Irreversible Phenomena and Dynamical Systems Analysis in Geosciences

NATO ASI Series

Advanced Science Institutes Series

A series presenting the results of activities sponsored by the NATO Science Committee, which aims at the dissemination of advanced scientific and technological knowledge, with a view to strengthening links between scientific communities.

The series is published by an international board of publishers in conjunction with the NATO Scientific Affairs Division

A Life Sciences B Physics	Plenum Publishing Corporation London and New York
C Mathematical and Physical Sciences	D. Reidel Publishing Company Dordrecht, Boston, Lancaster and Tokyo
D Behavioural and Social Sciences E Engineering and Materials Sciences	Martinus Nijhoff Publishers The Hague, Boston and Lancaster
F Computer and Systems Sciences G Ecological Sciences	Springer-Verlag Berlin, Heidelberg, New York and Tokyo

Series C: Mathematical and Physical Sciences Vol. 192

Irreversible Phenomena and Dynamical Systems Analysis in Geosciences

edited by

C. Nicolis

Institut d'Aéronomie Spatiale de Belgique,
Brussels, Belgium

and

G. Nicolis

Faculté des Sciences, Université Libre de Bruxelles,
Brussels, Belgium

D. Reidel Publishing Company

Dordrecht / Boston / Lancaster / Tokyo

Published in cooperation with NATO Scientific Affairs Division

Proceedings of the NATO Advanced Study Institute on
Irreversible Phenomena and Dynamical Systems Analysis in Geosciences
Crete, Greece
July 14-24, 1985

Library of Congress Cataloging in Publication Data

NATO Advanced Study Institute on Irreversible Phenomena and Dynamical Systems Analysis in Geosciences (1985: Crete)
Irreversible phenomena and dynamical systems analysis in geosciences.

(NATO ASI series. Series C, Mathematical and physical sciences: vol. 192)
"Published in cooperation with NATO Scientific Affairs Division."
"Proceedings of the NATO Advanced Study Institute on Irreversible Phenomena and Dynamical Systems Analysis in Geosciences, Crete, Greece, July 14–24, 1985"–T.p. verso.
Includes index.
1. Earth sciences–Mathematics–Congresses. 2. Dynamics–Congresses. 3. Stochastic processes–Congresses. I. Nicolis, Cathy. II. Nicolis, G.,1939- III. North Atlantic Treaty Organization. Scientific Affairs Division. IV. Title. V. Series: NATO ASI series. Series C, Mathematical and physical sciences; vol. 192.
QE33.2.M3N37 1985 550′.1′51 86–21985
ISBN 90–277–2363–X

Published by D. Reidel Publishing Company
P.O. Box 17, 3300 AA Dordrecht, Holland

Sold and distributed in the U.S.A. and Canada
by Kluwer Academic Publishers,
101 Philip Drive, Assinippi Park, Norwell, MA 02061, U.S.A.

In all other countries, sold and distributed
by Kluwer Academic Publishers Group,
P.O. Box 322, 3300 AH Dordrecht, Holland

D. Reidel Publishing Company is a member of the Kluwer Academic Publishers Group

---

Printed in The Netherlands

TABLE OF CONTENTS

FOREWORD

On July 14-24, 1985 an Advanced Study Institute on "Irreversible Phenomena and Dynamical Systems Analysis in Geosciences" was held in Crete, Greece under the auspices of the Greek Ministry of Science and Technology. It was sponsored by NATO, the International Solvay Institutes of Physics and Chemistry, and I.B.M. Europe. Thanks are due to Professor A. Berger who originally launched the idea of such an Institute, Dr. L. da Cunha of the NATO Scientific Affairs Division for his interest and cooperation and to Professor N. Flytzanis for his help in the organization.

The present Volume includes most of the material of the invited lectures delivered in the Institute as well as material from some seminars whose content was directly related to the themes of the invited lectures.

When dealing with the complex and diverse phenomena encountered in the various branches of geosciences, a multitude of approaches and tools is quite naturally envisaged. Many of them are highly specific to the particular branch considered. Thus, there is little in common between say, the numerical methods of short term weather prediction, and the monitoring of earthquakes. But from time to time one witnesses a process of transfer of knowledge, whereby methods hitherto limited to a certain field allow major progress in seemingly unrelated problems. The dating techniques of carbon and oxygen isotope ratios have, for instance, introduced a revolution in geology some decades ago. More unexpectedly perhaps, this had in turn important repercussions in climatology, since it allowed one to infer temperature or ice volume values of past geological periods within the last 10^6 years or so.

The principal goal of the present Advanced Study Institute was to contribute to the awareness of geoscientists about the existence and usefulness of another major tool of interdisciplinary research, namely, the theory of nonlinear dynamical systems. Unlike the isotope dating and many other similar examples, the immediate objective here is not to develop a new experimental technique enabling one to carry more accurate measurements. Rather, dynamical systems theory provides one with new ways of looking at a physical phenomenon, is at the basis of new concepts and tools of primary importance in the art of modelling, and helps to raise new questions which otherwise could not even be formulated. Its vocabulary and its methods are characterized by a great flexibility, as they can be adapted to such diverse problems as the study of catalytic chemical reactions in the laboratory scale, quantum optics, hydrodynamic instabilities, or large scale phenomena of interest in physics, geology, atmospheric dynamics and climatology.

We expect that the present Volume will acquaint researchers and advanced students in geosciences with the concepts and tools of dynamical systems and will thus serve as a jumping off point for new developments and new points of view.

Brussels, July 1986.
C. Nicolis and G. Nicolis.

ORGANIZING COMMITTEE

Benzi R., I.B.M. Scientific Center, Via Giorgione 129, 00147 Roma, Italy.

Berger A., Institut d'Astronomie et de Géophysique Georges Lemaître, Université Catholique de Louvain, Chemin du Cyclotron 2, 1348 Louvain-la-Neuve, Belgium.

Flytzanis N., Physics Department, University of Crete, Iraklion, Crete, Greece.

Ghil M., Department of Atmospheric Sciences and Institute of Geophysics and Planetary Physics, University of California, Los Angeles, CA 90024, U.S.A.

Nicolis C., Institut d'Aéronomie Spatiale de Belgique, Avenue Circulaire 3, 1180 Bruxelles, Belgium.

Nicolis G., Faculté des Sciences, Université Libre de Bruxelles, Campus Plaine, C.P. 226, Boulevard du Triomphe, 1050 Bruxelles, Belgium.

Velarde M.G., U.N.E.D.-Ciencias, Apdo Correos 50487, 28080 Madrid, Spain.

LIST OF PARTICIPANTS

André J.C.
Centre National de Recherches
Météorologiques,
Avenue G. Coriolis 42
31057 Toulouse Cedex
France

Bellemin L.
CCE DG XII
Square de Méeus 8
1049 Bruxelles
Belgium

Benzi R.
I.B.M. Scientific Center
Via Giorgione 129
00147 Roma
Italy

Bernardet P.
Centre National de Recherches
Météorologiques
Avenue G. Coriolis 42
31057 Toulouse Cedex
France

Birchfield G.E.
Department of Geological Sciences
Northwestern University
Evanston, Illinois 60201
U.S.A.

Brenig L.
Faculté des Sciences
Université Libre de Bruxelles
Campus Plaine, C.P. 231
Boulevard du Triomphe
1050 Bruxelles
Belgium

Chadam J.
Department of Mathematics
Indiana University
Bloomington, Indiana 47405
U.S.A.

da Cunha L.V.
Science Administrator
NATO Scientific Affairs Division
1110 Bruxelles
Belgium

Dalfes N.
National Center for
Atmospheric Research
P.O. Box 3000
Boulder, Colorado 80307
U.S.A.

Dalloubeix Ch.
Laboratoire de Géophysique
et Géodynamique Interne
Université de Paris Sud
91405 Orsay Cedex
France

Dalu G.A.
Instituto di Fisica della
Atmosfera
Piazza Luigi Sturzo 31
00144 Roma
Italy

Ruiz de Elvira A.
Departamento de Fisica
Universidad de Alcalà de Henares
Apdo. 20
Alcalà de Henares
Madrid
Spain

De Gregorio S.
Dipartimento di Matematica
Universita degli Studi di Roma
"La Sapienza"
Piazzale Aldo Moro 2
00185 Roma
Italy

de Miranda P.M.
Centro de Geofisica das
Universidades de Lisboa
Faculdade de Ciencias de Lisboa
Rua da Escola Politécnica 58
1200 Lisboa
Portugal

de Swart H.
Center for Mathematics and
Computer Science, dpt. AM.
P.O.B. Box 4079
1009 AB Amsterdam
The Netherlands

Deprit A.
Mathematical Analysis Division
National Bureau of Standards
Admin. Bldg., Room A-302
Washington, D.C. 20234
U.S.A.

Do Cêo Marques M.
Laboratorio de Fisica
Faculdade de Ciencias
Universidade do Porto
4000 Porto
Portugal

Egger J.
Meteorologisches Institut
Universität München
Theresienstrasse 37
8 München 2
West Germany

Flytzanis N.
Physics Department
University of Crete
Iraklion, Crete
Greece

Garcia A.
Faculté des Sciences
Université Libre de Bruxelles
Campus Plaine, C.P. 231
Boulevard du Triomphe
1050 Bruxelles
Belgium

Ghil M.
Department of Atmospheric Sciences
and Institute of Geophysics and
Planetary Physics
University of California
Los Angeles, CA 90024
U.S.A.

Gruendler J.
Mathematical Analysis Division
National Bureau of Standards
Gaithersburg
Maryland 20899
U.S.A

Guy B.
Ecole des Mines
158 Cours Fauriel
42023 Saint-Etienne Cedex
France

Hallet B.
Quarternary Research Center AK-60
University of Washington
Seattle, Washington 98195
U.S.A.

Harvey L.D.
National Center for
Atmospheric Research
P.O. Box 3000
Boulder, Colorado 80307
U.S.A.

Hense A.
Meteorologisches Institut
der Universität Bonn
Auf dem Hügel 20
5300 Bonn 1
West Germany

Hide R.
Geophysical Fluid Dynamics Laboratory
Geophysical Office (Met 0 21)
London Road, Bracknell
Berkshire, RG 12 2SZ
United Kingdom

Hyde W.
Department of Physics
University of Toronto
Ontario M5S 1A7
Canada

Iarlori S.
IBM Scientific Center
Via Giorgione 129
00147 Roma
Italy

Ioannidou E.I.
Tsakalov 17
Kolonaki
Athens
Greece

Jentsch V.
Max-Planck-Institut
für Meteorologie
Bundesstrasse 55
2000 Hamburg 13
West Germany

Kockarts G.
Institut d'Aéronomie
Spatiale de Belgique
Avenue Circulaire 3
1180 Bruxelles
Belgium

Lalas J.
Greek Meteorological Institute
Athens
Greece

Lemke P.
Max-Planck-Institut
für Meteorologie
Bundesstrasse 55
2000 Hamburg 13
West Germany

Lin Ch.
Department of Physics
Univervity of Toronto
Toronto, Ontario M5S 1A7
Canada

Lippolis G.
IBM Scientific Center
Via Giorgione 129
00147 Roma
Italy

Lorenz E.N.
Department of Meteorology and
Physical Oceanography
M.I.T.
Cambridge, Mass. 02139
U.S.A.

Maasch K.
Department of Geology and Geophysics
Kline Geology Laboratory
Yale University
New Haven, Connecticut 06511
U.S.A.

Madariaga R.
Institut de Physique du Globe
Université Pierre et Marie Curie
4 Place Jussieu, Tour 14
75230 Paris Cedex 05
France

Matteucci G.
Department of Geology and Geophysics
Kline Geology Laboratory
Yale University
P.O. Box 6666
New Haven, Connecticut 06511
U.S.A.

Michaelides S.C.
Meteorological Service
Larnaca Airport
Larnaca
Cyprus

Nicolis C.
Institut d'Aéronomie
Spatiale de Belgique
Avenue Circulaire 3
1180 Bruxelles
Belgium

Nicolis G.
Faculté des Sciences
Université Libre de Bruxelles
Campus Plaine, C.P. 226
Boulevard du Triomphe
1050 Bruxelles
Belgium

Nye J.
H.H. Wills Physics Laboratory
University of Bristol
Tyndall Avenue
Bristol BS8 1TL
United Kingdom

Garcia Ortiz J.M.
Departamento de Fisica
Universidad de Alcala de Henares
Madrid
Spain

Ortiz Bevia M.J.
Departamento de Fisica
Universidad de Alcala de Henares
Madrid
Spain

Ortoleva P.
Department of Chemistry
Indiana University
Bloomington, Indiana 47401
U.S.A.

Paladin G.
Dipartimento di Fisica
Universita degli Studi di Roma
"La Sapienza"
Piazzale Aldo Moro 2
00185 Roma
Italy

Pandolfo L.
Department of Geology
and Geophysics
Kline Geology Laboratory
Yale University
P.O. Box 6666
New Haven, Connecticut 06511
U.S.A.

Peltier W.R.
Department of Physics
University of Toronto
Ontario, M5S 1A7
Canada

Penland C.
Max-Planck Institut für Meteorologie
Bundesstrasse 55
2000 Hamburg 13
West Germany

Pnevmatikos S.
Physics Department
University of Crete
Iraklion, Crete
Greece

Prigogine I.
Faculté des Sciences
Université Libre de Bruxelles
Campus Plaine, C.P. 231
Boulevard du Triomphe
1050 Bruxelles
Belgium

Prigogine M.
Faculté des Sciences
Université Libre de Bruxelles
Campus Plaine, C.P. 238
Boulevard du Triomphe
1050 Bruxelles
Belgium

Provenzale A.
Instituto di Fisica Generale
Universita di Torino
C. so Massimo d'Azeglio 46
Torino
Italy

Putterman S.
School of Mathematics
University of
Newcastle upon Tyne
NE17RU
United Kingdom

Rebhan E.
Institut für Theoretische
Physik der Universität Düsseldorf
Universitätstrasse 1
Geb. 2532
4000 Düsseldorf
West Germany

Rees D.E.
Department of Applied Mathematics
University of Sydney
Sydney N.S.W. 2006
Australia

Roberts P.H.
Institute of Geophysics &
Planetary Physics
University of California
Los Angeles
California 90024
U.S.A.

Roumeliotis G.
Department of Applied Mathematics
The University of Sydney
Sydney N.S.W. 2006
Australia

Ronday F.
Institut de Physique
Université de Liège
Sart-Tilman B-6
4000 Liège
Belgium

Saltzman B.
Department of Geology
and Geophysics
Yale University
P.O. Box 6666
New Haven, Connecticut 06511
U.S.A.

Schilling H.D.
Meteorologisches Institut
Universität München
Theresienstrasse 37
8 München 2
West Germany

Speranza A.
FISBAT - CNR
Institute of Physics
University of Bologna
40126 Bologna
Italy

Spithas E.
Department of Research
and Technology
Greek Ministry of Industry
Energy and Technology
Athens
Greece

Tavantzis J.
New Jersey Institute of
Technology
Newark, N.J. 07102
U.S.A.

Van Delden A.
Inst. of Meteorology and
Oceanography
Rijksuniversiteit Utrecht
Princetonplein 5
2506 Utrecht
The Netherlands

Van der Mersch I.
Institut d'Astronomie et de
Géophysique Georges Lemaître
Université Catholique de
Louvain
Chemin du Cyclotron 2
1348 Louvain-La-Neuve
Belgium

Velarde M.G.
U.N.E.D. - Ciencias
Apdo Correos 50487
28080 Madrid
Spain

Verkley W.T.M.
Royal Netherlands Meteorological
Institute
P.O. Box 201
3730 AE de Bilt
The Netherlands

Viterbo A.P.
Centro de Geofisica das
Universidades de Lisboa
Facultade de Ciencias de Lisboa
Rua da Escola Politecnica 58
Portugal

Vulpiani A.
Dipartimento di Fisica
Universita degli Studi di Roma
"La Sapienza"
Piazzale Aldo Moro 2
00185 ROMA
Italy

Wheeler M.F.S.
St. Bede's College
Alexandra Park
Manchester
M16 8HX
United Kingdom

Wang X.J.
Faculté des Sciences
Université Libre de Bruxelles
Campus Plaine, C.P. 231
Boulevard du Triomphe
1050 Bruxelles
Belgium

Xanthopoulos V.
Physics Department
University of Crete
Iraklion, Crete
Greece

Garcia-Ybarra P.
Dept. Fisica Fundamental
U.N.E.D. - Fisica
Apdo Correos 50487
28080 Madrid
Spain

Wellens S. (Secretaty of the Conference)
Faculté des Sciences
Université Libre de Bruxelles
Campus Plaine, C.P. 231
Boulevard du Triomphe
1050 Bruxelles
Belgium

INTRODUCTORY REMARKS

Ilya Prigogine
Chimie Physique II
CP 231 Université Libre de Bruxelles
B 1050 Bruxelles, Belgique

It is a privilege to introduce this NATO Advanced Study Institute dealing with dynamical systems and their applications to Geophysics, Atmospheric Science, Climatology and Geology.

The very object of this conference shows how much progress has been achieved in applying the recent explosion of knowledge about the behavior of dynamical systems to problems which are of fundamental importance for the understanding of our natural environment.

This transfer of knowledge has in fact been reciprocal. On one side, the theory of bifurcations, the discovery of strange attractors and progress in the the description of stochastic processes have led to a number of far-reaching consequences, when applied to a wide variety of subjects in Earth sciences, Atmospheric sciences, biology and other fields; however, conversely, it is hardly necessary to remind here that it was actually the study of instabilities in atmospheric dynamics which had a decisive effect on recent research in dynamical systems. It is very fortunate for the all of us that E. N. Lorenz, whose work had such a decisive impact, is here to participate in our discussions.

For me, the most surprising feature is the amazing variety of behaviors we have now found to be the case in dynamical systems. In order to visualize this variety, it may be convenient to use the language of symbolic dynamics. Let us introduce a partition on some phase space. A "trajectory" can then be characterized by a succession of symbols, corresponding to the regions through which the point representing the state of the system passes.

For the sake of simplicity, let us assume we have adopted a partition in two regions, coded by the symbols {0, 1}. Any "trajectory" is then a succession of 0's and 1's. We may now consider two limiting situations: some periodical behavior, coded by a sequence such as {0101010101...}; or random-like behavior, coded by a sequence such as {011010001011...}

C. Nicolis and G. Nicolis (eds.), Irreversible Phenomena and Dynamical Systems Analysis in Geosciences, xvii–xix.

in which, at each step, the "choice" of the symbol satisfies a Bernoulli process. These two examples correspond to the classification of dynamical processes suggested by Ford, Eckhardt and Vivaldi(1):

-For **Algorithmically (A-) integrable systems**, the evolution operator is predictable enough for the trajectory of a point to be computable with arbitrary precision over any time lapse from prescribed initial data. Examples of such systems include all analytically integrable systems like the pendulum.

-In contrary, **Kolmogorov (K-) systems** exhibit a complex behavior of individual trajectories, like exponential divergence, As a result, no finite algorithm can reliably compute the motion of phase space points sensibly faster than the dynamics itself. Systems of this kind include geodesic flow on surfaces with negative curvature, hard spheres in three dimensions, the planar Lorentz gas model, etc.

The new, unexpected situation is the increased importance of the K-systems, as compared to the A-integrable systems. Obviously, here a new approach becomes necessary, as we can never deal with an infinite amount of information, which would correspond to the knowledge of an infinite amount of digits needed to locate a point in phase space. Indeed, our knowledge corresponds to a finite window, whatever its size. An alternative formulation would be to state that our knowledge corresponds to a set of partitions, but not to the observation of individual points.

How then to formulate classical dynamics? Over the recent years, our group in Brussels and Austin has been much involved in this problem; In the theoretical description of K-flows, contributions by Misra, Courbage, Tirapegui, Elskens and others have been of special importance. In short, the main conclusion of recent work is that the elimination of the ideal, unobservable information leads to a semi-group representation of dynamics, including irreversibility.(2)

The usual group representation corresponding to the Liouville formulation is only recovered in the singular limit in which the size of the partitions becomes strictly equal to zero. This suggests a amazing analogy with the transition from quantum mechanics to classical mechanics, where also classical representation is only recovered as a singular limit when Planck constant h tends to zero.

It is a characteristic feature of the ongoing scientific revolution that the theory of unstable dynamical systems is of importance even in a paragon field of classical mechanics such as celestial dynamics. Recent work by Wisdom and Petrosky(3) over the asteroid belt or the formation of the cometary clouds is an excellent example. In both cases, the individual prediction of the motion of a celestial body becomes impossible, and has to be replaced by a statistical description. Quantum mechanics has introduced probability at the microscopic level. We see now the role of randomness and probability appearing in the heart of what seemed to be a definitive possession of classical science.

1 B. Eckhardt, J. Ford and P. Vivaldi, 'Analytically Solvable Dynamical Systems which are not Integrable' *Physica* **13D** (1984) 339-356.

2 see for instance I. Prigogine, *From Being to Becoming,* Freeman, San Francisco 1980,

3 J. Wisdom 'The Origin of the Kirkwood Gap: a Mapping of the Asteroidal Motion near the 3/1 Commensurability', *Astronom. J.,* **87** (1982) 577-593.

PART I

DYNAMICAL SYSTEMS, STABILITY AND BIFURCATION

BIFURCATION AND STOCHASTIC ANALYSIS OF NONLINEAR SYSTEMS : AN INTRODUCTION.

G. Nicolis
Faculté des Sciences
Université Libre de Bruxelles
1050 Bruxelles
Belgique

ABSTRACT.

The dynamical systems approach to the study of complex phenomena of relevance in geosciences is outlined. The basic concepts and tools of the theory of dynamical systems are surveyed, with special emphasis on the phenomena of instability and bifurcation. Next, the theory is enlarged to take into account the effect of internal fluctuations or of external random disturbances. The connection between random noise and the complex behaviors related to chaotic dynamics is finally discussed.

1. INTRODUCTION

The basic philosophy underlying the dynamical systems approach to complex phenomena is to set up, before any attempt at a detailed quantitative description is undertaken, a qualitative study in which the general trends and potentialities of a system are sorted out. This is achieved by analyzing the way in which the solutions of an underlying set of equations of evolution are affected when various control parameters built in it are varied. The main reason that makes this task difficult to accomplish is that in nature one deals, as a rule, with systems subjected to a multitude of constraints and evolving according to nonlinear laws of interaction. The most dramatic consequence of these nonlinearities is to generate, under certain conditions, a multiplicity of solutions some of which may represent dynamical regimes characterized by very complex space or time dependencies. A brute force numerical approach will as a rule, miss these potentialities-whence the need for a qualitative approach.

We shall write the rate of change of the variables X_i, describing the state of such a system in the form

$$\frac{\partial X_i}{\partial t} = F_i(X_1, \ldots, X_n; \lambda, \mu, \ldots) \quad ; \quad i = 1, \ldots, n \qquad (1)$$

C. Nicolis and G. Nicolis (eds.), Irreversible Phenomena and Dynamical Systems Analysis in Geosciences, 3–29.

where F_i stand for the laws of evolution and λ, μ etc ... for the parameters. Let us give some typical examples of relevance in geosciences.

(i) Energy-balance models at the planetary scale.

In this class of models, which are discussed further in the Chapters by M. Ghil and C. Nicolis the evolution of the mean temperature T of the planet earth is studied as the result of the balance between incoming solar energy and outgoing infrared energy :

$$C \frac{dT}{dt} = Q[1 - a(T)] - \epsilon \sigma T^4 \tag{2}$$

where C is the heat capacity, Q the solar constant, σ the Stefan constant, ϵ the emissivity and a(T) the albedo. Because of the surface albedo feedback a depends on temperature T in a highly nonlinear fashion. This gives rise to multiple steady state solutions of Eq. (2).

(ii) Energy balance coupled to the cryosphere.

As discussed in detail in the Chapters by C. Nicolis and B. Saltzman, here one takes into account the negative feedback of ice extent on temperature and the positive feedback of temperature T on ice extent in some range of temperatures. The resulting equations

$$\frac{dT}{dt} = f(T, \ell)$$

$$\frac{d\ell}{dt} = g(T, \ell) \tag{3}$$

give rise to self-oscillations which are of considerable importance in climate dynamics.

(iii) The vorticity equation.

The importance of this equation in atmospheric dynamics is discussed in the Chapters by J.Egger, R. Benzi, A. Speranza and M. Ghil. One starts with the momentum balance in a rotating frame, takes the curl of both sides and neglects, to a first approximation, contributions coming from the vertical motion. The result is :

$$\left(\frac{\partial}{\partial x} + \frac{\partial}{\partial y}\right)(\zeta + f) = -(\zeta + f)\left(\frac{\partial u}{\partial x} + \frac{\partial v}{\partial y}\right) \tag{4a}$$

Here x and y are the horizontal coordinates, u and v the corresponding velocity components, ζ the vorticity

$$\zeta = \frac{\partial v}{\partial x} - \frac{\partial u}{\partial y} \tag{4b}$$

and f the Coriolis parameter,

$$f = 2\,\Omega \sin\theta \qquad (4c)$$

where Ω is the rotation velocity and θ the latitude. We see that, because of (4b), Eq. (4a) is highly nonlinear. Again, this nonlinearity is at the basis of the complexity of atmospheric circulation.

We are now sufficiently motivated to start our analysis. We shall first present (Section 2) the rudiments of the phase space description which, in a way, provides us with a "geometric view" of dynamical systems. A more algebraic approach, based on stability and bifurcation analysis will be outlined in Sections 3 and 4. Finally, in Section 5 the description of nonlinear dynamical systems will be enlarged to include the effect of stochastic perturbations.

2. PHASE SPACE DESCRIPTION : A GEOMETRIC VIEW OF DYNAMICAL SYSTEMS

Let us restrict ourselves for the time being, to problems amenable to a finite number of space-independent variables. Such problems arise quite frequently in geosciences through, for instance, truncation of the expansion of a space-dependent property into a finite number of modes. They can also arise independently of any heuristic approximation, by a systematic perturbative expansion. Whetever their origin, eqs. (1) simplify by the fact that the variables X_i depend solely on time, which allows us to write the evolution laws in the form

$$\frac{dX_i}{dt} = F_i(X_1, \cdots, X_n; \lambda, \mu, \cdots) \qquad (5)$$

or, introducing a column vector $\underline{X}$ whose components are equal to $X_1, \ldots, X_n$,

$$\frac{d\underline{X}}{dt} = \underline{F}(\underline{X}, \lambda) \qquad (6)$$

We now embed the evolution of the system in a space spanned by the variables $\{X_i\}$, which we call the phase space [1],[2]. An instantaneous state is represented in this space by a point (and vice-versa), whereas a succession of such states defines a curve, the phase space trajectory. Fig. 1 depicts a typical trajectory,C.
A degenerate case is represented by the fixed point S, for which $\underline{F}$ is identically zero and thus $\underline{X}$ does not evolve in time. The set of all regular phase space trajectories and of all fixed points is the phase portrait of a dynamical system. Its structure is determined to a large extent, by the following two very important elements :
(i) The uniqueness theorem of solutions of the system of eqs (5), (6) which asserts that under very mild conditions on $\underline{F}$, a system starting initially in a state different from a fixed point will evolve to one and only one state after a sufficiently long lapse of time $t < T_{max}$, where T_{max} depends on the structure of the equations. One can easily check that this theorem precludes intersections between trajectories or of a trajectory with itself.

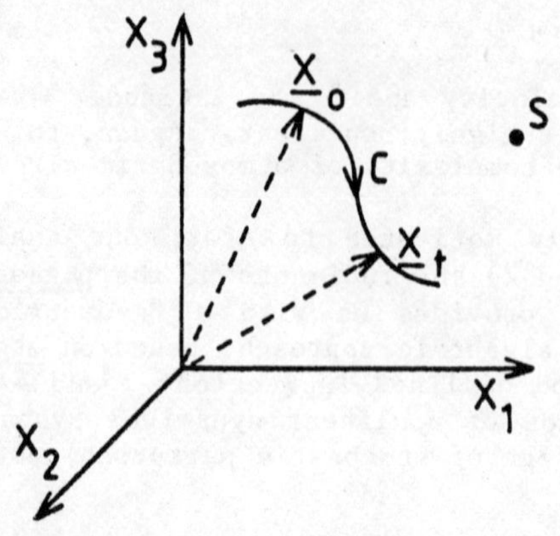

Fig. 1 Visualization of the evolution of a dynamical system in phase space, C : regular phase space trajectory. S : Singular phase space trajectory (fixed point).

(ii) The nature of the invariant sets that is to say, those bounded sets of points in phase space which are transformed into themselves during the evolution.

A very important type of invariant set is the set representing the regime reached by a dynamical system as time grows and all transients die out. We call this regime the attractor. The two simplest attractors are depicted in Fig. 2.

In Fig. 2a the phase space trajectories converge to a fixed point S, representing a time-independent solution of the equations of evolution. We call S a point attractor. In Fig. 2b, despite the existence of a fixed point S, the trajectories converge to a closed curve C, representing a time-periodic behavior. We call C a limit cycle. Notice that the very existence of attractors implies that physical

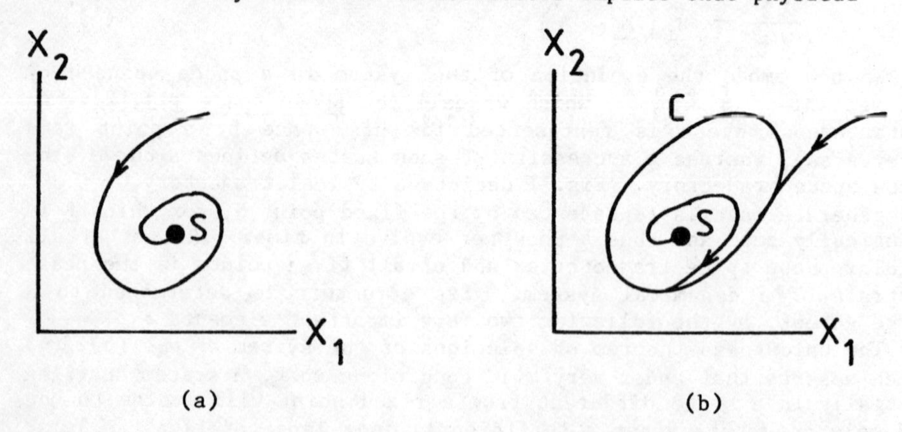

Fig. 2 The simplest attractors of a dissipative dynamical system : (a) point attractor attained through a monotonic evolution toward a steady state ; (b) limit cycle attractor describing an asymptotically stable periodic behavior in time.

systems are capable of damping perturbations and of forgetting initial conditions. This is the property of asymptotic stability, to which we shall come back in more detail later. Suffice it to stress, at this stage, that only dissipative dynamical systems, giving rise to irreversible phenomena and obeying to the second law of thermodynamics, can possess attractors and enjoy asymptotic stability. This rules out simple mechanical devices like the pendulum, for which conservation precludes irreversibility and dissipation. In the sequel we shall concentrate on dissipative systems, which constitute the vast majority of systems encountered in nature.

The next level of complexity relates to the coexistence of multiple attractors. This immediately leads to a topological problem, namely, how to delimit the relative basins of attraction, that is to say, the set of initial states in phase space that will evolve to either of the attractors. The solution of this problem, depicted in Fig. 3, involves necessarily an intermediate unstable state as well as a family of orbits remaining invariant under the flow known as separatrices. Clearly, the coexistence of multiple attractors constitutes the natural model of systems capable of performing transitions between various modes of behavior. The implications of this possibility in climate dynamics are discussed in the Chapter by C. Nicolis.

The existence of 1-dimensional attractors suggests the possibility of higher-dimensional attracting objects whose cross section along different phase space coordinates would give a limit cycle. Such objects, can indeed be shown to exist. They have the topology of a torus, and model multiperiodic behavior. We do not discuss these attractors in detail. Instead, we jump directly to what is undoubtetly the most complex and challenging attracting object known to date, namely a chaotic attractor [3]. An example constructed from a model system involving three variables [4],[5] is shown in Fig. 4. We observe two opposing trends. On the one side an instability of the motion tending to remove the phase space trajectory away from the steady state solution, which here turns out to be the state $x = y = z = 0$; and the bending of the outgoing trajectories followed by their reinjection back to the vicinity of the steady state. Two ingenious tricks of nature allow to reconcile these contradictory tendencies. One is the adoption of a fractal geometry [6], that is to say, the existence of attracting objects of non-integer dimensionality, which are intermediate between a surface and a volume. And the other is the sensitivity of the trajectories on the attractor to small changes in the initial conditions, as a result of which two nearby initial states can diverge, momentarily, in an exponential fashion. Quantitatively this is reflected by the existence of at least one positive Lyapounov exponent, σ_L [2], [3], providing an average of the rates of this exponential divergence over all points on the attractor. Clearly then, chaotic attractors provide the archetype of natural phenomena characterized by a limited predictability. Some obvious examples are atmospheric and climatic variability, discussed further

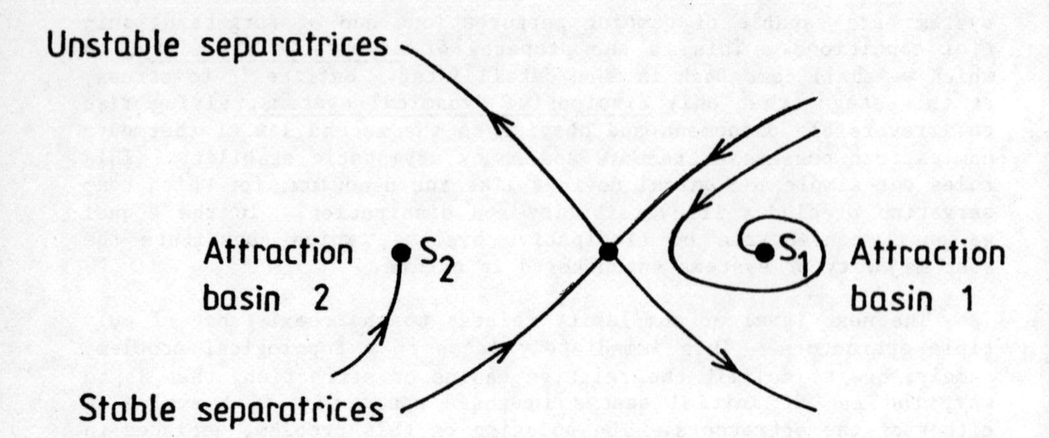

Fig. 3 Geometry of phase space corresponding to the coexistence of multiple attractors.

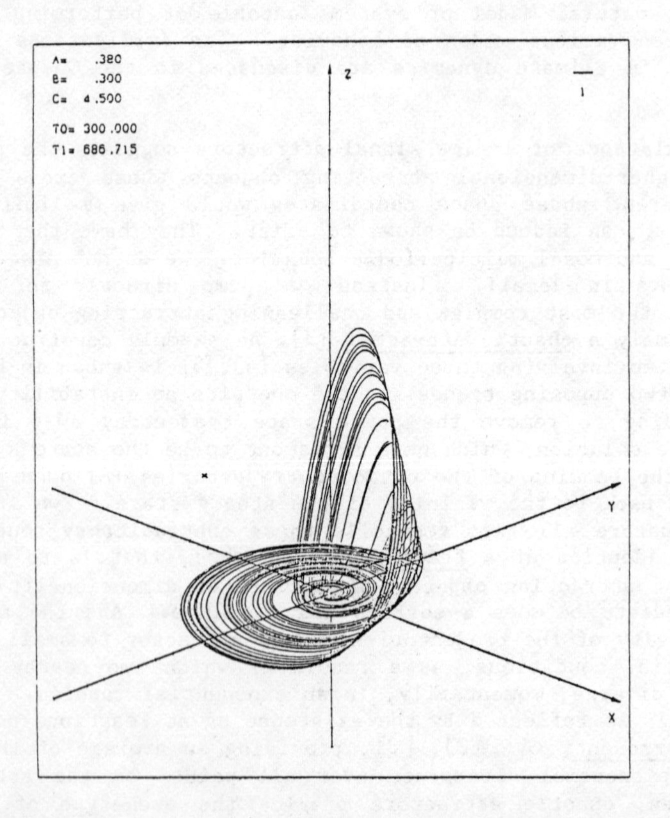

Fig. 4 Chaotic attractor constructed from a model system involving three variables and a single nonlinearity of second degree.

in the Chapters by M. Ghil, E. Lorenz and C. Nicolis. We discuss chaotic dynamics more extensively in Section 4.

One might be tempted to deduce, from the above discussion, that a given system can only be modelled by a particular type of attractor. This is not so, however. As a matter of fact the most exciting aspect of physical phenomena is that the same system can show a great variety of behaviors, each corresponding to a different attractor. The mechanism which is at the origin of this diversification is the instability of a reference state and the subsequent bifurcation of new branches of states as the paramaters $\lambda_1, \ldots, \lambda_m$ built in the system are varied [1], [2].

The simplest bifurcation is depicted in Fig. 5. We represent in a graph the way a state variable of the system X (a component of the horizontal velocity in an atmospheric problem, the ice extent in a problem pertaining to long term climatic variability, etc) is affected by a single control parameter λ(the pressure difference, the ice-albedo feedback parameter etc). We obtain in this way a bifurcation diagram.

For a range of values of λ only one solution is accessible. It enjoys the property of asymptotic stability, since in this range the system is capable of damping internal fluctuations or external disturbances. But beyond a critical value, denoted by λ_c in Fig. 5, we find that the states on this branch become unstable : the effect of fluctuations or of small external perturbations is no longer damped. The system acts like an amplifier, moves away from the reference

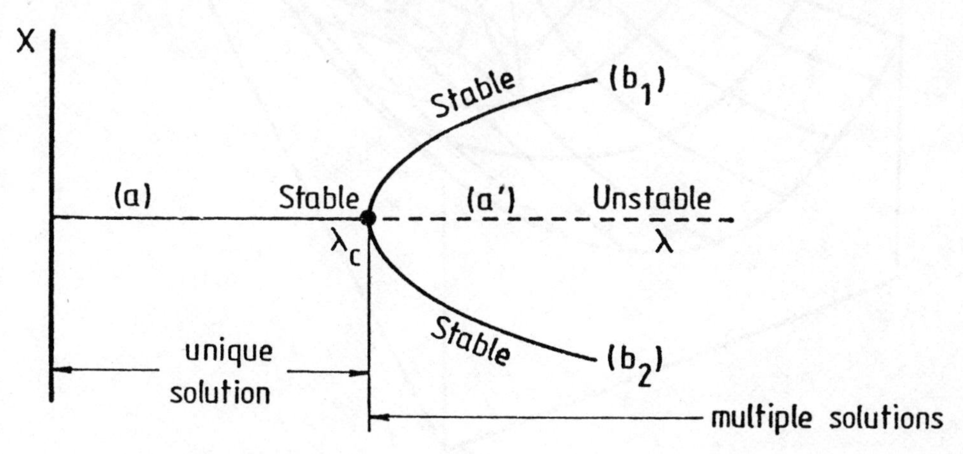

Fig. 5 Bifurcation diagram showing how a state variable x is affected when the control parameter varies. A unique solution (a), loses its stability at λ_c. At this value of the control parameter new branches of solutions (b_1, b_2) which are stable in the example chosen, are generated.

state and evolves to a new regime (which may be a state of self-sustained oscillations in the case of the climate problem). The two regimes coalesce at $\lambda = \lambda_c$, but are differentiated for $\lambda \neq \lambda_c$. This is the phenomenon of bifurcation.

Fig. 6 represents a mechanical analog of the phenomenon. A ball moves in a valley (branch (a)) which at a particular point becomes branched and leads to either of two new valleys (branches b1 and b2) separated by a hill. Although it is always perillous to draw analogies and extrapolations, it is nevertheless thought-provoking to imagine for a moment that instead of the ball in Fig. 6 one could have the planet earth sitting there prior to the onset of glaciations of the quaternary era!

A bifurcation phenomenon can be local, as in the examples of Figs 5 and 6, or global, as in the example of Fig. 7. In the first case the bifurcating branches of solutions appear in the vicinity of

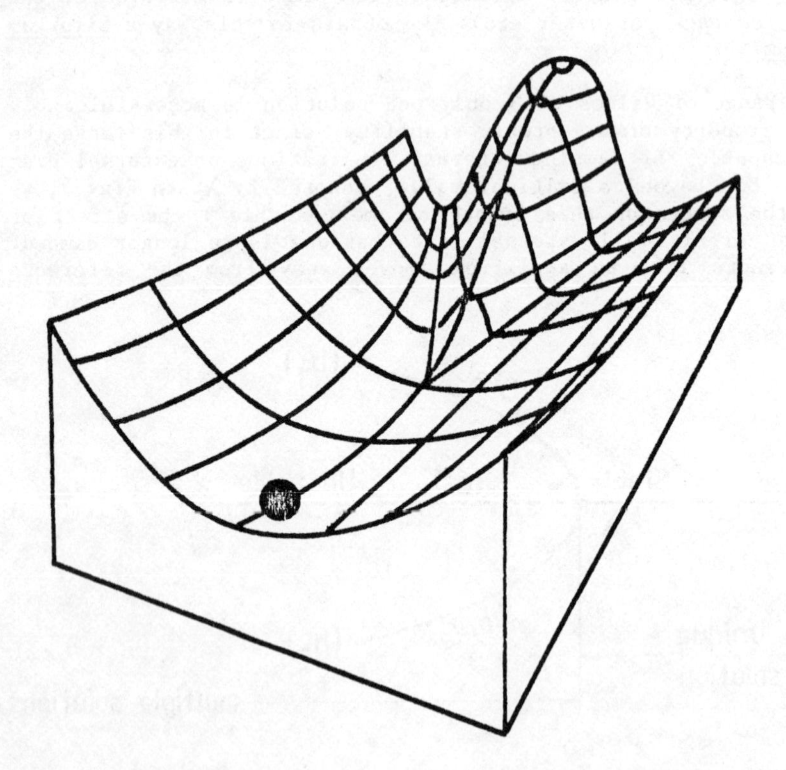

Fig. 6 Mechanical illustration of the phenomenon of bifurcation

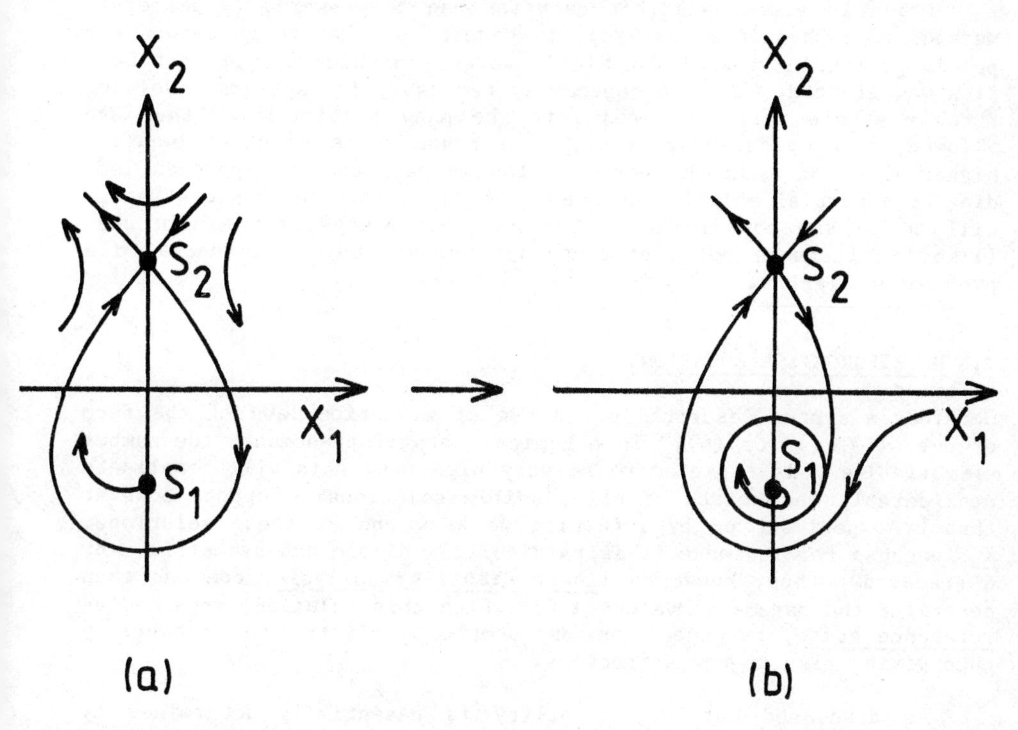

Fig. 7 Example of global bifurcation. As the parameters vary a homoclinic orbit (closed loop in Fig. 7a) opens up and gives rise to a stable limit cycle (Fig. 7b).

a fixed point when λ is close to λ_c. The situation in the second case is different : for a certain combination of parameter values the separatrices of an unstable fixed point S_2 merge to form a closed loop (Fig. 7a), known as homoclinic orbit ; on further variation of parameters a limit cycle is seen to bifurcate from the separatrix loop (Fig. 7b), and is thus formed at a finite distance from the fixed point S_1 [2], [7]. In that respect it constitutes a global bifurcation, whose period becomes as large as desired if one looks sufficiently near the domain of existence of the separatrix loop. The implications of such global phenomena in atmospheric and climate dynamics are discussed in the Chapters by M. Ghil and C. Nicolis.

In large classes of systems the diversification of behavior through bifurcation does not stop just at the first transition - referred to as primary bifurcation. One observes, instead, complicated bifurcation cascades leading to secondary, tertiary etc branches and culminating in certain cases to the bifurcation of chaotic attractors. This raises naturally the question of the structure of bifurcation diagrams and of the possibility of providing a list of the attractors of a given dynamical system.

While in some cases this question can be answered by geometric methods (See Chapter by J. Nye), in general one has to go beyond the purely geometric view of dynamical systems to which the present Section was limited. This is especially necessary for systems involving three ore more variables, owing to the many possibilities that are allowed, topologically speaking, in a phase space of dimensionality higher than two. In the next Section we describe one approach leading to a partial solution of these difficulties. The basic strategy will be to see how complex attractors can emerge from simple ones (like for instance point attractors), whose study is amenable to a problem of algebra.

3. THE BIFURCATION EQUATIONS

Consider a system described by a set of evolution laws of the form of eqs. (1), (5) or (6). In a typical natural phenomenon the number of variables n is expected to be very high, and this will complicate considerably the search of all possible solutions. Suppose however that by experiment or by intuition we know one of these solutions, $\underline{X}_s$, because for instance it is particularly simple and symmetric. By a standard method, known as linear stability analysis, one can then determine the parameter values λ for which this solution, regarded as reference state, switches from asymptotic stability to instability thus giving rise to new attractors.

As discussed earlier, stability is essentially determined by the response of the system to perturbations. It is therefore natural to cast the dynamical laws in a form in which the perturbations appear explicitly. Setting

$$X_i(t) = X_{i,s} + x_i(t) \tag{7}$$

substituting into Eqs.(5) and taking into account that $X_{i,s}$ is also a solution of these equations, we arrive at

$$\frac{dx_i}{dt} = F_i(\{X_{i,s}+x_i\},\lambda) - F_i(\{X_{i,s}\},\lambda)$$

These equations are homogeneous in the sense that the right hand side vanishes if all $x_i = 0$. To get a more transparent form of this homogeneous system we expand $F_i(\{X_{i,s} + x_i\}, \lambda)$ around $X_{i,s}$, and write explicitly the part of the result that is linear in x_j, plus a non linear correction whose structure need not be specified at this stage :

$$\frac{dx_i}{dt} = \sum_j L_{ij}(\lambda)x_j + h_i(\{x_j\},\lambda) \qquad (i=1,\ldots,n) \tag{8}$$

L_{ij} are the coefficients of the linear part and h_i the non linear contributions. The set of L_{ij} defines and operator (n x n matrix in our case), depending on the reference state X_s and on the parameters λ.

Now a basic result of stability theory [1], [2], [7], establishes that the asymptoptic stability or instability of the reference state ($\underline{X} = \underline{X}_s$ or $\underline{x} = 0$) of system (8) are identical to those of its linearized part.

$$\frac{dx_i}{dt} = \sum_j L_{ij}(\lambda)\, x_j \qquad ; \quad (i = 1, \ldots, n) \tag{9}$$

Stability reduces in this way to a linear problem, which is soluble by methods of elementary calculus.

Fig.8 summarizes the typical outcome of a stability analysis carried out according to this procedure. What is achieved is the computation of the rate of growth of the perturbations γ as a function of a control parameter. If $\gamma < 0$ (as it happens in Fig. 8 branch (1) if $\lambda < \lambda_c$) the reference state is asymptotically stable, and if $\gamma > 0$ ($\lambda > \lambda_c$ for branch (1) of Fig. 9) it is unstable. At $\lambda = \lambda_c$ one has a state of marginal stability , the frontier between asymptotic stability and instability.

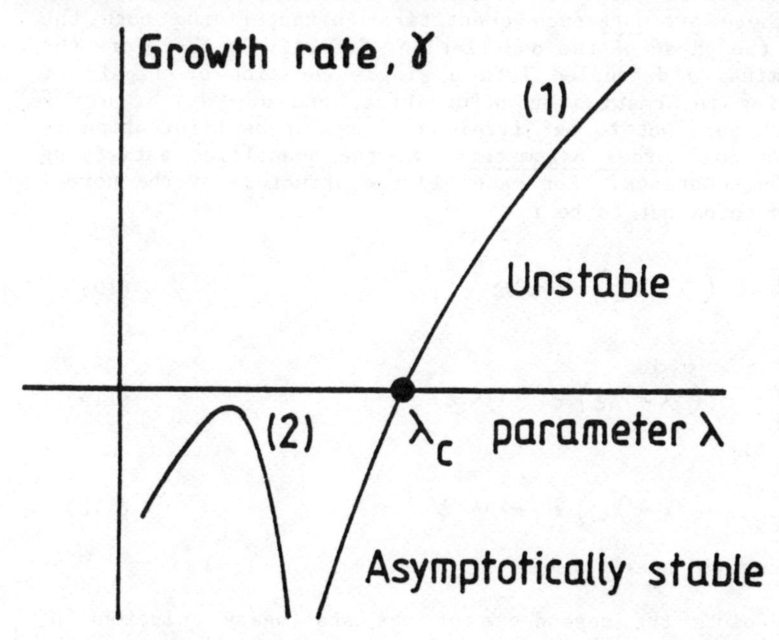

Fig. 8 Rate of growth of perturbations, γ as a function of the control parameter, λ deduced from linear stability analysis (cf. eq. (9)). Curve (1) : the reference state is asymptotically stable for $\lambda < \lambda_c$ and becomes unstable for $\lambda > \lambda_c$, where λ_c is the critical value of marginal stability. Curve (2) : the reference state remains asymptotically stable for all values of λ.

In general a multivariable system gives rise to a whole spectrum of γ, just like a crystal has a multitude of vibration modes. One will have therefore several γ versus λ curves in Fig. 8. Suppose first that of all these curves only one (curve (1)) crosses the λ axis, while all others are below it. Under well-defined mild conditions, one can then show that at $\lambda = \lambda_c$ a bifurcation of new branches of solutions takes place. Two cases can be distinguished.

(i) if at $\lambda = \lambda_c$ the perturbations are non-oscillatory, the bifurcating branches will correspond to steady-state solutions

(ii) if at $\lambda = \lambda_c$ the perturbations are oscillatory the bifurcating branches will correspond to time-periodic solutions in the form of limit cycles.

In either case a most remarkable point is that a suitable set of quantities can be defined which obey to a closed set of equations having a universal, normal form [2] if the parameters remain close to their critical values λ_c (see also Chapter by J. Chadam). In case (i) it turns out that there is only one relevant quantity, which measures the amplitude of the bifurcating branches. In case (ii), on the other hand, there are two such quantities characterizing both the amplitude and the phase of the oscillation. Effectively therefore the original dynamics is decoupled into a single equation or a pair of equations giving information on bifurcation, and n - 1 or n - 2 equations which turn out to be "irrelevant" as far as bifurcation is concerned. We call order parameters, z, the quantities satisfying the bifurcation equations. For case (i) the structure of the normal form equations turns out to be :

$$\frac{dz}{dt} = (\lambda - \lambda_c) - uz^2 \tag{10}$$

or

$$\frac{dz}{dt} = (\lambda - \lambda_c)z - uz^3 \tag{11a}$$

or

$$\frac{dz}{dt} = (\lambda - \lambda_c)z - uz^2 \tag{11b}$$

where u is a certain coefficient

Fig. 9 depicts the dependence of the stationary solutions of eq. (10) on the parameter λ. These solutions are given by :

$$z_s : z_{\pm} = \pm(\lambda - \lambda_c)^{1/2} \tag{12}$$

A standard stability analysis shows that branch z_- is unstable, whereas branch z_+ is asymptotically stable. We notice that as λ decreases toward λ_c the stable and unstable branches "collide" at $\lambda = \lambda_c$

and subsequently are annihilated. For this reason we call $\lambda=\lambda_c$ a limit point. Its presence signals the emergence of new branches of solutions. We have here one of the simplest examples of the phenomenon of bifurcation. Notice the singular (nonanalytic) dependence of the bifurcating solutions on the parameter $\lambda-\lambda_c$, even though the evolution equation (eq. (10)) depends smoothly on this parameter.

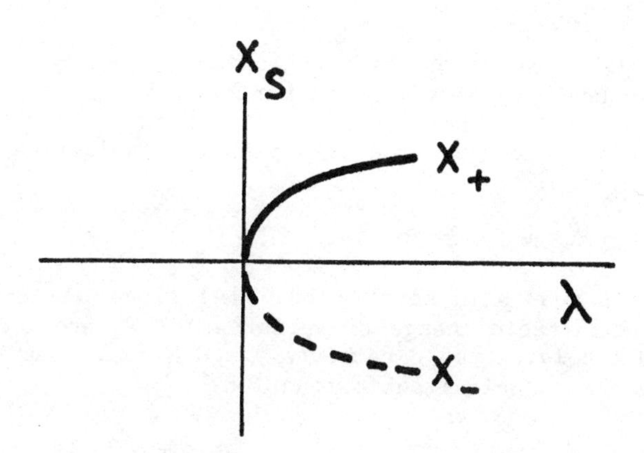

Fig. 9 Limit point bifurcation.

Figs. 10a-b depict the dependence of the stationary solutions of eqs. (11 a-b) on λ. In both cases the trivial solution $z = 0$ exists for all values of λ. It loses its stability at $\lambda=\lambda_c$ and gives rise to symmetric (Fig. 10a) or to transcritical (Fig 10b) bifurcating solution branches. We speak, respectively, of pitchfork and transcritical bifurcations :

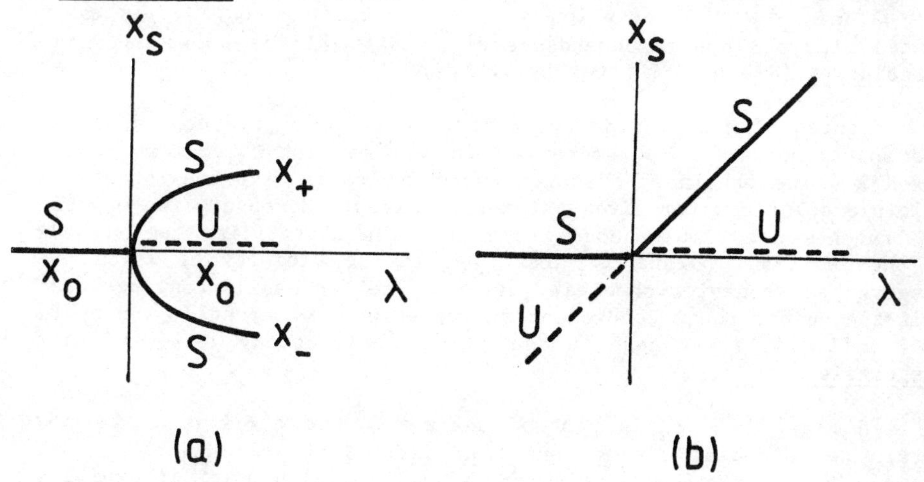

Fig. 10 Pitchfork (a) and transcritical (b) bifurcation.

$$z_s : \begin{cases} z_0 = 0 \\ z_\pm = \pm(\lambda - \lambda_c)^{1/2} \end{cases} \quad \text{(pitchfork bifurcation)} \tag{13a}$$

$$z_s : \begin{cases} z_0 = 0 \\ z = \lambda - \lambda_c \end{cases} \quad \text{(transcritical bifurcation)} \tag{13b}$$

As in the limit point case, the bifurcation is marked by a singular dependence of the branches on the parameter λ.

Let us now combine our previous examples by considering the dissipative system

$$\frac{dz}{dt} = -z^3 + \lambda z + \mu \tag{14}$$

(we do not add the quadratic term of Eq. (14) since one can always eliminate it by a suitable change of variables). We are led in this way to a problem involving two parameters λ and μ. The fixed points are now given by the canonical cubic equation

$$-z_s^3 + \lambda z_s + \mu = 0$$

From elementary algebra we know that this equation can have up to three real solutions. Moreover, as the parameters vary the three solutions merge, and one is left with only one solution. One can define regions in parameter space separating these two regimes, specifically

$$4\lambda^3 + 27\mu^2 = 0 \tag{15}$$

The value $\lambda = \mu = 0$ for which all these solutions coalesce is associated with a singular dependence of λ on μ. This is known as <u>cusp singularity</u> (See also Chapter by J. Nye).

Figures 11 a-b provide two different views of the dependence of the solutions on the parameters. In 11a we plot z_s against μ for fixed λ. We obtain an S-shaped curve indicating the coexistence of multiple solutions for given parameter values. Moreover, two of these branches are simultaneously stable. The bistability region ends at the two limit points μ_1 and μ_2 in the vicinity of which one observes the behavior shown in Fig. 9 One can easily convince oneself that under these conditions an increase of μ beyond μ_1 and up to μ_2, followed by a change in the opposite direction, will lead to a <u>hysteresis cycle</u>.

In Fig. 11 b z_s is plotted against λ for fixed μ. We now obtain two disjoint curves, one (c_1) defined for all values of λ, and another (c_2), defined only for $\lambda \geq \bar{\lambda}$ in which point it exhibits a limit point singularity. For $\lambda < \bar{\lambda}$ only one stable solution is

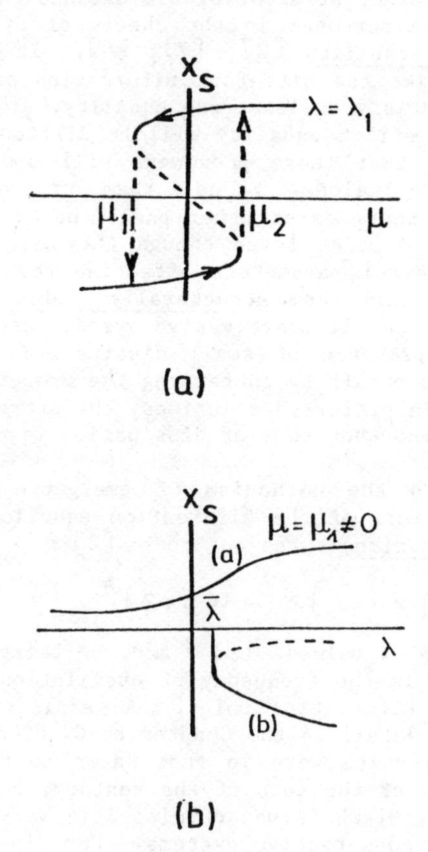

Fig. 11 Effect of parameters in the bifurcation of steady state solutions of eq. (14) (a) : hysteretic behavior of the solution at fixed λ, as the parameter μ is varied ; (b) : destruction of the pitchfork bifurcation when the parameter μ , acting as an imperfection, is not identically zero.

available, but for $\lambda > \bar{\lambda}$ one has bistability as before.

It is important to realize that for no non-vanishing μ, however small, can the pitchfork bifurcation depicted in Fig. 10 be observed. μ acts therefore like an "imperfection" destroying bifurcation. On the other hand, the limit point, Fig. 9 proves to be robust, in the sense that it is recovered in both Fig. 11a and 11b. But if both λ and μ are varied simultaneously, there will always be a particular combination of values ($\mu = 0$, λ variable in our case) for which the bifurcation will be recovered, as the system will be able to traverse the cusp singularity in a symmetric fashion.

Actually the above rather straightforward discussion illustrates a deep concept of great importance in the theory of dynamical systems, namely structural stability [2], [7], [8]. It shows indeed that certain phenomena like the pitchfork bifurcation occur only if the parameters present satisfy at least one equality. Inasmuch as in a physical system such a strict equality will be difficult to achieve, we expect therefore that these phenomena will disappear under slight changes of parameter values. We call them structurally unstable. On the other hand there exist other phenomena, like the limit point bifurcation, which subsist (even though they may be shifted) under changes of the control parameters affecting the structure of the evolution laws. We call them structurally stable. Note that according to this definition all conservative systems are structurally unstable, since the presence of small dissipative terms alters qualitatively the phase portrait by conferring the property of asymptotic stability to certain preferred solutions, the attractors. This brings out clearly the important role of dissipation in nature.

Let us now turn to the mechanism of emergence of periodic attractors. The normal form of the bifurcation equations (referred to usually as Hopf bifurcation) turns out to be [2] :

$$\frac{dz}{dt} = [(\lambda-\lambda_c) + i\omega_0]z - uz|z|^2 \tag{16}$$

where z, u, are now complex valued, $|z|^2 = zz^*$, z^* being the complex conjuguate of z, and ω_0 is the frequency of oscillation of the perturbations right at the bifurcation point. An explicit example of eq. (16) is discussed in detail in the Chapter by C. Nicolis.

The above powerful results were in fact known to Poincaré, the great mathematical genius of the turn of the century, but it was the American mathematician G. Birkhoff who developed it very far for the study of bifurcations in conservative systems. For dissipative systems the idea of reduction to a few order parameters is frequently associated to the name of the Soviet physicist L. Landau, in connection to his theory of phase transitions. In the mathematical literature it is frequently referred to as the Lyapounov-Schimidt procedure and, more recently, as the center manifold theory [2].

More intricate situations can also be envisaged in which several branches cross the λ axis in Fig. 8. This leads to the interaction between bifurcating solutions generating secondary or even tertiary bifurcation phenomena. The above results carry through, in the sense that one can guarantee that the part of the dynamics that gives information on the bifurcating branches takes place in a phase space of reduced dimensionality. The explicit construction of the normal form becomes, however, much more involved. In addition, their universality can no longer be guaranteed. For instance, in addition to fixed points, limit cycles or tori, new attractors can be generated by global bifurcation mechanisms which are intruding the normal form without having signaled their existence at the level of the stability analysis or of the detailed construction of the normal form. If the latter contains at least three coupled order parameters these global bifurcations may lead to chaotic dynamics, a flavor of which was already given in the preceeding Section.

It should be noticed that the reduction of the dynamics to a normal form remains possible in certain classes of bifurcations leading to space patterns in spatially distributed systems [1], [10]. In such cases the order parameters represent combinations of amplitudes of the dominant modes appearing in an expansion of the state variables in Fourier series or, more generally, in a series of linearly independent functions compatible with the symmetry properties of the system and the boundary conditions. As a matter of fact, such "low order models" are widely used in atmospheric physics even far away from any bifurcation point. A very interesting question is therefore to assess their status outside of the range of validity of the normal form. Some comments on this problem are presented in the concluding remarks by E. Lorenz.

4. CHAOTIC DYNAMICS

When a dynamical system evolves in a phase space of three or more dimensions, the topological constraints arising from the non-intersection of the trajectories are greatly relaxed and, as a result, a new kind of behavior known as chaotic dynamics can arise. Turbulence, one of the universal properties of large scale flows, is the most familiar example of a well-defined dynamical system whose state variables show an apparently erratic behavior in space and/or time. In the last years however, an explosion of experimental results has occurred which has established the existence of chaotic, turbulent-like motions in chemistry, optics, electrical circuits or materials science among others. In this section we discuss some models and define some minimal conditions that must be satisfied for the onset of chaotic behavior. We refer to the monographs of Schuster [3] and Bergé, Pomeau and Vidal [11] for a detailed introduction on this subject.

The ideas prevailing until the 1960's suggested a rather discouraging picture for the emergence of turbulence : turbulence was thought to be the manifestation of the dynamics of an infinity of coupled modes, each of which is associated to a particular frequency. Moreover, in order that these modes become fully coupled, an infinity of transitions occurring beyond the loss of stability of the regular steady-state (laminar) flow appeared to be necessary. It was therefore with great surprise and excitement that subsequently scientists became aware of the possibility that dynamical systems involving only a few variables can present chaotic behavior after a few bifurcations from the reference state.

The first such example was discovered by Lorenz [12].It provides a simplified description of the thermal convection problem in which only three modes, each associated to a particular spatial wavelength, are retained. It gives rise to an attractor which captures the principal features of turbulent convection. Such a phenomenon occurs routinely in the earth's atmosphere, and is at the origin of the well-known difficulty of weather prediction. What is more, it suggests (See also Chapter by E. Lorenz and C. Nicolis) that weather and climate are fundamentally unpredictable, since they should share one of the basic properties of chaotic motion, which is to depend in a sensitive way on the initial conditions.

Lorenz's work stimulated a wealth of developments on chaotic dynamics. In the following we give a few selective illustrations of "prototype" equations giving rise to chaotic behavior, and summarize the principal properties of chaotic attractors.

4.1. Some prototypes giving rise to chaotic behavior

A very interesting mathematical model of chaotic behavior has been suggested by Rössler [4], [5]. It contains three variables and only one quadratic nonlinearity :

$$\begin{aligned} \frac{dx}{dt} &= -y - z \\ \frac{dy}{dt} &= x + ay \\ \frac{dz}{dt} &= bx - cz + xz \end{aligned} \tag{17}$$

a, b, c being positive constants. The equations of evolution have two fixed points P_0 and P_1, one of which is the trivial solution : $x_s = y_s = z_s = 0$, and the other the solution $x_s = c - ab$, $y_s = b - c/a$, $z_s = c/a - b$. Hereafter we discuss solely the phenomena occurring around the first fixed point.

For a large range of parameter values the linear stability analysis predicts that the trivial solution can be unstable. The behavior in its vicinity has the following peculiar features : The trajectories are repelled along a two-dimensional surface of phase space associated to a pair of unstable complex eigenvalues of the linear stability operator (See eq. (9)), and are attracted along a one-dimensional curve corresponding to a negative (real) eigenvalue of this operator. We call such a fixed point a saddle-focus. Such a configuration gives rise to instability of motion, a basic ingredient of chaotic behavior, but it also allows for the reinjection of the unstable trajectories in the vicinity of P_0 and thus for the eventual formation of a stable attractor. Fig. 4 depicts the chaotic attractor attained for a particular set of parameter values. Both of the features just mentioned are clearly present.

Another important feature shown clearly in Fig. 4 is the folding of the surface along which the unstable motion occurs. It is instructive to visualize it by cutting transversally the flow by the plane $y = 0$, $x < 0$, $z < 1$. This is illustrated in Fig. 12. Here one plots the position of the (n + 1)st intersection point between the above defined surface (referred to as Poincaré surface of section) and the phase space trajectory, as a function of its position at the nth intersection. The numerical construction shows that one obtains a smooth bell-shaped curve [5]. Note that the positions of the successive intersections on this curve are not given by consecutive points but, rather, by points which appear to be distributed randomly. This establishes quite naturally the connection between our continuous time model, eq. (17), and a second very important class of dynamical systems showing chaotic behavior. These are discrete time models defined by the iterative equation

$$x_{n+1} = f_{n}(x_{n}) \tag{18a}$$

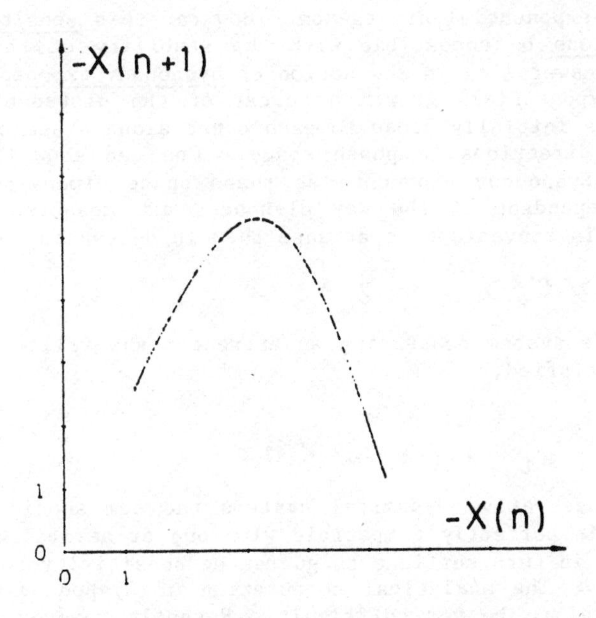

Fig. 12 Discrete time dynamics induced by the system of eqs (17) possessing the chaotic attractor of Fig. 4 on a Poincaré surface of section. The trajectories are cut by the plane ($y=0$, $x < 0$, $z < 1$). When the value of $-x$ at the (n+1)st intersection is plotted against its value at the nth intersection the bell-shaped curve shown in the Figure is obtained.

A typical bell-shaped curved similar to Fig. 12 is obtained by choosing

$$f = 1 - \lambda x^2 \tag{18b}$$

Despite their apparent simplicity, Eqs. (18) show an incredibly rich behavior ranging from simple fixed points to periodic or chaotic solutions, reviewed in the Chapter by E. Lorenz.

4.2. Characterization of chaotic attractors.

Asymptotic stability with respect to nearby initial conditions, and instability of motion on the attracting set itself are the principal features of chaotic attractors. Their very coexistence implies that these attractors should enjoy certain peculiar mathematical properties which we now briefly summarize.

In the first place, the instability of motion implies that if one probes at random pairs of nearby states on the attractor and follows their subsequent evolution, one should find that in most cases there is exponential divergence. How can this sensitivity to initial conditions be compatible with the stability of the overall system ? The answer lies in the notion of Lyapounov exponents, which measure the (exponential) growth or decay of the distance bet ween two trajectories initially close to each other along a set of linearly independent directions in phase space. One can show that there exist as many Lyapounov exponents as phase space dimensions, whose values are independent of the way distances are measured in phase space [2]. It is convenient to arrange them in decreasing order,

$$\sigma_1 \geq \sigma_2 \geq \cdots\cdots \geq \sigma_n \tag{19}$$

In a dissipative system possessing an attractor the following condition must be satisfied,

$$\sigma_1 + \sigma_2 + \cdots\cdots + \sigma_n < 0 \tag{20}$$

whereas for conservative dynamical systems the sum should be zero. This condition is perfectly compatible with one or several σ's being positive. This in turn suffices to guarantee sensitivity to the initial conditions. The analytical computation of Lyapounov exponents proves in general to be very difficult. Recently however, interesting numerical algorithms have been constructed which allow for the computation of the first few largest exponents [13].

The coexistence of instability of motion and attractivity forces on chaotic attractors a very peculiar geometrical structure. For instance, the folding and reinjection patterns found in the Rössler attractor (Fig. 4) turn out to be typical. Moreover, the discovery of chaotic dynamics in systems of three variables, and the fact that a two-dimensinal attractor cannot present instability of motion, imply that a large class of chaotic attractors should have a fractal geometry [6]. The key mathematical concept characterizing fractal objects is the Hausdorff dimension D expressing essentially the increase of the number N_ϵ of hypercubes of size ϵ necessary to cover the set as ϵ goes to zero. By its very definition D does not distinguish between uniform attractors enjoying self-similarity on all scales and highly non-uniform ones. For this reason a number of alternatives has been proposed [14]. Just like for Lyapounov exponents these dimensions are very difficult to compute analytically. An interesting algorithm [15] allows however the determination of D from the knowledge of a time series pertaining to the evolution of a single variable, as seen further in the Chapter by C. Nicolis.

It is amazing that seemingly "static" properties like D can be related to "dynamical" properties like Lyapounov exponents. The link is provided by the Kaplan-Yorke conjecture [16]

$$D = j + \frac{\sum_{i=1}^{j} \sigma_i}{|\sigma_{j+1}|} \tag{21}$$

where j is the largest integer for which $\sum_{i=1}^{j} \sigma_i \geq 0$.

An alternative way to characterize chaotic attractors, which is at the heart of the very nature of chaotic dynamics, is the probabilistic description. The motivation is obvious : owing to the sensitivity to initial conditions the concept of trajectory is no longer operationally meaningful. Instead, it becomes natural to argue in terms of the probability of observing the system in the vicinity of a certain state. Consider, for instance, the simple one-dimensional map [17].

$$x_{n+1} = f(x_n) = 2x_n \bmod 1$$
$$0 \leq x \leq 1 \; ; \quad n = 0, 1, 2, \ldots \tag{22}$$

If the initial condition is written in binary representation,

$$x_0 = \sum_{n=0}^{\infty} a_n \frac{1}{2^n} \; ; \quad a_n = 0 \text{ or } 1 \tag{23}$$

then one immediately sees that the action of f is to delete the first digit and shift the remaining sequence of digits to the left. Because of this Bernoulli property, one can always find an initial x_0 such that the successive iterates $f^{(1)}(x_0) \cdots, f^{(k)}(x_0)$ correspond to the random sequence generated, say, by coin tossing. We can thus label the two possible outcomes of each "trial" by the "symbols" L and R, according as the representative point is in the left or in the right half of the unit interval, and study the symbolic dynamics of L's and R's instead of eq. (23). Implicit in the emergence of the above strong properties is the fact that one deals, typically, with irrational numbers. However, even if one approximates such a number arbitrary well by a finite sequence of binary digits, the "error" is amplified exponentially by successive iterations (the Lyapounov exponent of (22) is $\sigma = \ln 2$) and, because of this, the trajectories will come as close as desired to any given state in the course of time.

The fact that the output of a system undergoing chaotic dynamics can be viewed as a succession of symbols released with a certain probability suggests that chaos should be intimately related to information. Let Pi be the a priori probability of state i. The entropy of the underlying stochastic process [18]

$$S = -\sum_{i=1}^{N} P_i \ln P_i \tag{24}$$

describes the information needed to locate the system in a particular state. The Kolmogorov-Sinai entropy, h on the other hand [3] measures the rate at which information about the state of a dynamical system is lost in the course of time. A system on a chaotic attractor should therefore possess a positive Kolmogorov-Sinai entropy. As it turns out, the latter can be related to the average over the positive Lyapounov exponents.

5. STOCHASTIC DESCRIPTION OF DYNAMICAL SYSTEMS.

In many circumstances the environment impinging on a physical system is a complex, noisy one. For instance, atmospheric and climate dynamics are affected by solar variability arising from fluctuations of the solar constant, or by changes of atmospheric composition arising from volcanic eruptions or other unpredictable events. In addition, even if the environment remains constant, complex physical systems generate spontaneously random variations from the mean values of their state variables. As an example, a local imbalance in the energy budget of a small area of the earth's surface can arise at any moment even though on a somewhat larger scale one would still characterize the state of the system by space and time-averaged quantities. Last but not least chaotic dynamics which is ubiquitous in geosciences if only because of atmospheric turbulence, provides a universal mechanism of variability. Depending on the level of description, this variability can be perceived as a fluctuation characterized by certain statistical properties, or as a deterministic event.

Whatever their origin might be, fluctuations are likely to affect the evolution of our dynamical system and must therefore be incorporated in the analysis. This is achieved by adopting a probabilistic description in which the deterministic dynamics is continuously "perturbed" by the fluctuations, modelled as a random noise. Considering for simplicity one-variable systems, this yields a stochastic differential equation [19], [20] of the form

$$\frac{dX}{dt} = F(X, \lambda) + G(X) f(t) \tag{25}$$

in which $F(X, \lambda)$ is the deterministic rate law (cf. eq. (6)), f(t) is a random force describing the fluctuations and G(X) a function expressing the coupling of the fluctuations to the system's dynamics.

In many cases f(t) can be modelled as a Gaussian white noise. This means that if we take moments of f over many possible evolutions (technically speaking "realizations"), moments of order higher than two are expressed in terms of those of order two, the latter being uncorrelated in time. More quantitatively :

$$\langle f(t) \rangle = 0$$

$$\langle f(t) f(t') \rangle = q^2 \delta(t-t') \qquad (26)$$

where the brackets denote averaging and q^2 is the <u>variance</u> of the fluctuations.

One may of course wonder whether the Gaussian white noise idealization is legitimate. In so far as the value of the quantity giving rise to the fluctuating source f(t) arises primarily from the superposition of a large number of loosely coupled variables (think for instance of a global quantity being the sum of local variables), this idealization will be reasonable : on the one side the central limit theorem, a major result of probability theory [20], gives under these conditions to the Gaussian distribution the privileged role it plays in statistics ; and on the other side, the loose coupling of the local variables will ensure statistical independence of the successive values of the global variable for different times. But if the fluctuations happen to be manifestations of some underlying chaotic dynamics, the situation may be quite different. For instance, the fluctuations of the variables of the Lorenz model [12] are neither Gaussian nor white.

At this point it may be instructive to allude briefly to a <u>credo</u> prevailing in geoscience literature, according to which fluctuations arise from the elimination of the "fast" variables. This belief is, at best, an oversimplified view of a much subtler reality. Let $\underline{F}_s$, $\underline{F}_f$ be the evolution laws for the "slow" and "fast" variables $\underline{X}$ and $\underline{Y}$ respectively. The equations for these variables are then of the form

$$\frac{d\underline{X}}{dt} = \underline{F}_s(\underline{X}, \underline{Y}, \mu)$$

$$\mu \frac{d\underline{Y}}{dt} = \underline{F}_f(\underline{X}, \underline{Y}, \mu) \qquad (27)$$

where the presence of the parameter $\mu \ll 1$ expresses the fact that the time scale of variation of $\underline{Y}$, $\tau = t/\mu$ is much faster than t. In the limit of a complete scale separation, $\mu \rightarrow 0$, an important theorem of analysis due to Tikhonov [21] asserts that under certain conditions on $\underline{F}_f$ the second equation (27) becomes

$$\underline{F}_f(\underline{X}, \underline{Y}, 0) = 0$$

from which Y can be expressed as a function of X, Y = h(X). Substituting into the first equation (27) one obtains then a closed set of equations for the slow variables,

$$\frac{d\underline{X}}{dt} = F_s(\underline{X}, \underline{h}(\underline{X})) \tag{28}$$

in which there is no trace whatsoever of a randomness associated with the fast variables. True, the reduced system of equations for X may have chaotic solutions due to the extra nonlinearity introduced by $\underline{h}(\underline{X})$, but this is a completely different mechanism. What may also

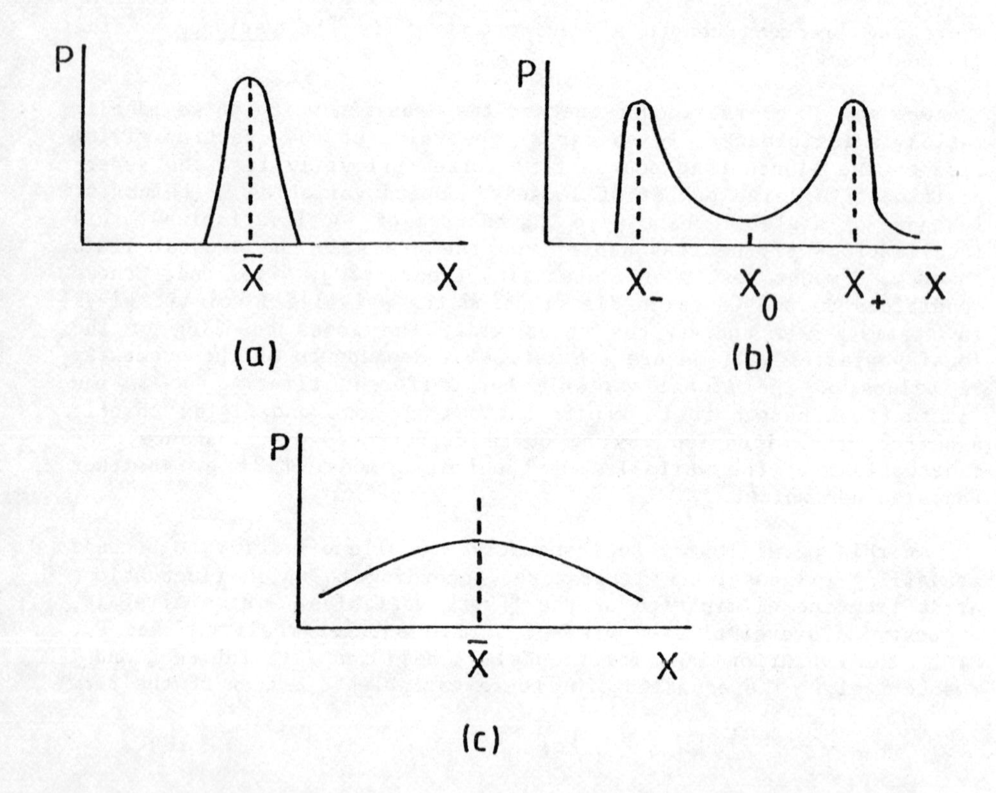

Fig. 13 Stochastic counterpart of bifurcation. As the system crosses the critical value λ_c of a control parameter λ the probability function switches from a unimodal form peaked sharply on a unique attractor ($\lambda < \lambda_c$, Fig. 13a) to a multihumped distribution whose maxima coincide with the new attractors emerging beyond bifurcation ($\lambda > \lambda_c$, Fig. 13c). At the critical value $\lambda = \lambda_c$ the probability, although one-humped, is considerably flattened (Fig. 13b). This reflects the ability of the system to visit large parts of state space with appreciable probability.

happen is that, to begin with, the full set of eqs (27) runs on a chaotic attractor. In this case it will in general be impossible to eliminate $\underline{Y}$, despite the appearance of μ , because the other conditions of validity of Tiknonov's theorem will not be secured. In short, chaotic dynamics is the ultimate mechanism of generation of random behavior. This is true even in simple laboratory problems like brownian motion : the jitery motion of a heavy particule in a fluid arises because the molecules undergo a random (chaotic) motion owing to the encounters between them.

Be it as it may, let us use the simple model described by eqs (25) - (26) as an illustration and proceed to a brief presentation of the techniques of solving stochastic differential equations (SDE). The main point is that SDE forced by white noise define a class of random processes known as diffusion processes, whose probability distribution P(X, t) obeys to the Fokker-Planck equation [19], [20]. We hereafter write this equation explicitly in the case of additive fluctuations for which the coupling function G(X) in eq. (25) reduces to a constant :

$$\frac{\partial P}{\partial t} = -\frac{\partial}{\partial X} F(X, \lambda) P + \frac{q^2}{2} \frac{\partial^2}{\partial X^2} P \tag{29}$$

Notice that if G(X) is not a constant (we speak then of multiplicative fluctuations), eq. (25) needs to be supplemented with a prescription extending the rules of classical calculus.

Let us seek for time-independent solutions of eq. (29). Integrating once over X we obtain

$$- F(X, \lambda) P_s + \frac{q^2}{2} \frac{\partial P_s}{\partial X} = \text{const}$$

Now, when X reaches its extreme values (referred to as "the boundaries" of the process), we expect that P_S and all its finite order derivatives will vanish. The constant in the right hand side is therefore equal to zero and

$$- F(X, \lambda) P_s + \frac{q^2}{2} \frac{d P_s}{d X} = 0 \tag{30}$$

In a sense, eq. (30) expresses the vanishing of the "probability flux" traversing the system. By analogy with Statistical Mechanics we say that we have in this case a "generalized detailed balance".

The solution of eq. (30) is straightforward :

$$P_s \sim \exp\left\{-\frac{2}{q^2} U(X)\right\} \tag{31a}$$

where the kinetic potential U(X) is defined by

$$U = - \int F(X, \lambda)\, dX \tag{31b}$$

If, for example , F is given by eq. (11a) U becomes

$$U = -\left[\frac{\lambda - \lambda_c}{2} X^2 - \frac{u}{4} X^4\right] \qquad (31c)$$

Eqs (31a-c) allow us to set up the stochastic counterpart of stability and bifurcation, as depicted in Fig. 13. To be specific, suppose that the dynamics gives rise to a unique globally stable attractor (Fig. 13a). This is what happens, typically, below a bifurcation point ($\lambda < \lambda_c$ in eq. (11a)). The probability P_s presents then a single maximum which is very sharp if q is small, and is well-approximated by a Gaussian : the system spends most of its time around the stable state, but is nevertheless capable of exploring the remaining of the phase space as well. We call this property metric transtivity.

Let now many states be available (Fig. 13c). This will happen, typically, beyond bifurcation ($\lambda > \lambda_c$ in eq. (11a)). The probability distribution becomes then multi-humped, and displays a very deep minimum between two sharp maxima. The statistics of X is very complex in this case and corresponds to what Lorenz has called metric almost-intransitivity [22], since for q small transitions across the minimum would take an exceedigly long time. Some implications of this behavior in climate dynamics are discussed in the Chapter by C. Nicolis.

Fig. 13b depicts the borderline case in which $\lambda = \lambda_c$: there is a unique macrostate available which is still stable, but the kind of stability it enjoys is weaker : the perturbations are not damped exponentially as when $\lambda < \lambda_c$, but decay as an inverse power of time. Correspondingly the distribution P_S, although one-humped, is considerably flatter than in Fig. 13a, and allows the system to visit large parts of phase space with appreciable probability.

The above results can be extended to multi-variable nonlinear systems operating near bifurcation. Some results on other kinds of noise and on the time-dependent properties of the probability distribution are also available [23]. A very important problem is also to set clearcut limits between random noise and the kind of variability connected with chaotic dynamics. For further comments on this basic question we refer to the Chapters by E. Lorenz and C. Nicolis.

REFERENCES

1. Nicolis G. and Prigogine I., Self-Organization in Nonequilibrium Systems, Wiley, New York, (1977).
2. Guckenheimer J. and Holmes Ph., Nonlinear Oscillations, Dynamical Systems and Bifurcations of Vector Fields, Springer, Berlin (1983).
3. Schuster H.G., Deterministic Chaos, Physik-Verlag, Weinheim (1984).
4. Rössler O., in Bifurcation Theory and its Application to Scientific Disciplines, Ann. N.Y. Acad. Sci. Vol. 316, New York (1979).
5. Gaspard P. and Nicolis G., J. Stat. Phys. 31, 499 (1983).
6. Mandelbrot B., Fractals : Form, Chance, Dimension, Freeman, San Francisco (1977).
7. Andronov A., Vit. A. and Khaikin C., Theory of Oscillators, Pergamon, Oxford (1966).
8. Thom R., Stabilité Structurelle et Morphogénèse, Benjamin, New York (1972).
9. Landau L. and Lifshitz E.M., Statistical Physics, 3rd edition, Pergamon, Oxford (1980).
10. Nicolis G., Dissipative Systems, Repts Progress in Physics, in press.
11. Bergé P., Pomeau Y. and Vidal C., L'ordre dans le chaos, Herman, Paris (1984).
12. Lorenz E. , J. Atmos. Sci. 20, 130 (1963).
13. Wolf A., Swift J., Swinney H. and Vastano J., Physica 16D, 185 (1985).
14. Farmer J., Z. Naturf. 37a, 1304 (1982).
15. Grassberger P. and Procaccia I., Phys. Rev. Lett. 50, 346 (1983).
16. Kaplan J. and Yorke J., in Functional Differential Equations and Approximations of Fixed Points, Springer, Berlin (1979).
17. Lichtenberg A. and Lieberman M., Regular and Stochastic Motion, Springer, Berlin (1983)
18. Khinchine A., Mathematical Foundations of Information Theory, Dover New York (1957).
19. Stratonovitch R., Topics in the Theory of Random Noise, Vol. II, Gordon and Breach, New York (1967).
20. Soong T., "Random Differential Equations in Science and Engineering", Academic Press, New York (1973).
21. Wasow W., "Asymptotic Expansions for Ordinary Differential Equations", Wiley, New York (1965).
22. Lorenz E., J. Appl. Meteor. 9, 325 (1970).
23. See, for instance, Gardiner C., "Handbook of Stochastic Methods", Springer, Berlin (1983).

CATASTROPHE THEORY IN GEOPHYSICS

J. F. Nye
Department of Physics
University of Bristol
Tyndall Avenue
Bristol BS8 lTL U.K.

ABSTRACT. Two lectures were given on this topic, and what is reproduced here are the introductory notes provided for the students. Lecture 1 discussed some of this basic material and illustrated it with optical examples drawn from Nye (1978), Nye (1979), Berry, Nye and Wright (1979) and Berry and Upstill (1980). Lecture 2 applied the results of catastrophe theory summarised here to caustics formed in the Earth by seismic events, as fully described in Nye (1985). Three examples were discussed : (1) caustics between 10° and 30° from the event, associated with velocity changes in the transition zone of the mantle; (2) the 143° caustic caused by seismic rays that have been refracted twice at the core-mantle boundary; (3) the caustic formed at the antipodal point from the event. All these caustics are affected by the non-sphericity of the Earth's structure in ways that are predictable in general outline by catastrophe theory. In particular, it is suggested that the well-known 143° caustic is actually cusped because of bumps on the core-mantle boundary.

1. CATASTROPHE OPTICS AND SEISMOLOGY

The lectures discuss how catastrophe theory is useful in understanding seismic wave phenomena. This is an unconventional approach to seismic waves not in the text books. Most of the discussion in the literature about catastrophe theory, applied to waves, has been about light rays (geometrical optics, catastrophe optics) but it applies equally well to seismic rays. The Earth is an inhomogeneous refracting medium that bends seismic rays, just as an inhomogeneous optically transparent medium bends light rays. There are two differences to notice: (1) the solid earth transmits shear waves (S waves) and compression waves (P waves) at different velocities even if it is assumed elastically isotopic, while an isotropic optical medium transmits just one kind of light wave; (2) seismic disturbances travelling through the body of the earth (body waves) consist of short pulses that show very little dispersion, while in optics one usually considers continuous (monochromatic) waves. (Surface seismic waves are dispersive.) Apart from these theoret-

C. Nicolis and G. Nicolis (eds.), Irreversible Phenomena and Dynamical Systems Analysis in Geosciences, 31–40.

ical differences there is the practical difference that in optics one commonly observes spatial patterns of light, while with seismic waves observations are made at points (seismic observatories) as a function of time. None of these differences seriously affects the underlying theory, at the first level of approximation we are concerned with, which is ray theory. At the next level of approximation we shall consider scalar diffraction theory, and again this is equally applicable to both kinds of waves. Consequently, in the theoretical development we can consider seismic and optical rays largely interchangeably.

Catastrophe optics is about the foci of rays (caustics) in the absence of symmetry. It differs in this respect from conventional geometrical optics, which is about lenses, with cylindrical symmetry, and about departures from the ideal point focus (aberrations). Thus it deals with the optics of Nature. It is thus particularly appropriate in seismology, where nearly all problems, except the simplest idealised ones, do in fact lack symmetry. Development of the subject has been stimulated by a remarkable mathematical theorem (Thom's theorem) which is part of singularity theory. This classifies the ways in which the extrema of functions can coalesce. Rays are the extrema of an optical distance function (Fermat's Principle). Focusing occurs on the envelopes (caustics) where neighbouring rays touch, and so is precisely the coalescence of these extrema. Thus Thom's theorem provides a classification of 'generic' or 'structurally stable' caustics.

As an illustration, the ideal point focus of a perfect lens is highly unstable; the slightest departure from perfection in the lens destroys it. On the other hand, the light caustic you see in your coffee cup on a sunny day (fig. 1), with its bright fold line and cusp, is structurally stable; the cup need not be circular at all, just

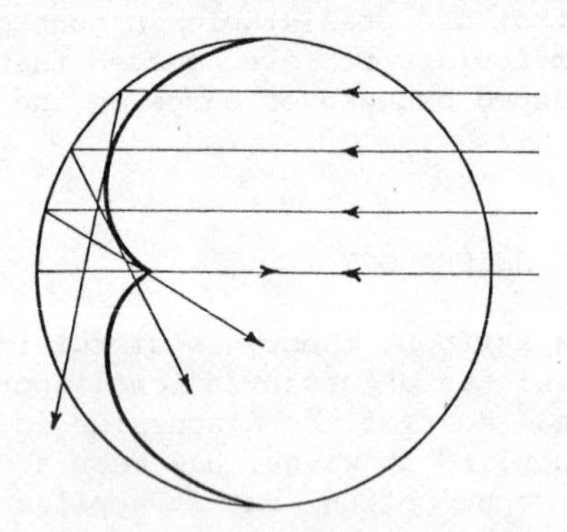

Fig. 1. Caustic in a coffee cup

smooth and reflecting (try it with a plastic cup and try deforming the cup). The fold and the cusp are in fact the only structurally stable caustics in two dimensions (the surface of your coffee).

The structurally stable caustics are classified by their codimension, that is, the number of parameters needed to specify them. For example, the fold has codimension 1 and so appears as a point on a line, as a smooth curve in a plane (the surface of the coffee), or as a surface in space (which could be explored by drinking some of the coffee). The cusp has codimension 2 and so does not appear on a line; it appears as a point in a plane, or as a line (a sharp edge on the fold surface) in space. The complete list up to codimension 3 is shown in Table I. This list is exhaustive (Thom's theorem). There are no other

Table I
List of caustics

Caustic in space	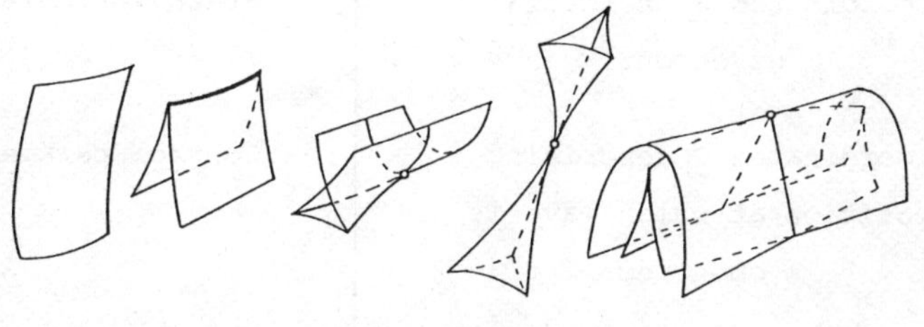				
NAME	Fold	Cusp	Swallowtail	Elliptic umbilic	Hyperbolic umbilic
Codimension	1	2	3	3	3

stable caustics in three dimensions. All others would break up, if the system were perturbed, into a combination of these elementary forms. These forms are the 'atoms'.

Examples

In optics: rays from a point source when refracted by smooth objects (glass, water drops) will produce folds and cusps on screens, photographic plates or the retina. In space they will produce swallowtails, elliptic and hyperbolic umbilics.

Seismic: rays from an earthquake will produce, in general, folds and cusps on the surface of the Earth. Within the Earth they will produce swallowtails, elliptic and hyperbolic umbilics.

Notice that the caustic (catastrophe) of codimension 2 (the cusp) contains the caustic of codimension 1 (the fold). Similarly the three caustics of codimension 3 contain the caustics of lower codimension. We say that the point singularities (foci) at the centres of these caustics "unfold" into an array of singularities of lower codimension.

2. CONNEXION BETWEEN THE MATHEMATICS AND THE PHYSICS

The glossary in Table II indicates the correspondence between the physics of ray theory and the mathematics of catastrophe theory. The

Table II

Glossary

Ray theory (physics)	Catastrophe theory (mathematics)
Coordinates $\underline{S}$ on initial wavefront	State variables $\underline{S}$
Coordinates $\underline{C}$ describing position at which wave is observed	Control parameters $\underline{C}$
Optical distance $\underline{S}$ to $\underline{C}$	Potential function $\phi(\underline{S};\underline{C})$
Fermat's Principle gives coordinates $\underline{S}$ from which rays travel to given $\underline{C}$	Gradient map $\nabla_{\underline{S}}\phi(\underline{S};\underline{C})=0$ gives extrema in $\underline{S}$ space
Caustics in $\underline{C}$ i.e. envelope of rays	Locus of $\underline{C}$ for which ϕ has degenerate extrema in $\underline{S}$
Moving on to a caustic in $\underline{C}$	Changing parameters $\underline{C}$ so that extrema in $\underline{S}$ coalesce
Structurally stable caustic	Elementary catastrophe in $\underline{C}$ i.e. singularity of gradient map

After Berry and Upstill (1980)

easiest way of starting is to imagine the rays starting out perpendicular to some initial wavefront (but we shall find later that this is not the most convenient formulation for the seismic application).

Table II uses general coordinates: $\underline{S}$ in state space, which in this case is the initial wavefront, and $\underline{C}$ in control space, which in this case is the three-dimensional space in which the caustics are observed. Thom's theorem tells us, remarkably, that the potential function

$\phi(\underline{S}; \underline{C})$ is always (for structurally stable caustics) reducible to one of five standard polynomial forms (Table III). Reducible here means that

Table III

Standard polynomials

Name	Potential $\phi(\underline{S}; \underline{C})$
fold	$S_1^3 + C_1 S_1$
cusp	$S_1^4 + C_2 S_1^2 + C_1 S_1$
swallowtail	$S_1^5 + C_3 S_1^3 + C_2 S_1^2 + C_1 S_1$
elliptic umbilic	$S_1^3 - 3S_1 S_2^2 + C_3 (S_1^2 + S_2^2) + C_2 S_2 + C_1 S_1$
hyperbolic umbilic	$S_1^3 + S_2^3 + C_3 S_1 S_2 + C_2 S_2 + C_1 S_1$
Note: To all these potentials may be added squared terms of either sign in the remaining ("inessential") state variables. For example, for the fold $\phi(\underline{S};\underline{C}) = S_1^3 + C_1 S_1 + S_2^2 + S_3^2 - S_4^2$	

the coordinates have to be appropriately chosen. Thus smooth, reversible coordinate changes are allowed in the state variables

$$\underline{S} \to \underline{S}'$$

$$\underline{C} \to \underline{C}'.$$

2.1 Example for homogeneous medium (straight line rays)

The following example of a coordinate scheme will help to make matters clearer. Fig. 2 shows an initial wavefront W specified by its height f(x, y) above a reference plane and let its slope be small: $\partial f/\partial x \ll 1$, $\partial y \ll 1$. So a general point Q on the wavefront has coordinates

(x,y,f). The distance ℓ between Q and a general point P in the space above the wavefront (X,Y,Z) is given by

$$\ell^2 = (X-x)^2 + (Y-y)^2 + (Z-f)^2. \qquad (1)$$

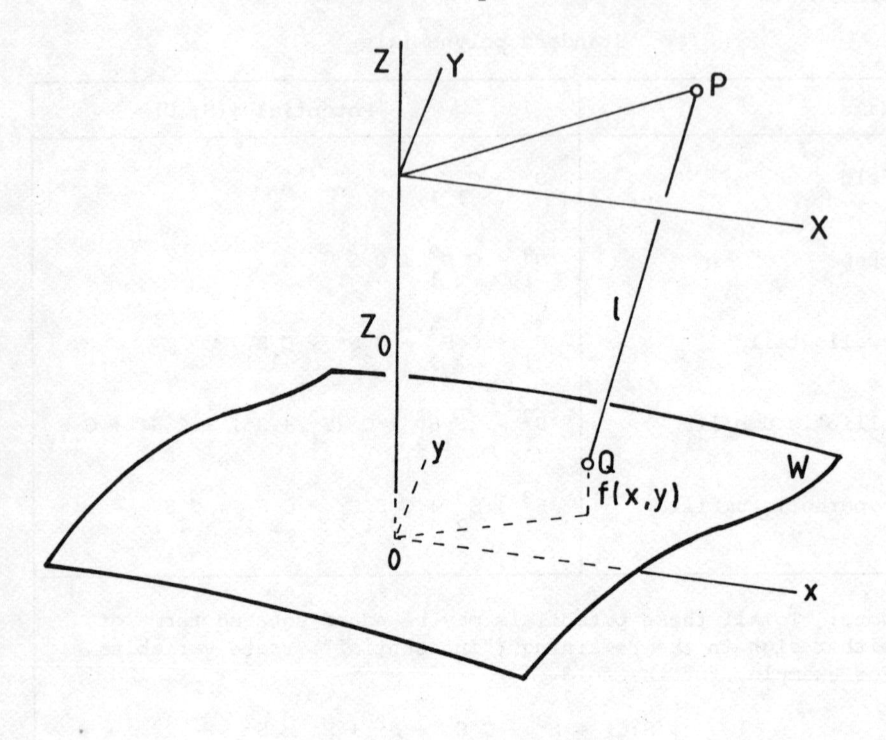

Fig.2. Showing coordinates

In this example we shall suppose that the rays travel in straight lines (which is not true for the Earth), so to find the ray through P we keep (X,Y,Z) fixed and look for extrema of ℓ, or ℓ^2, with respect to Q. So we can discard the part of ℓ^2 that does not depend on (x,y), and it is convenient to divide the remainder by $-2Z$. Thus we define our distance function, $\phi(\underline{S};\ \underline{C})$, in Table II, by

$$\phi(x,y;\ X,Y,Z) = f(x,y) - \frac{x^2+y^2}{2Z} + \frac{xX+yY}{Z} - \frac{\{f(x,y)\}^2}{2Z}. \qquad (2)$$

As a check, let us verify that the extrema of ϕ do indeed give the rays. Keeping (X,Y,Z) fixed and varying (x,y) to satisfy $\partial\phi/\partial x = \partial\phi/\partial y = 0$ gives

$$\frac{\partial f}{\partial x} = \frac{x-X}{Z-f}, \quad \frac{\partial f}{\partial y} = \frac{y-Y}{Z-f}. \qquad (3)$$

These are just the conditions that PQ is normal to the wavefront, that is, that PQ is a ray through P. The rays provide a gradient mapping from (x,y) to (X,Y,Z).

Suppose the wavefront can now be expressed as

$$f(x,y) \equiv g(x,y) + \frac{x^2+y^2}{2Z_o}, \tag{4}$$

where g(x,y) is quadratic or higher. Then choose the point $(0,0,Z_o)$ as a new origin (focus) in control space, put

$$X = X_1, \quad Y = Y_1, \quad Z = Z_o + Z_1 \tag{5}$$

and take X_1, Y_1, Z_1 to be small. We then obtain (assuming f<<Z)

$$\phi(x,y;\ X,Y,Z) = g(x,y) + \frac{Z_1}{2Z_o^2}(x^2+y^2) + \frac{X_1 x}{Z_o} + \frac{Y_1 y}{Z_o}. \tag{6}$$

Example

Let $g(x,y) \equiv a(x^3-3xy^2)$. Then reference to Table III shows that, after appropriate scaling of the variables, ϕ is the potential for the elliptic umbilic. We can then deduce that the caustic from this wavefront will be as sketched in Table I.

The caustics are the singularities of the mapping $\underline{S}\rightarrow\underline{C}$. Thus with one dimension x of state space and one dimension X of control space we could have figure 3(a). On one side of the caustic (fold) F in X there are 2 rays (2 values of x) and on the other no rays. The fold caustic separates regions of different "multiplicity".

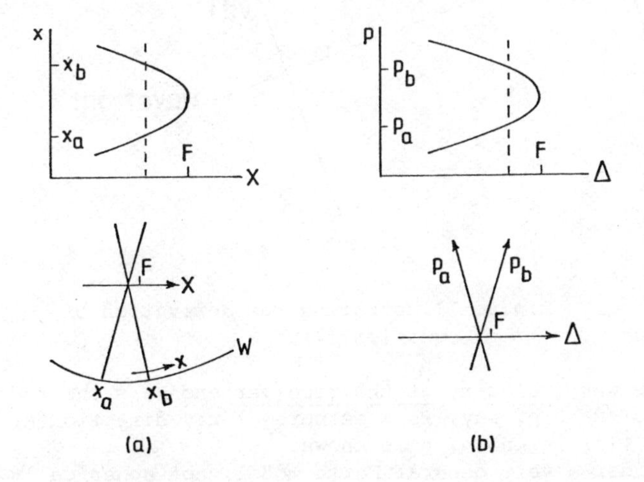

Fig. 3. Mapping between state space and control space.
F is the fold caustic in control space.

2.2 Inhomogeneous medium (curved rays)

When the medium is inhomogeneous, like the Earth, the rays are not straight. To deal with this we do not use the idea of an initial wavefront for defining the state coordinates (labelling the rays), but instead set up a theory which uses exclusively quantities measured in the neighbourhood of the observing point (eg. Nye and Hannay 1984). We do this by labelling the rays with their directions p. Thus figure 3(b) shows a mapping from p, a state variable, to Δ, a control variable, with a singularity F (fold caustic). On one side of the caustic there are 2 p's for each Δ and on the other side no p's.

Now apply this to the Earth. Consider a seismic event at the surface of a spherically symmetric Earth and let $T(\Delta)$ denote the travel time of a body wave to a receiving station on the surface at an angular distance Δ. The angle α between a ray reaching the receiver and the Earth's surface is given by (figure 4)

$$\cos \alpha = \frac{v}{R} \frac{dT}{d\Delta}, \tag{7}$$

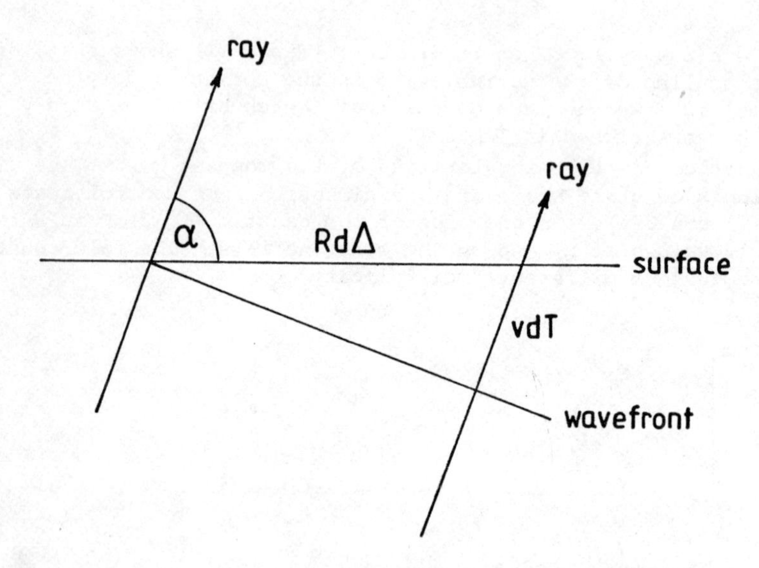

Fig.4. Illustrating the derivation of equation (7).

where v is the wave velocity at the receiver and R is the radius of the Earth. Thus $dT/d\Delta \equiv p$, say, is a measure of ray direction at the receiver. If $T(\Delta)$ is known, p is known.

Now consider a very general Earth model, not spherically symmetric, inhomogeneous and anisotropic, where there is a single seismic event at any depth. At a chosen receiving station take surface coordinates

Δ_1, Δ_2 and assume the travel-time function $T(\Delta_1,\Delta_2)$ is known. Then $(\partial T/\partial\Delta_1, \partial T/\partial\Delta_2) \equiv (p_1,p_2)$, say, provides a measure of the direction of the arriving ray for each point (Δ_1,Δ_2). Arrivals by different paths mean that $T(\Delta_1,\Delta_2)$ is multivalued. We want to construct a potential function that is single-valued. To do this first construct the single-valued time function

$$\tau(p_1,p_2) \equiv T(\Delta_1(p_1,p_2),\Delta_2(p_1,p_2)) - \Delta_1(p_1,p_2)p_1 - \Delta_2(p_1,p_2)p_2.$$

Then note that, by differentiation

$$\frac{\partial\tau}{\partial p_1} = -\Delta_1, \qquad \frac{\partial\tau}{\partial p_2} = -\Delta_2 . \tag{8}$$

τ is analogous to f in (2) and (3).

Next define a potential by

$$\phi(p_1,p_2;\ \Delta_1,\Delta_2) \equiv \tau(p_1,p_2) + \Delta_1 p_1 + \Delta_2 p_2, \tag{9}$$

with p_1,p_2 as state variables, and Δ_1,Δ_2 as controls. The gradient condition

$$\frac{\partial\phi}{\partial p_1} = \frac{\partial\phi}{\partial p_2} = 0 \qquad (\Delta_1,\Delta_2 \text{ constant})$$

generates (8), as we require (just as the gradient condition on (2) generated the ray equations (3)). Thus the potential (9), constructed from $T(\Delta_1,\Delta_2)$, does indeed generate the gradient mapping $\underline{p}\rightarrow\underline{\Delta}$ that we want and Thom's theorem tells us what stable caustics to expect.

REFERENCES

A general textbook on catastrophe theory:
T. Poston and I. Stewart 1978. Catastrophe theory and its applications. Pitman : London.

Review articles on catastrophe theory applied to optics:
M.V. Berry 1976. 'Waves and Thom's theorem'. Advances in Physics, 25, 1-26.
M.V. Berry and C. Upstill 1980. 'Catastrophe optics : morphologies of caustics and their diffraction patterns'. Prog. in Optics, 18, 257-346.

For background material on the Earth and seismic waves:
M.H.P. Bott 1982. The interior of the Earth : its structure, constitution and evolution. 2nd ed. Arnold : London.

Other applications of catastrophe theory to seismic waves:
J.A. Rial 1984. 'Caustics and focussing produced by sedimentary basins : applications of catastrophe theory to earthquake seismology'. Geophys. J.R. astr. Soc. 79, 923-938.

Other references in text:
J.F. Nye 1978. 'Optical caustics in the near field from liquid drops'. Proc. Roy. Soc. A361, 21-41.
J.F. Nye 1979. 'Optical caustics from liquid drops under gravity'. Phil. Trans. Roy. Soc. A292, 25-44.
M.V. Berry, J.F. Nye and F.J. Wright 1979. 'The elliptic umbilic diffraction catastrophe'. Phil. Trans. Roy. Soc. A291, 453-484.
J.F. Nye and J.H. Hannay 1984. 'The orientations and distortions of caustics in geometrical optics'. Opt. Acta. 31, 115-130. See §8 for caustics in an inhomogeneous medium.
J.F. Nye 1985. 'Caustics in seismology'. Geophys. J.R. astr. Soc. 83, 477-486.

PART II

GEOPHYSICAL FLUID DYNAMICS

PATTERNS OF HEAT TRANSFER BY THERMAL CONVECTION IN A ROTATING FLUID SUBJECT TO A HORIZONTAL TEMPERATURE GRADIENT : A SUMMARY

Raymond Hide
Geophysical Fluid Dynamics Laboratory
Meteorological Office (Met O 21)
Bracknell, Berkshire RG12 2SZ
England, UK

1. INTRODUCTION

Laboratory studies and associated analytical and numerical investigations of thermally-induced motions in a rapidly-rotating fluid of low viscosity are of intrinsic interest in fluid dynamics, and they also bear on a wide variety of natural flow phenomena in atmospheres and oceans. Advective heat flow perpendicular to the rotation axis $r = 0$ (where (r, ϕ, z) are the cylindrical polar co-ordinates of a general point in a frame of reference that rotates with angular speed Ω about $r = 0$, requires inter alia that u_r, the r-component of the Eulerian relative flow velocity $\underline{u}$, shall be non-zero. Since $u_r \doteqdot (2\Omega\bar{\rho}r)^{-1}\partial p/\partial\phi$ (where p denotes pressure and $\bar{\rho}$ mean density) nearly everywhere when Ω is so large that the geostrophic relationship (see equation (2) below) holds throughout most of parts of the fluid, in rapidly rotating systems advective heat transfer must be associated with departures from axial symmetry in the pattern of flow. Theory indicates and experiments amply confirm that strong departures from axial symmetry are 'generic' features of the patterns of thermally-induced flow in rapidly rotating fluids, even when the boundary conditions are axisymmetric.

There have been many experiments on thermal convection in a fluid annulus that rotates about a vertical axis and is subject to axisymmetric impressed differential heating and cooling with strong horizontal components and, in many cases, no impressed vertical component. These include careful determinations of the dependence of heat transfer across the annulus on Ω and other factors, including sloping endwalls which produce effects that are analogous in some respects to the latitudinal variational of the vertical component of Coriolis parameter that features prominently in theories of large-scale flow phenomena in atmospheres and oceans. The principal findings of these experiments provide a paradigm for understanding large-scale flows in planetary atmospheres, some of which are highly chaotic, and others (e.g. long-lived eddies in the atmospheres of Jupiter and Saturn) exhibit a high degree of order.

We note here in passing that systematic studies of effects due to

C. Nicolis and G. Nicolis (eds.), Irreversible Phenomena and Dynamical Systems Analysis in Geosciences, 43–51.

departures from axial symmetry in the boundary conditions have also been instructive. So far as heat transfer is concerned, the most dramatic enhancement is produced simply by introducing a rigid radial barrier connecting the inner wall of the annulus to the outer wall at all levels. Any tendency for rotation to reduce heat transfer is largely annulled by the presence of the barrier, across which at any station (r, z) can be supported the non-zero geostrophic pressure drop

$$\int_{\phi_2}^{2\pi-(\phi_2-\phi_1)} 2\Omega \bar{\rho} u_r r \, d\phi \tag{1}$$

(where $\phi_1 \leq \phi \leq \phi_2$ is the range of azimuth occupied by the barrier) and concomitant temperature drop, associated with non-zero geostrophic values of u_r that do not depend strongly on ϕ. Systematic studies of concomitant velocity and temperature fields are shedding light on the detailed dynamical processes involved.

In a single lecture it was impossible to do more than indicate in general terms how rotation affects thermal convection in a rotating fluid subject to a horizontal temperature gradient. Further details and applications to problems in geophysical and astrophysical fluid dynamics can be found by consulting the literature cited at the end of this summary (which differs in minor details only from my contribution to the forthcoming proceedings of another Advanced Study Institute at which I was invited to talk about this work, namely on Large-Scale Processes in the Oceans and Atmosphere, held in Les Houches, France).

2. GEOSTROPHY

It is useful to consider certain general properties of the motion of a fluid of low viscosity that departs but little from solid body rotation with steady angular velocity $\underset{\sim}{\Omega}$ when typical timescales of the relative motion greatly exceed $(2\Omega)^{-1}$. Such motion is 'geostrophic' nearly everywhere, satisfying

$$2\rho \underset{\sim}{\Omega} \times \underset{\sim}{u} + \nabla p - \rho \nabla V = 0 \tag{2}$$

Here $\underset{\sim}{u}$ is the Eulerian relative flow velocity, ρ denotes density, p pressure, and ∇V is the acceleration due to gravity and centripetal effects. Equation (2) is the leading approximation to the full equation of motion

$$2\rho \underset{\sim}{\Omega} \times \underset{\sim}{u} + \nabla p - \rho \nabla V = \underset{\sim}{A} \tag{3}$$

where

$$\underset{\sim}{A} \equiv -\rho D\underset{\sim}{u}/Dt - \rho \underset{\sim}{r} \times d\underset{\sim}{\Omega}/dt + \underset{\sim}{F} \tag{4}$$

the 'ageostrophic' contribution. It is valid in regions where the

Coriolis term $2\rho\boldsymbol{\Omega}\times\mathbf{u}$ greatly exceeds the relative acceleration term $\rho D\mathbf{u}/Dt \equiv (\partial\mathbf{u}/\partial t + (\mathbf{u}.\nabla)\mathbf{u})$ (where t denotes time), the 'precessional' term $\rho\,\mathbf{r}\times d\boldsymbol{\Omega}/dt$ (where $\mathbf{r}$ is the position vector of a general point in a frame which rotates with angular velocity $\boldsymbol{\Omega}$ relative to an inertial frame), and the term $\mathbf{F}$ which represents the viscous and other forces (e.g. Lorentz forces in magnetohydrodynamic systems) per unit volume.

Now equation (2) is mathematically degenerate, for its order is lower than that of the full equation of motion, (3), and it cannot therefore be solved under the complete set of boundary conditions. For this to be possible it is necessary to include ageostrophic terms in the analysis, so that the flow cannot be geostrophic everywhere. Hence:

<u>regions of highly ageostrophic flow occurring not only on the boundaries of the system but also in localised regions (detached shear layers, jet streams, etc) of the main body of the fluid are necessary concomitants of geostrophic motion.</u> (5)

Within these highly ageostrophic regions, $\mathbf{A}$ is comparable in magnitude with $2\rho\boldsymbol{\Omega}\times\mathbf{u}$ and the corresponding relative vorticity $\boldsymbol{\xi} \equiv \nabla\times\mathbf{u}$ can be comparable with or even exceed $2\boldsymbol{\Omega}$ in magnitude. Many examples of such vorticity concentrations are found in Nature and in laboratory systems, such as those considered below. They are often associated with steep gradients of temperature ('thermal fronts'), as in the case of jet-streams and western boundary currents found in atmospheres and oceans.

Equations (2) and (3) lead directly to another important finding (see above), namely that:

<u>the motion of a fluid of low viscosity that departs only slightly from steady rapid rigid-body rotation will not in general be symmetric about the rotation axis, even when the boundary conditions are axisymmetric.</u> (6)

This simple result elucidates of the occurrence of large-scale non-axisymmetric disturbances in the Earth's atmosphere and other natural systems, and it receives direct verification from the experiments outlined below. It can be deduced as follows. In cylindrical co-ordinates (r,ϕ,z) where $\boldsymbol{\Omega} = (0,0,\Omega)$, the ϕ-component of equation (3) gives

$$\rho u_r = (2\Omega)^{-1}\left\{-r^{-1}\partial p/\partial\phi + A_\phi\right\} \qquad (7)$$

if $\mathbf{u} = (u_r, u_\phi, u_z)$ and $\mathbf{A} = (A_r, A_\phi, A_z)$, since $\partial V/\partial\phi = 0$ by the assumption of axial symmetry in the boundary conditions. Now over any cylindrical surface of radius r, the rate of advective transport $M(r,t;Q)$ of any quantity Q (per unit mass), such as heat, angular momentum, etc. is given by

$$M(r,t;Q) \equiv \int_{z_1}^{z_2}\int_0^{2\pi} \rho r u_r Q \, d\phi \, dz$$

$$= \frac{1}{2\Omega}\int_{z_1}^{z_2}\int_0^{2\pi}\left\{-\frac{\partial p}{\partial\phi} + rA_\phi\right\} Q \, d\phi \, dz \tag{8}$$

where $z_1 \leq z \leq z_2$ is the axial extent of the surface. Since the ageostrophic contribution A_ϕ to equation (7) decreases rapidly with increasing Ω, advective transport perpendicular to the axis of rotation, as measured by $M(r,t;Q)$, will be negligible unless the flow pattern departs significantly from axial symmetry. In the axisymmetric case we have $\partial p/\partial\phi = 0$, and $M(r,t;Q)$ is consequently of the order of the small ageostrophic contribution.

This argument is the basis of (6). There may be singular cases where the flow remains axisymmetric and in consequence any advective transport perpendicular to the rotation axis is negligible. Indeed, by manipulating the axisymmetric boundary conditions it is possible to construct 'untypical' systems for which the flow pattern is also axisymmetric and M therefore close to zero. But 'generic' (i.e. typical) systems should have M significantly greater than zero and display pronounced departures from axial symmetry in their flow patterns.

Order of magnitude estimates of $\underline{A}$ given by equation (4) in cases when $d\underline{\Omega}/dt = 0$ and $\underline{F} = \nu\rho\nabla^2\underline{u}$, where ν denotes kinematic viscosity, indicate that geostrophic balance can be expected in thermally-driven systems when Ω is sufficiently large to satisfy both

$$\Omega \gg \nu/\ell^2 \quad \text{and} \quad \Omega \gg (g\alpha\Delta T/\ell)^{1/2}(\nu/\kappa)^{1/2}. \tag{9a, b}$$

Here ℓ is a typical length, ΔT a typical impressed temperature contrast, α the thermal coefficient of cubical expansion, and κ the coefficient of thermometric conductivity or its radiative equivalent.

We note here in passing that the result expressed by (6) bears on the role of rotation in the generation of magnetic fields by fluid motions. In the Earth's liquid metallic core for example, Coriolis forces influence the so-called geomagnetic dynamo process by promoting crucial departures from axial symmetry in the velocity field, for (by Cowling's theorem, see article by P H Roberts in this volume, and its recent extensions) axisymmetric magnetic fields cannot be maintained by fluid motions.

3. RÉGIMES OF THERMAL CONVECTION IN A ROTATING FLUID ANNULUS

The earliest laboratory experiments on thermal convection in a rotating liquid cylindrical annulus which rotates with angular speed Ω about its vertical axis of symmetry and is subject to axisymmetric applied differential heating showed that the general character of the flow is largely determined by two dimensionless parameters, namely

$$\Theta \equiv g d \Delta\rho / \bar{\rho}\, \Omega^2 (b-a)^2 \quad (10)$$

and

$$\mathcal{J} \equiv 4\Omega^2 L^4 / \nu^2. \quad (11)$$

Here g denotes the acceleration of gravity, d the fluid depth, b and a are the radii of the outer and inner side-walls; L depends on b, a and d and has the dimensions of length (and is equal to $(b-a)^{5/4}/d^{1/4}$ over a wide range of conditions); $\Delta\rho$ is a measure of the impressed horizontal density contrast associated with the applied differential heating and cooling, which can be taken as the difference between the maximum and the minimum density associated with the impressed temperature field $T_0(r)$.

The side-walls are held at temperatures T_b and T_a respectively and the fluid is subject to diabatic heating per unit volume at a rate $q\bar{\rho}$ times the specific heat c. The radial variation of the impressed temperature $T_0(r)$ takes the simple form

$$T_0(r) = (b-a)^{-1}\left[(r-a)T_b + (b-r)T_a\right] - q(r-a)(r-b)/2\kappa \quad (12)$$

(where κ is the thermometric conductivity of the fluid) when $2(b-a) \ll (b+a)$ and q is constant. When there are no heat sources or sinks within the fluid, we have the most extensively studied case of all, often referred to as the 'wall heated' case. Then equation (12) becomes

$$T_0(r) = (b-a)^{-1}\left[(r-a)T_b + (b-r)T_a\right], \ \{\text{wall heated, wall cooled}\} \quad (13)$$

and $\Delta\rho = |\rho(T_b) - \rho(T_a)|$.

When $q \neq 0$ we have "internally heated" or "internally cooled" systems according as $q > 0$ or $q < 0$. There are three particularly interesting limiting cases of internally heated systems, namely the 'inner wall cooled' case when the outer wall is a thermal insulator, so that $dT_0/dr = 0$ at $r = b$ and equation (12) becomes

$$T_0(r) = T_a + q(r-a)(2b-a-r)/2\kappa, \ \{\text{inner wall cooled}\}; \quad (14)$$

the 'outer wall cooled' case, when the inner wall is a thermal insulator so that $dT_0/dr = 0$ at $r = a$ and equation (12) becomes

$$T_0(r) = T_a - q(r-a)^2/2\kappa \quad , \quad \{\text{outer wall cooled}\}; \quad (15)$$

and the 'both walls cooled' case, when heat is removed at the same rate via both outer and inner side-walls simultaneously, so that dT_0/dr at $r=a$ is equal to $-dT_0/dr$ at $r=b$ and equation (12) becomes

$$T_0(r) = T_a - q(r-a)(r-b)/2K, \quad \{\text{both walls cooled}\}. \quad (16)$$

In the first two of these internally heated cases, $\Delta\rho$ is the same as in the wall-heated case, $|\rho(T_b) - \rho(T_a)|$, but in the 'both walls cooled' case we have $\Delta\rho = |\rho(T_0(\frac{1}{2}(a+b))) - \rho(T_a)|$.

Careful studies of the principal spatial and temporal characteristics of flows over a wide range of precisely specified and carefully controlled experimental conditions led to the discovery of several fundamentally different free types of flow, only one of which is symmetrical about the axis of rotation. When $\mathcal{J}$ is less than a certain critical value of about 2×10^5, viscosity ensures that the motion is essentially axisymmetric for all values of Θ. However, when $\mathcal{J}$ exceeds this critical value there exists a range of Θ, namely $\Theta_R > \Theta > \Theta_L$ (where Θ_R and Θ_L depend on $\mathcal{J}$), within which highly non-axisymmetric flow occurs. These non-axisymmetric motions are either 'regular' or 'irregular' depending on the values of Θ and $\mathcal{J}$. The regular flows, which occur when $\Theta_R > \Theta > \Theta_I$, are spatially periodic in ϕ and often exhibit periodic temporal 'vacillation' in amplitude, shape or even wavenumber, but under certain conditions these periodic variations are so slight that, apart from a steady azimuthal drift of the flow pattern relative to the walls of the apparatus, the flow is virtually steady. In sharp contrast to this behaviour, irregular flows (which occur within the range $\Theta_I > \Theta > \Theta_L$) exhibit complicated aperiodic fluctuations in both space and time. Axisymmetric flow occurs when $\Theta > \Theta_R$ or $\Theta < \Theta_L$.

The non-axisymmetric flow regimes are manifestations of 'sloping' or 'slantwise' convection and the observed dependence of Θ_R and Θ_L on $\mathcal{J}$ and other parameters has been successfully interpreted on the basis of baroclinic instability theory suitably modified to take viscous boundary layers into account. Barotropic instability is the likely cause of the transition from regular to irregular non-axisymmetric flow that occurs at $\Theta = \Theta_I$.

Heat flow determinations show that the tendency for advective heat transfer to decrease with increasing Ω is pronounced within the axisymmetric and irregular non-axisymmetric flow regimes found, respectively, where $\Theta > \Theta_R$ and where $\Theta_I > \Theta > \Theta_L$. However within the regular non-axisymmetric regime (where $\Theta_R > \Theta > \Theta_I$) advective heat transfer is virtually independent of Ω and typically about 0.8 times the 'non-rotating' value.

4. PATTERNS OF REGULAR NON-AXISYMMETRIC FLOW

The mathematic equations governing the flows we are considering are the equation of motion, equation (3), together with the equations of continuity and state for a liquid, respectively

$$\nabla \cdot \underset{\sim}{u} = 0 \quad (17)$$

and

$$\rho = \hat{\rho}\left[1 - \alpha\left(T - \hat{T}\right)\right] \tag{18}$$

(where T denotes temperature, α thermal coefficient of cubical expansion (for convenience here taken as constant), and $\hat{\rho}$ is the density at the reference temperature $\hat{T}$), and the equation of heat transfer

$$\partial T/\partial t + \underline{u}\cdot\nabla T = \kappa\nabla^2 T + q \tag{19}$$

Equations (12) to (16) are, of course, solutions of equation (19) when $\underline{u} = 0$. Across any cylindrical vertical surface $r =$ constant, the rate of heat transfer (made up of a conductive component and a convective (or 'advective') component when radiative effects are negligible) is given by

$$H(r,t) = \int_{-\frac{1}{2}d}^{\frac{1}{2}d}\int_{0}^{2\pi} \rho c\left[\kappa\frac{\partial T}{\partial r} + u_r T\right] r\, d\phi\, dz \tag{20}$$

if the fluid extends in the axial direction from $z = -\frac{1}{2}d$ to $z = \frac{1}{2}d$. As anticipated above (see equation (8)), the geostrophic contribution to the advective heat flow term on the right-hand side of equation (20) would vanish if the flow were axisymmetric, since the geostrophic part of u_r is proportional to $\partial p/\partial\phi$. As we have already indicated, this result points to the raison d'être of the non-axisymmetric regimes of flow found when Ω is sufficiently large; geostrophic flow cannot convey heat perpendicularly to the axis of rotation unless the flow is non-axisymmetric!

The boundary conditions on $\underline{u}$ under which the governing equations must be satisfied are that $\underline{u} = 0$ at a rigid bounding surface, and that the stress should vanish at a free surface. The thermal boundary conditions require continuity of heat flow which, at a bounding surface, is purely conductive and proportional to $\kappa\nabla T$. When the boundary conditions on the side-walls at $r = r^*$ where $r^* = a$ or $r^* = b$ are combined with the geostrophic relationship given by equation (2) and used in conjunction with the standard relationship for the radial flow in the Ekman boundary layers on $z = -\frac{1}{2}d$ and $z = \frac{1}{2}d$ to evaluate the radial heat flow at $r = r^*$ (see equation (20)), expressions for $H(r^*, t)$ can be obtained. These show that

$$H(r^*,t) = (\text{negative definite quantity}) \times \Gamma(r^*, \tfrac{1}{2}d, t) \tag{21}$$

if

$$\Gamma(r^*,t) \equiv \int_{0}^{2\pi} U_\phi(r^*,\phi,z,t)\, r\, d\phi \tag{22}$$

where $U_\phi(r^*,\phi,z,t)$ is the value of u_ϕ evaluated just outside the viscous boundary layer on $r = a$ or $r = b$ as the case may be.

This relationship between the heat flow at a side-wall and the line integral of the tangential velocity near the side-wall embodies the arguments that have been used to provide a general interpretation of the upper-level flow pattern in the case when $q = 0$ everywhere (see equation (13)), heat being introduced into the system via one of the side-walls and removed via the other side-wall. The corresponding impressed radial temperature gradient has the same sign at all values of r and the upper level pattern of motion in the regular flow regime consists of a single jet-stream meandering in a wavy pattern between the bounding cylinders, with a positive (ie 'westerly') azimuthal component when heat enters via the outer side-wall and leaves via the inner side-wall (so that the impressed radial temperature gradient is positive), and negative ('easterly') when the overall radial heat transfer is in the opposite direction, from the inner to the outer cylinder.

As a further test of equation (21), experiments have been carried out using internal heating, so that the term q in equation (19) is positive. This was done by passing an alternating electric current through the fluid. Heat could be removed via the inner side-wall, the outer side-wall, or both side-walls. The observed upper surface flow patterns were found to be in good agreement with predictions for these three cases made on the basis of equation (22). In the cases where heat is removed via one side-wall only, $H(r^*, t)$ vanishes at the other side-wall and, by equation (22), the quantity $\Gamma(r^*, \frac{1}{2}d, t)$ must vanish. For this to happen, $U_\phi(r^*, \phi, \frac{1}{2}d, t)$ will be positive at some values of ϕ and negative at others (or zero at all values of ϕ, as in the axisymmetric regime that occurs when $\Theta > \Theta_R$). This requirement can be satisfied by adding closed eddies to the wavy pattern found in the cases when $q = 0$. The most striking case of all studied in the experiments is that when heat is removed via both side-walls, implying that the impressed radial temperature gradient changes sign near mid-radius. The corresponding upper flow level consists of one or several separate closed eddies, each circulating anti-cyclonically (since $q > 0$), in accordance with equation (22), with the horizontal flow confined to a narrow jet-stream at the periphery of each eddy.

REFERENCES

The results summarized in this lecture are but a few of the extensive findings of many investigations of thermal convection in rotating fluids carried out during the past thirty-five years and described in numerous reports in the literature. References to these reports and to related geophysical studies can be found in the bibliographies of the following recent papers:

Hide, R. 1982, "On the role of rotation in the generation of magnetic fields by fluid motions". Philos, Trans. Roy. Soc. London A306, 223-234.

Hide, R. 1984, 'Presidential address: the giant planets; Galileo Galilei to Project Galileo', Quart. Journal Roy. Astron. Soc. 25, 232-247.

Hide, R. 1985, 'Thermal convection in a rotating fluid subject to a horizontal temperature gradient', pp 159-171 in Turbulence and predictability in geophysical fluid dynamics and climate dynamics'. (Ed M. Ghil, R. Benzi and G. Parisi), Societa Italiana di Fisica, Bologna, Italy.

Hignett, P. 1985, 'Characteristics of amplitude vacillation in a differentially heated rotating fluid annulus', Geophys. Astrophys. Fluid Dyn. 31, 247-281.

Hignett, P., White, A.A., Carter, R.D., Jackson, W.D.N. and Small, R.M., 1985 'A comparison of laboratory measurements and numerical simulations of baroclinic wave flows in a rotating cylindrical annulus', Quart. Journ. Royal Meteorological Soc. 111, 131-154.

Read, P.L. and Hide, R., 1984, 'An isolated baroclinic eddy as a laboratory analogue of the Great Red Spot on Jupiter', Nature 308, 45-48.

[illegible] 1981. [illegible] Soc. [illegible]

[illegible] 1983. [illegible] temperature [illegible] and [illegible] In: [illegible] Institute [illegible]

[illegible] 1988. [illegible] population [illegible] tropics. [illegible]

[illegible] and [illegible] In: [illegible]

[illegible] isolated [illegible]

CONVECTION IN SPHERICAL SYSTEMS

P.H. Roberts
Institute of Geophysics & Planetary Physics
University of California
Los Angeles
California 90024
U.S.A.

ABSTRACT. Aspects of convection theory, expounded elsewhere in this book for plane layers, are here generalized to spherical geometry. Particular attention is focussed on the important example of convection in the Earth's mantle, to the theory of which an introduction is given. Finally attention is transferred to the Earth's core. Aspects of magnetoconvection in a rotating sphere are described, these being extensions to spherical geometry of the plane layer models discussed by the author in his chapter on dynamo theory elsewhere in this book.

1. INTRODUCTION

Although the salient characteristics of convection can be understood through simple planar models, we should never lose sight of the fact that naturally occurring convection on cosmic scales almost invariably takes place in bodies that are, to a first approximation, radially symmetric, this being the natural state taken up by a large body like the Earth under its own gravitational forces. Other effects (rotation, magnetic fields, chemical inhomogenities, and even the convective motions themselves) produce only small departures from this radial symmetry, and these can be included if need be at second order. It is clearly pertinent to enquire how convective motions are most conveniently described in spherical geometry, and to ask what significant differences are introduced by the curvature of the system.

The topic is a large one, and we have time only to study two systems: the Earth's mantle (§3) and the Earth's core (§4). In each case the models are over-idealized, but more realism is not possible in the space available. It is hoped nevertheless that the flavour of the subjects, in their current states of development, has not been lost. The discussion of §4 is essentially a continuation of §§5.2 and 5.3 of the chapter 'Dynamo Theory' (referred to here as 'DT') elsewhere in this book. We start however by developing a mathematical technique that is useful for flows in spherical systems. (See also DT §2.1.)

C. Nicolis and G. Nicolis (eds.), Irreversible Phenomena and Dynamical Systems Analysis in Geosciences, 53–71.

2. KINEMATICS

2.1. Applicability of Oberbeck-Boussinesq Approximation.

The rheology of the Earth's mantle is poorly known. Even if the composition were precisely specified, its behaviour under the great temperatures and pressures prevailing in the deep mantle would be quite uncertain, as would be its response to stresses that persistently act on it over millennia. Faced with such uncertainties, it seems reasonable to adopt a simple constitutive model: a Newtonian fluid. Even then important information is still lacking, e.g. precisely how the viscosity of that fluid should be chosen as a function of temperature. We shall also suppose that the density scale-height, H, of the mantle is large compared with its depth, so that we may suppose it a Boussinesq fluid:

$$\underline{\nabla}.\underline{v} = 0. \qquad (2.1)$$

Such an approximation is highly questionable for other spherical convective systems, such as the solar convection zone. It is even dubious for the Earth's mantle. If we take $g = 10\ ms^{-2}$, $\bar{\alpha} = 3.10^{-5}\ K^{-1}$, $d = 3\ 10^6 m$, $C_p = 1.2\ 10^3\ J\,kg^{-1}\,K^{-1}$, we obtain $H = C_p/g\bar{\alpha} = 4\ 10^6 m$ and $d/H = 0.75$. This really suggests that the so-called 'anelastic approximation', (e.g. Glatzmaier & Gilman, 1981) in which (2.1) is replaced by $\underline{\nabla}.(\rho_0\underline{v}) = 0$ with ρ_0 the horizontally averaged density, would be preferable to (2.1). Although it is still possible to make progress, by expanding $\rho_0\underline{v}$ rather than $\underline{v}$ in (2.2) below, the added complications do not seem to be justified here in view of the great uncertainties in the rheology.

2.2. Lamb's Representation.

Since the three components of $\underline{v}$ are subject to one constraint (2.1), we expect that only two independent scalar fields will be needed to represent $\underline{v}$, and in spherical polar coordinates (r,θ,ϕ) the most convenient choice is that of Lamb (1881). We write

$$\underline{v} = \underline{v}_T + \underline{v}_P, \qquad (2.2)$$

where

$$\underline{v}_T = \underline{\nabla}\times(T\underline{r}) = (0, \frac{1}{\sin\theta}\frac{\partial T}{\partial\phi}, -\frac{\partial T}{\partial\theta}), \qquad (2.3)$$

$$\underline{v}_P = \underline{\nabla}\times\underline{\nabla}\times(S\underline{r}) = [\frac{L^2 S}{r}, \frac{1}{r}\frac{\partial}{\partial\theta}(\frac{\partial(rS)}{\partial r}), \frac{1}{r\sin\theta}\frac{\partial}{\partial\phi}(\frac{\partial(rS)}{\partial r})], \qquad (2.4)$$

$\underline{r}$ is the radius vector, and

$$L^2 = -[\frac{1}{\sin\theta}\frac{\partial}{\partial\theta}(\sin\theta\frac{\partial}{\partial\theta}) + \frac{1}{\sin^2\theta}\frac{\partial^2}{\partial\phi^2}] \qquad (2.5)$$

is the angular momentum operator. The canonical examples of a toroidal vector, $\underline{v}_T$, and a poloidal vector, $\underline{v}_P$, are obtained by taking their defining scalars, T and S, to be axisymmetric with respect to the preferred axis, $\theta = 0$, of the coordinates (the geographic axis in most cases below). When T is independent of ϕ, the only non-zero component of $\underline{v}_T$ is $v_{T\phi}$, i.e. the flow is purely zonal, along lines of latitude: when S is independent of ϕ, $v_{P\phi}$ vanishes and the flow lines lie in meridian planes. Any toroidal vector is orthogonal to any poloidal vector over a sphere of constant r, in the sense that

$$\int \underline{v}_T \cdot \underline{v}_P \, d\omega = \frac{1}{r} \iint \frac{\partial(\partial(rS)/\partial r, T)}{\partial(\theta,\phi)} \, d\theta d\phi = 0. \quad (2.6)$$

Here $d\omega = \sin\theta \, d\theta d\phi$ is the element of solid angle and the integrations (as later) are over the entire surface $0 \le \theta \le \pi$, $0 \le \phi < 2\pi$.

Since L^2 annihilates functions that depend on r alone, there is no way of including a radially symmetric flow into scheme (2.1) - (2.5): such a flow would in any case require a source or sink of mass at $r = 0$, somewhat contrary to the spirit of (2.1). Also, since $L^2 f(r) = 0$, we may without loss of generality subtract from any T and S its horizontal average:

$$\langle T \rangle = \frac{1}{4\pi} \int T(r,\theta,\phi) d\omega \, . \quad (2.7)$$

Thus, in future we may assume that

$$\langle T \rangle = 0 \, , \qquad \langle S \rangle = 0 \, . \quad (2.8)$$

It is easy to invert $L^2 T$ or $L^2 S$ for such functions. One uses the fact that, if $P_\ell^m(\theta)$ is the Associated Legendre function,

$$L^2 [P_\ell^m(\theta) e^{im\phi}] = \ell(\ell+1) [P_\ell^m(\theta) e^{im\phi}]. \quad (2.9)$$

Since $\ell = 0$ terms are excluded by (2.8),

$$F = \sum_{\ell=1}^{\infty} \sum_{m=-\ell}^{\ell} F_\ell^m(r) \, [P^m(\theta) e^{im\phi}] \quad (2.10)$$

implies

$$L^{-2} F = \sum_{\ell=1}^{\infty} \sum_{m=-\ell}^{\ell} [F_\ell^m / \ell(\ell+1)] [P_\ell^m(\theta) e^{im\phi}] \, . \quad (2.11)$$

Convergence of (2.10) assures that of (2.11). Inversion of L^2 is thus merely a matter of spherical harmonic analysis, and division of the resulting terms of order ℓ by $\ell(\ell+1)$.

The vorticity $\underline{\omega}$ of the flow (2.2) is easily obtained:

$$\underline{\omega} \equiv \underline{\nabla} \times \underline{v} = \underline{\omega}_T + \underline{\omega}_P , \tag{2.12}$$

where

$$\underline{\omega}_T = \underline{\nabla} \times (-\nabla^2 S\underline{r}) , \qquad \underline{\omega}_P = \underline{\nabla} \times \underline{\nabla} \times (T\underline{r}). \tag{2.13}$$

Thus toroidal flow has poloidal vorticity and, more unexpectedly, poloidal flow has toroidal vorticity. The curling process can easily be continued as often as is necessary, e.g.

$$\nabla^2 \underline{v} = \underline{\nabla} \times (\nabla^2 T\ \underline{r}) + \underline{\nabla} \times \underline{\nabla} \times (\nabla^2 S\ \underline{r}) , \text{ etc.} \tag{2.14}$$

Toroidal vectors have no radial components according to (2.3) and the poloidal defining scalar of any vector can therefore easily be obtained from its radial part; see (2.4). In this way we see that

$$\underline{r}.\underline{v} = rv_r = rv_{Pr} = L^2 S , \qquad \underline{r}.\underline{\omega} = r\omega_r = r\omega_{Pr} = L^2 T ,$$

or

$$T = L^{-2}(\underline{r}.\underline{\nabla} \times \underline{v}) , \qquad S = L^{-2}(\underline{r}.\underline{v}). \tag{2.15}$$

For given $\underline{v}$, these expressions are quickly evaluated by using (2.11).

2.3. Surface Motions.

If the fluid is confined to a sphere ($r = a$, say), v_r must vanish on its surface so that by $(2.15)_2$

$$S(a,\theta,\phi) = 0 . \tag{2.16}$$

Unless the fluid is in contact with a stationary no-slip surface [see (2.26) below], there is no reason why its surface motion

$$\underline{v}_H \equiv (v_\theta, v_\phi) = \left[\frac{1}{\sin\theta}\frac{\partial T}{\partial \phi} + \frac{\partial S'}{\partial \theta} ,\ -\frac{\partial T}{\partial \theta} + \frac{1}{\sin\theta}\frac{\partial S'}{\partial \phi}\right] , \tag{2.17}$$

where

$$S'(\theta,\phi) = \left[\frac{\partial S(r,\theta,\phi)}{\partial r}\right]_{r=a} , \tag{2.18}$$

should vanish. It may be, as in the case of the Earth's mantle, that $\underline{v}_H$ is especially interesting as providing the only readily accessible source of information about $\underline{v}$. It is therefore worth considering $\underline{v}_H$ further.

As (2.17) shows, $\underline{v}_H$ is the sum of a toroidal part, $\underline{v}_{HT}$, and a poloidal part $\underline{v}_{HP}$. Though $(2.15)_1$ still provides the means of extracting $\underline{v}_{HT}$, $(2.15)_2$ is vacuous as it stands. We note however that by (2.1) and (2.16)

$$\left(\frac{\partial v_r}{\partial r}\right)_{r=a} + \underline{\nabla}_H \cdot \underline{v}_H = 0 , \qquad (2.19)$$

where

$$\underline{\nabla}_H \cdot \underline{v} = \frac{1}{a \sin\theta} \left[\frac{\partial}{\partial\theta} (v_\theta \sin\theta) + \frac{\partial v_\phi}{\partial\phi}\right] \qquad (2.20)$$

is the "horizontal divergence". And now by differentiating $(2.15)_2$ with respect to r and using (2.19) we obtain

$$S' = -L^{-2} a \underline{\nabla}_H \cdot \underline{v}_H . \qquad (2.21)$$

A toroidal motion is horizontally divergenceless; a poloidal motion has no vertical vorticity, as we saw in §2.2. The following illustrations may be helpful. First recall the canonical poloidal motion, the flow in meridian planes, and imagine a "dipolar-type" circulation which has a northward component deep within the sphere, and which returns as southward motion on and near its surface. Then $\underline{\nabla}_H \cdot \underline{v}_H$ is positive (negative) in the northern (southern) hemisphere of $r = a$, corresponding to upwelling (downwelling) of fluid from (to) depths beneath $r = a$. Geologically such upwellings (downwellings) might be associated with ridges (trenches) on the Earth's surface. Second recall the canonical toroidal motion, the zonal flow along lines of latitude, and imagine that the two hemispheres turn in opposite directions about the axis $\theta = 0$. There will be a shear on the equator which, geologically, would probably be classed as a strike-slip fault.

It is easy to find the spherical harmonic parts of $\underline{v}_{HT}$ and $\underline{v}_{HP}$, by using the orthogonality relation,

$$\int [P_\ell^m(\theta) e^{im\phi}][P_{\ell'}^{m'}(\theta) e^{-im'\phi}] d\omega = N_\ell^m \delta_{\ell\ell'} \delta_{mm'} , \qquad (2.22)$$

for surface harmonics, where N_ℓ^m depends on the normalization selected for P_ℓ^m. By $(2.15)_1$ and (2.21) we quickly obtain the coefficients, $T_\ell^m(a)$ and $S_\ell'^m$, for expansions of $T(a,\theta,\phi)$ and $S'(\theta,\phi)$ of the form (2.10):

$$T_\ell^m(a) = \frac{1}{\ell(\ell+1) N_\ell^m} \int \{\underline{v}_H \times \underline{\nabla}[P_\ell^m(\theta)^{-im\phi}]\} . \hat{\underline{r}} d\omega , \qquad (2.23)$$

$$S_\ell'^m = \frac{1}{\ell(\ell+1) N_\ell^m} \int \{\underline{v}_H \cdot \underline{\nabla}[P_\ell^m(\theta) e^{-im\phi}]\} d\omega . \qquad (2.24)$$

The vertical vorticity and horizontal divergence of $\underline{v}_H$ are then easily computed if need be:

$$\omega_r = \frac{1}{a} \sum_{\ell=1}^{\infty} \sum_{m=-\ell}^{\ell} \ell(\ell+1) T_\ell^m(a) P_\ell^m(\theta) e^{im\phi} , \tag{2.25}$$

$$\underline{\nabla}_H . \underline{v} = -\frac{1}{a} \sum_{\ell=1}^{\infty} \sum_{m=-\ell}^{\ell} \ell(\ell+1) S_\ell'^m P_\ell^m(\theta) e^{im\phi} . \tag{2.26}$$

Finally, in preparation for §3 we make one dynamical point. Experiments on convection in the laboratory often use no-slip walls for which $\underline{v}_H = 0$. If $r = a$ is such a wall, $(2.15)_1$, (2.16) and (2.19) require

$$S = \partial S/\partial r = T = 0 , \quad \text{on} \quad r = a. \tag{2.27}$$

In studying mantle convection, it is more reasonable to suppose that the stresses associated with the normal (here $\hat{\underline{r}}$) vanish, and this leads to

$$S = \frac{\partial^2 S}{\partial r^2} = \frac{\partial}{\partial r}\left(\frac{T}{r}\right) = 0 , \quad \text{on } r = a . \tag{2.28}$$

3. MANTLE CONVECTION

3.1. Earth's Surface Motions

The magnitude and direction of the horizontal motion of the surface of the Earth's mantle at the present geological epoch is shown in Fig. 1. Convergent boundaries are trenches, divergent boundaries are ridges, and both are associated with poloidal flow through $\underline{\nabla}_H . \underline{v}$. Strike-slip faulting may also be seen, and this stems from toroidal flow through ω_r. It is interesting to ask how much of each type of motion is present, and how each is divided amongst the different spherical harmonics. The kinetic energy (per unit depth) of the surface motions is, by (2.6) and (2.22),

$$K = \tfrac{1}{2}\rho a^2 \int_0^\pi \int_0^{2\pi} \underline{v}_H{}^2 \sin\theta \, d\theta \, d\phi = \tfrac{1}{2}\rho a^2 \sum_{\ell=1}^{\infty} (\sigma_{T\ell}^2 + \sigma_{P\ell}^2) , \tag{3.1}$$

where ρ is density and $\sigma_{T\ell}^2$ and $\sigma_{P\ell}^2$ are variances,

$$\sigma_{T\ell}^2 = \sum_{m=-\ell}^{\ell} N_\ell^m |T_\ell^m|^2 , \quad \sigma_{P\ell}^2 = \sum_{m=-\ell}^{\ell} N_\ell^m |S_\ell'^m|^2 , \tag{3.2}$$

that encapsulate the importance of the order ℓ harmonics to each type of motion. It is clear from the results, shown in Fig. 2, that there is at each ℓ an approximate equipartition of energy between the poloidal and toroidal motions. This fact is not easily explained by convection theory.

3.2. Apparent Conflict with Theory

To simplify the theory suppose that in the absence of convective motions the gravitational field and temperature gradient are

$$\underline{g} = -g\underline{r}/a , \qquad \underline{\nabla}\Theta_0 = -\beta\underline{r}/a , \tag{3.3}$$

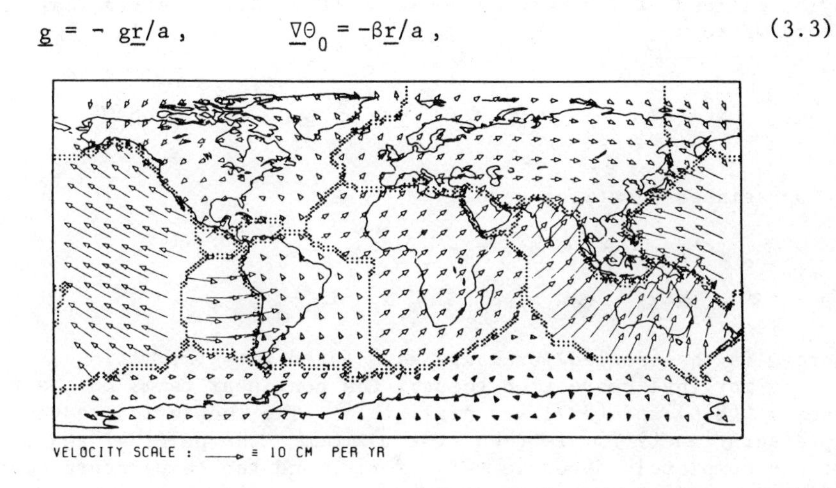

Figure 1. Plate Boundaries and Surface Velocities (from Peltier, 1985; Reproduced, with permission, from the Annual Reviews of Fluid Mechanics, Vol. 17, © 1985 by Annual Reviews Inc.)

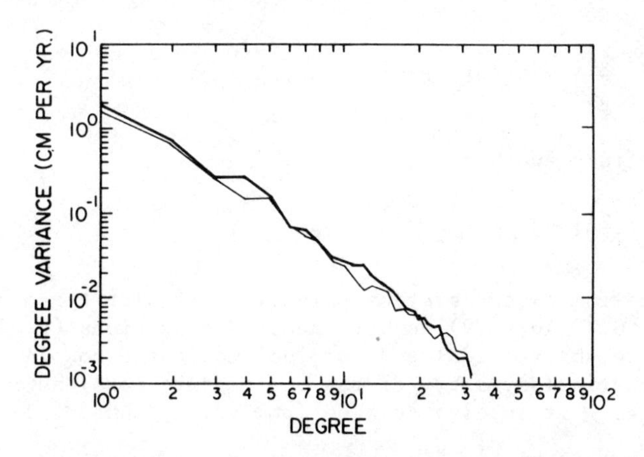

Figure 2. σ^2_{T1} and σ^2_{P1} as functions of harmonic degree 1 (from Peltier, 1985 ; reproduced with permission, from the Annual Reviews of Fluid Mechanics, Vol. 17, c 1985 by Annual Reviews Inc.) The thin line shows σ^2_{T1}, the thick σ^2_{P1}.

the constants $-g$ and $-\beta$ being the surface values of these quantities. These simple proportionalities in r are appropriate when $\underline{g}$ is the self-gravitation of a uniform density sphere, and when Θ_0 is the temperature created by a uniform distribution of heat sources within that sphere. More general cases can be treated without difficulty, and have many points of similarity. The Rayleigh and Prandtl numbers,

$$R = g\bar{\alpha}\beta a^4/\nu\kappa , \qquad p_r = \nu/\kappa , \tag{3.4}$$

may be defined. I use here my notation (DT§5.3.1). After scaling, according to

$$\underline{x} \to a\underline{x} , \quad t \to \frac{a^2}{\kappa} t , \quad \underline{v} \to \frac{\kappa}{a} \underline{v} , \quad \Theta - \Theta_0 \to \beta a \Theta' , \tag{3.5}$$

the Boussinesq equations become

$$\partial\Theta'/\partial t - \nabla^2\Theta' - \underline{v}.\underline{r} = - \underline{v}.\underline{\nabla}\Theta' , \tag{3.6}$$

$$p_r^{-1}\partial\underline{v}/\partial t - \nabla^2\underline{v} - R\Theta'\underline{r} + \underline{\nabla}\tilde{\omega} = p_r^{-1}\underline{v}.\underline{\nabla}\,\underline{v} , \tag{3.7}$$

where $\tilde{\omega}$ is the (scaled) pressure associated with the motions.

In marginal convection theory, the nonlinear terms on the right-hand sides of (3.6) and (3.7) are negligible. The theoretical advantages of representation (2.2) are then at a maximum: the poloidal and toroidal motions completely decouple. The former and the temperature perturbation define one closed system of equations

$$\left(\frac{\partial}{\partial t} - \nabla^2\right)\Theta' = L^2 S , \quad \left(\frac{1}{p_r}\frac{\partial}{\partial t} - \nabla^2\right)\nabla^2 S = -RL^2\Theta' ; \tag{3.8P}$$

$$\Theta' = S = \frac{\partial^2 S}{\partial r^2} = 0 \quad \text{at} \quad r = 1, \tag{3.9P}$$

and the latter defines another

$$\left(\frac{1}{p_r}\frac{\partial}{\partial t} - \nabla^2\right)T = 0 ; \quad \frac{\partial}{\partial r}\left(\frac{T}{r}\right) = 0 \text{ at } r = 1 . \tag{3.8T, 3.9T}$$

(Strictly these define closed systems only after regularity conditions are imposed at $r = 0$.) In (3.9) we have adopted conditions (2.28). Indeed not only are the toroidal and poloidal motions decoupled, so because of the spherical symmetry of the basic state are the individual spherical harmonics. It is easy to show from (3.8T) and (3.9T) that

$$T \to 0 , \qquad \text{as} \quad t \to \infty. \tag{3.10}$$

In the absence of a motor to drive the vertical vorticity, it must disappear in a time of order a^2/ν (i.e. 10^{-4}s for $\nu = 10^{17}\text{m}^2/\text{s}$).

3.3. Possible Explanations Explored

How then can we explain the strength of the toroidal surface motion? A first attempt might be through the nonlinear term in (3.7), which does indeed produce toroidal motion from poloidal motion. While it is probably true that convection in the mantle is highly supercritical, R being perhaps $10^5 - 10^7$ times its critical value R_c, the Prandtl number is extremely large (perhaps 10^{23}) and the Reynolds number of the flow is tiny (maybe 10^{-19}). Thus the nonlinearity of (3.6) is far more significant than that of (3.7). Equations $(3.8P)_2$ and $(3.8T)_1$ even with the P_r^{-1} terms omitted are acceptable in the nonlinear regime, and the T system remains decoupled from the P system of equations. [The converse is not true since toroidal motions contribute to the right-hand side of (3.6): in view of (3.10) however, this is not significant.] In short, to study finite amplitude convection we need only amend $(3.8P)_1$, and toroidal motions will be absent as before.

Before totally dismissing generation of toroidal flow in the bulk, we should (especially in view of §3.5.2 below) examine the effects of temperature dependent viscosity. The origins of the $\nabla^2\underline{v}$ term in (3.7) may be traced in $\rho\nabla^2 v_i = \partial\sigma_{ij}/\partial x_j$ where σ_{ij}, the stress tensor, is for a Newtonian fluid

$$\sigma_{ij} = \rho\nu\left[\frac{\partial v_i}{\partial x_j} + \frac{\partial v_j}{\partial x_i} - \frac{2}{3}\delta_{ij}\underline{\nabla}\cdot\underline{v}\right].$$

Taking $\nu = \nu(\Theta)$, ρ constant and divergenceless flow (2.1), we find for the linear stability problem not the $\nabla^2\underline{v}$ of (2.14) but

$$\underline{\nabla}\times\left([\nu_0\nabla^2 T + r\frac{d\nu_0}{dr}\frac{\partial}{\partial r}\left(\frac{T}{r}\right)]\underline{r}\right) + \underline{\nabla}\times\underline{\nabla}\times\left([\nu_0\nabla^2 S + 2r\frac{d\nu_0}{dr}\frac{\partial}{\partial r}\left(\frac{S}{r}\right)]\underline{r}\right),$$

where $\nu_0 = \nu_0(\Theta_0) = \nu_0(r)$; there is no mixing of modes. When however we look at finite amplitude convection, we discover that toroidal motion is created from poloidal motion through the departure of $\nu(\Theta_0 + \Theta')$ from radial symmetry via Θ'. Because this is a second order effect we shall ignore it, noting however that further investigation of the topic may be called for in the future.

Returning to the case of constant ν, we look next at the boundary conditions (3.9P) and (3.9T). The real mantle does not extend to $r = 0$, but has a lower surface at which further conditions must be applied, demands that replace the regularity conditions at $r = 0$ imposed above. While it is true in principle that the stresses exerted by core motions across the core-mantle boundary can drive the mantle into motion, even toroidal motion, it is unlikely that these flows are persistent enough, or the stresses large enough, to be effective.

The final possibility is that (2.27) and (2.28) are both incorrect, and should be replaced by conditions that couple the toroidal and poloidal flows. One may picture the continents as light scum floating on the heavier mantle, and being carried along with it in its convection.

When however the mantle is subducted at a trench the continents are too light to be subducted with it, and are therefore "pushed aside" along the trench, so generating a strike-slip fault. This deflection of poloidal motion into toroidal motion at the surface of the Earth may, because of the high viscosity of the upper mantle, drive toroidal flows to considerable depth.

Although these ideas may currently be somewhat speculative, they may ultimately provide the best explanation for the anomalously large toroidal components of motion shown in Figs. 1 and 2.

3.4. Some Correlations

It is interesting to try to relate the surface motions to the gravitational field of the Earth and to the heat flow from the Earth, and to attempt to discern a consistent overall picture. One might expect that the spreading ($\underline{\nabla}_H.\underline{v} > 0$) associated with a ridge indicates that here a convection cell is rising to the surface, and that therefore this will be a region of abnormally high heat from the Earth. Conversely, a trench ($\underline{\nabla}_H.\underline{v} < 0$) would be pictured as the cold descending current of a cell above which the heat flux would be comparatively small. Corroboration is difficult because of the presence of radioactive sources which tend to be concentrated in the continental crust, and whose heat flux is not readily separated from that associated with mantle convection. There does however appear to be a correlation between the locations of ridges and of regions of high heat flux: compare Fig. 3 with Fig. 1.

A correlation seems to exist between surface motions and anomalies in the geogravitational field. Two (competing) physical causes can be discerned. First, there is a topographical effect: an upwelling

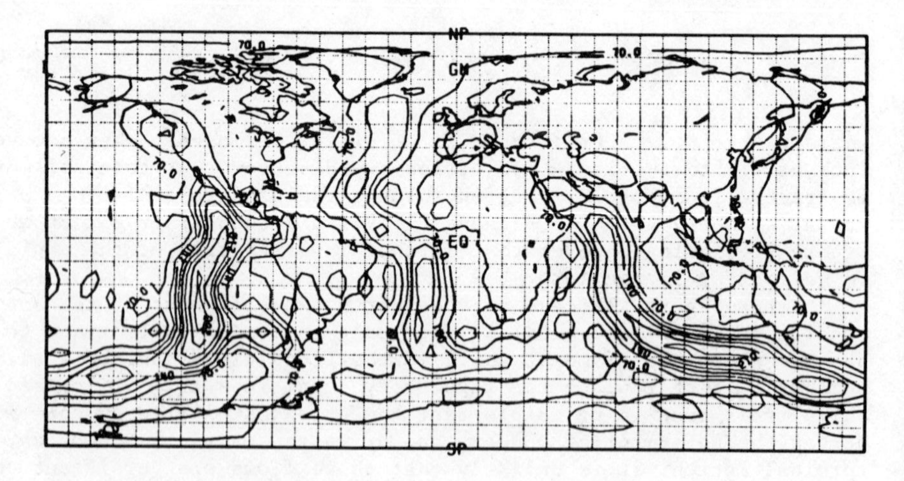

Figure 3. Surface Heat Flux. Contours of equal heat flux based on Chapman and Pollack (1980) using all spherical harmonics up to $\ell = 18$. Contour interval = 35m W/m^2. (Figure constructed by W.R. Peltier and reproduced with his permission.)

(downwelling) tends to elevate (depress) the free surface above (below) its equilibrium position, $r = a$. This suggests that a positive (negative) gravity anomaly should exist at an upwelling (downwelling). Second, there is an expansion effect brought about by variations of density with temperature: an upwelling (downwelling) is associated with a hot rising (cold falling) convective stream which is lighter (denser) than ρ_0, the density of the equilibrium state. This suggests a geoid high (low) over trenches (ridges).

In examining the topographic effect we should first recognize that (2.16) is an oversimplification. Suppose that a fluid element, situated at $\underline{x}$ in the equilibrium state, is carried to $\underline{x} + \underline{\xi}(\underline{x},t)$ by the convection; in particular, let the point $\underline{a} = (a,\theta,\phi)$ on the unperturbed free surface, S_0, be carried to $\underline{a} + \underline{\xi}(\underline{a},t)$ on the disturbed free surface, S. The stress associated with the normal, $\underline{n}$, to S (which is not in the same direction as the radial normal of S_0) must vanish. In particular

$$(p - 2\rho_0\nu\ \partial v_n/\partial n)_S = 0 .$$

In linear stability theory this reduces to

$$(p' + \xi_r \frac{dp_0}{dr} - 2\rho_0\nu_0 \frac{\partial v_r}{\partial r})_{S_0} = 0 , \qquad (3.11)$$

where $dp_0/dr = -g_0\rho_0$ by hydrostatic balance. The first and last terms in (3.11) are of the same order of magnitude so that, in steady convection;

$$\xi_r = 0(\frac{\nu_0}{g_0}\frac{\partial v_0}{\partial r})_{r=a} . \qquad (3.12)$$

Instead of (2.16), we should have made v_n vanish on $a + \xi_r$, where ξ_r is given by (3.12). But v_r and therefore ξ_r are perturbation quantities, and the resultant changes to the solution of the linear stability problem (3.8P) and (3.9P) would be quadratically small, i.e. negligible. In working out the gravitational field, $\underline{g}'_T$, produced by the topography, we must however use (3.12) and not (2.16). It is easy to see that $g'_T = 0(G\rho_0\xi_r)$, where G is the universal constant of gravitation, and since $g_0 = 0(G\rho_0 a)$ this is $g'_T = 0(g_0\xi_r/a) = 0(\nu_0\partial v_r/r\partial r)_{r=a}$ by (3.12). More precisely, for the ℓth harmonic using (3.9P) we obtain

$$g'_{Tr} = -\frac{3\nu_0(\ell+1)}{(2\ell+1)}\left[\frac{\partial^3 S}{\partial r^3} - \frac{3\ell(\ell+1)}{r^2}\frac{\partial S}{\partial r}\right]_{r=a} . \qquad (3.13)$$

The density changes associated with the buoyancy of the rising and falling convection currents are of order $\rho_0\alpha\Theta'$ and the expansion effect therefore produces on S a change, $\underline{g}'_E$, in gravitational acceleration of order $G\rho_0 a\alpha\Theta' = 0(\nu_0 v_r/a^2)$, i.e. of the same order as $\underline{g}'_T$. More precisely, for the ℓth harmonic using (3.8P) and (3.9P) we obtain

$$g'_{Er} = \frac{3\nu_0(\ell+1)}{(2\ell+1)}\left[\frac{\partial^3 S}{\partial r^3} - \frac{(\ell+1)(\ell+2)}{r^2}\frac{\partial S}{\partial r}\right]_{r=a}. \tag{3.14}$$

[We have omitted some working here. The analysis of Roberts (1965) may be helpful in obtaining (3.13) and (3.14).]

It is difficult to tell a priori whether g'_T predominates over g'_E or vice versa. In Table 1 we give, for the most easily excited of the first five harmonics, the values of g'_{Tr}/g'_{Er} and also the critical Rayleigh number, R_c. It will be seen that, except for $\ell = 1$ where they are equal and opposite, the effects of topography dominate.

Table 1
Rayleigh Number and Gravity Ratio

ℓ	1	2	3	4	5
R_c	3091.2	5224.1	8774.5	13981.	21204.
$-g'_{Tr}/g'_{Er}$	1	1.251	1.418	1.544	1.644

(In connection with $\ell = 1$ it may be noted that a pure lateral displacement of the entire sphere is a neutral eigenfunction.) It should be emphasized that these results depend on the choice (3.3) of model, the assumption of constant viscosity, and the linearization of the governing equations. There are however some indications that g'_T dominates g'_E for the Earth also, except possibly for very high harmonic numbers: there are indications that narrow regions near trenches are geoid highs, not lows.

3.5. Other Aspects of Mantle Convection Theory

The many ramifications of mantle convection now provides an industry for a small army of geoscientists, and it is obviously impossible here to present anything approaching a complete survey of its status today. A few further points of interest to fluid dynamicists of theoretical bent should however be adumbrated:

3.5.1. Mode of Heating. Once the full sphere has been replaced by the more realistic spherical shell, one can imagine that convection is driven not by internal heating but by heating from below, as in the canonical example of convection: the Bénard layer. The character of convection in a Bénard layer, and in a layer internally heated but thermally insulating at the bottom, are quite different. At large R, the convective roll in a Bénard layer acts as a cartwheel on whose surface rides the thin boundary layer connecting a hot boundary layer at the bottom, into which heat pours from the source, to a cold boundary layer at the top, out of which heat is drawn. In a less well ordered (turbulent) regime, the lower boundary layer will spasmodically release plumes to take the heat upwards. The cartwheel cell would not play such a useful role for the internally heated cell, and therefore it does not arise. No hot boundary layer will be present at the lower insulated walls which could feed heat to the cartwheel. Moreover, the heat sources in the interior parts of the cartwheel would continually release heat that could only inefficiently reach the surface by thermal conduction. The internally heated layer therefore adopts a difficult mode, in which the whole interior gradually rises to release its heat, apart from fast concentrated descending streams of fluid chilled in the upper boundary layer. (See Peckover & Hutchinson, 1974; McKenzie et al, 1974.) The preferred horizontal scale, $\mathcal{L}$, of the internally heated layer appears, at large R, to be significantly greater than its depth, whereas for the Bénard layer they are roughly equal.

It is not known whether heating from below or internal heating more closely resembles the convective state of the Earth's mantle. On the one hand the existence of ridges may favour driving from below; on the other the Earth is still cooling from its formation 4½ billion years ago, and this cooling sets up an adverse temperature gradient conducive to convection, akin to that of the internally heated model.

3.5.2. Temperature-dependent Viscosity. A constant-viscosity, Newtonian fluid is probably a rather poor model of the Earth's mantle, but its status as a rheological model is greatly improved if ν is allowed to depend strongly on the temperature, Θ. Such a viscosity introduces new features of even a qualitative nature. While the material is cold and effectively solid, it can transmit heat only inefficiently by conduction. Heat will therefore tend to be held back warming the material until it effectively melts, and can transmit heat efficiently by convection. In such a manner the thermal state of the convecting layer is essentially regulated by the $\nu(\Theta)$ law (Tozer, 1972; see also §4.4 of Peltier, 1985). Similar considerations show that the upper layers of the mantle, which are efficiently chilled by losing heat to space, will be effectively stagnant; and that the mobility evinced in plate tectonics has other causes.

3.5.3. Phase changes. Mantle properties change abruptly at a number of depths (420 km, 670 km,.....) where material of essentially the same composition takes up different crystalline forms. Convection currents distort such phase boundaries from the horizontal, but the boundaries are not to be thought of as barriers for the convecting material. They do have implications for the stability of the layer (see for example Schubert et al, 1975).

4. MAGNETOCONVECTION IN RAPIDLY ROTATING BODIES

4.1. Convection in a Rapidly Rotating Sphere

Central to Geophysical Fluid Dynamics (GFD), as the term is generally understood, is convection in thin stratified layers on a rotating sphere, representing the Earth's atmosphere and oceans. The resulting literature is truly vast, and quite beyond the scope of this lecture. We shall consider only some aspects of convection in a thick layer of almost constant density, representing the Earth's fluid core. As in GFD, technical difficulties arise because of the different orientations of $\underline{g}$ to the angular velocity, $\underline{\Omega}$, in different regions of the fluid. Also, the core is electrically conducting and is pervaded by magnetic field, $\underline{B}$, whose Lorentz force $\underline{J} \times \underline{B}$, where $\underline{J} = \mu_0^{-1} \underline{\nabla} \times \underline{B}$ is the electric current density, may not be negligible.

We shall first consider the case $\underline{B} = 0$, and try to extend the discussion (DT§5.2.2) of marginal Bénard convection in a rapidly rotating plane layer to marginal convection in a rapidly rotating sphere. The Taylor-Proudman ("TP") theorem holds for both systems and we may expect that the nascent convective motions will also be highly two-dimensional, with respect to the direction of $\underline{\Omega} = \Omega \hat{\underline{z}}$.

In DT§5.2.2 we considered two cases, namely (case I) $\underline{\Omega}$ vertical, and (case II) $\underline{\Omega}$ horizontal. In the latter, the constraints of the TP-theorem did not inhibit convection at all: every fluid column in the x-direction (parallel to $\underline{\Omega}$) moved together as a two dimensional convection roll that took heat out of the layer (parallel to $\underline{g}$). Coriolis forces were ineffective and the critical Rayleigh number, R_c, for marginal convection was independent of Ω. Since R_c increases rapidly with Ω when $\underline{g}$ and $\underline{\Omega}$ are parallel [see DT$(5.8)_2$], it is natural at first to suppose that the sphere will emulate case II by convecting in columns [now defined by constant s and ϕ, where (s,ϕ,z) are cylindrical coordinates and $\underline{\Omega} = \Omega \hat{\underline{z}}$] at the equator. Unfortunately however this argument has ignored a crucial geometrical factor. The plane layer in case II extended to infinity in the direction (x) parallel to $\underline{\Omega}$: in the putative cells near the equator of the sphere, the columns are cut-off at $r = a$, and must therefore vary considerably in length $2\sqrt{(a^2-s^2)}$ as they carry heat out of the sphere in their s-motion. Such cells are far from being two-dimensional.

Case II nevertheless gives insight into the spherical case. Columns located at intermediate s are not as drastically inhibited by geometry as are the equatorial cells ($s \doteqdot a$); they can convect heat in the s-direction by only slightly adjusting their length. We will presently look at the other extreme case, a cell close to the rotation axis, where the columns scarcely change their length at all. But first we note that, in any cell but the equatorial cell, $\underline{\Omega}$ and $\underline{g}$ are not perpendicular (except on the equatorial plane). Such cells have therefore some of the characteristics of case I. In particular we expect that their scale, in projection on the equatorial plane ($z = 0$), will be $O(aT^{-1/6})$ where

$$T = (2\Omega a^2/\nu)^2 \tag{4.1}$$

is the Taylor number, the usual non-dimensional measure of rotation in convection problems. We must also expect that $R_c = O(T^{2/3})$ as $T \to \infty$: see DT(5.8). Here a is the radius of the sphere.

Since the equatorial cell convects poorly, while the cells of smaller s convect better, we might look at the other extreme, a cell centred on the z-axis. It transpires that this (axisymmetric) motion is also not optimal, because it has lost all its case II character. The radial component, g_s, of $\underline{g}$ vanishes as $s \to 0$, and all small-s cells convect heat inefficiently in the s-direction.

In view of this discussion, it is not surprising that the preferred mode of marginal convection should be one concentrated in cells strung out round $s = s_c$, where s_c is intermediate between 0 and a. Since the dimensions of these cells are of order $aT^{-1/6}$ they correspond to a highly asymmetric mode. If the perturbation fields are proportional to $\exp[i(m\phi-\omega t)]$, we must expect [cf. DT(5.8)] that

$$m_c = O(T^{1/6}), \qquad R_c = O(T^{2/3}), \qquad T \to \infty. \tag{4.2}$$

This conclusion and arguments along the lines given above were first given by Roberts (1968), and were refined and corrected by Busse (1970) who provided the sketch shown in Fig. 4. Later Soward (1977) showed that it was strictly incorrect to refer to these modes of convection as 'marginal' in the usual sense of linear stability theory: we do not have space here to describe this subtle effect of curvature. We should however note another, more obvious, effect of curvature. In DT§5.2.2 we assumed that $p_r > 1$, so that we did not need to consider overstable convection. In the sphere asymmetric modes are necessarily time dependent and, in the case of (4.2) above, ω_c is real and $O(m_c\Omega)$ i.e. the instability pattern drifts round the rotation axis with an $O(\Omega)$ phase velocity. Since $\omega_c > 0$, it drifts eastwards.

Busse (1975) made this model the basis for a dynamo. In the notation of DT§4.2 it is a weak field (α^2) dynamo.

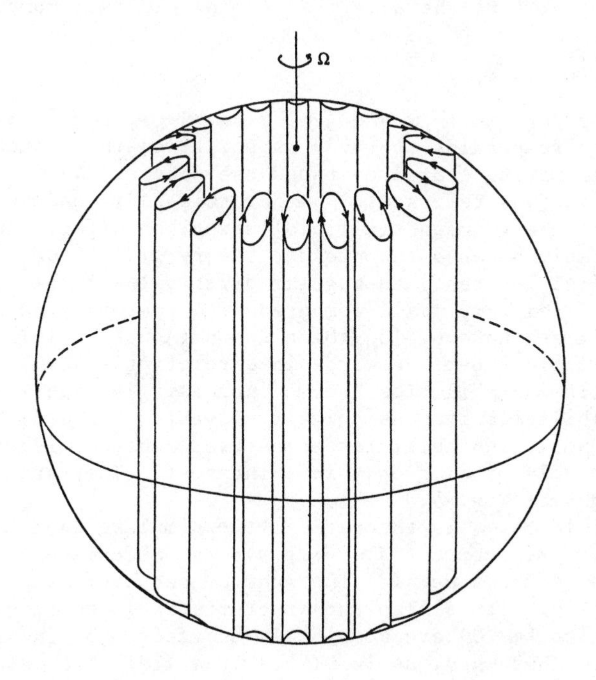

Figure 4. Qualitative sketch of marginal convection in an internally heated rotating sphere. (From Busse, 1970, with permission.)

4.2. Magnetoconvection in a Rapidly Rotating Sphere

DT§5.2 contained some surprising developments. In Case I of $\underline{\Omega}$ parallel to $\underline{g}$, a plane layer could be de-stabilized by a horizontal magnetic field, $\underline{B}_0$. In particular in the range where the Chandrasekhar number (a non-dimensional measure of B_0)

$$Q = \sigma B_0^2 a^2/\rho\nu \tag{4.3}$$

is of order $T^{1/2}$, and where in consequence the Coriolis and Lorentz forces are comparable, R_c is only of order $T^{1/2}$ rather than $T^{2/3}$: see DT(5.30).

It was natural to examine whether similar conclusions held in other geometries, and to generalize §4.1 by adding a zonal magnetic field $B_0\hat{\underline{\phi}}$. Assuming

$$\underline{B} = B_0(s/a)\hat{\underline{\phi}}, \qquad (B_0 = \text{const.}), \tag{4.4}$$

Fearn (1979) showed that, as for DT(5.30),

$$m_c = O(1), \quad R_c = O(T^{\frac{1}{2}}), \quad \text{for} \quad Q = O(T^{\frac{1}{2}}), \quad T\to\infty. \tag{4.5}$$

His results are summarized in Fig. 5, in which R_{CRIT} is not the critical value of R but of $R/T^{1/2}$; the abscissa is the Elsasser number

$$\Lambda = Q/T^{\frac{1}{2}}. \tag{4.6}$$

The results apply for finite κ and $\nu\to 0$ [i.e. $T\to\infty$; see (4.5)]. For small Λ the ω_c corresponding to (4.5) is positive, the instability pattern therefore moving eastwards round the sphere. As Λ is increased however this motion was reversed. Whether this bears on the observed westward drift of the geomagnetic field, described briefly in DT§1.4, is perhaps doubtful, because the present theory gives (for $\kappa \ll \eta$ the case of geophysical interest) an angular drift velocity of order only $\kappa/a^2 \doteqdot 10^{-18}s^{-1}$, i.e. very small compared with the observed motion ($2.10^{-11}s^{-1}$). It has however long been known (Acheson, 1972) that magnetic instabilities move westwards in a rotating system, and as Q increases magnetic instabilities become increasingly significant and convective instabilities increasingly irrelevant. [Indeed, for large Q very curious magnetic instabilities arise (see Acheson, 1980) which, however, are probably of no geophysical interest. They are indicated on the lower right of Fig. 5.]

Although field (4.4) is theoretically the simplest choice, the corresponding electric current $\underline{J}_0 = (B_0/\mu_0 a)\hat{\underline{z}}$ is uniform, and must therefore be driven by sources outside the sphere, geophysically a highly unrealistic premise. It is also uncharacteristic in that the **axisymmetric** modes of convection ($m = 0$) are completely unaffected by the presence of the field (4.4). Thus when, as in DT§5.2, the field becomes so large that axisymmetric motions are preferred (according to the two-dimensional analogue of the TP theorem), R_c/Q becomes independent of Q.

For these reasons, and also to see whether (4.5) is characteristic of a wider class of $\underline{B}_0$, Fearn and Proctor (1983) examined choices other than (4.4). They also took a significant step towards geophysical realism by adding a zonal shear. We have seen in DT§§2.4 and 4.2 that such a shear may play a crucial inductive role in the geodynamo, as well as providing through the anti-ω-effect (DT§5.1) a nonlinearity that limits its growth. More recently they have reported an attempt to build this programme into a larger study that produces a working dynamo of Taylor type, as adumbrated in DT§5.3.2 (Fearn and Proctor, 1984). Clearly a discussion of these matters is outside my remit, but I would like to stress that many questions remain to be answered in this active and exciting area of mathematical geophysics.

ACKNOWLEDGEMENTS

I am grateful to Professor W.R. Peltier for criticising a draft of §§1-3 of this Chapter and for permission to reproduce Figs. 1-3 above. Thanks are also due to the following: Annual Reviews Inc. for allowing me to reproduce Figs. 1 and 2; Cambridge University Press (publishers of the Journal of Fluid Mechanics) and Professor Fritz Busse for permission to copy Fig. 4; and Dr. David Fearn and Gordon and Breach Science Publishers for allowing me to reproduce Fig. 5.

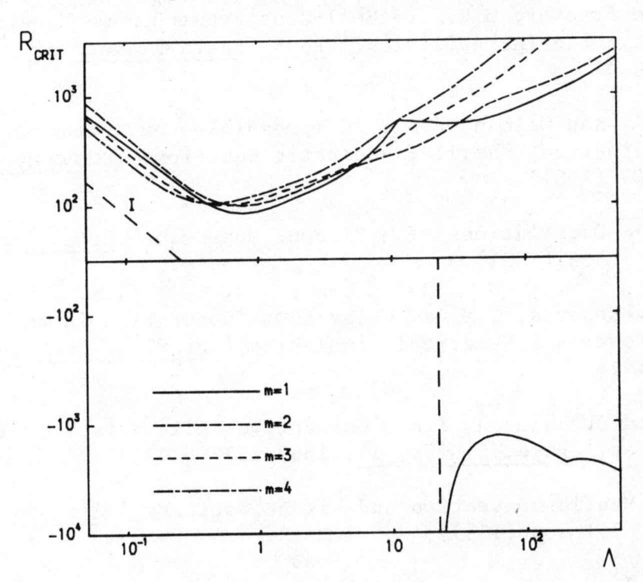

Figure 5. The critical Rayleigh number, R_{CRIT}, as a function of Elsasser number, Λ, for the first four values of the azimuthal wavenumber in the case $\nu = 0$. (From Fearn, 1979, with permission.) Here R is not (3.4) but $R/T^{\frac{1}{2}}$; Λ is $Q/T^{\frac{1}{2}}$ [see (4.5) and (4.6)]. The lower right of the diagram shows an instability of the field (4.4) catalysed by a bottom heavy density gradient (Acheson, 1980).

REFERENCES

Acheson, D.J., 'On the Hydromagnetic Stability of a Rotating Fluid Annulus,' J. Fluid Mech., 52, 529-541 (1972).

Acheson, D.J., "Stable' Density Stratification as a Catalyst to Instability,' J. Fluid Mech., 96, 723-733 (1980).

Busse, F.H., 'Thermal Instabilities in Rapidly Rotating Systems,' J. Fluid Mech. 44, 441-460 (1970).

Busse, F.H., 'A Model of the Geodynamo,' Geophys. J. R. Astr. Soc., 42, 437-459 (1975).

Chapman, D.S. and Pollack, H.N., 'Global heat flow: degree 18 spherical harmonic representation' (abstract). EOS Trans. AGU., 61,383 (1980).

Fearn, D.R. 'Thermal and Magnetic Instabilities in a Rapidly Rotating Fluid Sphere,' Geophys. Astrophys. Fluid Dynam., 14, 103-126 (1979).

Fearn, D.R. and Proctor, M.R.E., 'Hydromagnetic Waves in a Differentially Rotating Sphere,' J. Fluid Mech., 128, 1-20 (1983).

Fearn, D.R. and Proctor, M.R.E., 'Self-Consistent Dynamo Models Driven by Hydromagnetic Instabilities,' Phys. Earth Planet. Int., 36, 78-84 (1984).

Glatzmaier, G.A. and Gilman, P.A., 'Compressible Convection in a Rotating Spherical Shell. I. Anelastic Equations,' Astrophys. J. Suppl. 45, 335-349 (1981).

Lamb, H. 'On the Oscillations of a Viscous Spheroid,' Proc. Lond. Math. Soc., 13, 51-66 (1881).

McKenzie, D.P., Roberts, J.M., and Weiss N.O. 'Convection in the Earth's Mantle: Towards a Numerical Simulation,' J. Fluid Mech., 62, 465-538 (1974).

Peckover, R.S. and Hutchinson, I.H. 'Convective Rolls driven by Internal Heat Sources,' Phys. Fluids, 17, 1369-1372 (1974).

Peltier, W.R. 'Mantle Convection and Viscoelasticity,' Ann. Rev. Fluid Mech., 17, 561-608 (1985).

Roberts, P.H. 'Convection in a Self-Gravitating Fluid Sphere,' Mathematika, 12, 128-137 (1965).

Roberts, P.H. 'On the Thermal Instability of a Rotating Fluid Sphere Containing Heat Sources,' Phil. Trans. R. Soc. Lond. A, 263, 93-117 (1968).

Schubert, G., Yeun, D.A. and Turcotte, D.L. 'Role of Phase Transitions in a Dynamic Mantle', Geophys. J.R. Astron. Soc., 42,705-735 (1975).

Soward, A.M. 'On the Finite Amplitude Thermal Instability of a Rapidly Rotating Fluid Sphere', Geophys. Astrophys. Fluid Dyn., 9, 19-74 (1977).

Tozer, D.C. 'The Present Thermal State of the Terrestrial Planets', Phys. Earth Planet. Int., 6, 182-197 (1972).

DYNAMO THEORY

P.H. Roberts
Institute of Geophysics & Planetary Physics
University of California
Los Angeles
California 90024
U.S.A.

ABSTRACT. After a brief description of the geomagnetic field, past and present, the electrodynamics of moving conductors is described, including the skin effect and Alfvén's frozen flux theorem. The magnetic Reynolds number is introduced, poloidal and toroidal vectors are defined, and it is shown how large toroidal fields can be created in the Earth's core from poloidal fields by the ω-effect, which is the shearing of poloidal field lines by zonal differential motion. Such toroidal fields cannot be detected by magnetic observations at the Earth's surface, but may play a significant dynamical role. The kinematic dynamo problem is formally stated, and is followed by a proof of Cowling's theorem, which rules out the dynamo maintenance of axisymmetric magnetic fields. Ponomarenko's dynamo is exhibited, and is proved to work by a boundary layer argument. Induction by turbulently moving conductors is discussed as an introduction to helicity and the associated α-effect, which creates mean poloidal field from mean toroidal field by cyclonic motion. A similar effect produced by large-scale motions of large magnetic Reynolds number is briefly studied. The discussion of dynamical aspects of dynamo theory opens with a review of possible driving mechanisms, and continues with a study of marginal convection in rotating magnetic systems. Nonlinear aspects are briefly noted. Crude models for reversals are mentioned. Finally there is a brief guide to the literature.

1. DESCRIPTION OF THE GEOMAGNETIC FIELD

1.1. Internal Origin of Field

It was Gilbert (1540-1603), a Court physician to Queen Elizabeth I of England, who first noticed that the source of the Earth's magnetism lay within it. Gilbert had a sphere of lodestone made (his "terrella") and showed how the directions in which a freely suspended magnetic needle dipped towards it, as it was placed on its surface at different distances from the magnetic poles, resembled the directions at which a similar needle dipped when placed at different latitudes on the Earth's

C. Nicolis and G. Nicolis (eds.), Irreversible Phenomena and Dynamical Systems Analysis in Geosciences, 73–133.

surface. Gauss gave mathematical teeth to the idea in 1839, by using results of potential theory that are today very familiar, and by fitting those results to the observations. He posited

$$\underline{\nabla} \times \underline{B} = 0, \qquad \underline{\nabla} \cdot \underline{B} = 0. \tag{1.1,1.2}$$

The second of these is universal, and expresses the impossibility of magnetic monopoles. The first, applied above the Earth's surface, expressed his belief (borne out today) that extra-terrestrial electric currents are too feeble to affect the measured magnetic fields much. According to (1.1) a magnetostatic potential, V, exists such that

$$\underline{B} = -\underline{\nabla} V , \qquad \nabla^2 V = 0 , \tag{1.3,1.4}$$

where (1.4) follows from (1.2) and (1.3).

Let (r,θ,ϕ) be spherical coordinates, where r is distance from the geocentre, θ is geographic co-latitude and ϕ is geographic East longitude. The general solution of (1.4) may be written as

$$V = a \sum_{\ell=0}^{\infty} \sum_{m=0}^{\ell} [\{g_\ell^m (\frac{a}{r})^{\ell+1} + q_\ell^m (\frac{r}{a})^\ell\} \cos m\phi$$

$$+ \{h_\ell^m (\frac{a}{r})^{\ell+1} + s_\ell^m (\frac{r}{a})^\ell\} \sin m\phi] P_\ell^m(\theta) , \tag{1.5}$$

where P_ℓ^m is the associated Legendre function, and a (= 20,000 km/π) is the radius of the Earth. When Gauss fitted the resulting $\underline{B}$ to the observed field he discovered that, to an excellent approximation

$$q_\ell^m = s_\ell^m = 0 , \qquad \text{for all } \ell \text{ and } m. \tag{1.6}$$

All surviving terms in (1.5) decrease with increasing r. There are no sources of field "at infinity"; the field $\underline{B}$ is generated by sources beneath the surface of the Earth, $r = a$. All subsequent analyses of the main geomagnetic field have confirmed Gauss's finding (1.6). Since magnetic monopoles do not exist, the $\ell = 0$ term of (1.5) should also be omitted, allowing us to write

$$V = a \sum_{\ell=1}^{\infty} \sum_{m=0}^{\ell} [g_\ell^m \cos m\phi + h_\ell^m \sin m\phi](a/r)^{\ell+1} P_\ell^m(\theta) . \tag{1.7}$$

As r increases, the series (1.7) is increasingly dominated by the three $\ell = 1$ terms, g_1^0, g_1^1 and h_1^1 which correspond to dipoles of strength m_z, m_x and m_y at $r = 0$ with axes along $\theta = 0$, $\phi = 0$ and $\phi = \frac{1}{2}\pi$ respectively, the z, x and y axes. Together they make up the "centred dipole", $\underline{m} = (m_x, m_y, m_z)$, whose magnitude $m = |\underline{m}|$ is about $8\ 10^{15}\ \mathrm{T\,m^3}$, giving a polar field strength of about 0.6 gauss. The direction of $\underline{m}$ is "the magnetic axis" of the Earth. It meets the northern hemisphere at $(\theta,\phi) = (\theta_0,\phi_0)$ where

$$\theta_0 = \sin^{-1}(m_z/m) , \qquad \phi_0 = \tan^{-1}(m_y/m_x) .$$

This is the "North magnetic pole", and $(\theta,\phi) = (\pi - \theta_0, \pi + \phi_0)$ is the "South magnetic pole". Since at great distances only $\underline{m}$ is seen,

$$\underline{B} = O(r^{-3}), \quad \text{for } r \to \infty . \tag{1.8}$$

This fact is central to these lectures on dynamo theory, for it expresses the absence of external sources of geomagnetic field.

1.2. Internal Structure of Earth

Since we must seek an origin for the geomagnetic field within the Earth, it is appropriate here to draw together the few facts about the internal constitution of the Earth that we will need later. For our purposes, the Earth is spherically symmetric in all respects and consists of three parts:

(i) a light <u>mantle</u> occupying $a \geq r \geq c \doteqdot 3485$ km. Being composed of semi-conducting silicates, its electrical conductivity, σ, rises rapidly with depth but even at $r = c$, the core-mantle boundary (CMB), σ is probably only of order 3×10^3 mho/m. Often, the mantle may be assumed insulating without serious error.

(ii) an outer <u>fluid core</u> occupying $c \geq r \geq c_1 \doteqdot 1215$ km. Being iron-rich, this region is dense ($\rho \doteqdot 10^4$ kgm/m^3) and a good electrical conductor ($\sigma \doteqdot 3\times10^5$ mho/m);

(iii) an <u>inner body</u> $r \leq c_1$, which is dense and solid, possibly of a composition similar to the outer core, but frozen (see §5.1). If so, its conductivity σ will also be high.

As we shall usually be interested in fluid motions, we will pay no special attention to the inner core, usually using the word "core" to embrace (ii) + (iii), and regarding (iii) as a region of a fluid core which happens to move with the mantle.

If [see (i)] we ignore mantle conduction, we may apply (1.1) - (1.4) and (1.7) right down to the CMB. As r decreases the higher order terms in (1.7) become increasingly potent. Thus, even though g_ℓ^m and h_ℓ^m decrease rapidly with increasing ℓ, the corresponding coefficients $g_\ell^m(a/c)^{\ell+1}$ and $h_\ell^m(a/c)^{\ell+1}$ for V at the CMB do not. Thus, the (angular) structure of $\underline{B}$ on the CMB is on a smaller scale than at the Earth's surface. Some representative values (in nT) are given in Table I; they are from the Definitive Geomagnetic Reference Field for 1975. (The so-called "Schmidt-normalized" P_ℓ^m are used.)

TABLE I

Coefficient	g_1^0	g_1^1	h_1^1	g_2^0	g_2^1
Value at $r = a$	-30100	-2013	5675	-1902	3010
$(a/c)^{\ell+1}$ multiple	-100400	-6717	18940	-11590	18350

h_2^1	g_2^2	h_2^2	g_6^0	g_6^1	h_6^1	g_6^2	h_6^2
-2067	1632	68	45	66	-13	28	99
-12600	9948	-414	3100	4500	-880	1900	6700

g_6^3	h_6^3	g_6^4	h_6^4	g_6^5	h_6^5	g_6^6	h_6^6
-198	75	1	-41	6	-4	-111	11
13400	5100	70	-2800	400	200	-7530	750

1.3. Link with Earth's Rotation

We have seen that the dipole terms in (1.7) dominate V at the Earth's surface. Further, it is clear from Table I that the "axial dipole" m_z corresponding to g_1^0 has roughly ten times the strength of the "equatorial dipole" (m_x, m_y) corresponding to g_1^1 and h_1^1. This fact explains both the usefulness of the compass and the genesis of one of the earliest "explanations" of the compass's behaviour, namely that lodestone contains a "quality" (which can be transferred to iron) which attracts it to the Pole Star. We shall interpret the fact differently! We shall see in it compelling evidence that Coriolis forces strongly influence motions in the fluid core.

1.4. Secular Variation of Field: Westward Drift

The geomagnetic field depends on time, t. After rapid variations of external origin (e.g. magnetic storms) have been removed, a slow secular change remains: the g and h coefficients in (1.7) are functions of t. For instance, the centred dipole is not fixed in time. Its strength and direction change continuously. Of particular interest are the rate of change, $\dot{g}_1^0$, of the axial dipole moment (much the same as $\dot{m}$) which is shown in Fig. 1, and the location of the North magnetic pole which is shown in Fig. 2.

There is a suggestion in Fig. 2 that the magnetic pole has been drifting westwards since 1550. If the ϕ_0 displacement from 1780 to 1980 is representative, the pole will circumnavigate the globe in about 4500 years. Between 1780 and 1980 the "tilt", θ_0, of the magnetic axis has been about 11^0, although it seems to have been much less 400 years ago.

Other harmonics of the field also show secular change and westward drift. In Fig. 3 we see the secular rate of change of B_ϕ in 1983. The isolines are large, characteristically of continental dimensions. This is true also of other components of $\underline{B}$ and $\dot{\underline{B}}$. In Fig. 4 the magnetic

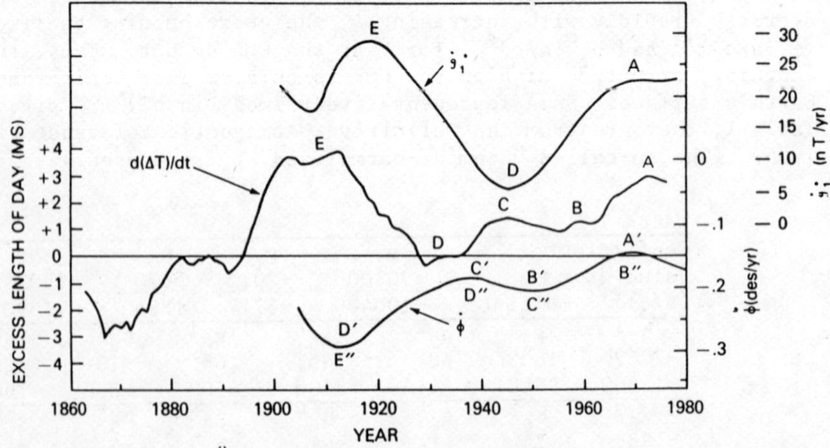

Figure 1. Comparison of $\dot{g}_1^0$, $\dot{\phi}$ and d(Δt)dt (from Langel, 1985, with permission).

declination is shown at six different epochs. While the successive field patterns have a different structure, they are recognizably distortions of each other, and the principal distortion is again a "westward drift". The drift is not a solid-body rotation of the pattern; the westward motion depends on latitude and even, in some places, on longitude. Said another way, the drift rate of each non-axial harmonic

$$\dot{\phi}_\ell^m = (g_\ell^m \dot{h}_\ell^m - h_\ell^m \dot{g}_\ell^m)/m[(g_\ell^m)^2 + (h_\ell^m)^2] \qquad (m \neq 0)$$

is different, with that of $\ell = 1$ being particularly small. For this reason $\ell = 1$ is often omitted; the drift of the "non-dipole field" based on an average of $\dot{\phi}_\ell^m$ over all harmonics with $2 \leq \ell \leq 5$ is shown in Fig. 1, where it is labelled $\dot{\phi}$. Assuming Figs. 1 and 2 are representative, the non-dipole field appears to include shorter timescales than the dipole field; $\dot{\phi}$ is characteristically -0.2^0/yr, suggesting circumnavigation of the Earth in less than 2000 years. Persuasive cases have been made for even a 60 year component in the secular variation.

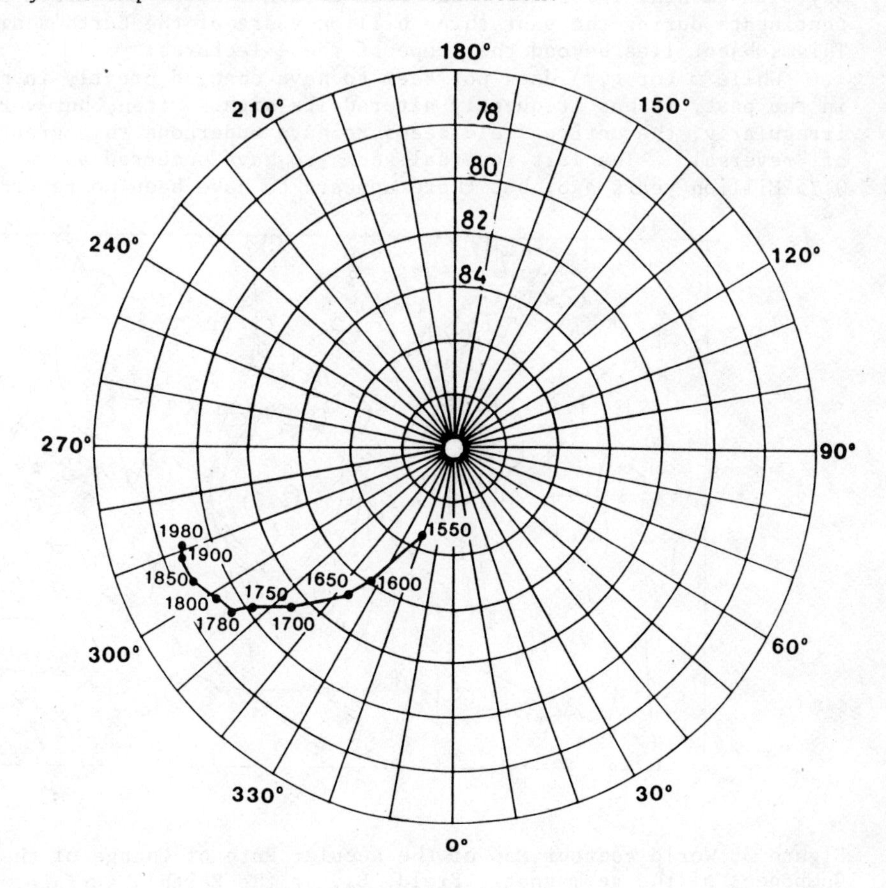

Figure 2. Position of geomagnetic North Pole as a Function of Time (from Langel, 1985, with permission).

1.5. Longevity of Field. Reversals.

When a rock is formed, be it sedimentary or igneous, it captures the prevailing magnetic field at the time of its birth. It is on the basic fact that palaeomagnetism rests, a science that has revealed startling and profound facts about the Earth's structure, dynamics and past history. Its power as a geological tool depends on the fact that the geomagnetic field is as old as the Earth.

From measurements of the fields fossilized in geologically recent lavas, erupted over times long compared with the secular variation scale, it has been shown that, if Fig. 2 were plotted over such long periods, the pole would fill a spherical cap of roughly 10° radius with its centre at the geographic pole. Assuming that this is true over all geological time, the palaeomagnetist can infer the position of the geographic pole (relative to his sample site for the relevant geological epoch) by taking the mean of the virtual geomagnetic poles they give. In this way, fundamental information has been obtained about the drift of continents during the past three billion years of the Earth's history. This subject lies beyond the scope of these lectures.

While m (or $g_1^{\,0}$) does not seem to have changed greatly in magnitude in the past, it has frequently altered its sign. Often, but very irregularly, the entire field seems to have undergone this great spasm of "reversal". The last reversal seems to have occurred about 0.75 Million years ago, but there appears to have been no reversal

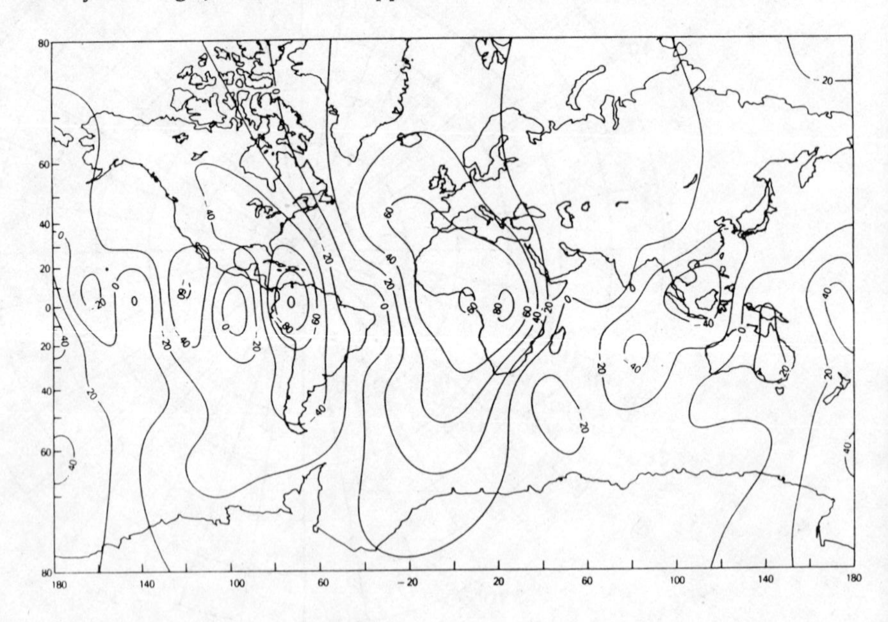

Figure 3. World contour Map of the secular Rate of Change of the East Component of the geomagnetic Field, $\dot{B}_\phi$, at the Earth's surface. Source: the 1983 model from the Goddard Space Flight Center (from Langel, 1985, with permission). Units are nT/yr.

during the 50-70 Million Years of Permian. Long periods of one polarity are called "polarity epochs"; short periods are "polarity events".

1.6. Variations in Length of Day

Changes occur in the length of the day for many reasons but, even after the seasonal variation and the slow tidal deceleration of the Earth have been subtracted, a secularly varying part remains, showing time-scales not unlike the geomagnetic field. The suggestion is strong that they are connected, though not perhaps in a very direct way. Let Δt be the change in the length of the day, compared with some standard value, then $d(\Delta t)/dt$ is the rate of change of the mantle's rotation period. This is plotted in Fig. 1, where the similarity with $\dot{g}_1^{\,0}$ and $\dot{\phi}$ can clearly be seen.

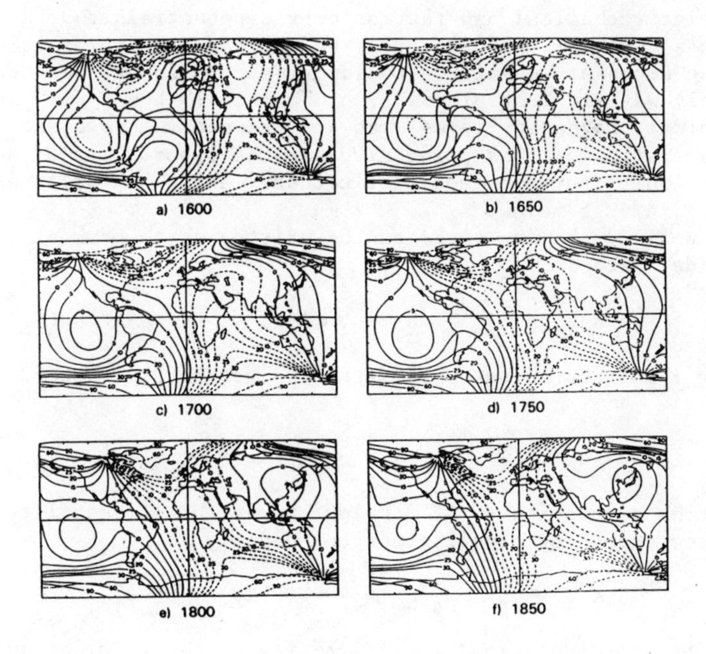

Figure 4. Global Maps of the magnetic Declination, $D = -\tan^{-1}(B_\phi/B_\theta)$, at six different Epochs (From Barraclough, 1974, with permission). Units are Degrees.

2. ELECTRODYNAMICS

2.1. Early Theories of Geomagnetism.

Except in its uppermost layers, the interior of the Earth is above the Curie point. It is impossible to explain the magnetic field of the Earth by the permanent magnetism of its interior. The geomagnetic field owes its existence to electric currents, and it "only" remains to explain how these currents have managed to persist over geological time.

Although currents in the mantle are required in order to explain the apparent connection between the geomagnetic field and the length of the day (§1.6), these currents are likely to be small compared with those flowing in the core, simply because (§1.2) the conductivity of the core is much greater than that of the mantle. We will be little in error if we suppose that the currents responsible for the geomagnetic field flow only in the core.

How are these currents generated? Various mechanisms have been proposed. For example, electric currents will certainly be driven by the electrochemical and thermoelectric potentials that exist in the Earth's core. It has, however, been found hard to explain how potentials of the necessary magnitude could arise. Moreover, the currents produced in this way are most plausibly of "poloidal" type, with attendant "toroidal" magnetic fields that are trapped in the core. Further consideration of this point gives us the opportunity of defining these terms, which pervade geomagnetic theory and which will be directly useful in §2.5 below.

A field obeying (1.2) can be written as the sum of toroidal and poloidal vectors:

$$\underline{B} = \underline{B}_T + \underline{B}_P , \quad \underline{B}_T = \underline{\nabla} \times (T\underline{r}) , \quad \underline{B}_P = \underline{\nabla} \times \underline{\nabla} \times (S\underline{r}) , \tag{2.1}$$

where $\underline{r}$ is the radius vector from the geocentre. According to Ampère's law

$$\underline{\nabla} \times \underline{B} = \mu_0 \underline{J} \tag{2.2}$$

[see the discussion of §2.3 below; μ_0 is the permeability of free space], the corresponding electric current density is

$$\underline{J} = \underline{J}_T + \underline{J}_P , \quad \mu_0 \underline{J}_P = \underline{\nabla} \times \underline{\nabla} \times (T\underline{r}) , \quad \mu_0 \underline{J}_T = \underline{\nabla} \times (-\nabla^2 S \underline{r}) , \tag{2.3}$$

i.e. the toroidal field $\underline{\nabla} \times (T\underline{r})$ is fed by the poloidal current $\underline{\nabla} \times \underline{\nabla} \times (T\underline{r})/\mu_0$, and the poloidal field $\underline{\nabla} \times \underline{\nabla} \times (S\underline{r})$ is fed by the toroidal current $\underline{\nabla} \times (-\nabla^2 S \underline{r})/\mu_0$. In an insulator ($\underline{J} = 0$) we have

$$T = 0 , \quad \nabla^2 S = 0 , \tag{2.4}$$

and in fact S is closely related to the potential V introduced in (1.3):

$$V = - \frac{\partial}{\partial r}(rS) . \tag{2.5}$$

Clearly, however, there is no toroidal field in an insulator.

If we temporarily ignore the motion $\underline{v}$ of the core and suppose that $\underline{B}$ is steadily maintained by electrochemical or thermoelectric potentials, Φ, we have by Ohm's law

$$\underline{J} = \sigma\,\underline{E} = -\,\sigma\underline{\nabla}\Phi\,, \tag{2.6}$$

where $\underline{E}$ is the electric field and σ is the electrical conductivity (assumed constant). We now deduce from (2.3) and (2.6) that

$$L^2 T = -\,\mu_0 \sigma\, r \frac{\partial\Phi}{\partial r}\,, \qquad \nabla^2 S = 0\,, \tag{2.7}$$

where

$$L^2 = -\left[\frac{1}{\sin\theta}\frac{\partial}{\partial\theta}\left(\sin\theta\frac{\partial}{\partial\theta}\right) + \frac{1}{\sin^2\theta}\frac{\partial^2}{\partial\phi^2}\right] \tag{2.8}$$

is the angular momentum operator. Without going further into the technical details of toroidal and poloidal vectors (but see §2 of my chapter on "Convection in spherical systems" in this book), we can appreciate that, since S is by $(2.4)_2$ and $(2.7)_2$ a potential field <u>everywhere</u>, is continuous on $r = R$, and according to (1.8) vanishes as r^{-2} at infinity,

$$S \equiv 0. \tag{2.9}$$

In reaching this conclusion we have included only potential differences that arise in the bulk, e.g. between a rising hot stream of fluid of one chemical composition and a falling cold stream of another. A more obvious source is the CMB where two very different media meet, an iron-rich core and a semi-conducting mantle. It is therefore pertinent to remark that (2.9) applies also in this case.

We have shown that, no matter how strong the electrochemical and thermoelectric effects, they can only steadily produce toroidal fields, and these are magnetically undetectable at the Earth's surface. To create the observed field, some further mechanism must be invoked that excites poloidal fields from toroidal fields. This is not a simple matter, but such a mechanism does exist: the α-effect, to be described in §4.2 below. The α-effect is a subtle form of electromagnetic induction.

2.2. Dynamo Hypothesis

Although electrochemical/thermoelectric potentials might, if large enough, generate the toroidal field of the core, there is an alternative and simpler process, the ω-effect (§2.4), which can efficiently create toroidal field from poloidal field. The ω-effect is a straightforward form of electromagnetic induction, and one is naturally led to wonder whether, instead of invoking electrochemical or thermoelectric effects of doubtful potency, it would not be simpler and more realistic to suppose that electromagnetic induction "does the entire job", both creating poloidal field from toroidal field and the reverse. This is the dynamo hypothesis for the origin of the geomagnetic field.

A self-excited dynamo may be pictured as follows: the moving part of the dynamo induce from the prevailing magnetic field, $\underline{B}$, an electromotive force $\underline{v} \times \underline{B}$ (where $\underline{v}$ is the velocity of the conductor) and this e.m.f. sets up electric currents which through Ampère's law (2.2) produce the very same inducing field $\underline{B}$ that created $\underline{J}$. Stated this way, the process seems suspiciously like "hoisting oneself up by his own boot straps", but the mechanism is essentially the one employed in commercial power stations and is in no sense a perpetual motion machine violating energy conservation. The simplest example, the homopolar dynamo, is illustrated in Figs. 5.

In Fig. 5(a), the conducting disk, D, rotates with angular velocity Ω in a uniform field $\underline{B}_0$ parallel to its axis, A'A. Positive charges (P) build up on the rim of D, negative charges (N) on A'A. Charge separation is "necessary" to create an electrostatic field radially inwards along the disk to cancel out exactly the equal and opposite electromotive force, $\underline{u} \times \underline{B}_0$, created by the motion $\underline{u} = \underline{\Omega} \times \underline{r}$ of the disk. The electric circuit is not complete, and no current flows. In Fig. 5(b), a stationary wire, W, is added to join A'A to D, electrical contact being made through sliding contacts (brushes) at S_1 and S_2. The charges, that build up in Fig. 5(a), are in Fig. 5(b) drawn off as electric current, I: the electromotive force is now motive. In Fig. 5(c) the wire, instead of taking a short path from D to A'A, is wound near the periphery of the disk, and in its plane. The direction of this winding is chosen so that the magnetic field, $\underline{b}$, produced by the current, I, flowing in the wire, cuts the disk in the same sense as the original uniform field. Of course, there was no special reason for choosing a uniform field in the first place. The dipole-type field created by the winding also induces current in the moving disk. It is plausible and true that, if the Ω is large enough, so that the e.m.f. $\underline{u} \times \underline{b}$ can create a sufficiently large I, the uniform field can be completely dispensed with, and the device will retain its fields and currents for as long as Ω is maintained. It is then a "self-excited dynamo".

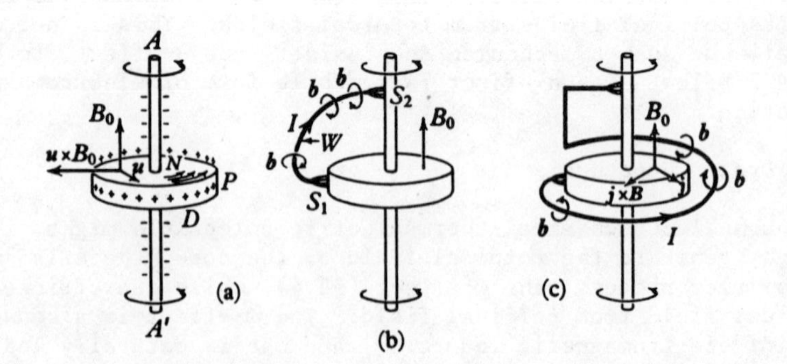

Figure 5. The Homopolar Dynamo.

There are ohmic losses in the circuitry of Fig. 5(c) which must be made good by the couple that maintains the angular velocity, $\underline{\Omega}$, of the disk. It is easy to see that the Lorentz force, $\underline{j} \times \underline{B}$, in the disk opposes its motion, as one would anticipate from Lenz's law. In the steady state the associated couple is equal and opposite to the applied torque $\underline{\Gamma}$ maintaining $\underline{\Omega}$. In this way, the available $\underline{\Gamma}$ determines the steady state strengths of field and current.

2.3. Basic Electromagnetism.

We will be interested only in conductors moving with non-relativistic speeds,

$$v \ll C, \tag{2.10}$$

on timescales long compared with the time taken by light (travelling at speed C) to cross the system considered. Under these circumstances we may neglect displacement currents, and use the pre-Maxwell equations:

$$\underline{\nabla} \times \underline{E} = -\partial \underline{B}/\partial t , \qquad \underline{\nabla} . \underline{B} = 0 , \tag{2.11,2.12}$$

$$\underline{\nabla} \times \underline{H} = \underline{J} , \qquad (\underline{\nabla} . \underline{D} = \vartheta) . \tag{2.13,2.14}$$

Here $\underline{D}$ is the electric displacement, $\underline{H}$ the magnetizing force, and ϑ the electric charge density. We must add a constitutive theory. Assuming isotropic materials and no intrinsic magnetization, we postulate

$$\underline{H} = \underline{B}/\mu_0 , \qquad (\underline{D} = \varepsilon \underline{E}) , \tag{2.15,2.16}$$

$$\underline{J} = \sigma(\underline{E} + \underline{v} \times \underline{B}) , \tag{2.17}$$

where ε is the permittivity. It was from (2.13) and (2.15) that we obtained (2.2). Ohm's law (2.17) is the appropriate generalization of the more familiar form (2.6) which applies only to stationary conductors: it includes the e.m.f. $\underline{v} \times \underline{B}$ discussed earlier.

From (2.11), (2.13), (2.15) and (2.17) we obtain "the induction equation":

$$\frac{\partial \underline{B}}{\partial t} = \underline{\nabla} \times [\underline{v} \times \underline{B} - \eta \underline{\nabla} \times \underline{B}] , \tag{2.18}$$

where

$$\eta = 1/\mu_0 \sigma \tag{2.19}$$

is "the magnetic diffusivity" (dimensions m^2/s). It implies that $(\partial/\partial t)\underline{\nabla}.\underline{B} = 0$, so that if (2.12) holds initially (as it must) it holds for ever. We shall be vitally concerned with the solution of (2.18) under appropriate boundary conditions, and with the $\underline{J}$ and $\underline{E}$ that result from (2.2) and (2.17). If need be we can then deduce $\underline{D}$ and ϑ from (2.16) and (2.14) but, through the neglect of displacement currents, these quantities play a passive role that leaves $\underline{B}$, $\underline{J}$ and $\underline{E}$ unaffected. For this reason (2.14) and (2.16) are uninfluential and have been placed in brackets. It may also be shown that, for consistency with (2.10), the energy density of the electric field and the electric stress tensor are both negligible compared with their magnetic counterparts, and that the advection $\vartheta \underline{v}$ of charge by motion is properly omitted from $\underline{J}$ in (2.17). Also, $\underline{B}$ is invariant under translation or rotation of frames: by (2.2) the same is true of $\underline{J}$, but by (2.17) not of $\underline{E}$.

Equation (2.18) is often written in its constant-η form:

$$\frac{\partial \underline{B}}{\partial t} = \underline{\nabla} \times (\underline{v} \times \underline{B}) + \eta \nabla^2 \underline{B} , \tag{2.20}$$

where we have used (2.12) and the identity $\underline{\nabla} \times \underline{\nabla} \times \underline{B} = \underline{\nabla}(\underline{\nabla}.\underline{B}) - \underline{\nabla}^2 \underline{B}$. A dimensionless form follows from the scaling

$$t \to (L^2/\eta)t , \quad \underline{x} \to L\underline{x} , \quad \underline{v} \to U\underline{u} , \tag{2.21}$$

where L is a characteristic length (e.g. c) and U is a characteristic fluid velocity:

$$\frac{\partial \underline{B}}{\partial t} = R_m \underline{\nabla} \times (\underline{u} \times \underline{B}) + \nabla^2 \underline{B} , \tag{2.22}$$

where

$$R_m = UL/\eta \tag{2.23}$$

is "the magnetic Reynolds number".

2.4. Skin Effect. Flux Ropes. The ω-Effect. Alfvén's Theorem.

The objective of this subsection is to gain relevant experience in the solution of the induction equation (2.20) by using simple models involving solid conductors.

An important example, quoted in every book on electromagnetism, illustrates "the skin effect". Imagine that a uniform stationary conductor fills the half-space $z \geq 0$, and that $z < 0$ is vacuum. Let an oscillating field,

$$\underline{B} = B_0 \hat{\underline{x}} \cos \omega t , \tag{2.24}$$

be present in $z \leq 0$, where $\hat{\underline{x}}$ is the unit vector in the x-direction. Since $\underline{v} = 0$ in the conductor, the field in $z \geq 0$ obeys according to (2.20) the vector diffusion equation,

$$\frac{\partial \underline{B}}{\partial t} = \eta \nabla^2 \underline{B} , \tag{2.25}$$

and using (2.24) as boundary condition on $z = 0$ we quickly find that in the conductor

$$\underline{B} = B_0 e^{-z/\delta} \hat{\underline{x}} \cos \omega(t - z/\delta) , \tag{2.26}$$

where

$$\delta = (2\eta/\omega)^{\frac{1}{2}} . \tag{2.27}$$

The currents induced in $z > 0$ by the field (2.24) flow in a skin on the surface of the conductor and screen its deep interior from that field; the amplitude of $\underline{B}$ decreases exponentially in the e-folding distance, δ, called "the skin depth".

We may apply this idea to determine the steady field, $\underline{b}$, induced by a rapidly rotating, electrically conducting cylinder, $s \leq a$, placed in a field $\underline{B}_0 = B_0 \hat{\underline{x}}$ perpendicular to its axis, Oz. [Here s, the distance from Oz, is one of the cylindrical coordinates (s,ϕ,z) used below.] The exterior of the cylinder is a uniform stationary conductor in perfect electrical contact across $s = a$ with the cylinder. The configuration, for a zero angular velocity, ω, of the cylinder, is sketched in Fig. 6(a). We shall concentrate on the rapidly rotating limit by which we mean that the magnetic Reynolds number,

$$R_m = a^2\omega/\eta , \tag{2.28}$$

is very large.

Imagine an observer fixed to the periphery, $s = a$, of the rotor. He will be subjected to a rapidly oscillating magnetic field which,

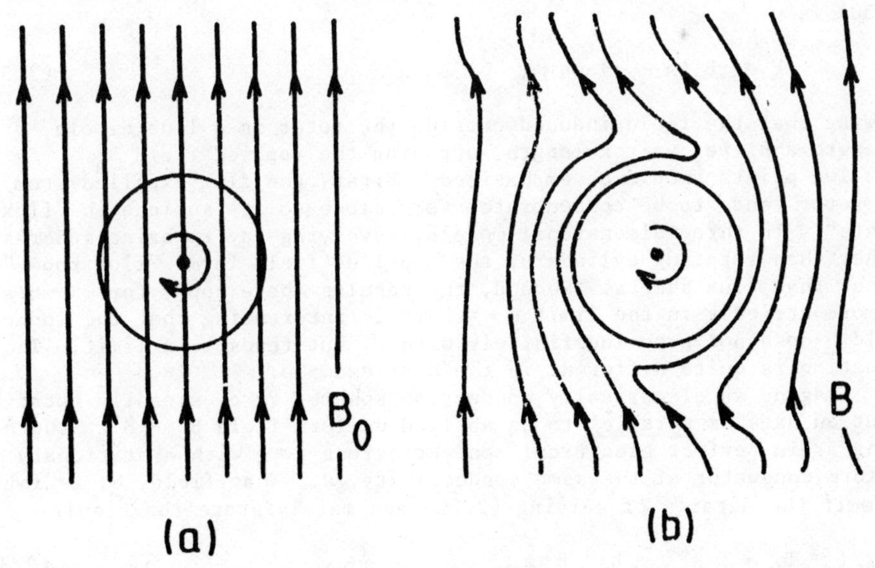

Figure 6. Induction by an infinite rotating cylinder in perfect electrical contact with a surrounding stationary conductor. In (a) the applied transverse field $\underline{B}_0$ is shown; in (b) the sum of applied and induced fields, $\underline{B}$, is shown.

therefore, penetrates as a thin boundary layer only a small distance δ (<<a) into the rotor. Thus, the rotor is almost completely diamagnetic: magnetic flux is expelled from it. [See Fig. 6(b).] To a good approximation therefore

$$B_s = 0 \,, \quad \text{on} \quad s = a \,. \tag{2.29}$$

Consider now the field in the surroundings. This must obey (2.29) and at great distances where $\underline{b}$ is negligible

$$\underline{B} \rightarrow \underline{B}_0 \,, \quad \text{as} \quad s \rightarrow \infty \,. \tag{2.30}$$

Since the state is steady and the conductor is uniform (2.11), (2.17) and (2.13) show that

$$\underline{\nabla} \times \underline{E} = 0 \,, \quad \underline{\nabla} . \underline{E} = \underline{\nabla} \,.\, \underline{J}/\sigma = 0 \,.$$

Thus $\underline{E}$ is an electrostatic potential field

$$\underline{E} = -\underline{\nabla}\Phi \,, \qquad \nabla^2\Phi = 0 \,.$$

It is however clear, from the direction of $\underline{v} \times \underline{B}$ in the rotor, that $\underline{E}$ can have only a z-component, i.e. that Φ and therefore E_z is a function of z alone. But E_z must vanish for $s \rightarrow \infty$, since no electric field has been applied. It follows that $\underline{E} = 0$, and $\underline{B}$ is therefore a potential field as in (1.3) and (1.4). The relevant V which obeys (2.29) and (2.30) is

$$V = -B_0\left(s + \frac{a^2}{s}\right)\cos\phi \,, \qquad s \geq a \,, \tag{2.31}$$

showing that the field induced outside the rotor is a line-dipole, strength a^2B_0 per unit z-length, opposing the applied field $\underline{B}_0$.

Two points should be emphasized. First, the flux expelled from the rotor tends to be concentrated very close to its surface in "flux sheets". In three dimensional models, involving say rotating spheres rather than rotating cylinders, the expelled field forms "flux ropes" rather than flux sheets. Second, the results above apply for $\omega >> \eta/a^2$ or more strictly in the limit $\omega \rightarrow \infty$. It is interesting that the induced field $\underline{b}$ does not grow indefinitely with ω, but tends to a limit. The situation is quite different in the next example.

Imagine an electrically conducting sphere, $r \leq c$, steadily rotating about an axes Oz <u>parallel</u> to an applied uniform field $\underline{B}_0 = B_0\hat{\underline{z}}$, there being again perfect electrical contact across $r = c$ with a stationary uniform conductor of the same conductivity, σ. What field, $\underline{b}$, is induced? We omit the details of solving (2.20) and merely state the result:

$$\underline{B} = \underline{B}_0 + \underline{b} \,, \qquad \underline{b} = B_\phi \hat{\underline{\phi}} \,, \tag{2.32}$$

where

$$B_\phi = \frac{1}{5}\frac{\omega c^2}{\eta}\left(\frac{c}{r}\right)^3 \sin\theta\cos\theta, \qquad r \geq c, \tag{2.33}$$

$$B_\phi = -\frac{1}{5}\frac{\omega c^2}{\eta}\left(\frac{r}{c}\right)^2 \sin\theta\cos\theta, \qquad r \leq c. \tag{2.34}$$

The current system induced by the motion of the rotor is indicated in Fig. 7(a). It may be seen that, since $\underline{b}$ and $\underline{v}$ are parallel, only the applied field $\underline{B}_0$ induces currents. Since (for $\omega > 0$) B_ϕ is negative in the northern hemisphere and positive in the southern hemisphere, the lines of force of the total field $\underline{B}$ are as sketched in Fig. 7(b). It seems as though the lines of force are dragged round in the direction of the rotor's motion. This may be understood by an appeal to Alfvén's frozen flux theorem.

Alfvén's theorem states that, in a perfectly conducting fluid (i.e. $\sigma = \infty$, $\eta = 0$, $R_m = \infty$), the magnetic tubes of force are material volumes, i.e. volumes that move with the fluid as though frozen to it. (For a proof, see for example §3.1 of Moffatt, 1978.) Our fluids are not perfectly conducting. (Indeed by §2.5 the dynamo problem has no meaning in a perfect conductor.) But our fluids will in some respects resemble perfect conductors when R_m is large. There will however always be some drift of the lines of force relative to the conductor proportional to their curvature and to the size of η. This drift tends to reduce the curvature, i.e. "straighten up" lines of force that are bent. Thus, the turning of the rotor tends by Alfvén's theorem to pull the lines of force round with it, as illustrated in Fig. 7(b), but this effect

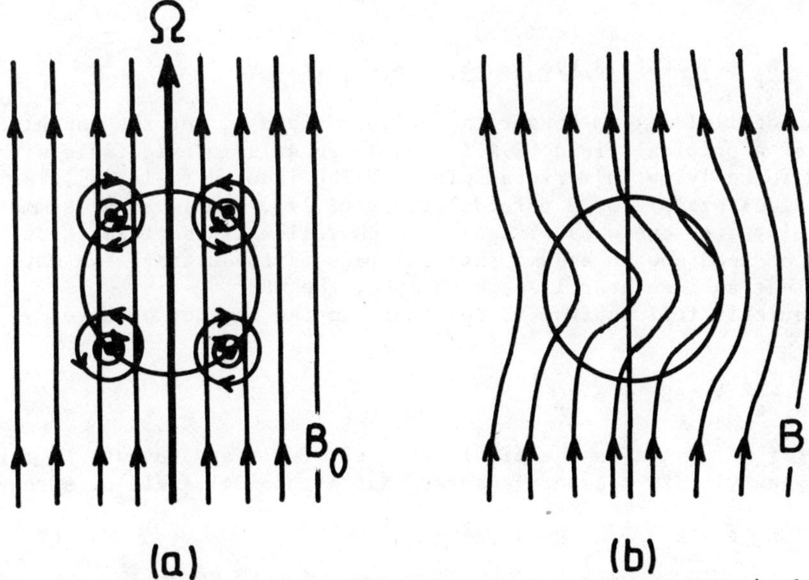

Figure 7. Induction by a rotating sphere in perfect electrical contact with a surrounding stationary conductor. In (a) the applied field, $\underline{B}_0$, parallel to the angular velocity $\underline{\Omega}$ is shown; also the current system induced is illustrated. In (b) the sum of applied and induced fields is shown.

gradually lessens as the curvature in the lines of force increases, and the lines of force increasingly drift through the conductor back towards their original positions. In the steady state this drift is equal and opposite to the drag associated with Alfvén's theorem. The larger the $\underline{v}$, the more the drag, the greater the distortion of $\underline{B}_0$, i.e. the larger the induced field $\underline{b}$. In confirmation we see from (2.33) and (2.34) that there is, unlike the cylindrical rotor considered earlier, no limit to the magnitude of B_ϕ, which increases linearly with ω as $\omega \to \infty$, i.e. $B_\phi \propto R_\omega B_0$ where $R_m = \omega c^2/\eta$ is the magnetic Reynolds number of motion; see also (2.39) below.

It should also be particularly noticed that it is the shear at $r = c$ that induces the field, not the velocity of the entire rotor. Because of the invariance of the induction equation (2.20) to rotation of the frame, the rotation of the entire conductor can no more induce field than can a stationary conductor, which is to say not at all.

The idea that shear can create zonal field B_ϕ from a meridional field $\underline{B}_P$ is, if generalized, a useful one in geomagnetism. Instead of supposing the shearing motion is confined to a single surface as above, we suppose that

$$\underline{v} = \omega(r,\theta)\, r \sin\theta\, \hat{\underline{\phi}} , \tag{2.35}$$

and in place of (2.32) we write

$$\underline{B} = \underline{B}_P + \underline{B}_T , \tag{2.36}$$

where

$$\underline{B}_P = \underline{\nabla} \times [A(r,\theta,t)\hat{\underline{\phi}}] , \quad \underline{B}_T = B_\phi(r,\theta,t)\hat{\underline{\phi}} . \tag{2.37}$$

The inducing field is no longer the uniform field $\underline{B}_0$ but the canonical example of a poloidal field $(2.1)_3$, namely an axisymmetric field with lines of force lying in meridian planes. The induced field, $\underline{B}_T$, is the canonical example of a toroidal field $(2.1)_2$, namely an axisymmetric zonal field, i.e. one whose lines of force follow lines of latitude. There is no need now to assume that all space is conducting. We may suppose that an insulator lies above $r = c$, the CMB.

It quickly follows from (2.20) that, in the conducting core,

$$\frac{\partial B_\phi}{\partial t} = \eta \Delta B_\phi + s\, \underline{B}_P \cdot \underline{\nabla}\omega , \tag{2.38}$$

where $\Delta = \nabla^2 - s^{-2}$. It is the final term, $s\underline{B}_P \cdot \underline{\nabla}\omega$, that prevents B_ϕ from diffusing away. Its presence ensures that a toroidal field of strength

$$B_T \doteqdot R_\omega B_P , \qquad R_\omega = \omega c^2/\eta , \tag{2.39}$$

is created from the poloidal field. This is the "ω-effect" referred to in §2.2, and which was there advocated as an attractive alternative to the dubious charms of electrochemical or thermoelectric excitation of $\underline{B}_T$. It should be noticed however that the ω-effect does not maintain

$\underline{B}_P$, and only generates $\underline{B}_T$ while $\underline{B}_P$ is present. Thus, while the ω-effect may temporarily create a toroidal field large compared with a parent poloidal field, unless a mechanism can be found to support $\underline{B}_P$ its ultimate fate will be to diffuse away, after which $\underline{B}_T$ must follow suit. The ω-effect cannot unaided sustain a self-exciting dynamo: see §3.2.

2.5. Decay Modes for a Sphere

We have had occasion more than once in the last section to point out that, unless supported by a source, a field will die away in time through ohmic resistance, i.e. through the finiteness of σ and η. This is an important point for the theory of geomagnetism for, were it true that the diffusion of field from the core is insignificant over the age of the Earth, there would be no need to search for a theory of geomagnetism: the geomagnetic field would, as much as the rocks, be a fossilised relic of the Earth's birth.

To examine this question, we suppose that the conducting region is a uniform sphere $r \leq c$ (the inner plus outer cores of the Earth), surrounded by an insulator (the mantle and the free space above it). We neglect motions, and therefore seek solutions of the vector diffusion equation (2.25) in $r \leq c$, and join it continuously at $r = c$ to a potential field of the form (1.7), which satisfies the crucial condition (1.8).

It is convenient to divide $\underline{B}$ into its toroidal and poloidal parts as in (2.1). It is easy to show using the techniques of §2.1 that (2.25) implies that the defining scalars, T and S, obey analogous scalar diffusion equations:

$$\frac{\partial T}{\partial t} = \eta \nabla^2 T , \qquad \frac{\partial S}{\partial t} = \eta \nabla^2 S , \qquad r \leq c ; \tag{2.40}$$

while above the conductor (2.4) holds. Continuity of $\underline{B}$

$$\langle \underline{B} \rangle = 0 , \qquad \text{at} \quad r = c , \tag{2.41}$$

requires

$$\langle T \rangle = 0 , \qquad \langle S \rangle = \langle \partial S/\partial r \rangle = 0 , \qquad \text{at} \quad r = c. \tag{2.42}$$

Here $\langle Q \rangle$ is the leap, $Q(c+) - Q(c-)$, in a quantity Q in crossing from vacuum to conductor.

This system of equations and boundary conditions may be conveniently solved by dividing T and S into their spherical harmonic components:

$$T = T_\ell^m(r) P_\ell^m(\theta) e^{im\phi - \lambda_T t} , \qquad S = S_\ell^m(r) P_\ell^m(\theta) e^{im\phi - \lambda_S t} . \tag{2.43}$$

Equation $(2.4)_2$ shows that since $\underline{B} \to 0$ as $r \to \infty$,

$$S_\ell^m \propto r^{-(\ell+1)} , \qquad r \to \infty , \tag{2.44}$$

so that everywhere in $r > c$

$$\frac{\partial S_\ell^m}{\partial r} + \frac{(\ell+1)}{r} S_\ell^m = 0 . \tag{2.45}$$

In particular, (2.45) holds on $r = c+$, and by (2.42) also on $r = c-$. Thus, using also $(2.4)_1$, we see that (2.40) must be solved subject to

$$T_\ell^m = 0 , \quad \frac{\partial S_\ell^m}{\partial r} + \frac{(\ell+1)}{c} S_\ell^m = 0 , \qquad r = c. \tag{2.46}$$

After substituting (2.43) into (2.40) we quickly find

$$T_\ell^m \propto j_\ell(k_\ell r/c) , \qquad S_\ell^m \propto j_\ell(k_{\ell-1} r/c) , \tag{2.47}$$

where

$$\lambda_{T\ell} = \eta k_\ell^{\,2}/c^2 , \qquad \lambda_{S\ell} = \eta k_{\ell-1}^{\,2}/c^2 , \tag{2.48}$$

$j_\ell(z)$ being the spherical Bessel function of order ℓ, and k_ℓ being one of its positive zeros. [The property $j_\ell'(z) + (\ell+1)j_\ell(z)/z = j_{\ell-1}(z)$ has been used in order to obtain $(2.47)_2$ from $(2.46)_2$.]

Equations (2.48) give the decay rates of the ℓth harmonics of toroidal and poloidal field. The longest lived modes are those corresponding to the smallest positive zero of $j_\ell(z)$, and we will confine attention to these (while recognizing that the entire spectrum of modes would be required to solve the general initial value problem for $\underline{B}$). These smallest k_ℓ are increasing functions of ℓ, the smallest being $k_0 = \pi$. [Note: $j_0(z) = \sin z/z$.] Thus the longest lived modes are the poloidal dipole fields S_1^0 and S_1^1, associated with $\underline{m}$ in §1.1. As ℓ increases, so do the decay rates, those of the modes T_ℓ^m always being the same as those of $S_{\ell+1}^m$, according to (2.48), and all being independent of m. The e-folding time for the centred dipole moment $\underline{m}$ is

$$\tau = c^2/\eta\pi^2 , \tag{2.49}$$

or about 8000 yrs for the Earth, using values quoted in §1.2. That τ should be proportional to c^2/η might have been anticipated on dimensional grounds, but the additional reduction of τ by a factor of nearly 10 could not. Of course, similar results hold for conductors that are not spherical, but the analysis is usually more difficult to perform.

3. RESULTS FROM KINEMATIC DYNAMO THEORY

3.1. Statement of Kinematic Dynamo Problem

Let $\mathcal{V}$ be a bounded volume of conductor moving internally with given (steady) velocity $\underline{v} = U\underline{u}$ (where $\underline{u}$ is dimensionless) but bounded by a fixed surface, $\mathcal{S}$, beyond which lies an insulator, $\hat{\mathcal{V}}$. We seek a magnetic field, $\underline{B}$, that obeys the induction equation in the conductor:

$$\frac{\partial \underline{B}}{\partial t} = R_m \, \underline{\nabla} \times (\underline{u} \times \underline{B}) + \nabla^2 \underline{B} \, , \qquad \text{in } \mathcal{V}, \tag{3.1}$$

and a magnetic field $\underline{\hat{B}}$ that is potential in $\hat{\mathcal{V}}$:

$$\underline{\hat{B}} = -\underline{\nabla}\hat{V} \, , \qquad \nabla^2 \hat{V} = 0 \, , \qquad \text{in } \hat{\mathcal{V}}; \tag{3.2}$$

is continuous with $\underline{B}$:

$$\langle \underline{B} \rangle = 0 \, , \quad \text{on } \mathcal{S} \, , \tag{3.3}$$

and which vanishes at great distances from an origin in $\mathcal{V}$:

$$\underline{\hat{B}} = O(r^{-3}) \, , \qquad r \to \infty \, . \tag{3.4}$$

The motions $\underline{v}$ are said to provide "a self-excited dynamo" if

$$\underline{B} \not\to 0 \, , \qquad t \to \infty \, . \tag{3.5}$$

We recognize in (3.1), (3.2), (3.3)* and (3.4), the equations (2.22), (1.3) and (1.4), (2.41) and (1.8) derived and discussed earlier. Condition (3.5) states that the system sustains its magnetism as did the homopolar dynamo of §2.1. Unlike the manmade and deliberately asymmetric structure of the homopolar dynamo, we have in mind the simple, almost spherical $\mathcal{V}$ of cosmic bodies such as the Earth, planets and stars, in which many paths exist to short-circuit currents, and for which the existence of solutions to (3.1) - (3.5) is by no means obvious. Indeed, some negative results are stated in §3.2.

Considered as a mathematical problem in the theory of vector partial differential equations, (3.1) - (3.4) is linear and not self-adjoint. As stated (for a finite body $\mathcal{V}$) we may expect that the eigenvalue spectrum is discrete and complete, i.e. a solution can be expressed as a sum of normal modes of the form

$$\underline{B}(\underline{x};t) = \underline{B}_0(\underline{x}) e^{\lambda t} \, . \tag{3.6}$$

There are basically two possibilities:

(a) if $R\ell(\lambda) < 0$ for all λ, $\underline{B}$ will not satisfy (3.5) and dynamo action does not occur;

(b) if $R\ell(\lambda) \geq 0$ for any λ, (3.5) is satisfied and $\underline{v}$ is "regenerative".

The eigenvalues λ are continuous functions of R_m and since we know (§2.5) that (a) obtains when $R_m = 0$ we may infer that (a) obtains for all

---

* We may note that, in addition to (3.3), $\langle \underline{n} \times \underline{E} \rangle$ must vanish on $\mathcal{S}$. This condition would be influential were $\hat{\mathcal{V}}$ a conductor, and it must be used, for instance, for the rotors of §2.4 or in §3.3 below. In an insulator, however, $\underline{\hat{E}}$ contains an electrostatic part, $-\underline{\nabla}\hat{\Phi}$, which can always be chosen both to vanish at infinity and to make $\langle \underline{n} \times \underline{E} \rangle = 0$ on $\mathcal{S}$. Thus, it is unnecessary to determine $\underline{E}$ in order to solve the problem stated: it may have been seen that $\underline{E}$ was considered in the analogous situation of §2.5.

sufficiently small R_m. If now we consider a sequence of states of increasing U, it may happen that a marginal state, $R_m = R_{mc}$, is reached where $R\ell(\lambda) = 0$ for one (or more) $\lambda = \lambda_1$, say. If $Im(\lambda_1) = 0$ we may, in analogy with convection theory, say that "the principle of the exchange of stabilities holds": a field can then be steadily maintained for $R_m = R_{mc}$. If $Im(\lambda_1) \neq 0$, we have "overstability", a field of constant amplitude can be obtained, but it is oscillatory. If we have reason to believe that steady solutions exist, we may seek the marginal R_{mc} in a different manner, by omitting $\partial \underline{B}/\partial t$ on the left of (3.1) and treating R_m as an "eigenvalue" to be determined.

3.2. Non-support of Axisymmetric Fields: Cowling's Theorem.

We here amplify remarks made in §2.4 and generalize them. We examine whether axisymmetric magnetic fields can be maintained by dynamo action. Clearly the container $\mathcal{V}$ and inducing velocities must be assumed to be axisymmetric also, since otherwise asymmetric fields would necessarily appear. We therefore write

$$\underline{v} = \underline{v}_T + \underline{v}_P, \quad \underline{v}_T = \omega(r,\theta) r \sin\theta \, \hat{\underline{\phi}}, \quad \underline{v}_P = \underline{\nabla} \times [\chi(r,\theta)\hat{\underline{\phi}}]. \quad (3.7)$$

This represents a generalization of (2.35) by the addition of a **meridional** motion $\underline{v}_P$, and (2.38) requires modification. The magnetic field is written in the form (2.36) and (2.37), as before.

The induction equation (3.1) gives

$$\frac{\partial B_\phi}{\partial t} + s\, \underline{v}_P . \underline{\nabla}\left(\frac{B_\phi}{s}\right) = \eta\, \Delta B_\phi + s\underline{B}_P . \underline{\nabla}\omega, \quad (3.8)$$

$$\frac{\partial A}{\partial t} + \frac{\underline{v}_P}{s} . \underline{\nabla}(sA) = \eta\, \Delta A, \quad (3.9)$$

where we have reverted to dimensional units. According to (3.2), (3.3) and (3.4) we have

$$\hat{B}_\phi = \Delta\hat{A} = 0, \quad \text{in } \hat{\mathcal{V}}. \quad (3.10)$$

$$\langle B_\phi \rangle = \langle A \rangle = \langle \underline{n}.\underline{\nabla}A \rangle = 0, \quad \text{on } \mathcal{S}, \quad (3.11)$$

$$\hat{A} = O(r^{-2}), \quad r \to \infty. \quad (3.12)$$

It should be particularly noted that, while the toroidal field has in (3.8) the ω-effect source to create it from $\underline{B}_P$, the poloidal field equation (3.9) has no source, the second test on the left-hand side being no more than part of the motional time derivative of sA following the poloidal motion. It is not surprising that, by multiplying (3.9) by s^2A and integrating over $\mathcal{V}$ one can with the help of (3.10) - (3.12) prove that

$$\frac{\partial}{\partial t}\int_{\mathcal{V}} (sA)^2 dV = -2\eta \int_{\mathcal{V}+\hat{\mathcal{V}}} [\underline{\nabla}(sA)]^2 dV < 0, \quad (3.13)$$

i.e. the poloidal field vanishes as $t \to 0$. Once $\underline{B}_P = 0$, there is no longer an ω-source in (3.8), and by multiplying that equation by B_ϕ/s^2 and integrating over $\mathcal{V}$ one can then by using (3.10) - (3.11) prove that

$$\frac{\partial}{\partial t}\int_{\mathcal{V}} \left(\frac{B_\phi}{s}\right)^2 dV = -2\eta \int_{\mathcal{V}} \left[\underline{\nabla}\left(\frac{B_\phi}{s}\right)\right]^2 dV < 0 , \tag{3.14}$$

so that the toroidal field also vanishes as $t \to \infty$.

We have given S.I. Braginskii's (1964) proof of T.G. Cowling's (1933) theorem: an axisymmetric magnetic field cannot be maintained by dynamo action. It should be noticed that the theorem in no way rules out axisymmetric motions as possible dynamos, but would show that such flows could only sustain asymmetric magnetic fields (see §3.3). The most celebrated anti-dynamo theorem concerning motions is that of E.C. Bullard and H. Gellman (1954): purely toroidal motions are not self-excited dynamos. These theorems have analogies in plane geometry: a two-dimensional magnetic field (by which we mean a field that is independent of one of the Cartesian coordinates but may have three nonzero components) cannot be maintained by dynamo action*; motions that have no component parallel to one of the axes of Cartesian coordinates are not self-excited dynamos. For proofs of these results and helpful insights see Chapter 6 of H.K. Moffatt's (1978) book.

In view of all these negative results, it is appropriate to supply a working dynamo, and this we shall do in §3.3. We conclude this subsection with an interesting extension of the Bullard-Gellman (1954) theorem due to F.H. Busse (1975).

One of the big uncertainties of geomagnetic theory is the size of the toroidal field, $\underline{B}_T$, in the core (see also §4.2 below). It is possible to assess roughly the strength of the poloidal field, $\underline{B}_P$, by extrapolating the observed field downwards, and estimates of the poloidal field energy, M_P, made in this way are unlikely to be grossly in error. It is however quite unsure how much of the total magnetic field energy, M, that this accounts for. F.H. Busse's bound concerns M_c, the field energy of the core

$$M_P = \frac{1}{2\mu_0}\int B_P^2 dV , \qquad M_c = \frac{1}{2\mu_0}\int_{\mathcal{V}} B^2 dV , \tag{3.15}$$

see (2.1); the first integral is over all space. Now $\underline{B}_P$ is created from $\underline{B}_T$ by some kind of induction process (§2.1). Recalling Alfvén's theorem (§2.4) we may picture the lines of force of the toroidal field, which lie totally on spherical surfaces, being distorted off these surfaces to supply a field with a radial component, i.e. $\underline{B}_P$. This evidently requires that $\underline{v}$ has a radial component. F.H. Busse's (1975) bound relates the maximum value of v_r in the core to M_P/M_c.

---

* Again, the non-existence of a two-dimensional dynamo field does not imply that a two-dimensional flow cannot regenerate three-dimensional fields. See §4.2 below, and discussion of the motion (4.14).

Busse makes use of the result

$$\int [\nabla(\underline{r}.\underline{B})]^2 dV \geq 2 \int \underline{B}_P{}^2 dV = 4\mu_0 M_P . \qquad (3.16)$$

Since the proof of this inequality given by Busse and several subsequent workers is fallacious, we should sketch the correct argument here. Defining $\underline{B}_P$ as in (2.1) and expanding S as in (2.43), i.e.

$$S = \sum_{\ell=1}^{\infty} \sum_{m=-\ell}^{\ell} S_\ell{}^m(r,t) P_\ell{}^m(\theta) e^{im\phi} , \qquad (3.17)$$

we can show, appealing to the orthonormality of the spherical harmonics, that

$$\int \underline{B}_P{}^2 dV = \sum_{\ell=1}^{\infty} \sum_{m=-\ell}^{\ell} \ell(\ell+1) \int_0^{\infty} [|\frac{\partial}{\partial r}(rS_\ell{}^m)|^2 + \ell(\ell+1)|S_\ell{}^m|^2] dr, \qquad (3.18)$$

$$\int [\underline{\nabla}(\underline{r}.\underline{B})]^2 dV = \int [\underline{\nabla}(\underline{r}.\underline{B}_P)]^2 dV =$$
$$= \sum_{\ell=1}^{\infty} \sum_{m=-\ell}^{\ell} [\ell(\ell+1)]^2 \int_0^{\infty} [r^2 |\frac{\partial S_\ell{}^m}{\partial r}|^2 + \ell(\ell+1)|S_\ell{}^m|^2] dr . \qquad (3.19)$$

Now

$$|\frac{\partial}{\partial r}(rS_\ell{}^m)|^2 - r^2 |\frac{\partial S_\ell{}^m}{\partial r}|^2 = \frac{\partial}{\partial r}[r|S_\ell{}^m|^2]$$

integrates to zero, since $S_\ell{}^m = O(r^{-2})$ for $r \to \infty$ and is continuous at $r = a$, and we therefore have

$$\int [\underline{\nabla}(\underline{r}.\underline{B})]^2 dV \geq 2 \sum_{\ell=1}^{\infty} \sum_{m=-\ell}^{\ell} \ell(\ell+1) \int_0^{\infty} [|\frac{\partial}{\partial r}(rS_\ell{}^m)|^2 + \ell(\ell+1)|S_\ell{}^m|^2] dr,$$

from which (3.16) follows, by (3.18). We also see that equality obtains in (3.16) if and only if $\underline{B}_P$ is a pure centred dipole ($\ell = 1$).

The scalar product of (2.18) with $\underline{r}$ may be written as

$$(\frac{\partial}{\partial t} + \underline{v}.\underline{\nabla})(\underline{r}.\underline{B}) - \eta\nabla^2(\underline{r}.\underline{B}) = \underline{B}.\underline{\nabla}(\underline{r}.\underline{v}) . \qquad (3.20)$$

This equation is analogous to that governing "temperature", $\underline{r}.\underline{B}$, in a moving fluid, with $\underline{B}.\underline{\nabla}(\underline{r}.\underline{v})$ representing "heat sources" internal to the fluid. If these are zero, the temperature will equalize unless sources are present on the boundary. In fact (3.2) - (3.4) ensure that such boundary sources are absent so that, if the flow is toroidal ($\underline{r}.\underline{v} = 0$), then $\underline{r}.\underline{B}$ (and so $\underline{B}_P$) will vanish as $t \to \infty$. This in fact proves "half" of the Bullard-Gellman theorem: the proof that $\underline{B}_T \to 0$ as $t \to \infty$ is more elaborate, and will not be attempted here (see H.K. Moffatt, 1978).

Multiply (3.20) by $\underline{r}.\underline{B}$, integrate over $r < a$, apply the divergence theorem, and use (3.3) to convert the surface integrals to integrals of $\hat{\underline{B}}$ over $\mathcal{S}$. With the help of (3.2) and (3.4), these can be converted into integrals of $\hat{\underline{B}}$ over $\hat{\mathcal{V}}$, leading to

$$\tfrac{1}{2}\frac{d}{dt}\int(\underline{r}.\underline{B})^2 dV + \eta\int[\underline{\nabla}(\underline{r}.\underline{B})]^2 dV = -\int_{\mathcal{V}}(\underline{r}.\underline{v})\underline{B}.\underline{\nabla}(\underline{r}.\underline{B})dV, \qquad (3.21)$$

the integrals on the left-hand side being over all space, $\mathcal{V}+\hat{\mathcal{V}}$. The integral on the right-hand side can be bounded, using the Schwarz inequality, as

$$-\int_{\mathcal{V}}(\underline{r}.\underline{v})\underline{B}.\underline{\nabla}(\underline{r}.\underline{B})dV \le (\underline{r}.\underline{v})_{max}\left[\int_{\mathcal{V}} B^2 dV\int_{\mathcal{V}}[\underline{\nabla}(\underline{r}.\underline{B})]^2 dV\right]^{\frac{1}{2}} \qquad (3.22)$$

$$\le (\underline{r}.\underline{v})_{max}\left[2\mu_0 M_c\int[\underline{\nabla}(\underline{r}.\underline{B})]^2 dV\right]^{\frac{1}{2}}.$$

[Since its integrand is positive, the last integral in (3.22) could be extended from $\mathcal{V}$ to all space.] We now have from (3.21) and (3.16)

$$\tfrac{1}{2}\frac{d}{dt}\int(\underline{r}.\underline{B})^2 dV \le [(2\mu_0 M_c)^{\frac{1}{2}}(\underline{r}.\underline{v})_{max} - \eta(4\mu_0 M_P)^{\frac{1}{2}}]\int[\underline{\nabla}(\underline{r}.\underline{B})]^2 dV. \qquad (3.23)$$

It now follows that if the poloidal field is to be maintained **indefinitely**

$$(\underline{r}.\underline{v})_{max}/\eta \ge (2M_P/M_c)^{\frac{1}{2}}, \qquad (3.24)$$

which is Busse's result. The uncertainty in $\underline{B}_T$ and M_c is in this way related to uncertainty in the radial velocity.

3.3. Helical Dynamo.

Probably the simplest of all "fluid" dynamos is the model devised by Yu. B. Ponomarenko (1973). From a geophysical viewpoint, it is too idealized to be realistic, for it departs from §3.1 in two significant respects. First, the inducing motion occurs in a circular cylinder, $\mathcal{V}$, extending to infinity. Second the exterior, $\hat{\mathcal{V}}$, of this cylinder is not a vacuum but is a material of the same electrical conductivity as the cylinder, with which it is in perfect electrical contact. Nevertheless we may believe that the model is a dynamo if (3.5) is obeyed and if [see (3.4)] $\underline{B} = O(s^{-2})$ for $s \to \infty$. It is possible that by wrapping the cylinder into a torus, and surrounding it by a finite conductor, the two over-idealizations could be removed without detriment to the dynamo's regenerative capability, but the attractive simplicity of the model would be lost, and no-one has so far thought this generalization worth undertaking. The flow assumed by Ponomarenko is particularly simple: a helical motion of the cylinder as a whole as though solid:

$$\underline{v} = \begin{cases} \omega\hat{\underline{\phi}} + U\hat{\underline{z}}\,, & s < a\,; \\ \underline{0}, & s > a\,, \end{cases} \tag{3.25}$$

ω and U being constants.

The mathematical problem is now that of solving (3.1) subject to

$$\langle \underline{B} \rangle = \langle E_\phi \rangle = \langle E_z \rangle = 0\,, \qquad s = a\,, \tag{3.26}$$

and of course (3.4). In $\mathcal{V}$, (3.1) takes the simple form

$$\left(\frac{\partial}{\partial t} + \Omega\frac{\partial_1}{\partial\phi} + U\frac{\partial}{\partial z}\right)\underline{B} = \eta\nabla^2\underline{B}\,, \qquad s < a\,, \tag{3.27}$$

where $\partial_1/\partial\phi$ denotes differentiation holding the unit vectors $\hat{\underline{s}}$ and $\hat{\underline{\phi}}$ fixed; we have reverted to dimensional equations. In $\hat{\mathcal{V}}$, (3.1) reduces to the vector diffusion equation (2.25). These equations and boundary conditions admit simple helical solutions of the form

$$\underline{B} = \underline{B}_0(s)\,e^{im\phi+ikz+\lambda t}\,, \tag{3.28}$$

where

$$\nabla^2\underline{B} = q^2\underline{B}\,, \qquad \underline{\nabla}.\underline{B} = 0\,, \qquad \text{in } \mathcal{V}; \tag{3.29}$$

$$\nabla^2\hat{\underline{B}} = \hat{q}^2\underline{B} \qquad \underline{\nabla}.\hat{\underline{B}} = 0\,, \qquad \text{in } \hat{\mathcal{V}}; \tag{3.30}$$

$$q^2 = k^2 + (\lambda + im\omega + ikU)/\eta\,, \qquad \hat{q}^2 = k^2 + \lambda/\eta\,, \tag{3.31}$$

and, without loss of generality, $\mathcal{R}\ell\, q \geq 0$ and $\mathcal{R}\ell\, \hat{q} \geq 0$.

We may solve (3.29) and (3.30) by precisely the same boundary-layer type methods that we used in §2.4 to solve induction by a cylindrical rotor in a transverse field. We suppose that $|qa| \gg 1$ and $|\hat{q}a| \gg 1$, and note that the ∇^2 in (3.29) and (3.30) reduces in the first approximation to $\partial^2/\partial s^2$, so that in cylindrical coordinates

$$\underline{B}^{(0)} = [0, A_\phi^{(0)}, A_z^{(0)}]e^{q(s-a)}\,, \quad \hat{\underline{B}}^{(0)} = [0, A_\phi^{(0)}, A_z^{(0)}]e^{-\hat{q}(s-a)}\,. \tag{3.32)$^{(0)}$, (3.33)$^{(0)}$}$$

The superfix (0) indicates that these are the first and largest terms of a series solution; we shall have to take the approximation two further steps. The constants $A_\phi^{(0)}$ and $A_z^{(0)}$ are the same for both $\underline{B}$ and $\hat{\underline{B}}$ in $(3.32)^{(0)}$ and $(3.33)^{(0)}$ because of $(3.26)_1$. The fields are confined to thin layers, of thickness $|q|^{-1}$ and $|\hat{q}|^{-1}$ on either side of $s = a$. The continuity equations $(3.29)_2$ and $(3.30)_2$ imply that in the next approximation $B = \underline{B}^{(0)} + \underline{B}^{(1)}$, where

$$B_s^{(1)} = -\frac{i}{qa}(mA_\phi^{(0)} + kaA_z^{(0)})e^{q(s-a)}\,, \quad \hat{B}_s^{(1)} = \frac{i}{\hat{q}a}(mA_\phi^{(0)} + kaA_z^{(0)})e^{-\hat{q}(s-a)}\,. \tag{3.34)$^{(0)}$, (3.35)$^{(0)}$}$$

But the continuity of B_s on $s = a$ then requires that

$$mA_\phi^{(0)} + ka\, A_z^{(0)} = 0\,, \qquad B_s^{(1)} = \hat{B}_s^{(1)} = 0. \qquad (3.36)^{(0)}$$

Since a non-zero B_s is essential for dynamo action, we must proceed to the next approximation in which $(3.29)_1$ and $(3.30)_1$ give

$$\underline{B}^{(1)} = [0, A_\phi^{(1)}, A_z^{(1)}]e^{q(s-a)} - \frac{1}{2a}(s-a)\,\underline{B}^{(0)}\,, \qquad (3.32)^{(1)}$$

$$\hat{\underline{B}}^{(1)} = [0, A_\phi^{(1)}, A_z^{(1)}]e^{-\hat{q}(s-a)} - \frac{1}{2a}(s-a)\,\hat{\underline{B}}^{(0)}\,. \qquad (3.33)^{(1)}$$

The final terms in $(3.32)^{(1)}$ and $(3.33)^{(1)}$ arise because of the curvature term $s^{-1}\partial/\partial s$ in the ∇^2 operators. They play no part in what follows. The curvature terms in $\underline{\nabla}.\underline{B}$ and $\underline{\nabla}.\hat{\underline{B}}$ are however crucial when we use $(3.29)_2$ and $(3.30)_2$ to find $B_s^{(2)}$ and $\hat{B}_s^{(2)}$. For example $(3.29)_2$ gives

$$\frac{dB_s^{(2)}}{ds} = \frac{im}{a^2}(s-a)A_\phi^{(0)}e^{q(s-a)} - \frac{i}{a}(mA_\phi^{(1)} + kaA_z^{(1)})e^{q(s-a)}\,, \qquad (3.37)$$

so that

$$B_s^{(2)} = \frac{im}{qa}\left[\frac{(s-a)}{a} - \frac{1}{qa}\right]A_\phi^{(0)}e^{q(s-a)} - \frac{i}{qa}(mA_\phi^{(1)} + kaA_z^{(1)})e^{q(s-a)} \qquad (3.34)^{(1)}$$

and similarly

$$\hat{B}_s^{(2)} = -\frac{im}{\hat{q}a}\left[\frac{(s-a)}{a} + \frac{1}{\hat{q}a}\right]A_\phi^{(0)}e^{-\hat{q}(s-a)} + \frac{i}{\hat{q}a}(mA_\phi^{(1)} + kaA_z^{(1)})e^{-\hat{q}(s-a)}\,. \qquad (3.35)^{(1)}$$

The continuity of B_s on $s = a$ now implies

$$mA_\phi^{(1)} + kaA_z^{(1)} = m\left(\frac{1}{\hat{q}a} - \frac{1}{qa}\right)A_\phi^{(0)}\,, \quad B_s^{(2)}(a) = \hat{B}_s^{(2)}(a) = \frac{im}{q\hat{q}a^2}A_\phi^{(0)}\,. \qquad (3.36)^{(1)}$$

Finally we must ensure continuity of $\underline{n} \times \underline{E}$ on $s = a$. Condition $(3.26)_2$ has already been met; we need apply only $(3.26)_3$ which requires

$$\eta < dB_\phi/ds > + B_s <u_\phi> = 0\,, \qquad s = a. \qquad (3.38)$$

Substituting from $(3.32)^{(0)}$, $(3.33)^{(0)}$ and $(3.36)_2^{(1)}$, we finally obtain

$$q + \hat{q} = \left(\frac{i\omega m}{\eta a}\right)\frac{1}{q\hat{q}}, \tag{3.39}$$

the required "dispersion relation" determining the complex growth rate λ. In solving (3.39), we should recall that, for self-consistency ($|qa| >> 1$, $|\hat{q}a| >> 1$), the magnetic Reynolds number $R_m = \omega a^2/\eta$ must be large.

The transcendental equation (3.39) is easily solved when

$$\Omega \equiv m\omega + kU \tag{3.40}$$

is zero. The significance of $\Omega = 0$ may be appreciated by observing that according to $(3.36)_1$ the field $(3.32)^{(0)}$ and $(3.33)^{(0)}$ lies on helices whose pitch $(sd\phi/dz)$ is $-ka/m$, a constant. The helix pitch defined by the motion (3.25) is variable: $\omega s/U$. These helices are parallel on the surface of the cylinder when $\Omega = 0$. When $\Omega = 0$, (3.31) and (3.39) give

$$\hat{q} = q \doteqdot (i\omega m/2\eta a)^{1/3}, \tag{3.41}$$

or

$$\lambda \doteqdot -\eta k^2 + \eta\left|\frac{m\omega}{2\eta a}\right|^{2/3}\left(\frac{1}{2} + \mathrm{sgn}(m\omega)\frac{\sqrt{3}}{2}i\right). \tag{3.42}$$

If

$$k^2 < 2^{-5/3}(m\omega/\eta a)^{2/3}, \tag{3.43}$$

λ has a positive real part, corresponding to dynamo action.

We expect (§3.2) from Cowling's theorem that the Ponomarenko dynamo will fail if $m = 0$; from the Cartesian antidynamo theorems we expect that it should also fail if $k = 0$ or $U = 0$. Such indeed is the case, and (3.43) demonstrates this for the present asymptotic limit. [Recall here that, if Ω given by (3.40) is zero, and k or U is zero, then $m\omega = 0$.]

If we want to ask the question, 'What is the smallest possible velocity that can make Ponomarenko's model regenerate field,' we must obtain the dispersion relationship in the general case, and search numerically for the smallest value of

$$\tilde{R}_m = a(U^2+\omega^2a^2)^{1/2}/\eta = R_m(1+\chi^2)^{1/2}, \quad \chi = U/\omega a. \tag{3.44}$$

[The maximum velocity, $(U^2+\omega^2a^2)^{1/2}$, which occurs on $s = a$, clearly makes $\tilde{R}_m$ more appropriate than R_m.] It is found that the minimum $\tilde{R}_m$ occurs for $\chi \doteqdot 1.314$, for the mode $m = 1$, $ka \doteqdot -0.38754$. The corresponding frequency is $\mathrm{Im}(\lambda) \doteqdot 0.41029961(\eta/a^2)$, and the minimum magnetic Reynolds number is

$$\tilde{R}_m = 17.7221175... \tag{3.45}$$

(These results were obtained through several overnight runs of the author's TI-59 programmable calculator.)

4. MEAN FIELD ELECTRODYNAMICS

4.1. Turbulent Diffusion.

The working dynamos we have so far encountered, in §2.1 and §3.3, are too artificial to be relevant to naturally occurring dynamos in planetary cores and stellar convection zones. When we have stepped too close to relevance, as in §2.2, we have failed to sustain fields. We have nevertheless acquired useful experience. The homopolar dynamo has taught us to value asymmetry of structure, and the Ponomarenko dynamo has told us that this asymmetry need not lie in the physical structure of the body but can instead reside in the (helical) structure of the motions. The axisymmetric dynamo, governed by (3.8) and (3.9), failed because it lacked a mechanism to create poloidal field from toroidal field. We may well wonder whether, by the addition of an asymmetric motion with a suitably helical structure, we can provide this missing ingredient. From this point of view we must be encouraged by the models of §3.3. And the mechanism sought exists; it is known as "the α-effect" (§4.2).

The easiest way to understand the α-effect is to consider induction by turbulence of known statistical properties. We must first establish a notation: by $\bar{Q}(\underline{x},t)$ we mean the average of the field Q, at location $\underline{x}$ and time t, over an ensemble of (identical) turbulent systems. We then denote by $Q' = Q - \bar{Q}$ the fluctuating remainder which, by definition, has a zero ensemble average. In short, we shall write

$$Q = \bar{Q} + Q' . \tag{4.1}$$

We shall at first suppose that the turbulent inducing velocity has no mean part:

$$\underline{v} = \underline{v}' . \tag{4.2}$$

Consider the turbulent diffusion of some passive scalar field, such as temperature, that obeys the equation

$$\frac{DQ}{Dt} \equiv \frac{\partial Q}{\partial t} + \underline{v}.\underline{\nabla} Q = D_m \nabla^2 Q , \tag{4.3}$$

where D_m is the molecular diffusivity; D/Dt is the motional derivative, i.e. the derivative following the motion of the fluid. When we substitute (4.1) and (4.2) into (4.3) and average (4.3) over ensembles, we obtain

$$\frac{\partial\bar{Q}}{\partial t} - D_m\nabla^2\bar{Q} = -\overline{\underline{v}'.\underline{\nabla}Q'} \;. \tag{4.4}$$

Subtracting this from (4.3) we deduce

$$\frac{\partial Q'}{\partial t} - D_m\nabla^2 Q' = -\underline{v}'.\underline{\nabla}\bar{Q} - [\underline{v}'.\underline{\nabla}Q' - \overline{\underline{v}'\underline{\nabla}Q'}] \;. \tag{4.5}$$

A crude approximation (that can sometimes be defended) omits the final terms, between square brackets, in (4.5). This is known as "first-order smoothing". We then obtain

$$\frac{\partial Q'}{\partial t} - D_m\nabla^2 Q' = -\underline{v}'.\underline{\nabla}\bar{Q} \;. \tag{4.6}$$

Regarding $\underline{v}'$ as known (see above), we may solve (4.6) as

$$Q' = \mathcal{F}(\underline{v}',\bar{Q}) \;, \tag{4.7}$$

where $\mathcal{F}$ is a linear functional of the unknown $\bar{Q}$. Substituting (4.7) back into (4.4) we obtain a closed linear equation for $\bar{Q}$:

$$\frac{\partial\bar{Q}}{\partial t} = D_m\nabla^2\bar{Q} - \overline{\underline{v}'.\underline{\nabla}\mathcal{F}(\underline{v}',\bar{Q})} \;. \tag{4.8}$$

Despite the apparent complexity of the last term in (4.8) we may make a plausible physical guess at its nature. Let us remind ourselves first how the term $D_m\nabla^2 Q$ in (4.3) could be motivated by a crude argument from kinetic theory. An imaginary surface, $\mathcal{S}$, is drawn through $\underline{x}$ perpendicular to $\underline{\nabla}Q$ $(=\hat{\underline{n}}|\underline{\nabla}Q|$, say), and it is noted that molecules within a mean-free-path, ℓ, on each side of $\mathcal{S}$ and moving towards it, probably cross $\mathcal{S}$ without suffering further collisions. Since they carry with them $Q(\underline{x}+\ell\hat{\underline{n}})$ or $Q(\underline{x}-\ell\hat{\underline{n}})$, depending on whether they start from the side of $\mathcal{S}$ with larger Q or smaller, a flux of Q is created, proportional to $-\bar{v}\ell\underline{\nabla}Q$ where v is the root-mean-square molecular speed. It is the divergence of this flux that gives the $D_m\nabla^2 Q$ term in (4.3), with $D_m \doteqdot \bar{v}\ell$ as diffusion coefficient.

We shall now recall that turbulent eddies behave, in some rough and ready sense, like molecules in transporting Q. We may not be greatly in error if we substitute "eddies" for "molecules", "correlation length" for "mean-free-path", "root-mean-square turbulent velocity" for "root-mean-square molecular speed", and $\bar{Q}$ for Q. And the last term in (4.8) becomes

$$\overline{-\underline{v}'.\underline{\nabla}\mathcal{F}(\underline{v}',\bar{Q})} \sim D_T\nabla^2\bar{Q} \;, \tag{4.9}$$

where D_T, the "turbulent diffusivity", is again of order $\bar{v}\ell$, but with the different meanings indicated. We can then write (4.8) as

$$\frac{\partial \bar{Q}}{\partial t} = D \nabla^2 \bar{Q}, \tag{4.10}$$

where

$$D = D_m + D_T \tag{4.11}$$

is the total diffusivity.

F. Krause and K.-H. Rädler (1980) have written a book that makes a special study of this and similar results based on first-order smoothing, an approximation that they consider is justified if either the microscale Reynolds number, $R_m = \bar{v}\ell/D_m$ is small or the Strouhal number $S = \bar{v}\tau/\ell$ is small, where τ is the correlation time of the turbulence. They indicate that (4.9) follows when either $R_m >> 1$ and $S << 1$ or when $R_m << 1$ and $S >> 1$. In the former case* $D_T = O(\bar{v}^2\tau)$, which is large compared with D_m if $R_m S >> 1$; in the latter case*, $D_T = O(\bar{v}^2\ell^2/D_m) << D_m$.

The idea that a term such as $-\overline{\underline{v}'.\nabla Q'}$ on the left of (4.4) could be represented by a simple diffusion term as on the right of (4.9) is very old. Osbert Reynolds suggested in the nineteenth century that the inertial term $-\underline{v}'.\nabla\underline{v}'$ in the equation governing the mean flow $\underline{\bar{v}}$ in a turbulent fluid could be represented by $\nu_T \nabla^2 \underline{\bar{v}}$, with ν_T a turbulent viscosity. And it is not surprising that historically the first attempt to understand electromagnetic induction in a turbulent, electrically conducting fluid emphasized strongly that the mean of the non-linear term $-\underline{v}.\nabla\underline{B}$, included in

$$\underline{\nabla} \times \underline{\mathcal{E}} \equiv \underline{\nabla} \times (\underline{v} \times \underline{B}) = \underline{B}.\underline{\nabla}\,\underline{v} - \underline{v}.\underline{\nabla}\,\underline{B} \tag{4.12}$$

in the induction equation (2.20), would behave like $\eta_T \nabla^2 \underline{\bar{B}}$, with a turbulent diffusivity, η_T possibly large compared with the molecular diffusivity**, η_m. This turned out to be an oversimplification, since it overlooked an effect that is very significant for field generation (see below). This is not to say that an enhanced (turbulent) diffusion of the mean magnetic field, $\underline{\bar{B}}$, does not arise. Indeed, the short (22 year) period of magnetic activity on the Sun (the solar cycle) could not be explained unless $\eta_T \doteqdot 10^4 \eta_m$. Rather, it means that in the transport of a passive vector field like $\underline{B}$, there is another consideration that may outweigh mere turbulent diffusion of $\underline{\bar{B}}$.

---

* Commonly $S = O(1)$ in a turbulent flow, so that $D_T = O(\bar{v}^2\tau)$ implies $D_T = O(\bar{v}\ell)$; if $R_m = O(1)$, the estimate $D_T = O(\bar{v}^2\ell^2/D_m)$ implies $D_T = O(\bar{v}\ell)$. In this way, within the framework of first order smoothing (but outside its validity), the earlier physical guess is recovered.

** In this section η will not denote the molecular diffusivity but the sum of turbulent and molecular diffusivities.

4.2. The Alpha Effect.

The oversight in early theories of magnetic induction arose through staying too close to the considerable body of knowledge about turbulence in laboratories and wind tunnels that had been built up since the time of Reynolds. These examples of turbulence are mirror-symmetric, i.e. all their statistical properties are unchanged by reflecting the turbulent motions in a mirror, or theoretically by replacing $\underline{x}$ by $-\underline{x}$. Rotation alone does not destroy mirror-symmetry, but rotation combined with other forces may do so. For example, turbulent convection in a rotating fluid layer is not mirror-symmetric (§5.2.2). Prior to the 1950's and 1960's when its importance in magnetohydrodynamics began to become apparent, non-mirror-symmetric turbulence was regarded as a curiosity, and was scarcely studied.

The lack of mirror-symmetry in a flow may be gauged by its "helicity",

$$H = \underline{v}.\underline{\nabla} \times \underline{v}\ , \tag{4.13}$$

or $H = \underline{v}.\underline{\omega}$, where $\underline{\omega} = \underline{\nabla} \times \underline{v}$ is the vorticity. An even better measure is perhaps $H/v\omega$, which by definition lies between -1 and +1, the extremes being realized in "Beltrami flows" ($\underline{\omega} \propto \underline{v}$) which are "maximally helical". We have already met a helical motion in §3.3. The Ponomarenko model (3.15) has positive helicity if $\omega U > 0$ and negative if $\omega U < 0$. It fails to work if $\omega U = 0$. The hint is strong that helicity is "good for field generation". Corroboration is provided by the two-dimensional Beltrami flow

$$\underline{v} = (\cos y - \cos z,\ \sin z, \sin y)\ , \tag{4.14}$$

(G.O. Roberts, 1972). The yz-streamlines of this cellular motion are sketched in Fig. 8, the + signs (- signs) being placed at the maxima (minima) of v_x, i.e. the component of $\underline{v}$ out of the paper. The model resembles an infinite square array of Ponomarenko motions with adjacent cylinders pointing parallel and antiparallel to Ox, each being in perfect electrical contact with its neighbours. Whether pointing "up" or "down" each cylinder contains positive helicity.

The way in which helical flows generate field can be qualitatively understood with the help of Alfvén's theorem (§2.4). We give two illustrations: first induction by flow (4.14). The lines of force of a field in the SW→NE direction in the plane of the paper ($x = 0$) are twisted in an anti-clockwise sense by the cells with positive v_x which lift the lines out of the paper, and in a clockwise sense by the cells with negative v_x which depress the lines below the plane of the paper ($x < 0$). In this way a field with a component in the SE→NW direction is created above $x = 0$, and a field with a component in the NW→SE direction is produced below $x = 0$. And so repeating the argument for every constant-x section of the model, we see how "average field", $\underline{\bar{B}}(x)$, perpendicular to Ox and turning slowly anticlockwise with increasing x, can be maintained. There will in addition be complicated fields $\underline{B}'$ varying on the yz-lengthscales. The one-dimensionality of $\underline{\bar{B}}$

in no way implies that $\underline{B} = \overline{\underline{B}} + \underline{B}'$ is one - or even two-dimensional. The anti-dynamo theorem of §3.2 is therefore not relevant and, through the two-dimensionality of the flow (4.14), we also provide an example of a two-dimensional motion that regenerates field [see footnote below (3.15)]. It may be noted that the mean current $\overline{\underline{J}} = \underline{\nabla}\times\overline{\underline{B}}/\mu_0$ is also in the yz-plane and anti-parallel to $\overline{\underline{B}}$.

Models like (4.14), which create magnetic field on a scale large compared with the motions, can be developed to a high degree of mathematical rigour, since use can be made of a powerful analytic technique: the "two-scale method". The technique is so powerful that it can even help answer dynamical questions (see §5.3).

A second and better known illustration, sketched in Fig. 9, concerns the effect of a turbulent eddy having positive helicity on a straight field line. The correlated flow $\underline{v}'$ and vorticity $\underline{\omega}'$ are thought of as a "wrench", i.e. a pull and a twist. The pull (from the top of the page) bends the field line into an Ω shape, as in Fig. 9(a). The twist then turns the Ω out of the plane of the paper, as in Fig. 9(b). The field depicted in Fig. 9(b) may be thought of as the original field line plus a circular loop. Such a loop is the kind of field that is, by Ampère's law, associated with a current $\underline{J}$ antiparallel to the

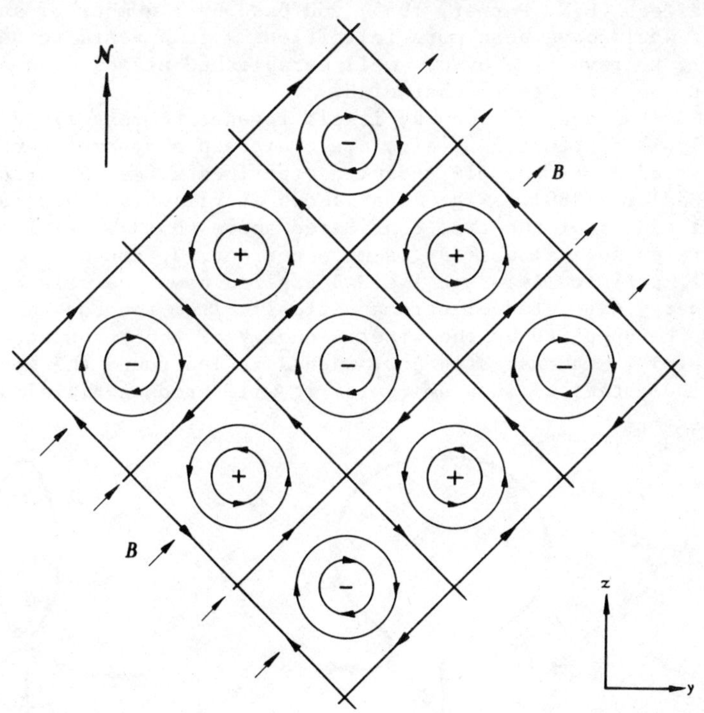

Figure 8. Illustration of the Flow (4.14), and of its Inductive Effects. Figures 8,9,10 and 11 are reprinted from Lectures in Applied Mathematics (1971) "Dynamo Theory" P.H. Roberts, Volume 14 (figures 2,10,11 and 12) by permission of the American Mathematical Society.

starting field, $\underline{B}$. One now imagines turbulence consisting of a statistical array of such eddies having all possible orientations and locations, but all having positive H. The average effect on the mean field $\overline{\underline{B}}$ will be the creation of a mean current, $\overline{\underline{J}}$, antiparallel to $\overline{\underline{B}}$, i.e. $\overline{\underline{J}} = \sigma\alpha\overline{\underline{B}}$, where $\alpha(<0)$ is a constant of proportionality, of dimensions L/T. More precisely, in terms of the mean of the turbulent e.m.f. $\underline{v} \times \underline{B}$,

$$\overline{\mathcal{E}} \equiv \overline{\underline{v}' \times \underline{B}'}\,, \tag{4.15}$$

we have

$$\overline{\mathcal{E}} \doteqdot \alpha\overline{\underline{B}} - \eta_T\, \underline{\nabla} \times \overline{\underline{B}}\,. \tag{4.16}$$

[We have restored in (4.16) the turbulent diffusion identified earlier; see (4.12).]

The creation of a mean e.m.f. with a component parallel to $\overline{\underline{B}}$ is called "the α-effect" for no better reason than the choice of letter used in a very influential paper on the subject (M. Steenbeck, F. Krause and K.-H. Rädler, 1966) to describe the constant of proportionality. The choice (Γ, i.e. 'G' for generation) employed by the discoverer of the α-effect (E.N. Parker, 1955) and used by a number of subsequent authors would have been more felicitious. It appears to be too late, however, to reverse a by now well-established usage. The sign of α tends to be opposite to that of H.

That the α-effect can by itself regenerate mean field (and therefore also a fluctuating field, i.e. maintain a dynamo) can be elegantly illustrated by two simple gedanken experiments (see F. Krause and K.-H. Rädler, 1980). Fig. 10(a) shows a cylindrical box full of helical eddies of the type considered above ($\underline{H} > 0$). Fig. 10(b) illustrates how, if desired, a mean current, $\overline{\underline{J}}$, could be drawn out of the box, antiparallel to a field $\overline{\underline{B}}$ applied down the axis of the cylinder. (The energy extracted electromagnetically from the box would, of course, have to be supplied by the kinetic energy of the turbulence.) Fig. 10(c) demonstrates how, when $\overline{\underline{J}}$ is led round the box (in the sense indicated) as in a solenoid, it will produce a field that

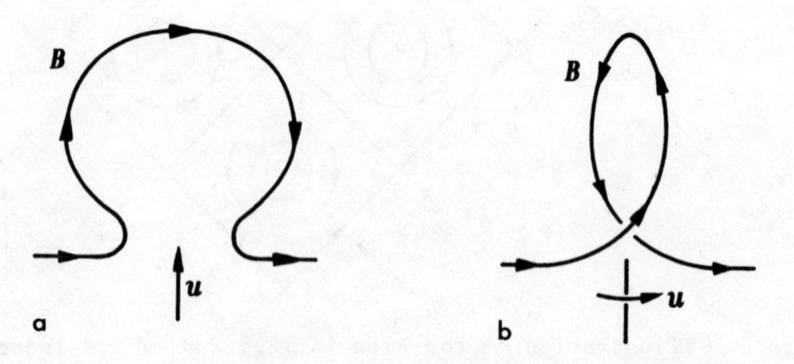

Figure 9. The α-effect of a helical eddy on a straight field line.

re-inforces the applied field, $\bar{B}$. When the applied field can be removed without causing the other fields and currents to collapse, the device is a self-exciting dynamo reminiscent of the homopolar dynamo (§2.2, Fig. 5).

In Fig. 11, two hollow rings full of helical turbulence ($H > 0$) thread each other, each being perpendicular to the plane of the other as it passes through the centre of that ring. The mean field $\bar{\underline{B}}$ in each ring generates (through the α-effect) the mean current $\bar{\underline{J}}$ required (by Ampère's law) to produce the mean field in the other.

More important for us is the way the α-effect saves the axisymmetric dynamo governed by (3.8) - (3.12). First, we abandon (4.2), and include a mean axisymmetric motion (3.7). Second, as indicated by (4.16), we add $\alpha\bar{\underline{B}}$ to $\bar{\underline{E}}$ when we ensemble average (2.17). Finally, we treat the resulting induction equation for $\bar{\underline{B}}$ in the way that led to (3.8) and (3.9). Omitting the bars for simplicity, we obtain

$$\frac{\partial B_\phi}{\partial t} + s\underline{v}_P\cdot\underline{\nabla}\left(\frac{B_\phi}{s}\right) = \eta\Delta B_\phi + s\underline{B}_P\cdot\underline{\nabla}\omega + \left[\alpha\Delta A + \frac{1}{s}\underline{\nabla}\alpha\cdot\underline{\nabla}(sA)\right], \tag{4.17}$$

$$\frac{\partial A}{\partial t} + \frac{1}{s}\underline{v}_P\cdot\underline{\nabla}(sA) = \eta\Delta A + \alpha B_\phi, \tag{4.18}$$

while (3.10) - (3.12) are unaltered. Of course, the crucial fact is that (4.18) has acquired a source. A way has been found to complete the regenerative cycle: poloidal field generates toroidal field <u>and</u> the reverse. The zonal field now has two sources, (4.17) containing an

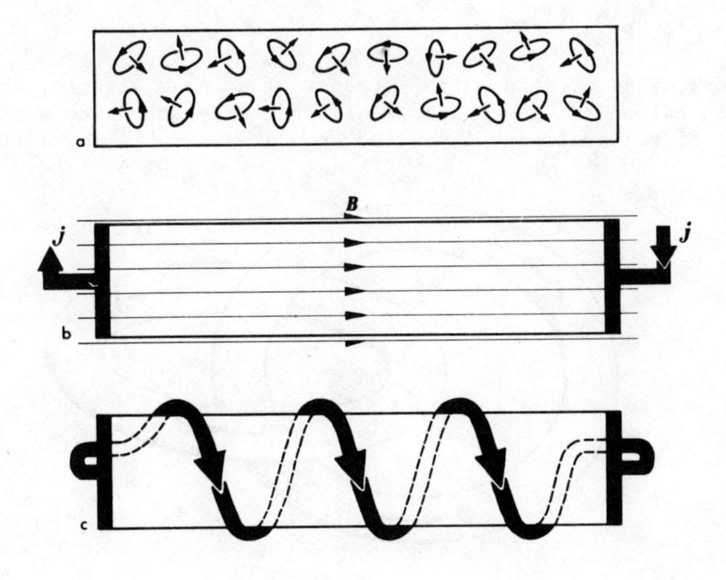

Figure 10. An α-Dynamo created by Turbulence in a Box.

α-effect term as well as the old ω-effect term. In fact, it is possible to dispense with the mean flow ($\omega = 0$) and still drive a dynamo, as the models shown in Figs. 10 and 11 demonstrate. The ω-effect is, however, such a natural and simple mechanism, and zonal flows are so easily set up in rotating systems, that one should be cautious before omitting ω.

Introduce two Reynolds numbers, the ω-effect Reynolds number $(2.39)_2$ and an α-effect Reynolds number:

$$R_\omega = \omega c^2/\eta , \qquad R_\alpha = \alpha c/\eta . \tag{4.19}$$

We may distinguish two extremes:

(S) $R_\omega >> R_\alpha$. The final term in square brackets in (4.17) is negligible, so that, as in $(2.39)_1$, $B_\phi = O(R_\omega B_P)$. Since $B_P = O(R_\alpha B_\phi)$ by (4.18), regeneration requires that "the dynamo number",

$$D \equiv R_\omega R_\alpha = \alpha\omega c^3/\eta^2 , \tag{4.20S}$$

should be O(1). In the resulting mean field,

$$B_\phi/B_P = O(R_\omega/R_\alpha)^{1/2} >> 1 ; \tag{4.21S}$$

(W) $R_\alpha >> R_\omega$. The ω-effect is negligible, so that both $B_\phi = O(R_\alpha B_P)$ and $B_P = O(R_\alpha B_\phi)$. Regeneration requires that R_α should be O(1), and in the resulting mean field

$$B_\phi/B_P = O(1). \tag{4.21W}$$

Working as it does on the product of α and ω effects, the S-dynamo is often called an "αω-dynamo", while the W-dynamo which works on a product of two α-effects is called an "α²-dynamo". The letters 'S'

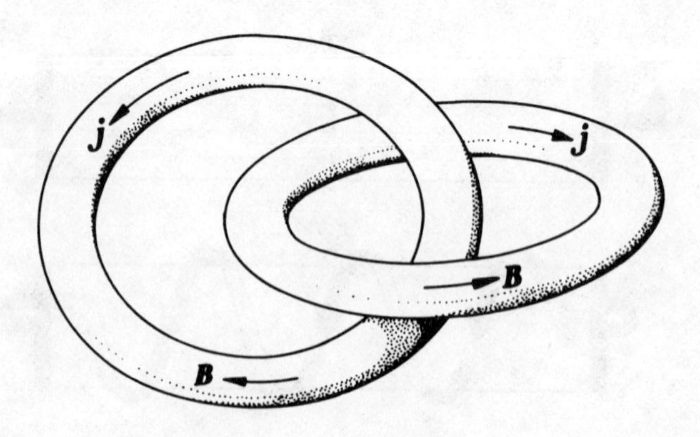

Figure 11. The double-Ring α-Dynamo.

and 'W' stand for 'strong' and 'weak', and arise because only $\underline{B}_P$ can be seen outside a star or a planet; B_ϕ is trapped inside it. Thus $\underline{B}_P$ provides the scale, and the question remains only whether $\overline{B}$ in the generating regions is strong or weak compared with $\underline{B}_P$. In reality dynamos occurring in Nature are unlikely to fall into either extreme category. In particular, the Lorentz forces associated with an amplifying α^2-dynamo field may create an ω-shear to limit its growth (§5.1): a kind of anti-ω-effect. The dynamo then makes use of all sources in (4.17) and (4.18); it is an "$\alpha^2\omega$-dynamo".

4.3. Dynamo Waves.

The αω-dynamo equations

$$\frac{\partial B_\phi}{\partial t} + s\underline{v}_P.\underline{\nabla}\left(\frac{B_\phi}{s}\right) = \eta\Delta B_\phi + s\underline{B}_P.\underline{\nabla}\left(\frac{v_\phi}{s}\right), \tag{4.22}$$

$$\frac{\partial A}{\partial t} + \frac{1}{s}\,\underline{v}_P.\underline{\nabla}(sA) = \ddot{\eta}\Delta A + \alpha B_\phi, \tag{4.23}$$

may have oscillatory solutions. These are most easily illustrated in an infinite fluid, without poloidal flow and without curvature effects, Cartesian coordinates (y,x,z) replacing spherical coordinates (r,θ,ϕ). It is further supposed for simplicity that

$$\alpha = \text{constant}, \qquad \omega = \partial\bar{v}_z/\partial y = \text{constant}. \tag{4.24}$$

On seeking solutions of the form

$$\underline{B} = \underline{B}(x,t), \qquad (B_x = 0), \tag{4.25}$$

we obtain from (4.24) and the Cartesian analogues of (4.22) and (4.23)

$$\frac{\partial B_z}{\partial t} = \eta\frac{\partial^2 B_z}{\partial x^2} + \omega B_y, \tag{4.26}$$

$$\frac{\partial B_y}{\partial t} = \eta\frac{\partial^2 B_y}{\partial x^2} - \alpha B_z, \tag{4.27}$$

in which B_y corresponds to the previous poloidal field, and B_z to the toroidal. These admit periodic solutions

$$\underline{B} = \underline{B}_0\, e^{i(kx-\lambda t)}, \tag{4.28}$$

where k, λ and $\underline{B}_0$ are constants, provided

$$\eta k^2 - i\lambda = \begin{cases} (1-i)(\tfrac{1}{2}k\alpha\omega)^{1/2}, & \text{if } k\alpha\omega > 0, \\ (1+i)(-\tfrac{1}{2}k\alpha\omega)^{1/2}, & \text{if } k\alpha\omega < 0. \end{cases} \tag{4.29}$$

These states are marginal $[R\ell(\lambda) = 0]$ if

$$D \equiv \omega\alpha/k^3\eta^2 = \pm 2, \tag{4.30}$$

and (4.28) is then a progressive wave travelling in the x-direction with phase velocity $\pm\eta k$, and dynamo-maintaining its amplitude against ohmic diffusion as it does so.

Solutions exist in spherical geometry that are similarly oscillatory and resemble progressive waves travelling from poles to equator or the reverse. They do not travel from pole to pole, since there are good dynamical reasons to believe that α has opposite signs in opposite hemispheres, and that (4.24) should more realistically be replaced by

$$\alpha = \alpha_0 \cos\theta, \quad \omega = \bar{v}_\phi/s = \omega_0 r, \tag{4.31}$$

where α_0 and ω_0 are constants. The waves then progress from poles to equator for $\alpha_0\omega_0 < 0$ and the reverse for $\alpha_0\omega_0 > 0$. A half-cycle of a marginal state is shown in Figs. 12. The progression in magnetic activity from poles to equator is obvious, as is the reversal of the polarity from figure (a) to figure (h). A further eight figures, identical with (a) - (h) but with reversed fields, would complete the cycle and return the system to state (a). Elaborations of these ideas form the basis of the preferred explanation of the solar cycle today. It would not be appropriate to take these matters further here. For further reading, see §6.

4.4. Nearly Symmetric Dynamos.

The ideas described in §§4.1 and 4.2 although, like nearly all aspects of turbulence theory, hard to put on a secure mathematical footing, have the merit of being comparatively easy to illustrate. It is doubtful however whether small-scale turbulence really has much inductive effect in the Earth's core. To draw an analogy, turbulence is so important in the solar convection zone that η is effectively 10^4 times its "molecular" value, and correspondingly [see (4.29)] the period of the solar cycle is reduced from about 10^5 years to the order of a decade, consistent with the observations. If η were similarly enhanced in the Earth's core from its "molecular" value, the timescales of the secular variation would be drastically reduced, to months rather than centuries. Since there is no evidence for this, one concludes that η_T is not large compared with η_m. (This does not mean that other diffusion coefficients, such as the kinematic viscosity or the thermal diffusivity of the core, whose molecular values are much smaller than η_m, are not greatly enhanced by turbulence.)

Despite the absence of small-scale turbulence, the α-effect can be

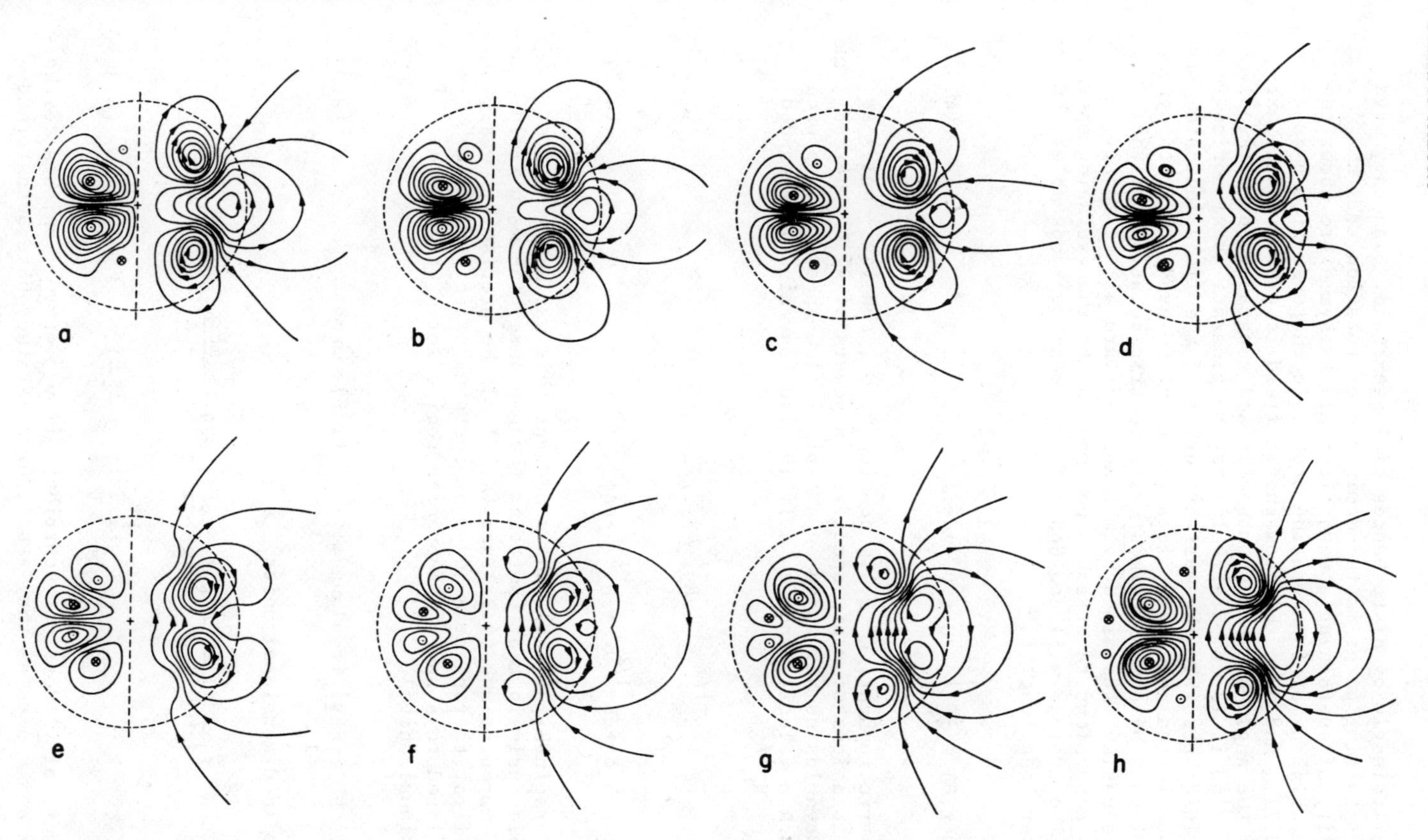

Figure 12. A half-cycle of a marginal $\alpha\omega$-dynamo field in a sphere when $\alpha_0\omega_0 < 0$. On the left of each figure contours of equal B_ϕ strength are shown; on the right the poloidal field lines are plotted. (From Roberts 1972 with permission of Royal Society of London.)

produced efficiently by disturbances of large-scale, planetary wave type. The starting point is the dominance of rotation and the impossibility of sustaining a magnetic field axisymmetric about the geographic axis, or of regenerating any field whatever with a zonal motion (§3.2). And yet, the geomagnetic field is predominantly axisymmetric; the angle between geomagnetic and geographic axes is only about 10°. And it is comparatively easy to generate zonal motion in a rotating fluid; possibly the westward drift of the field indicates its predominance in the core. The idea is, then, to evade the anti-dynamo theorems of §3.2 by seeking almost axisymmetric solutions of the induction equation, sustained by motions that are almost zonal.

It is convenient to attach a new meaning to the averaging operators of §4.1. The overbar will now denote not a mean over ensembles, but an average over longitude:

$$\bar{Q}(r,\theta) = \frac{1}{2\pi}\int_0^{2\pi} Q(r,\theta,\phi)d\phi . \tag{4.32}$$

The asymmetric remnant, $Q-\bar{Q}$, is denoted by Q', and (4.1) holds as before. Expansions of $\underline{v}$ are specified in powers of $R_m^{-1/2}$ where R_m, the magnetic Reynolds number based on a typical flow speed, is supposed large: such a velocity induces a field also conveniently expanded in powers of $R_m^{-1/2}$:

$$\underline{v} = \bar{v}\hat{\underline{\phi}} + R_m^{-1/2}\underline{v}' + R_m^{-1}\underline{\nabla}\times[\bar{\chi}\hat{\underline{\phi}}] , \tag{4.33}$$

$$\underline{B} = \bar{B}_\phi\hat{\underline{\phi}} + R_m^{-1/2}\underline{B}' + R_m^{-1}\underline{\nabla}\times[\bar{A}\hat{\underline{\phi}}] . \tag{4.34}$$

S.I. Braginskii (1964), whose concept this was, showed after considerable analysis that (omitting the overbars) the mean field obeyed the $\alpha\omega$-equations (4.22) and (4.23). To be precise, it was necessary first to replace $\bar{A}$ and $\bar{\chi}$ by certain "effective fields" closely related to $\bar{A}$ and $\bar{\chi}$. A precise recipe for α was deduced: if (in dimensional units)

$$\underline{v}' = \sum_{m=1}^{\infty} [\underline{v}'_{mc}(r,\theta)\cos m\phi + \underline{v}'_{ms}(r,\theta)\sin m\phi] , \tag{4.35}$$

then (also in dimensional units)

$$\alpha = \eta \sum_{m=1}^{\infty} \frac{1}{m}\Big[\frac{(1-m^2)}{s\bar{v}^2}(\underline{v}'_{ms}\times\underline{v}'_{mc})_\phi - \underline{\nabla}\Big(\frac{rv'_{mcr}}{\bar{v}}\Big).\underline{\nabla}\Big(\frac{v'_{msz}}{\bar{v}}\Big) + \underline{\nabla}\Big(\frac{rv'_{msr}}{\bar{v}}\Big).\underline{\nabla}\Big(\frac{v'_{mcz}}{\bar{v}}\Big)\Big] . \tag{4.36}$$

An m-harmonic of $\underline{v}'$ does not contribute if $\underline{v}'_{mc}$ or $\underline{v}'_{ms}$ vanishes (either with the x and y axes as they stand, or after they have been turned about Oz by any angle): the "waves", $\underline{v}'$, must be "tilted" before they can regenerate. Is the observed asymmetry of the geomagnetic field, including the tilt of the geomagnetic axis, a manifestation of $\underline{v}'$ motions necessary to make the geodynamo work? Are these asymmetric motions and fields indicative of a basic instability of an otherwise axisymmetric state? The correct answer to both questions is probably, "Yes", and we shall discuss how such instabilities may arise in §5. Meanwhile we must answer an obvious criticism: why, if the Earth is an $\alpha\omega$-dynamo, does it not exhibit a regular oscillation like the solar cycle and the dynamo wave (§4.3)? The answer seems to lie with $\underline{v}_P$, which was ignored in the models of §4.3. This was reasonable for the Sun on which no appreciable meridional motion is observed. But theoretical models that possess a meridional flow of order η/c are found typically to be steady dynamos rather than oscillatory. This suggests that $\underline{v}_P$ may not be negligible in the Earth's core.

5. DYNAMICAL ASPECTS OF DYNAMO THEORY

5.1. Driving Mechanisms.

So far we have concentrated on the kinematic dynamo problem (§3.1), in which $\underline{v}$ is specified and an everlasting field, $\underline{B}$, is sought which obeys (2.20). Because this equation is linear in $\underline{B}$, considerable mathematical progress can be made, but in this very linearity lies serious physical unrealism. If the magnetic Reynolds number, R_m, exceeds the critical value, R_{mc}, defined in §3.1, the induction equation will generally show that $|\underline{B}|$ increases without limit as $t \to \infty$. The reality is different. Even if $|\underline{B}|$ does increase initially in obedience to kinematic theory, the Lorentz force, $\underline{J} \times \underline{B}$, will eventually become too great for $\underline{v}$ to be maintained by any driving mechanism whatever. Given a specified driving mechanism, the Lorentz force modifies $\underline{v}$ and reduces its amplitude until eventually a new marginal kinematic state is reached. Equilibration is not always straightforward. For example, it may happen that at first the Lorentz forces increase $|\underline{v}|$ and so produce "runaway field growth" (§§5.2.4,5.3). Ultimately however a balance will be struck.

It is clearly of interest to specify a driving mechanism and to attempt the simultaneous solution of the induction equation (2.20) and the equations of fluid motion. This is a nonlinear system for $\underline{B}$ and $\underline{v}$, solutions are hard to find, and progress is slow. We will have space here to describe only a few aspects of this "dynamic dynamo problem", or "fully self-consistent dynamo problem" as it is often called.

One attack on the dynamic theory has been in the context of turbulence although, when the purely hydrodynamic situation presents such intractible difficulties (e.g. D.C. Leslie, 1973), it is not surprising that approaches to the magnetohydrodynamic problem have often been crude. Behind these approaches lies the idea that Lorentz forces acting on the microscale will reduce the vigour of the turbulent motion, and therefore weakens the α-effect, so bringing about the equilibration of the mean field strength.

Equilibration on the macroscale is another possibility. The mean magnetic field and its Lorentz force can create mean motion that limits field growth. In particular, as has been remarked already, zonal flows are easy to excite in a rotating fluid, and those created by the Lorentz force may provide a balance through an anti-ω-effect (§4.2); more of this in §5.2.

What of the driving mechanism on which the system relies to maintain fields against ohmic losses of energy? After examination of many alternatives, there is today a concensus that favours buoyancy. (In fairness it should also be stated that the driving of core motion by the luni-solar precession has not yet been utterly discredited.) Whether the density differences that provide buoyancy are produced by differences in temperature or in composition is not unequivocally decided. Thermal convection is simpler, and better understood than compositional convection. It suffers from two deficiencies. First, it is thermodynamically less efficient than compositional convection, and it may be that the heat flux from the core is too small to be consistent with thermal driving of the geodynamo. Second, the presence or absence of adequate radioactivity in the core is not easily decided, and geochemists have at various times given different answers. (For a recent proposal, see R.C. Feber _et al_, 1984.)

The proponents of compositional convection start from the usually uncontested fact that the density, ρ, of the outer fluid core is significantly less than that of iron at the same pressure, and that this betrays the presence of light elements in significant percentages. There is no general agreement on which light constituent predominates; Si, S and O all have their advocates. Whatever the composition of the iron alloy, it is likely to be different on liquidus and solidus. If, as has been suggested (§1.2), the solid core is the result of freezing the outer core during the general cooling of the Earth, the inner core boundary, $r = c_1$, is a gradually rising, freezing interface at which light constituent is released as the more iron-rich inner core accretes solid. This light constituent rises to stir the outer core into a state of compositional convection.

In addition to the latent heat released during freezing at the inner core boundary there is, despite the probable small change in composition between solidus and liquidus, abundent gravitational energy available, from this sinking of heavy iron in the core, to drive the geodynamo throughout geological times (S.I. Braginskii, 1963).

Although possibly a majority of geophysicists today believe compositional convection is more likely than thermal convection as the cause of core motion, we shall consider below only thermal convection. We do so for three reasons: (a) thermal convection and compositional convection have many points of similarity; (b) the theory of thermal convection is significantly simpler; and (c) it is more in the spirit of this meeting.

5.2. Marginal Convection in Rotating Magnetic Systems

5.2.1. _Bénard Convection_. The classic example of convection is the

Bénard layer. Only brief reminders need be given here, and a definition of our notation. An infinite horizontal layer, $0 < z < d$, of homogeneous fluid is bounded at $z = 0$ and $z = d$ by walls that are perfect thermal conductors and which are maintained at temperatures Θ_r and $\Theta_r - \Delta\Theta$, respectively; it lies in a uniform gravitational field $-g\hat{\underline{z}}$. The coefficient of thermal expansion of the fluid is $\bar{\alpha}$, its isothermal compressibility is $\bar{\beta}$, its kinematic viscosity is ν, its thermal diffusivity is κ, and its mean density is ρ, all assumed constant. It is supposed that $d << 1/g\rho\bar{\beta}$, the density scale height; the adiabatic gradient in the layer is then effectively zero, the applied superadiabatic gradient is $\Delta\Theta/d = \beta$ (say) and is constant because κ is constant. We expect that if β is "large enough" the static "conduction state"

$$\underline{v} = 0 , \quad \Theta = \Theta_0 \equiv \Theta_r - \beta z , \tag{5.1}$$

will become convectively unstable.

To determine when this happens we naturally see what dimensionless groupings we can form from the relevant **parameters**. We find

$$\bar{\alpha}\Delta T , \quad gd^3/\nu\kappa , \quad p_r = \nu/\kappa , \tag{5.2}$$

where p_r is the Prandtl number. The solutions of the equations of fluid motion and thermal conduction are simplest if we take the Oberbeck-Boussinesq limit: this is

$$\bar{\alpha}\Delta T \to 0 , \quad gd^3/\nu\kappa \to \infty , \tag{5.3}$$

with their product, the Rayleigh number,

$$R = g\bar{\alpha}d^3\Delta T/\nu\kappa , \tag{5.4}$$

finite. By (5.3), the density of the fluid can then be assumed constant, except in the buoyancy force. Marginal states are sought by linearizing these equations about the conduction solution (5.1). They then admit solutions proportional to

$$\exp [i(\ell x + my - \omega t)] . \tag{5.5}$$

Mode (5.5) becomes marginally stable when $\mathrm{Im}(\omega) = 0$, and we consider only the smallest possible value ($\bar{R}$, say) of R for this to occur. It may be shown that the marginal state is steady $[\mathrm{R}\ell(\omega) = 0]$, and $\bar{R}$ is therefore independent of p_r. In fact, $\bar{R}$ depends on ℓ and m only in the combination $a = \sqrt{(\ell^2+m^2)}$, the horizontal wavenumber; this is hardly surprising since no horizontal direction is preferred in the original system. It may also be shown that $\bar{R}(a)$ is proportional to a^{-2} as $a \to 0$, is asymptotic to $(ad)^4$ for $a \to \infty$, and between these extremes $\bar{R}(a)$ has a single minimum, R_c, the critical minimum Rayleigh number at which convection can occur. If the boundaries are stress-free (an experimentally unrealizable case but nevertheless a qualitatively representative one), the critical state can be found analytically:

$$a_c d = \pi/\sqrt{2}\,, \qquad R_c = 27\pi^4/4\,. \tag{5.6f}$$

By $(5.6f)_1$, the horizontal dimension $\mathcal{L} = \pi/a_c$ of a convective roll is $d\sqrt{2}$; roughly speaking, the convection cells are "as broad as they are deep". This statement is even closer to the truth when no-slip conditions are applied at the walls when, by numerical methods,

$$a_c d = 3.1162\ldots, \quad R_c = 1707.76\ldots \tag{5.6r}$$

5.2.2. Rotating Bénard Layer. Now suppose that the problem is generalized to the case where the boundaries both rotate about the vertical with angular velocity $\underline{\Omega} = \Omega\hat{\underline{z}}$. In the frame of reference rotating with the walls - the frame we now employ - (5.1) defines the conduction solution whose stability is to be examined. In the Boussinesq approximation the centrifugal force $\rho\underline{\Omega}\times(\underline{\Omega}\times\underline{r})$ can be absorbed into the pressure as $\underline{\nabla}(\frac{1}{2}\rho|\underline{\Omega}\times\underline{r}|^2)$. It therefore has no bearing on convection in the layer; only the Coriolis force $2\rho\underline{\Omega}\times\underline{v}$ is influential. In convection theory this is usually measured by the "Taylor Number"

$$T = \frac{4\Omega^2 d^4}{\nu^2} = \frac{1}{E^2}\,, \qquad E = \frac{\nu}{2\Omega d^2}\,, \tag{5.7}$$

although in other branches of rotating fluid theory it is the Ekman number, E, that is usually employed.

Proceeding as for $\underline{\Omega} = 0$ above, one seeks solutions of the form (5.5). If $p_r > 1$ (which we shall assume, for simplicity), it may be shown again that marginal states are steady; $\bar{R}$ is independent of p_r; it is also independent of ℓ/m, i.e. $\bar{R} = \bar{R}(a,T)$. The effective Taylor number, $4\Omega^2\mathcal{L}^4/\nu^2 = T(\mathcal{L}/d)^4$, for cells of large horizontal scale ($\mathcal{L} = \pi/a$) is large, and we may therefore expect that $\bar{R}$ depends strongly on rotation as $a\to 0$. It can be shown however that $\bar{R}\sim(ad)^4$ as $a\to\infty$, as before. The situation is sketched in Fig. 13, where it is apparent that both $a_c d$ and R_c are greater than their $\Omega = 0$ values (5.6). The increase is monotonic with T and

$$a_c d = O(T^{1/6})\,, \qquad R_c = O(T^{2/3})\,, \qquad T\to\infty\,. \tag{5.8}$$

The increasing inhibition of convection implied by (5.8) can be understood by reference to the Taylor-Proudman ("TP") theorem, which concerns the slow, steady motion of an inviscid fluid. In our context of Boussinesq convection for which, in dimensional units,

$$\frac{\partial \underline{v}}{\partial t} + \underline{v}.\nabla \underline{v} + 2\underline{\Omega} \times \underline{v} = -\nabla(\frac{p}{\rho} + \tfrac{1}{2}\Omega^2 s^2) + g\bar{\alpha}\Theta'\hat{\underline{z}} + \nu\nabla^2\underline{v}\,, \tag{5.9}$$

$$\nabla.\underline{v} = 0\,, \tag{5.10}$$

where $\Theta' = \Theta - \Theta_0$ is the temperature perturbation. The steadiness of the marginal state justifies the omission of $\partial\underline{v}/\partial t$, slowness (the neglect of $\underline{v}.\nabla\underline{v}$) is assured by the infinitesimal amplitude of the convection considered, and $\nu\nabla^2\underline{v}$ apparently becomes increasingly unimportant as $T \to \infty$. Thus (5.9) leads us to

$$2\Omega\hat{\underline{z}} \times \underline{v} = -\nabla\tilde{\omega} + g\bar{\alpha}\Theta'\hat{\underline{z}}\,, \tag{5.11}$$

and, operating with $\nabla\times$ and using (5.10), we obtain

$$\frac{\partial \underline{v}}{\partial z} = \frac{g\bar{\alpha}}{2\Omega}\,\hat{\underline{z}} \times \nabla\Theta'\,. \tag{5.12}$$

For a homogeneous fluid ($\bar{\alpha} = 0$), (5.12) implies

$$\underline{v} = \underline{v}(x,y)\,, \tag{5.13}$$

i.e. the motion is two-dimensional, the TP theorem. Even when buoyancy is present, (5.12) gives

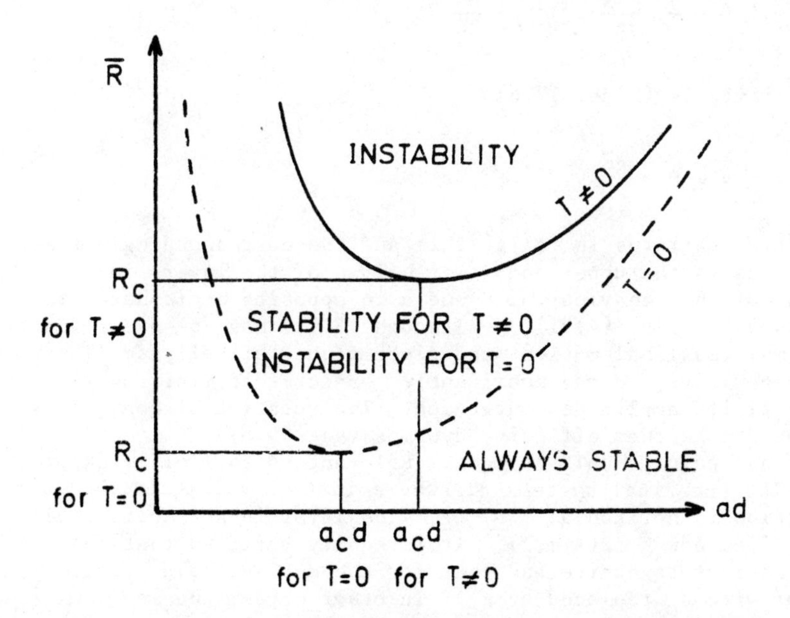

Figure 13. $\bar{R}(a)$ is shown for fixed T and zero magnetic field. The curve for $T = 0$ and fixed M is qualitatively similar (§5.2.3). For comparison the curve for $T=M=0$ is shown dashed (§5.2.1).

$$v_z = v_z(x,y) . \tag{5.14}$$

Since v_z is zero on the walls $z = 0,d$, (5.14) requires

$$v_z \equiv 0 . \tag{5.15}$$

There are no rising and falling motions; convection of heat cannot, apparently, occur.

Convection is nevertheless possible by violating the conditions of the TP theorem, by making one or more of the terms neglected in (5.9) important. In steady marginal convection, the only possibility is the $\nu\nabla^2\underline{v}$ term and indeed, by adopting a small horizontal scale, $\mathcal{L} = O(T^{-1/6}d)$ according to $(5.8)_1$, the viscous forces are so enhanced that $\nu\nabla^2\underline{v}$ cannot be neglected. The rate of viscous energy loss is very large for motions on such small scales, and convection cannot occur until it can be matched by the rate at which buoyancy does work; this demands the large R of $(5.8)_2$.

After the TP theorem has been violated in this way, vertical convective motions arise, but they are still small compared with the horizontal motions implied by (5.12), and so the vorticity is approximately

$$\underline{\omega} \doteqdot \left(-\frac{\partial v_y}{\partial z}, \frac{\partial v_x}{\partial z}, 0\right) = -\frac{g\bar{\alpha}}{2\Omega}\underline{\nabla}\Theta' , \tag{5.16}$$

and the helicity is (Hide, 1976)

$$H \equiv \underline{v}.\underline{\omega} \doteqdot -\frac{g\bar{\alpha}}{2\Omega}\underline{v}.\underline{\nabla}\Theta' . \tag{5.17}$$

For convection patterns in rolls, this and the corresponding α take opposite signs in the upper and lower halves of the layer, in a way reminiscent of the behaviour of H and α in opposite hemispheres in the model of §4.3, cf. eq. $(4.31)_1$. Although (5.17) has been derived from infinitesimal (marginal) motion and gives a quadratically small α that is non-regenerative, we may confidently expect that finite α is created by finite amplitude convection. The rotation of convecting fluids should make them effective dynamos (see §5.3).

One final point should be made. Relevant to §4.1 of my chapter on convecting spherical systems elsewhere in this Volume is the case where rotation is horizontal: $\underline{\Omega} = \Omega\hat{\underline{x}}$. The layer is not horizontally isotropic, i.e. now $\bar{R} = \bar{R}(\ell,m,T)$. It is easily verified that the $\ell = 0$ modes are easiest to excite and that for these (5.6) hold precisely as before, but with a_c replaced by m_c. In other words, the layer minimizes (indeed removes!) the constraint of the TP theorem by arranging to convect in rolls with axes parallel to $\underline{\Omega}$.

5.2.3. Magnetoconvection. Now generalize §5.2.1 in a different way. Suppose that the fluid is electrically conducting and that a vertical uniform field $\underline{B}_0 = B_0\hat{\underline{z}}$ is applied by external sources. Further dimensionless parameters are required:

$$Q = \frac{B_0^2 d^2}{\mu_0 \rho \nu \eta} = M^2 , \qquad M = B_0 d \left(\frac{\sigma}{\rho\nu}\right)^{\frac{1}{2}} , \qquad q = \frac{\kappa}{\eta} . \qquad (5.18)$$

In convection theory the strength of $\underline{B}_0$ is usually measured by the Chandrasekhar number, Q, rather than by the older Hartmann number, M. The ratio of diffusivities, q, has no name, although $p_r q = \nu/\eta$ is usually called "the magnetic Prandtl number": to avoid again the complications of overstability, we will suppose $q < 1$, which is usually easily satisfied in planetary cores. The remarks made below (5.7) apply once more, but with $\bar{R} = \bar{R}(a,Q)$. Also, ohmic dissipation is more significant than viscous dissipation in the energy budget, especially as $Q \to \infty$. The situation sketched in Fig. 13 is qualitatively repeated, but in place of (5.8) we have

$$a_c d = O(Q^{1/3}), \qquad R_c = O(Q) , \qquad Q \to \infty . \qquad (5.19)$$

An analogue to the TP theorem explains the inhibition of convection implied by (5.19). Instead of (5.9) we now have

$$\frac{\partial \underline{v}}{\partial t} + \underline{v}.\underline{\nabla}\,\underline{v} = -\underline{\nabla}\left(\frac{p}{\rho}\right) + g\bar{\alpha}\Theta'\hat{\underline{z}} + \frac{1}{\rho}\,\underline{J}' \times \underline{B}_0 + \nu\nabla^2\underline{v} , \qquad (5.20)$$

where $\underline{J}' = \underline{\nabla} \times \underline{B}'/\mu_0$ is the electric current density and $\underline{B}' = \underline{B} - \underline{B}_0$ is the field perturbation. Assuming steadiness, slowness, and lack of viscosity as before, we reduce (5.20) to

$$0 = -\underline{\nabla}\Pi + g\bar{\alpha}\Theta'\hat{\underline{z}} + \frac{B_0}{\mu_0\rho}\frac{\partial \underline{B}'}{\partial z} , \qquad (5.21)$$

where $\rho\Pi$ is the total pressure, $p + B^2/2\mu_0$. Operating with $\underline{\nabla}\times$, we obtain

$$\frac{\partial \underline{J}'}{\partial z} = \frac{g\bar{\alpha}\rho}{B_0}\,\hat{\underline{z}} \times \underline{\nabla}\Theta' . \qquad (5.22)$$

For clarity, we specialize to the case of convection in rolls in the xz-plane: electrodynamically the situation is analogous to that discussed below (2.27) but with an infinite horizontal array of parallel cylinders (the convection rolls) and with coordinates x and z interchanged. As in that discussion, $\underline{E}$ is zero and Ohm's law (2.17) gives

$$\underline{J}' = \sigma\underline{v} \times \underline{B}_0 . \qquad (5.23)$$

[Since $\underline{B}'$ vanishes with $\underline{v}$, the term $\sigma\underline{v} \times \underline{B}'$ omitted from (5.23) is quadratically small, i.e. negligible.] By (5.22) and (5.23) we have

$$\frac{\partial v_x}{\partial z} = - \frac{g\alpha\bar{\rho}}{\sigma B_0^2} \frac{\partial \Theta'}{\partial x} . \tag{5.24}$$

The analogue of the neglect of viscosity in the TP theorem is the **neglect** of resistivity in (5.24). This leads to

$$v_x = v_x(x) , \quad v_z = -z\, v_x(x) + \tilde{v}(x) , \tag{5.25}$$

where we have used (5.10). To make $v_z = 0$ on $z = 0, d$ we require $v_x = \tilde{v} = 0$, and then (5.15) follows, ruling out convection.

The resolution is much as before. By favouring cells of small horizontal scale, $\mathcal{L}$, the right-hand side of (5.24) becomes the product of the small $g\alpha\bar{\rho}/\sigma B_0^2$ with a large $\partial\Theta'/\partial x = O(\Theta'/\mathcal{L})$, and is finite: (5.25) is then false. It may be noted that the heat conduction **equation,**

$$\frac{\partial \Theta'}{\partial t} + \underline{v}.\underline{\nabla}\Theta' - \beta v_z = \kappa\nabla^2\Theta' , \tag{5.26}$$

gives, for the slow steady case,

$$-\beta v_z = \kappa\nabla^2\Theta' , \tag{5.27}$$

so that by (5.24) and (5.10)

$$R \frac{\partial^2\Theta'}{\partial x^2} + Qd^2 \frac{\partial^2}{\partial z^2} \nabla^2\Theta' = 0. \tag{5.28}$$

For $\mathcal{L} \gg d$ this becomes

$$\frac{\partial^2}{\partial x^2} [Qd^2 \frac{\partial^2\Theta'}{\partial z^2} + R] \doteqdot 0 , \tag{5.29}$$

from which the origin of $(5.19)_2$ becomes obvious.

The rate of ohmic energy loss is very large for motions on scale $(5.19)_1$, and convection cannot occur until it has been matched by the rate at which buoyancy does work; this demands the large R of $(5.19)_2$.

Relevant to §5.2.4 below is the case when the applied field is horizontal: $\underline{B} = B_0\hat{\underline{x}}$. This system is not horizontally isotropic, i.e. now $\bar{R} = \bar{R}(\ell,m,Q)$. It is easily verified that $\ell = 0$ modes are the easiest to excite and that for these (5.6) hold precisely as before, but with a_c replaced by m_c. In other words, the layer minimizes (indeed removes!) the constraint imposed by the applied field, $\underline{B}_0$, by arranging to convect in rolls with axes parallel to $\underline{B}_0$.

5.2.4. Magnetoconvection in a Rotating Layer. We now consider (for $p_r > 1$, $q < 1$) a Bénard layer rotating about the vertical in a horizontal applied field: $\underline{\Omega} = \Omega\hat{\underline{z}}$, $\underline{B}_0 = B_0\hat{\underline{x}}$. Most other orientations would lead to results similar to those that follow, but the absence of horizontal isotropy is a didactic advantage.

Suppose that T is fixed and large. Consider the effect of increasing Q gradually from zero. For $Q \ll T^{1/3}$, the critical R is again approximately $(5.8)_2$ but the preferred mode is a roll perpendicular to $\underline{B}_0$. This is at first sight surprising, in view of the model that concluded §5.2.3. By selecting this mode, however, the magnetic field is able partially to offset the constraint of rotation. As Q increases to $O(T^{1/3}) < Q < O(T^{1/2})$, R_c becomes of order T/Q though $\mathcal{L}$ remains small. For $Q = O(T^{1/2})$, R_c is $O(T^{1/2})$. Finally, for $Q = O(T^{1/2})$, the rolls are almost parallel to $\underline{B}_0$ in order to minimize the field constraint, although the Coriolis forces cannot reduce R_c below $O(Q)$.

The range $Q = O(T^{1/2})$ is particularly interesting, for as $Q/T^{1/2}$ increases the direction of the rolls gradually swings round from being perpendicular to $\underline{B}_0$ to being parallel. Also in this range

$$a_c d = O(1), \quad R_c = O(T^{1/2}) \qquad \text{for } Q = O(T^{1/2}), \quad T \to \infty. \tag{5.30}$$

The constraints of rotation and magnetic field offset one another, and as in the case $\underline{\Omega} = \underline{B}_0 = 0$, the convection cells are "as broad as they are deep" (see §5.2.1). In this way the system is also able to reduce its dissipative losses, so that the critical Rayleigh number $(5.30)_2$ is small compared with its value $(5.8)_2$ for the same rotation rate but for zero applied field. It may be noted that we may rewrite (5.30) as

$$a_c d = O(1), \qquad \hat{R} = O(1) \quad \text{for } \Lambda = O(1), \quad \nu \to 0, \tag{5.31}$$

where

$$\hat{R} = \frac{g\bar{\alpha}\beta d^2}{2\Omega\kappa}, \qquad \Lambda = \frac{B_0^2}{2\Omega\mu_0\rho\eta}, \tag{5.32}$$

are the modified Rayleigh number and Λ is the "Elsasser number". According to (5.31), thermal and magnetic diffusion are far more influential than viscosity in lightly damped convective systems. When $\Lambda = O(1)$, Coriolis and Lorentz forces are equal in magnitude. That this is generally an interesting parameter range is the conclusion of Eltayeb (1972), who looked at cases where $\underline{\Omega}$ and $\underline{B}_0$ are both vertical, and where they are both horizontal but perpendicular to one another. Similar results may also hold when curvature effects are present: see §4 of the chapter on 'Convection in Spherical Systems', elsewhere in this Volume.

5.2.5. Thermodynamic Reasons for Dynamos. The results of §5.2.4 support a heuristic principle proposed many years ago by Willem Malkus: "A convecting layer of electrically conducting fluid adopts that mode of convection that maximizes the heat transport across it (for given applied temperature contrast $\Delta\Theta$). In particular a rotating layer may set up a dynamo, so that its magnetic field can offset the constraint of rotation." (Malkus, unpublished.)

Although examples have been given that show Malkus's idea cannot be applied too literally, there is some feeling that it contains a qualitative truth. One begins to think that the geomagnetic field may not be an incidental bye-product of convection in the core, but may play an integral part in the cooling of the Earth. The question in §4.2, of a weak or strong geodynamo, also becomes more intriguing. At first sight, one might have thought that a large toroidal field is very unlikely, since it greatly enhances the ohmic energy loss, so that buoyancy forces must be correspondingly greater before convection can occur. We now see however that a large toroidal field also increases the scale of convective motions and so reduces energy losses on the microscale. Which effect is more important is not easy to answer: the models of §§5.2.3 and 5.2.4 are too prejudicial since they totally ignore the energy losses of the (external) mechanism creating $\underline{B}_0$. But there is at least a possibility that strong toroidal fields will be preferred.

One can imagine a sequence of increasing R in which first a marginal R_{c1} is reached, where the convective motions are infinitesimal and therefore non-regenerative (§3.1). Next, R attains a second critical value, R_{c2}, at which field regeneration first occurs: the necessary helicity is present because of rotation. As R rises further, $|\underline{B}|$ increases, as does the Lorentz force, a balance being struck of the kind described in §5.1. We may expect from §5.2.4 that the scale of the motions will increase, and that the dissipative energy losses may at some stage begin to decrease. If so, "runaway field growth" will occur, in which fields increase rapidly until a regime like (5.31) is set up, in which Lorentz forces and Coriolis forces come into approximate balance, and in which the motions and fields, instead of being on a small lengthscale, have a comparable scale to that of the body, with a correspondingly smaller dissipative energy loss. In §5.3, a convective dynamo is described in which more than an inkling of runaway growth has been detected. (See also Stevenson, 1984.)

5.3. Nonlinear Models.

5.3.1. An α^2-Model. Attacks on the dynamo problem along the lines of §§4.1 and 4.2 succeed, and can often be made rigorous, by availing themselves of the power of the two-scale method. The small scale evinced by $(5.8)_1$ suggested to S. Childress and A.M. Soward (1972) that the problem of §5.2.4 could be solved again, with the applied field, $\underline{B}_0$, replaced by a non-uniform field, in fact the $\underline{B}$ generated by the convection itself, acting as a dynamo. The layer has, of course, to be in a nonlinear supercritical state $R > R_{c2}$ (§5.2.5) but, provided $R - R_{c2}$ is not too large, the horizontal scale $\mathcal{L}$ of the motions remains small compared with d, and the requisite finite amplitude solution can be obtained by expansion using two-scale methods. The dynamo is of weak field (α^2) type: see §4.2.

Space is not available here to examine the Childress-Soward dynamo in detail, but one interesting fact should be reported. A.M. Soward (1974) found clear indications of runaway field growth. It was not possible to determine the ultimate fate of the system, because (§5.2.4) runaway growth is accompanied by an increase in horizontal scale, $\mathcal{L}$, the loss of the two-scale method, and a concomitant failure of the analytic approach (but see Y. Fautrelle and S. Childress, 1982).

5.3.2. αω-Models. The difficulty of matching the analytic success of α^2-dynamos by a comparable success for αω-theory has proved frustrating to those who see the αω-dynamo as physically the more plausible mechanism, providing in fact the α to ω of all planetary and stellar dynamo questions! Proponents of αω-dynamos have had to proceed more slowly, and to rely more heavily on large scale computation. But a few key ideas have emerged.

It has been mentioned in §5.2.4 that a Bénard layer convects particularly easily if $Q = 0(T^{1/2})$ and that the corresponding Lorentz force is then of the same size as the Coriolis force, with viscous and inertial forces negligible. That such a balance is the one struck in the Earth's core has often been mooted. It is based on the idea that the westward drift (of angular velocity 0.2°/year) may indicate a zonal core velocity of order $U \doteqdot 10^{-4}$ m/s, and (using values of §1.2) a magnetic Reynolds number of order $R_\omega \doteqdot 100$. Assuming $B_P \doteqdot 4$ Gauss, we see that the ω-effect should by (2.39) create $B_T \doteqdot 400$ Gauss. The corresponding Lorentz force density, taken to be of order $B_T^2/\mu_0 c \doteqdot 4\,10^{-4}$ N/m³, is now seen to be of the same order as the Coriolis force density $2\Omega\rho U \doteqdot 10^{-4}$ N/m³.

These order of magnitude estimates imply that the equation of motion,

$$\frac{\partial \underline{v}}{\partial t} + \underline{v}.\underline{\nabla}\,\underline{v} + 2\underline{\Omega}\times\underline{v} = -\underline{\nabla}\left(\frac{p}{\rho}\right) + g\bar{\alpha}\Theta\hat{\underline{z}} + \frac{1}{\rho}\,\underline{J}\times\underline{B} + \nu\nabla^2\underline{v}\,, \tag{5.33}$$

can, as in the transition from (5.20) to (5.21) above, be replaced by

$$2\underline{\Omega}\times\underline{v} = -\underline{\nabla}\left(\frac{p}{\rho}\right) + g\alpha\Theta\hat{\underline{r}} + \frac{1}{\rho}\underline{J}\times\underline{B}. \tag{5.34}$$

(Even the buoyancy term appears negligible compared with the rest, but it is energetically necessary and plays no role in the following argument. We are considering a spherical Earth with buoyancy radially upwards.)

J.B. Taylor (1963) noticed that, when the φ-component of (5.34) was integrated over a cylinder, $\mathcal{C}(s)$, of radius s (see Fig. 14) all terms except the last vanished, i.e.

$$\int_{\mathscr{C}(s)} (\underline{J} \times \underline{B})_\phi \, d\phi dz = 0 , \quad \text{for} \quad 0 < s < c. \tag{5.35}$$

If $\underline{B}$ initially did not satisfy (5.35) we would have to conclude that (5.34) had failed and, at the very least, the $\partial \underline{v}/\partial t$ term in (5.33) would have to be restored. Taylor proposed that this would quickly adjust $\underline{v}$ until (5.35) and the balance (5.34) were realized. He pointed out that, to add to $\underline{v}$ in (5.34) a "geostrophic flow", $v_G(s)\hat{\underline{\phi}}$, which is zonal and constant on $\mathscr{C}(s)$, does not affect the balance (5.34) although it does considerably alter the solution, $\underline{B}$, of the induction equation, especially through the ω-effect of v_G. Thus he visualized that, when the initial $\underline{B}$ did not satisfy (5.35), the geostrophic flow would quickly readjust to create a new field $\underline{B} + \underline{B}_G$ that satisfied (5.35) and (5.34). As $\underline{B}$ evolved subsequently, this balance would be maintained through an accompanying change in v_G. Indeed, J.B. Taylor (1963) showed how his concept could be used to time-step $\underline{v}$ and $\underline{B}$ efficiently. The feeling was general that a complete and satisfactory solution awaited only the advent of new generations of computers. Before this happened, however, a new idea emerged.

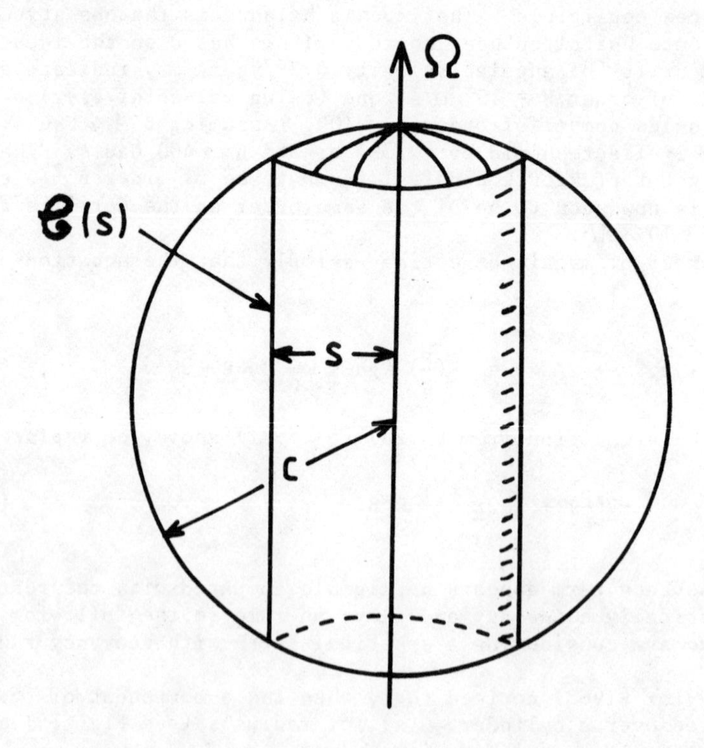

Figure 14. The cylinder $\mathscr{C}(s)$ in J.B. Taylor's (1963) argument.

In 1975 Braginskii proposed quite a different magnetohydrodynamic state in the core, which he named "model-Z" because the meridional field over most of the core would be principally in the z-direction. The process of readjustment envisaged by Taylor, which relied on B_s, would then be largely suppressed, so that (5.35) would not hold.

To test this idea, S.I. Braginskii (1978) examined numerically an axisymmetric model of given α, and with an ω-shear partly provided by buoyancy as a "thermal wind" but also including a geostrophic, ω_G, created by the model, and a contribution from the Lorentz force. He found that his models did not evolve to the Taylor state, but to a model-Z.

Today, attempts are being made to discover reasons why a Taylor state arises in one circumstance, but a model-Z in another, apparently similar, case. We do not have space to take matters further here. We have taken the reader up to the forefront of dynamo theory. In §6 we give references to current work.

5.4. Crude Models for Reversals.

Some really simple magnetic systems can exhibit some really complicated behaviours. No better illustration exists than the Rikitake two-disk dynamo illustrated in Fig. 15, where two homopolar dynamos of the type illustrated in Fig. 5 are coupled together, the current induced by one disk being fed to the wire loop that creates the inducing field for the other.

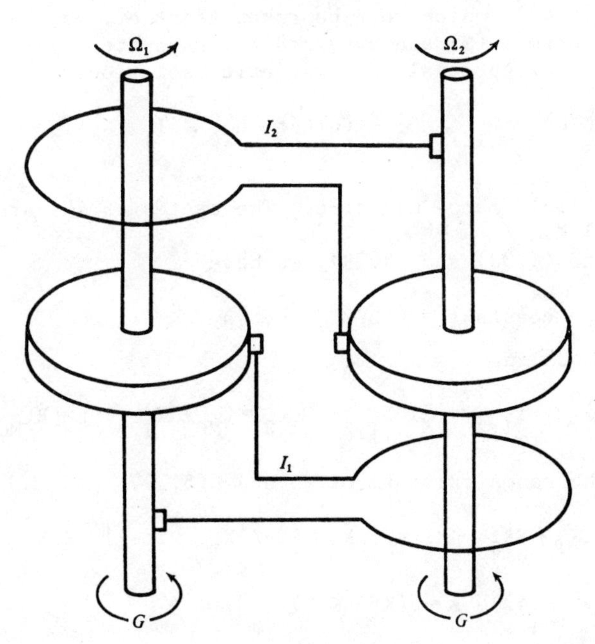

Figure 15. The Rikitake Coupled Disk Dynamos.

The equations governing the angular velocities Ω_1, Ω_2 of disks D_1, D_2 and the electric currents I_1, I_2 induced by that motion take the form

$$L \frac{dI_1}{dt} + RI_1 = M\Omega_1 I_2 , \qquad L \frac{dI_2}{dt} + RI_2 = M\Omega_2 I_1 , \qquad (5.36,\ 5.37)$$

$$C \frac{d\Omega_1}{dt} = G - MI_1 I_2 , \qquad C \frac{d\Omega_2}{dt} = G - MI_2 I_1 , \qquad (5.38,\ 5.39)$$

where L is the self-inductance of either electric circuit (assumed identical), R is its resistance, M is proportional to the mutual inductance between the circuits, C is the moment of inertia of either disk, and G is the identical couple with which each is driven. Equations (5.38) and (5.39) are equations of motion; (5.36) and (5.37) are induction equations.

The system (5.36) - (5.39) admits an infinity of steady-state "magnetic" solutions, namely

$$\Omega_1 = K^2 \frac{R}{M}, \quad \Omega_2 = K^{-2} \frac{R}{M}, \quad I_1 = \pm K\left(\frac{G}{M}\right)^{\frac{1}{2}}, \quad I_2 = \pm K^{-1} \left(\frac{G}{M}\right)^{\frac{1}{2}} . \qquad (5.40)$$

The ± signs in (5.40), which must be taken together, may be thought of as states of normal (N) and reversed (R) magnetic polarity (§1.5). There is also the exceptional "non-magnetic" solution:

$$\Omega_1 = (G/C)t + \Omega_{10} , \quad \Omega_2 = (G/C)t + \Omega_{20} , \quad I_1 = 0, \quad I_2 = 0, \qquad (5.41)$$

which any stray field makes magnetic. The states (5.40) are likewise unstable for all K.

According to (5.38) and (5.39), we have

$$\Omega_1 - \Omega_2 = \text{constant} = (GL/MC)^{\frac{1}{2}} A , \text{ say} . \qquad (5.42)$$

Introducing the scaling

$$t \to \left(\frac{CL}{MG}\right)^{\frac{1}{2}} t, \quad I_{1,2} = \left(\frac{G}{M}\right)^{\frac{1}{2}} X_{1,2}, \quad \Omega_{1,2} = \left(\frac{GL}{MC}\right)^{\frac{1}{2}} Y_{1,2}, \quad \mu = R\left(\frac{C}{MGL}\right)^{\frac{1}{2}}, \qquad (5.43)$$

we now obtain the canonical form of (5.36) - (5.39)

$$\dot{X}_1 = -\mu X_1 + YX_2 , \quad \dot{X}_2 = -\mu X_2 + (Y-A)X_1 , \quad \dot{Y} = 1 - X_1 X_2 , \qquad (5.44)$$

where $Y = Y_1$. By (5.42), $A = \mu(K^2 - K^{-2})$.

The third order dynamical system (5.44) exhibits chaotic behaviour, and possesses a strange attractor. Fig. 16 shows Y_1 in a typical integration. The current I_1 and its associated magnetic field oscillate with growing amplitude about one of the states (5.40) and then suddenly switch to the other; the process is then repeated with the reversed field until the original polarity is restored. The intervals of time between reversals is quite irregular. Sometimes they are long ("polarity epochs"); sometimes short ("polarity events").

Plotted in (X_1, X_2, Y) space, the trajectories continually approach a finite surface, the attractor, as illustrated in X_2Y-projection in Fig. 17. This surface consists of two sheets connected along the dotted lines; each sheet is associated with one of the solutions (5.40) but round each such solution there is a hole, indicating that the trajectory does not approach either solution during its convergence to the sheet. The solution trajectory circles the N-hole (R-hole) in a clockwise (anticlockwise) sense with increasing amplitude until eventually it meets the left-hand (right-hand) dotted line. It then crosses to the other sheet.

Many other reversing dynamos have been examined since Rikitake made his discovery, but his model is still probably the one most thoroughly studied. Speculative comparisons have been made with the geomagnetic

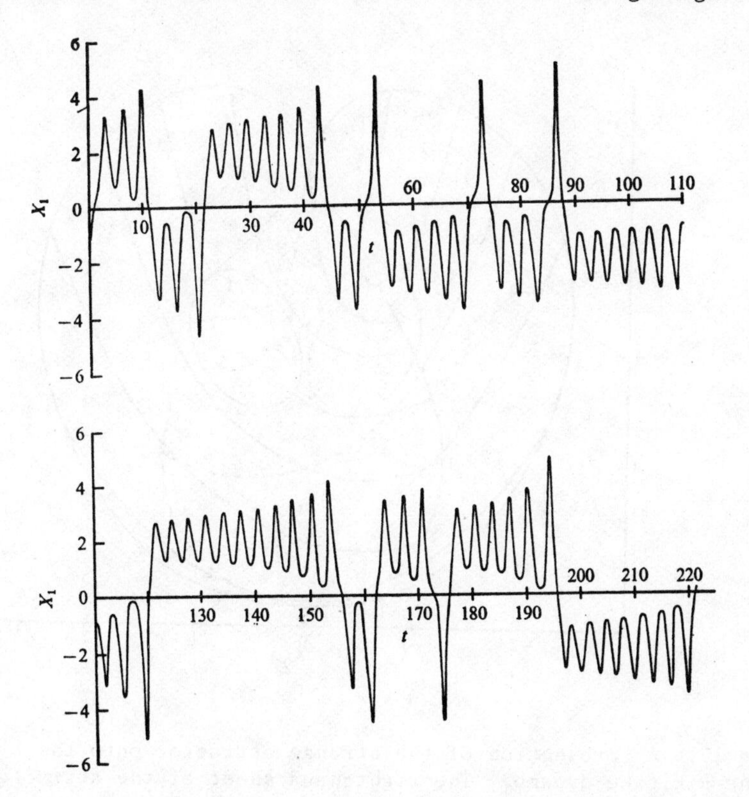

Figure 16. A plot of $X_1(t)$ for a typical integration of the Rikitake System ($\mu = 1$, $K = 2$). (From A.E. Cook and P.H. Roberts, 1970).

field (see R.T. Merrill and M.W. McElhinny, 1983). Further references are given in §6.

6. FURTHER READING

General Sources. No attempt will be made below to give a complete bibliography; the subject is today too large and too rapidly expanding to make this practicable. In addition to the citations of the text, the following reference list contains the locations of recent papers through which much interesting work can be traced.

The following books, or substantial sections of them, are devoted to dynamo theory: V. Bucha (1983), F. Krause and K.-H. Rädler (1980), R.T. Merrill and M.W. McElhinny (1983), H.K. Moffatt (1978), E.N. Parker (1979), E.R. Priest (1984), P.H. Roberts (1967), P.H. Roberts and A.M. Soward (1978), A.M. Soward (1983) and Ya.B. Zeldovich et al (1983). Reviews have been written by F.H. Busse (1978), P.H. Roberts (1971a,b), P.H. Roberts and A.M. Soward (1972) and A.M. Soward and P.H. Roberts (1977). Work of special interest to each section is mentioned below.

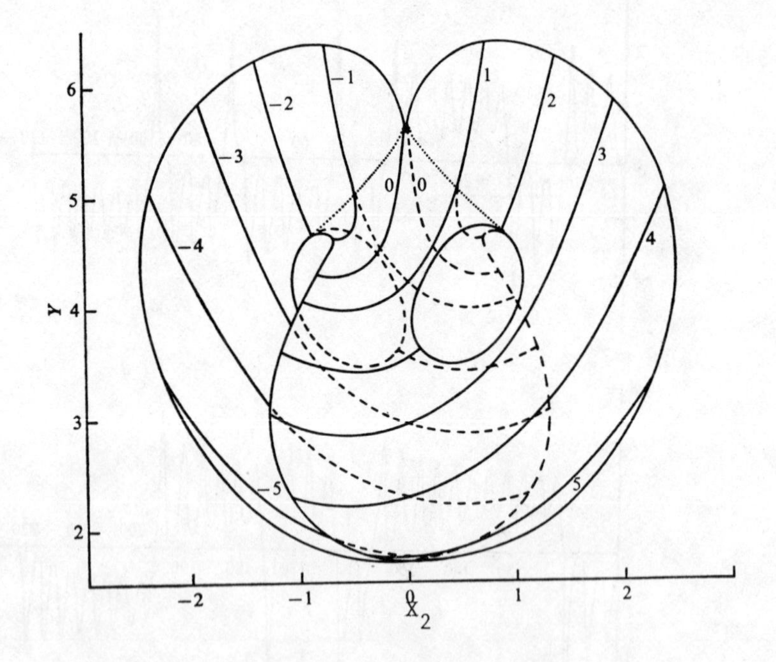

Figure 17. A projection of the strange attractor onto the X_2Y-plane for the Rikitake dynamo. The right-hand sheet of the attractor lies above $X_1 = 0$, the left-hand sheet below. The dashed lines show the edges hidden by an overlying sheet: the dotted lines show the apparent joins between the sheets. (From A.E. Cook and P.H. Roberts, 1970.)

Section 1. S. Chapman and J. Bartels (1940, Ch. 26) outlines the history of geomagnetic discovery up to 1940. R.A. Langel (1985) has recently reviewed observations and their analysis. J.A. Jacobs (1975) describes the structure of the Earth, and particularly the core. M. Stix and P.H. Roberts (1984) have made one of the most recent theoretical attempts to link variations in the length of the day with the secular variation.

Section 2. To H. Lamb (1881) goes the credit of discovering the toroidal-poloidal representation of a divergenceless field. Perhaps the most persuasive case for a thermoelectric origin to the geomagnetic field was made by S.K. Runcorn (1954): the electrochemical theory seems never to have been seriously explored. The first suggestion that a large fluid body owes its field to dynamo action came from J. Larmor (1919), who had in mind solar magnetism. One discussion of the pre-Maxwell equations has been given by myself (P.H. Roberts, 1967) but there are many others.

Perhaps the most complete discussion of induction by rotating solid conductors is that of A. Herzenberg and F.J. Lowes (1957). The creation of flux ropes by convection is a topical issue [e.g. See Chapter 2 of A.M. Soward (1983)]. Proofs of Alfvén's theorem can be found in many places; for a recent account, see H.K. Moffatt (1978). W.M. Elsasser (1947) first noticed the ω-effect and emphasized its importance for dynamo theory. There are many discussions of the decay modes of a sphere, all following the lines of §2.5.

Section 3. T.G. Cowling's (1933) theorem provided a sequence of papers, some extending his result to time-dependent fields (as in S.I. Braginskii's proof given in §3.2), others to compressible fluids, and some (e.g. D.J. Ivers and R.W. James, 1984, who also cite earlier work) searching for a higher degree of mathematical rigour.

Ramifications of Ponomarenko's model are considered by A. Gailitis and Ya. Freiberg (1979) and by Ya.B. Zeldovich et al (1983). The model is, in the terminology of §3.1, an overstable dynamo. A motion devised by D. Lortz (1968), also for a cylinder, is a steady dynamo. The motions are again helical but of a more complicated type than (3.15). In compensation, only mathematical analysis is required to demonstrate dynamo action; numerical work is not entailed. The Lortz dynamo is the subject of a recent paper by P. Chen and J.L. Milovich (1984) where earlier work is cited. D. Lortz (1972) examines the possibility of bending his cylindrical model into a torus.

Section 4. It was impossible in §4.1 to do more than hint at what is now a large topic. Indeed, it forms the theme of F. Krause and K.-H. Rädler's (1980) book. Those interested should also consult H.K. Moffatt (1978), E.N. Parker (1979) and Ya.B. Zeldovich et al (1983). G.O. Roberts [reported in P.H. Roberts (1971b)] has adapted his motion (4.14) to a sphere, and by convincing numerical work demonstrated an asymmetric magnetic field sustained by axisymmetric motion. As we saw in §3.2, this does not violate T.G. Cowling's theorem.

Dynamo waves were discovered and interpreted by E.N. Parker (1955) in his epoch-making paper that first isolated the α-effect; he was also the first (E.N. Parker, 1957) to see in these waves a viable explanation of the solar cycle. Figure 12, taken from a numerical integration by

P.H. Roberts (1972), supports his idea. There has been much subsequent work (see E.N. Parker, 1979; H. Yoshimura, 1983).

Braginskii's concept of the nearly symmetric dynamo (§4.4) has been elucidated and developed by A.M. Soward (1972) and H.K. Moffatt (1978).
Section 5. The theory of compositional convection in the Earth's core is reviewed by D.E. Loper and P.H. Roberts (1983).

The early parts of §5.2 are essentially all included in S. Chandrasekhar's book (1963), although there have since been many (and at least one curious) developments; see D.J. Acheson (1980), where other papers are referenced. In addition to the unrealism of uniform $\underline{B}_0$ mentioned in §5.2.5, the models of §5.2 have the second weakness of ignoring the effect of shear on the convective motions. Some efforts have been made to include shear: see my chapter in the present book on "Convection in spherical systems". The model of §5.2.4 has been taken into the nonlinear regime in several papers, most recently by A.M. Soward (1986), where earlier work is cited.

The Childress-Soward dynamo was slightly pre-dated by the convective $\alpha\omega$-dynamo of F.H. Busse (1973), the ω-shear being maintained by external means. Busse has also constructed convective α^2-dynamos for spherical geometry. The basic ideas are explained in his chapter of P.H. Roberts and A.M. Soward (1978).

Of the current attempts to understand the relationships between Braginskii's model-Z and Taylor's concept, we should mention particularly A.M. Soward and C.A. Jones (1983) and G.R. Ierley (1986), papers where other literature is referenced.

Since Rikitake's discovery, many other self-reversing systems have been isolated. The idea of an $\alpha\omega$-dynamo limited by a geostrophic flow created by the Lorentz force (the anti-dynamo effect of §5.1) led F. Krause and P.H. Roberts (1981) to study the system

$$\dot{P} = -P + \alpha T\,, \qquad \dot{T} = -T + \omega P\,, \qquad \dot{\omega} = \kappa^2(1 - PT)\,, \tag{6.1}$$

in which T and P are the toroidal and poloidal fields and ω is the zonal shear. A related model has been thoroughly studied by K.A. Robbins (1977). Recently D. Crossley, O. Jenson and J.A. Jacobs (1986) have added a stochastic element to the integration of such systems.

Evidently, models like (5.44) or (6.1) are extremely crude representations of the geomagnetic field. Similar models for the solar field (see Ya.B. Zeldovich et al, 1983) have been generalized by a slightly less severe truncation. N.O. Weiss et al (1984) studied a fifth order system (the complex Lorenz equations) which even exhibited periods of low magnetic activity like the Maunder minimum of the Sun.

ACKNOWLEDGEMENTS

I am grateful to the following for permission to reproduce figures: Figures 1-3: Dr. R.A. Langel and NASA Goddard Space Flight Center; Figure 4, Dr. D.R. Barraclough and the Royal Astronomical Society; Figures 8-11, the American Mathematical Society; Figure 12, the Royal Society of London; Figures 15-17, Dr. A.E. Cook and the Cambridge

University Press (as publishers of the Proceedings of the Cambridge Philosophical Society).

REFERENCES

Acheson, D.J. ''Stable' Density Stratification as a Catalyst for Instability,' J. Fluid Mech. 96, 723-733 (1980).

Barraclough, D.R. 'Spherical Harmonic Analyses of the Geomagnetic Field for Eight Epochs between 1600 and 1910,' Geophys. J.R. Astr. Soc., 36, 497-513 (1974).

Braginskii, S.I. 'Structure of the F-Layer and Reasons for Convection in the Earth's Core,' Soviet Phys.-Dokl., 149, 1311-1314 (1963).

Braginskii, S.I. 'Self-Excitation of a Magnetic Field during the Motion of a Highly-Conducting Fluid,' Soviet Phys.-JETP, 20, 726-735 (1965). Russian Original 1964.

Braginskii, S.I. 'Nearly Axially Symmetric Model of the Hydromagnetic Dynamo of the Earth I,' Geomagn. Aeron., 15, 122-128 (1975).

Braginskii, S.I. 'Nearly Axially Symmetric Model of the Hydromagnetic Dynamo of the Earth II,' Geomagn. Aeron., 18, 225-231 (1978).

Bucha, V. (Editor) Magnetic Field and the Processes in the Earth's Interior, Academia, Prague (1983).

Bullard, E.C. and Gellman, H. 'Homogeneous Dynamos and Terrestrial Magnetism,' Phil. Trans. R. Soc. Lond. A, 247, 213-278 (1954).

Busse, F.H. 'Generation of Magnetic Fields by Convection,' J. Fluid Mech., 57, 529-544 (1973).

Busse, F.H. 'A Necessary Condition for the Geodynamo,' J. Geophys. Res., 80, 278-280 (1975).

Busse, F.H. 'Magnetohydrodynamics of the Earth's Dynamo,' Ann. Rev. Fluid Mech, 10, 435-462 (1978).

Chandrasekhar, S. Hydrodynamic and Hydromagnetic Stability, Clarendon Press (1963).

Chapman, S. and Bartels, J. Geomagnetism. II. Analysis of the Data and Physical Theories, Clarendon Press (1940).

Chen, P. and Milovich, J.L. 'An Explicit Solution for Static Unbounded Helical Dynamos,' Geophys. Astrophys. Fluid Dynam., 30, 343-354 (1984).

Childress, S. and Soward, A.M. 'Convection-driven Hydromagnetic Dynamos,' Phys. Rev. Lett., 29, 837-839 (1972).

Cook, A.E. and Roberts, P.H. 'The Rikitake Two-Disc Dynamo System,' Proc. Camb. Phil. Soc., 68, 547-569 (1970).

Cowling, T.G. 'The Magnetic Field of Sunspots,' Mon. Not. R. Astr. Soc., 94, 39-48 (1933).

Crossley, D., Jenson, O. and Jacobs, J. 'Geomagnetic Reversals - Stochastic Excitation or Chaotic Behaviour?' to appear (1986).

Elsasser, W.M. 'Inductive Effects in Terrestrial Magnetism. Part III. Electric Modes,' Phys. Rev., 72, 821-823 (1947).

Eltayeb, I.A. 'Hydromagnetic Convection in a Rapidly Rotating Fluid layer, 'Proc. R. Soc. Lond. A, 326, 229-254 (1972).

Fautrelle, Y. and Childress, S. 'Convective Dynamos with Intermediate and Strong Fields,' Geophys. Astrophys. Fluid Dynam., 22, 235-279 (1982).

Feber, R.C., Wallace, T.C. and Libby, L.M. 'Uranium in the Earth's Core,' EOS, 65, 785 (1984).

Gailitis, A. and Freiberg Ya. 'Nature of the Instability of a Turbulent Dynamo', Magnetohydrodynam., 16, 116-122 (1980). Russian Original 1979.

Herzenberg, A. and Lowes, F.J. 'Electromagnetic Induction in Rotating Conductors,' Phil. Trans. R. Soc. Lond. A, 249, 507-584 (1957).

Hide, R. 'A Note on Helicity,' Geophys. Astrophys. Fluid Dynam., 7, 157-161 (1976).

Ierley, G.R. 'Macrodynamics of α^2-Dynamos,' Geophys. Astrophys. Fluid Dynam., 34, 143-173 (1986).

Ivers, D.J. and James, R.W. 'Axisymmetric Antidynamo Theorems in Compressible Non-Uniform Conducting Fluids,' Phil. Trans. R. Soc. Lond. A., 312, 179-218 (1984).

Jacobs, J.A. The Earth's Core, Academic Press (1975).

Krause, F. and Roberts, P.H. 'Strange Attractor Character of Large-Scale Nonlinear Dynamos,' Adv. Space Res., 1, 231-240 (1981).

Krause, F. and Rädler K.-H. Mean-Field Magnetohydrodynamics and Dynamo Theory, Pergamon (1980).

Lamb, H. 'On the Oscillations of a Viscous Spheroid,' Proc. Lond. Math. Soc., 13, 51-66 (1881).

Langel, R.A. Main Field, NASA Publication X-622-85-8 (1985).

Larmor, J. 'How could a Rotating Body Such as the Sun become Magnetic?' Rep. Brit. Assoc. Adv. Sci., 159-160 (1919).

Leslie, D.C. Developments in Turbulence Theory, Clarendon Press (1973).

Loper, D.E. and Roberts, P.H. 'Compositional Convection and the Gravitationally Powered Dynamo,' pp297-327 in Soward (1983), loc. cit.

Lortz, D. 'Exact Solution of the Hydromagnetic Dynamo Problem,' Plasma Phys., 10, 967-972 (1968).

Lortz, D. 'A Simple Stationary Dynamo Model,' Z. Naturforsch., 27A, 4-8 (1972).

Merrill, R.T. and McElhinny, M.W. The Earth's Magnetic Field. Its History, Origin and Planetary Perspective, Academic Press (1983).

Moffatt, H.K. Magnetic Field Generation in Electrically Conducting Fluids, Cambridge University Press (1978).

Parker, E.N. 'Hydromagnetic Dynamo Models,' Astrophys. J., 122, 293-314 (1955).

Parker, E.N. 'The Solar Hydromagnetic Dynamo,' Proc. Nat. Acad. Sci. USA, 43, 8-14 (1957).

Parker, E.N. Cosmical Magnetic Fields. Their Origin and Activity, Clarendon Press (1979).

Ponomarenko, Yu. B. 'On the Theory of Hydromagnetic Dynamo,' Zh. Prikl. Mech. Tech. Fiz., (USSR) 6, 775-778 (1973).

Priest, E.R. Solar Magnetohydrodynamics, Reidel (1984).

Reid, W.H. (Editor) Mathematical Problems in the Geophysical Sciences 2. Inverse Problems, Dynamo Theory and Tides. Lectures in Applied Mathematics, 14, Amer. Math. Soc. (1971).

Robbins, K.A. 'A New Approach to Subcritical Instability and Turbulent Transition in a Simple Dynamo,' Math. Proc. Camb. Phil. Soc., 82, 309-325 (1977).

Roberts, G.O. 'Dynamo Action of Fluid Motions with Two-Dimensional Periodicity,' Phil. Trans. R. Soc. Lond. A, 271, 411-454 (1972).

Roberts, P.H. An Introduction to Magnetohydrodynamics, Longmans (1967).

Roberts, P.H. 'Dynamo Theory,' pp129-206 in W.H. Reid (1971) loc. cit. (1971a).

Roberts, P.H. 'Dynamo Theory of Geomagnetism,' pp123-131 in Zmuda (1971), loc. cit. (1971b).

Roberts, P.H. 'Kinematic Dynamo Models,' Phil. Trans. R. Soc. Lond. A, 272, 663-703 (1972).

Roberts, P.H. and Soward, A.M. 'Magnetohydrodynamics of the Earth's Core', Ann. Rev. Fluid. Mech., 4, 117-154 (1972).

Roberts, P.H. and Soward, A.M. (Editors) Rotating Fluids in Geophysics, Academic Press (1978).

Roberts, P.H. and Stix, M. The Turbulent Dynamo. A Translation of a Series of Papers by F. Krause, K.-H. Rädler and M. Steenbeck, NCAR Technical Note IA-60 (1971).

Runcorn, S.K. 'The Earth's Core,' Trans. Am. Geophys. Un., 35, 49-78 (1954).

Soward, A.M. 'A Kinetic Theory of Large Magnetic Reynolds Number Dynamos', Phil. Trans. R. Soc. Lond. A, 272, 431-462 (1972).

Soward, A.M. 'A Convection Driven Dynamo I. The Weak Field Case,' Phil. Trans. R. Soc. Lond. A, 275, 611-651 (1974).

Soward, A.M. (Editor) Stellar and Planetary Magnetism, Gordon and Breach (1983).

Soward, A.M. 'Nonlinear Marginal Convection in a Rotating Magnetic System,' Geophys. Astrophys. Fluid Dynam., to appear (1986).

Soward, A.M. and Jones, C.A. 'α^2-Dynamos and Taylor's Constraint,' Geophys. Astrophys. Fluid Dynam., 27, 87-122 (1983).

Soward, A.M. and Roberts, P.H. 'Recent Developments in Dynamo Theory,' Magnetohydrodynam., 12, 1-36 (1977).

Steenbeck, M., Krause, F. and Rädler, K.-H. 'Berechnung der mittlerer Lorenz-Feldstärke $\underline{v} \times \underline{B}$ für ein elektrisch leitendes Medium in turbulenten, durch Coriolis-Kräfte beeinflusster Bewegung,' Z. Naturforsch., 21, 369-376 (1966). [For translation into English see Roberts, P.H. and Stix, M. (1971) loc. cit.]

Stevenson, D.J. 'The Energy Flux Number and Three Types of Planetary Dynamo,' Astron. Nachr., 305, 257-264 (1984).

Stix, M. and Roberts, P.H. 'Time Dependent Electromagnetic Core-Mantle Coupling,' Phys. Earth Planet. Int., 36, 49-60 (1984).

Taylor, J.B. 'The Magnetohydrodynamics of a Rotating Fluid and the Earth's Dynamo Problem,' Proc. R. Soc. Lond. A, 274, 274-283 (1963).

Weiss, N.O., Cattaneo, F. and Jones, C.A. 'Periodic and Aperiodic Dynamo Waves,' Geophys. Astrophys. Fluid Dynam., 30, 305-341 (1984).

Yoshimura, H. 'Dynamo Generation of Magnetic Fields in Three Dimensional Space: Solar Cycle Main Flux Tube Formation and Reversals,' Astrophys. J. Suppl., 52, 363-385 (1983).

Zeldovich, Ya. B., Ruzmaikin, A.A. and Sokoloff, D.D. Magnetic Fields in Astrophysics, Gordon and Breach (1983).

Zmuda, A.J. (Editor) World Magnetic Survey 1957-1969, IAGA Bulletin No. 28 (1971).

NONPROPAGATING SOLITONS AND EDGE WAVES

S. Putterman
Physics Department
University of California
Los Angeles, California 90024

ABSTRACT. In addition to the well known KdV and envelope solitons a channel can also support a nonpropagating soliton. The cause of localization of this soliton is similar to that of edge waves except that its nonlinear properties enable it to exist far from the boundary.

1. INTRODUCTION

Solitons are interesting because they are states of matter where energy is localized and does not spread out on a long time scale. Many investigations have focused on solitons which are localized in one spatial direction and include the KdV and envelope solitons.[1] For fluid in a channel the KdV soliton is a smooth localized elevation of the liquid level which does not spread as it moves down the channel with approximately the speed of a long gravity wave. The envelope soliton is a smoothly modulated wave packet which moves approximately with the group velocity of deep water waves having the frequency of the waves in the packet. Since water waves are highly dispersive non-spreading motions cannot be solutions of the linearized theory. The shift in frequency due to nonlinear processes is the key effect which balances dispersion and makes possible these localized states. The KdV soliton is the localized nonlinear state corresponding to a pulse in the linear theory and the envelope soliton is the localized nonlinear state corresponding to a wave packet in the linear theory. Recently it was discovered that there can exist another type of physical soliton which is the localized nonlinear state corresponding to an evanescent wave in the linear theory.[2)3)4)] Evanescent waves such as edge waves can only exist at the boundaries of a system. Nonlinearities enable this soliton to become unstuck from the boundary and exist in the bulk of the fluid. Thus the nonpropagating soliton has a particularly important physical property, its existence is a manifestation of broken translational invariance. Like an edge wave this soliton can exist with its center of mass at rest. It is also similar to edge waves in that the optimum manner of excitation is parametric.

C. Nicolis and G. Nicolis (eds.), Irreversible Phenomena and Dynamical Systems Analysis in Geosciences, 135–138.

2. SOLITON EQUATIONS OF MOTION

The basic fluid equations are reviewed in another paper of these proceedings.[5] In the linear approximation these equations yield the basic dispersion law for plane progressive surface waves

$$\bar{\omega}^2 = gk\tanh kd \cong gdk^2 - \frac{1}{3}gd^3k^4 + \ldots \tag{1}$$

If one seeks a solution to the incompressible, undamped ($\tau = 0$) equations of motion in the limit where terms of leading order in both nonlinearity and dispersion are retained then one finds for the surface height $\xi(x,t)$:

$$\frac{\partial^2\xi}{\partial t^2} - c_o^2\frac{\partial^2\xi}{\partial x^2} + \gamma\frac{\partial^4\xi}{\partial x^4} - \frac{c_o^2 G}{d}\frac{\partial^2(\xi^2)}{\partial x^2} = 0 \tag{2}$$

where the nonlinear coefficient and dispersion are

$$G = 3/2 \qquad \gamma = \frac{1}{2}(d^2c^2/dk^2)_o \tag{3}$$

and $c = \bar{\omega}/k$ so that the subscript "o" means evaluated at $k \to 0$. Looking for a solution to (2) of the form $\xi(x-ut)$ yields the KdV soliton

$$\xi = \frac{3}{2}\frac{(u^2-c^2)d}{c^2G}\operatorname{sech}^2\frac{1}{2}\left[\frac{c^2-u^2}{\gamma}\right]^{1/2}((x-ut) \tag{4}$$

From (1) we see that these long wavelength solitons travel at a speed higher than c since $\gamma = -gd^3/3$. These solitons were first observed in the nineteenth century by J. Scott Russell. If in (2) one sets $\xi = \xi(x-ut, t)$ and neglects terms involving $(\partial^2\xi/\partial t^2)$ where x-ut is held fixed then one is led to the KdV equation.

A wave packet modulated along the direction of the channel has the form

$$\xi = \xi'(\bar{x},t)\exp(ikx + i\omega t) \tag{5}$$

where $\bar{x} = x + (d\bar{\omega}/dk)t$. Due to dispersion such a packet would in the linear approximation spread out. if, however, we again retain the leading nonlinear order but neglect

$$(\partial^2\xi'/\partial t^2)_{\bar{x}} \tag{6}$$

one finds a nonlinear Schroedinger equation (NLS) for ξ' ;[6]

$$2i\omega(\partial\xi'/\partial t)_{\bar{x}} + (\bar{\omega}^2 - \omega^2)\xi' - \omega(d^2\bar{\omega}/dk^2)\partial^2\xi'/\partial\bar{x}^2 + $$
$$+ \beta|\xi'|^2|\xi' = 0 \tag{7}$$

where the nonlinear coefficient β in the deep water limit is ω^2k^2. The stationary localized solution to this equation is:

$$\xi = b_o\operatorname{sech} b_1\bar{x} \tag{8}$$

where $b_1^2 = (\bar{\omega}^2 - \omega^2)/\omega d^2\omega/dk^2$

$b_o^2 = -2(\bar{\omega}^2 - \omega^2)/\beta k^2$

Consider next a channel which is driven in its first crosswise mode; the free surface will be

$$\xi = \xi'(x) \cos(\pi y/\ell) e^{i\omega t} \tag{9}$$

where ℓ is the width of the channel. If $\omega < \bar{\omega}(\pi/\ell)$ then ξ' will be an exponentially decreasing function of x. That is, according to the linear theory, the disturbance will be trapped at the entrance to the channel in much the same fashion as an edge wave is trapped to the shore. Note that the edge wave is trapped when its frequency of motion is less than $\bar{\omega}(k_m)$ where k_m is the wavenumber of modulation parallel to the edge.

Once again nonlinear processes make it possible for a soliton to exist, but its role is quite different. The linear solution of the form (9) is already localized; the nonlinear softening of the frequency makes it possible for the soliton to leave the edge and move to any location in the channel. Letting $\xi' = \xi'(x,t)$ the leading order equation is:[3)]

$$2i\omega\partial\xi'/\partial t - (\omega^2-\bar{\omega}^2)\xi' - v^2\partial^2\xi'/\partial x^2 - |\xi|^2\xi\omega^2 A\pi^2/8\ell^2T^2 = 0 \tag{10}$$

where $T = \mathrm{Tanh}\,\pi d/\ell$

$$v^2 = \frac{1}{2} g\left[d(1-T^2) + \ell T/\pi\right]$$

and the nonlinear coefficient is

$$A = 9T^{-2} - 16 + 5T^2 - 6T^4 \tag{11}$$

The solution to (10) is again of the form (8) but with $\bar{x}$ replaced by x as the group velocity can be zero. Actually the velocity of this soliton is a free parameter.

3. DISCUSSION

Although the envelope and nonpropagating solitons are both described by the NLS they are physically different states of fluid flow. The envelope soliton exists as a result of the nonlinear hardening of the frequency of a travelling wave in deep water. The non-propagating soliton exists as a result of the softening of the standing (crosswise) wave also in deep water; in fact from (11) we find that $A > 0$ when $T > .77$. Although we have emphasized the equilibrium properties of the KdV and NLS these solutions are stable as can be demonstrated by the inverse scattering transform. Since the nonpropagating soliton has internal motion on a fast time scale $\approx 1/\omega$ it is strongly coupled to by a parametric drive at a frequency 2ω [2)] in analogy with edge waves.

Observations[7] in the Andaman sea strongly suggest that solitons can be stable for very long periods of time in the ocean. These solitons are about 100 km long and their stability against transversal break up gives good cause to consider the possibility of the existence of solitons in higher dimension. The localization of energy is dramatic with their measured power being about 2000 megawatts. The Andaman sea solitons are probably some generalization of KdV type solitons. Whether envelope or nonpropagating solitons play a role in geophysical phenomena remains to be seen.

ACKNOWLEDGEMENT

It is a pleasure to thank the organizers of this Conference for making possible these stimulating exchanges.

REFERENCES

1. Proceedings Int'l School of Physics, Course LXXX, 'Topics in Ocean Physics,' ed. A.R. Osborne, P. Malanotte Rizzoli, 1982, North-Holland; H. Segur p.235.

2. J. Wu., R. Keolian, I. Rudnick, Phys. Rev. Lett. **52**. 442 (1984).

3. A. Larraza, S. Putterman, Phys. Lett. **103A**, 15 (1984); J. Fluid Mech. **148**, 443 (1984).

4. J.W. Miles, J. Fluid Mech. **148**, 450 (1984).

5. A. Larraza, S. Putterman; 'Universal Power Spectra for Wave Turbulence.'

6. H.C. Yuen, p. 205, loc. cit., Proceedings.

7. A.R. Osborne, T.L. Burch, p. 312, loc. cit. Proceedings.

UNIVERSAL POWER SPECTRA FOR WAVE TURBULENCE; APPLICATIONS TO WIND WAVES AND 1/f NOISE

A. Larraza and S. Putterman
Physics Department
University of California
Los Angeles, California 90024

ABSTRACT. Nonlinear processes lead to a scattering of waves by waves and can produce a stationary state power spectrum with self-similar properties. At lowest nonlinear order one obtains the results of weak wave turbulence theory. At infinite nonlinear order the power spectrum accumulates at 1/f noise for non-dispersive waves and $1/\omega^5$ noise for deep water waves. Effects resulting from a renormalization of the phase velocity are also discussed.

1. INTRODUCTION

Consider a medium which receives a large input of energy at some wave number $k_o = 1/\ell_o$. By large it is meant that effects due to the reversible nonlinear terms in the equations of motion dominate the damping due to the linear irreversible terms such as molecular viscosity (inertial motion dominates transport). Since nonlinearities create sum and difference, frequencies ω and wave numbers k, the medium will soon have motion present at higher harmonics of k_o given by pk_o where p is an integer.

Kolmogorov[1] in pioneering work considered energy in the form of rotational or vortex motion which was input to a fluid at a constant rate Q. He argued that in the steady state Q must be balanced by the rate at which nonlinearities cause the energy to cascade to shorter and shorter length scales or equivalently higher and higher harmonics. Using dimensional analysis to balance these rates he was led to a power spectrum of turbulent motion proportional to $k^{-5/3}$ for energy between k and k+dk and to $\omega^{-2}d\omega$ for energy between ω and $\omega+d\omega$. Experiments[2] on tidal motion support the Kolmogorov spectrum.

Spectra proportional to a power of the frequency are also found for other off equilibrium systems such as wind driven ocean waves in deep water and voltage fluctuations across a resistor through which an externally imposed current is driven. For ocean waves the power spectrum has for years been matched to the Phillips[3] spectrum proportional to ω^{-5}, and voltage fluctuations[4] follow the virtually omnipresent[5] 1/f noise law, that appears to characterize most systems in the limit of very low frequency.

C. Nicolis and G. Nicolis (eds.), Irreversible Phenomena and Dynamical Systems Analysis in Geosciences, 139–144.

Deep water waves, sound waves, Alfven waves, etc., differ from rotational motion in that wave number and frequency are connected by a dispersion law. For water waves $\omega^2 = gk$ and for sound waves $\omega = ck$ where c is the sound velocity. For these systems the Kolmogorov picture can still be applied but the resonant nature of the wave-wave interactions changes the stationary state power spectrum to which the system is driven by the nonlinearities. For water waves[6)] the leading order nonlinearities yield a power spectrum proportional to ω^{-4} whereas for sound[7)] the off equilibrium spectrum is $\omega^{-3/2}$.

As the strength of the external drive increases one expects that higher order nonlinearities become important in determining the steady state. These higher order terms modify the power spectrum and in the limit of infinite nonlinearity lead to a saturation phenomenon where the spectrum of deep water waves goes as ω^{-5}, and the spectrum of dispersionless waves (sound) goes as ω^{-1} [8)]. The weak and strong nonlinear limits are developed in Section 3. In section 2 the basic underlying nonlinear equations are discussed.

2. NONLINEAR FLUID MECHANICS

For a barotropic fluid conservation of mass and Newton's law take the form:

$$\partial\rho/\partial t + \vec{\nabla}\cdot\rho\vec{v} = 0 \tag{1}$$

$$\rho[\partial\vec{v}/\partial t + (\vec{v}\cdot\vec{\nabla})\vec{v}] = -\vec{\nabla}p(\rho) + \rho\vec{g} - \eta\vec{\nabla}\times\vec{\nabla}\times\vec{v} + (\tfrac{4}{3}\eta+\zeta)\vec{\nabla}(\vec{\nabla}\cdot\vec{v}) \tag{2}$$

where ρ, v, p are density, velocity and pressure and where the first and second viscosities η, ζ are assumed constant. The irreversible viscous forces are derivable from the divergence of a viscous stress tensor τ_{ij}

$$-\partial\tau_{ij}/\partial r_j = -\eta[\vec{\nabla}\times\vec{\nabla}\times\vec{v}]_i + (\tfrac{4}{3}\eta+\zeta)[\vec{\nabla}(\vec{\nabla}\cdot\vec{v})]_i \tag{3}$$

For incompressible flow ρ is constant and

$$\vec{\nabla}\cdot\vec{v} = 0 \tag{4}$$

The curl of (2) then yields

$$\partial(\vec{\nabla}\times\vec{v})/\partial t - \eta\nabla^2\vec{\nabla}\times\vec{v} = \vec{\nabla}\times[\vec{v}\times(\vec{\nabla}\times\vec{v})] \tag{5}$$

If in addition the flow is irrotational

$$\vec{\nabla}\times\vec{v} = 0 \text{ or } \vec{v} = \vec{\nabla}\phi \tag{6}$$

and (2) yields

$$\partial\phi/\partial t + (1/2)(\vec{\nabla}\phi)^2 + gz + (p/\rho) = \text{constant} \tag{7}$$

where we take gravity in the z direction. This form of Bernoulli's law should be supplemented by two boundary conditions at the free surface $z = \xi(x,y,t)$. One is the continuity of stress

$$p + \tau_{\perp\perp} + \sigma \left[\frac{\xi_{xx}(1+\xi_y^2) + \xi_{yy}(1+\xi_x^2) - 2\xi_{xy}\xi_x\xi_y}{(1 + \xi_x^2 + \xi_y^2)^{3/2}} \right] = p_o \tag{8}$$

where the subscripts "x,y" indicate partial differentiation and $\perp$ denotes the component perpendicular to the orientation of the free surface and p_o is the constant atmospheric pressure. The other condition states that particles on the surface move with the local fluid velocity

$$\partial\xi/\partial t + \vec{\nabla}\phi\cdot\vec{\nabla}\xi = \partial\phi/\partial z \tag{9}$$

For sound waves the motion is irrotational yet compressible. Neglecting g and taking the time derivative of (1) and subtracting the divergence of (2) yields

$$\partial^2\rho/\partial t^2 - \nabla^2 p = \partial^2(\rho v_i v_j + \tau_{ij})/\partial r_i \partial r_j \tag{10}$$

Expanding p in a taylor series

$$p = p_o + c^2\delta\rho + \tfrac{1}{2}(\partial^2 p/\partial\rho^2)_o(\delta\rho)^2 + \ldots$$

where $\rho = \rho_o + \delta\rho$ enables (10) to be rewritten in the form

$$\frac{\partial^2\delta\rho}{\partial t^2} - c^2\nabla^2\delta\rho = \frac{\partial^2}{\partial r_i \partial r_j}(\rho v_i v_j + \tau_{ij}) + \nabla^2 \sum_{\alpha=2}^{\infty} \frac{1}{\alpha!} \frac{\partial^\alpha p}{\partial\rho^\alpha}(\delta\rho)^\alpha \tag{11}$$

Rotational turbulence (the original Kolmogorov problem) should be characterized by (4,5). Compressible or acoustic wave turbulence should be characterized by (11) and turbulence in deep water waves should be described by (4,6,7,8,9). A solution of these nonlinear equations supplemented by some consistent stochastic hypothesis should yield the stationary power spectra. To our knowledge this procedure can only be carried out successfully in the case of wave turbulence where the underlying Boltzmann equation for the spectral intensity or wave action can be derived. The reason for the difference between wave turbulence and vortex turbulence is that in the former case there exists a dispersion law and hence an adiabatic invariant of the motion. Furthermore for acoustic turbulence one can consider an expansion in the mach number $\delta\rho/\rho_o$ where ρ_o is the equilibrium density. For the vortex turbulence there is no material property corresponding to ρ_o and hence the mach number for that case is always large.

The inertial regime occurs when the amplitudes are so large that τ_{ij} can be neglected compared to the nonlinear terms.

3. POWER SPECTRA FOR WAVE TURBULENCE

Although kinetic equations for wave turbulence can be derived we will follow here the simpler Kolmogorov dimensional approach and study the process whereby wave energy cascades from one length scale to the next. Label the properties of successive scales with subscript n so that we have for the wave numbers

$$k_n = 2^n/\ell_o \tag{12}$$

For sound waves the energy per unit volume on this length scale is:

$$E_n = (c^2/\rho_o)(\delta\rho_n)^2 \tag{13}$$

The key to the cascade argument is that the rate at which energy rolls over from one length scale to the next is a function of the energy contained in that length scale (locality). For sound waves the lack of dispersion implies that the basic nonlinear interaction is a three wave resonance so that waves with frequencies ω_1 and ω_2 scatter to produce waves with frequency $\omega_3 = \omega_1 \pm \omega_2$. To leading order this effect is produced by the term in (11) with $\alpha = 2$ and therefore yields a rollover time t_n for the wave energy given by

$$1/t_n \cong \omega_n G^2 E_n/\rho c^2 \tag{14}$$

where $G = 1 + (\rho/c)dc/d\rho$ is the macroscopic Gruneison coefficient (the symbol $\cong$ means equality except for a numeric). The stationary state then follows from setting the rollover rate equal to the input rate Q or

$$E_n/t_n = Q \tag{15}$$

The discrete stationary spectrum then is

$$E_n \cong [Q\rho c^2/\omega_n G^2]^{1/2} \tag{16}$$

so that the continuous power spectral density is

$$e(\omega) = [Q\rho c^2/G^2]^{1/2}/\omega^{3/2} \tag{17}$$

The energy per unit area of water waves is

$$U_n = \rho g \xi_n^2 \tag{18}$$

These waves differ from sound wave in that the strong downward dispersion requires that the leading interaction effect be produced by a four wave process. Thus instead of (14) one has

$$\frac{1}{t_n} \cong \omega_n k_n^4 \xi_n^4 \tag{19}$$

Seeking a steady state with

$$U_n/t_n = q \tag{20}$$

yields the power spectral density

$$u(\omega) \cong g^2[q\rho^2]^{1/3}/\omega^4 \tag{21}$$

For weak wave turbulence one finds (17) for acoustics and (21) for surface waves on deep water. For many years ocean waves have been described by the Phillips spectrum wherein

$$u(\omega) \cong \rho g^3/\omega^5 \tag{22}$$

which differs from (21) as regards the power of ω and its independence of the value of the energy input q. For this reason (22) is thought of as corresponding to some kind of saturation regime. Numerous technical causes for saturation have been proposed especially white-capping. We would now like to argue that the passage from (21) to (22) can be understood in terms of higher order nonlinearities, i.e. in terms of higher order wave processes.

Although (19) yields the rollover time for a four wave process the cascade time for an m wave process will be

$$1/t_n \cong \omega_n k_n^{2(m-2)} \xi_n^{2(m-2)}$$

which with (18,20) yields (22) in the limit $m \to \infty$.

This suggests the following picture for the transition from (21) to (22). As the power input to wave motion is increased higher nonlinear effects come into play and shift the exponent of ω in the steady state power spectrum. In practice one should observe a response somewhere between (21) and (22) depending upon which range of m dominates.

In the acoustic case an m wave (i.e. m phonon) process leads to

$$1/t_n \cong \omega_n G_m^2 (E_n/\rho c^2)^{m-2}$$

where G_m is determined by the nonlinear equation of state in (11). Now one finds

$$e_m(\omega) \cong \frac{\rho c^2}{\omega} \left[\frac{Q}{\rho c^2 \omega G_m^2} \right]^{1/(m-1)}$$

which in the limit of large m goes over to 1/f noise.[8)]

4. CONCLUSIONS

For nondispersive waves the redistribution of externally imposed energy by high order nonlinearities leads to 1/f noise. The type of nonlinear effect which saturates the nondispersive spectrum at 1/f noise also saturates the deep water spectrum at ω^{-5} noise (22). However, even in a dispersive medium the response at very low frequency can shift to 1/f noise due to the renormalization of the dispersion law. The elasticity of the energy contained in the spectrum can for sufficiently small k lead to a contribution to the dispersion law for ω that is linear in k and therefore nondispersive. If the external injection of energy takes place at or below this region 1/f noise will result even though the quiescent medium is dispersive. It appears that 1/f noise is as fundamental for off-equilibrium response as equipartition is to equilibrium.

Recently Phillips[9] has called into question use of the saturated spectrum (22). He implies that the fit of experimental data taken over many years to (22) rather than (21) may have been motivated by a lack of appreciation of the theory of energy balance of various processes in the weak turbulence limit. To this extent it should be emphasized that the generalization of Hasselmann's[10] kinetic equation to higher order nonlinear effects will yield spectra with exponents between those of (21) and (22).

ACKNOWLEDGMENT

Many thanks are due to A. Provenzale for critical comments aimed at elucidating an extension of the work on 1/f noise to other systems

REFERENCES

1. A.N. Kolmogorov, Dokl. Akad. Nauk. SSR **30**, 299 (1941) [Sov. Phys. Dokl. **10**, 734 (1968)].
2. H.L. Grant, R.W. Stewart, A. Moilliet; J. Fluid Mech. **12**, 241 (1962).
3. O.M. Phillips, Dynamics of the Upper Ocean, Cambridge University Press, 1977.
4. P. Dutta, P.M. Horn, Rev. Mod. Phys. **53**, 497 (1981).
5. W.H. Press, Comments Astrophys. Space Phys, **7**, 103 (1978).
6. R.Z. Sagdeev, Rev. Mod. Phys., **51**, 1 (1979).
7. R. Kraichnan, Phys. Fluids **11**, 266 (1968); V.E. Zakharov, R.Z. Sagdeev, Dokl. Akad. Nauk SSR, **192**, 297 (1970) [Sov. Phys. Dokl. **15**, 439 (1970)]
8. A. Larraza, S. Putterman, P.H. Roberts, Phys. Rev. Lett., **55**, 897 (1985).
9. O.M. Phillips, J. Fluid Mech. **156**, 505 (1985).
10. K. Hasselmann, J. Fluid Mech. **12**, 481 (1962); **15**, 273, 385 (1963); Proc. Roy. Soc. **A299**, 77 (1967).

Fluctuations and Dissipation in Problems of Geophysical Fluid Dynamics*

Katja Lindenberg, Bruce J. West** and J. Kottalam
Department of Chemistry, B-014
University of California
San Diego, CA 92093 U.S.A.

ABSTRACT. Truncated descriptions of geophysical flow fields are often used to model the most important macroscopic features of the atmosphere. Fluctuations are typically added to these descriptions as a heuristic way of including the effect of the degrees of freedom that have been omitted in the truncated description. For example, truncated geophysical hydrodynamic systems with multiple equilibria have been used as models of atmospheric blocking. The fluctuations phenomenologically added to these models provide a mechanism for the transition between equilibrium states. Using a systematic reduction procedure we *arrive* at such a model including the effects of fluctuations *and* dissipation arising from small scale structure in the atmosphere.

1. INTRODUCTION

A common characteristic of geophysical flow fields is the absence of a clear separation in the spatial and/or temporal scales of fluid motions. Thus, when the flow field is described by a Fourier decomposition, the hydrodynamic equations give rise to an infinite set of coupled nonlinear rate equations for the Fourier coefficients ψ_j. It is nevertheless a common procedure to separate out a set of a small number of "significant" modes which, it is hoped, adequately represent the macroscopic phenomenon of interest. The customary way of proceeding is to select a subset of modes, ignore all the other modes, and test whether this truncated description exhibits the macroscopic behavior of interest. Examples of this strategy can be found in the modeling of Bénard convection (Lorenz, 1963), fluid turbulence (Kraichnan and Montgomery, 1980), meteorological predictability (Lorenz, 1969) and atmospheric blocking (Charney and DeVore, 1979). In general the selected mode amplitudes which might correspond to potential vorticity, temperature, etc. are assumed to obey equations of the form

*Supported in part by NSF grants ATM 85-07820 and ATM-85-07821
**Center for Studies of Nonlinear Dynamics, La Jolla Institute, La Jolla, California, 92038.

C. Nicolis and G. Nicolis (eds.), Irreversible Phenomena and Dynamical Systems Analysis in Geosciences, 145–156.

$$\dot{\psi}_j = F_j(\{\psi_l\}) \quad ; \quad j, l = 1, 2, \cdots n \tag{1}$$

where the description has been truncated to n modes. The form of the function F_j is determined by the geophysical hydrodynamic equations.

Even when it is believed that a highly truncated subset of modes is an adequate representation of *average* (macroscopic) properties of the flow, it is clear that fluctuations around these averages give rise to uncertainties that one associates with small scale variability of the flow field, e.g. turbulence. Thus, although the scales of these fluctuations may be unresolved in global circulation models, their effects are certainly manifest on the scale of the resolved modes. It is generally recognized that small scale turbulence is at least in part a manifestation of the degrees of freedom that have been ignored in the truncated description.

A traditional way of dealing with fluctuations is to expand the truncated set (1) so as to include the effects of small scale turbulence by phenomenologically adding fluctuations with assumed statistical properties (Thompson, 1957; Landau and Lifshitz, 1959; Leith, 1971; Hasselmann, 1976; Holloway and Hendershott, 1977; Egger, 1982): $\dot{\psi}_j = F_j(\{\psi_l\}) + f_j(t)$. The $f_j(t)$ are usually taken to be stationary, zero-centered, Gaussian and delta-correlated random variables that are independent one from the other. The statistical properties of the $f_j(t)$ are assumed to be independent of the mode amplitudes. The above set of stochastic nonlinear differential equations is assumed to be a coarse-grained representation of the underlying primitive equations in which the effect of the modes $\psi_{l>n}$ on ψ_j is "represented" by the noise $f_j(t)$. Although this connection has been recognized, quantitative models of it had not previously been developed (West and Lindenberg, 1984a, 1984b; Lindenberg and West, 1984; Kottalam, West and Lindenberg, 1985).

We have established a research program to explore the quantitative connection between the observed fluctuations and the coarse graining procedure. In outline, this is how we have proceeded:

1. We begin the analysis with a set of primitive hydrodynamic equations describing a geophysical flow field. These equations could be as simple as the potential vorticity equations on a β−plane or they might include coupling of the vorticity to a temperature field, etc.

2. We transform the set of primitive equations to some appropriate modal description wherein the modal amplitudes ψ_j satisfy the set of coupled mode rate equations $\dot{\psi}_j = F_j(\{\psi_l\}), l, j = 1, 2, ..., N \rightarrow \infty$.

3. Instead of truncating this equation to obtain (1), as is often done, we systematically eliminate the unresolved modes l, j = n+1, n+2, ..., N by explicit integration. The solutions thereby obtained allow us to derive a stochastic differential equation for the observed modes of the form

$$\dot{\psi}_j = \mathrm{F}_j(\{\psi_l\}) + \textit{fluctuations} + \textit{dissipation}, \quad l, j = 1, 2, \cdots, N \tag{2}$$

with functional forms for the fluctuations and dissipation that depend on the particular problem being analyzed, i.e. on the functional form of F_j.

4. Since the fluctuations and dissipation arise from the same eliminated modes, one would expect them to be related to one another. We establish this relation as a natural consequence of the coarse-graining procedure.

2. A GEOPHYSICAL FLOW FIELD

Although quite general, it is simpler to discuss the consequences of the coarse-graining procedure in the context of a particular geophysical problem. A problem of topical interest is that of atmospheric blocking (Baur, Hess and Nagel, 1944, 1951; Holloway and West, 1984) and seems to be a natural vehicle for presenting our ideas.

A model of this phenomenon introduced by Charney and DeVore (1979) is that of a homogeneous β−plane barotropic atmosphere confined between zonal walls. We accept this model without comment as to its geophysical validity since our intent is to illustrate our approach to any such model. It may turn out subsequently that our coarse graining procedure will enable us to comment on the physical merits of such modeling, but we defer any such comments until the future. We therefore turn to the primitive equation used by Charney and DeVore, that for the conservation of potential vorticity. The stream function $\psi(x,y,t)$ satisfies the equation (Pedlosky, 1982)

$$\frac{\partial}{\partial t}(\nabla^2-g^2)\,\psi + J\,(\psi,\nabla^2\psi+h) + \beta\frac{\partial}{\partial x}\psi + \gamma\nabla^2(\psi-\psi^*) = 0\,. \tag{3}$$

Here J is the Jacobian determinant, β is the Coriolis parameter, g is a parameter containing the acceleration due to gravity, γ is proportional to the Ekman depth, h (x,y) is the effect of the lower boundary elevation in the β−plane, and the last term arises from the frictionally-induced time-independent vorticity source $\gamma\nabla^2\psi^*(x,y)$ and vorticity sink $\gamma\nabla^2\psi(x,y,t)$.

A modal decomposition of $\psi(x,y,t)$, $h\,(x,y)$ and $\psi^*(x,y)$ subject to appropriate boundary conditions in the channel lead to the Fourier series $\psi(x,y,t) = \sum_k \psi_k(t)\,G_k(x,y)$ where the G_k are trigonometric functions. In terms of the amplitudes $\psi_k(t)$, the conservation equation (3) is re-expressed as

$$\dot{\psi}_k(t) = \sum_q a_{kq}\,(\psi_q(t)-\psi_q^*) + \sum_{q,q'} b_{kqq'}\,\psi_q(t)\,\psi_{q'}(t) \tag{4}$$

where the a_{kq} and $b_{kqq'}$ are coefficients determined by the form (3) (Kottalam, West and Lindenberg, 1985). In the Charney and DeVore (1979) model one truncates this system by retaining only the modes $G_A = \cos y$, $G_K = \cos nx \sin y$ and $G_L = \sin nx \sin y$, where n is determined by the topography. All other modes are ignored. The resulting set of equations for $\dot{\psi}_A$, $\dot{\psi}_K$ and $\dot{\psi}_L$ is of the form

$$\dot{\psi}_j = F_j(\psi_A, \psi_K, \psi_L)\;, \quad j = A, K, L \tag{5}$$

and is the minimal set giving rise to multiple equilibria of the flow field. Thus, in the phase space spanned by ψ_A, ψ_K and ψ_L one finds three solutions to the steady state equations $F_j = 0$, $j = A, K, L$ (cf. Fig. 1). Two of these are stable while one is unstable. Any initial condition (excluding the unstable state) is in the basin of attraction of either one or the other of the stable fixed points, and will therefore approach that state asymptotically.

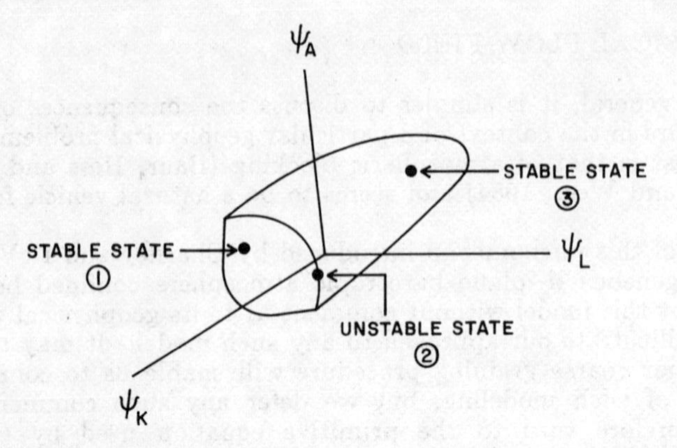

Figure 1. Phase space for the Charney-De Vore model of atmospheric blocking. The dots are the solution to the steady state equations $\dot{\psi}_j = 0$.

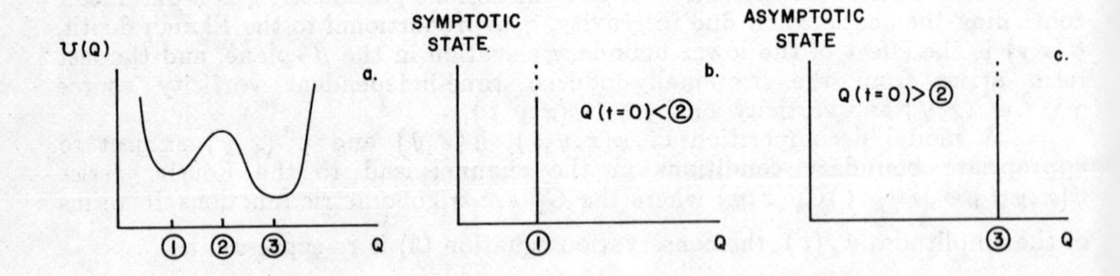

Figure 2. One-dimensional bistable system. a. Bimodal potential, with stable states at (1) and (3) and an unstable state at (2). b. Asymptotic state for any initial condition to the left of state (2). c. Asymptotic state for any initial condition to the right of state (2).

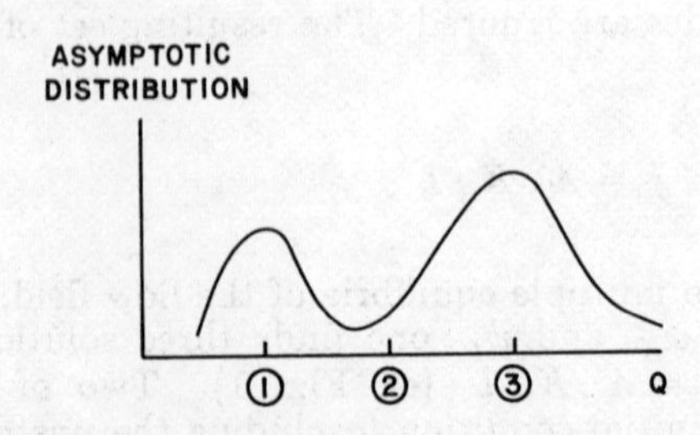

Figure 3. Asymptotic distribution in the presence of fluctuations.

An analogical description of this behavior can be obtained from a one-dimensional system with a bimodal potential $U(Q)$ such as shown in Fig. 2a. The system evolves according to the rate equation $\dot{Q} = U'(Q)$ and has two stable and one unstable fixed points. Any initial condition to the left of the unstable fixed point will asymptotically coalesce with state 1 (cf. Fig. 2b), while starting at the right of this point will lead to state 3 (cf Fig. 2c). Figures 2b and 2c represent the phase space distribution function for the asymptotic trajectory of the system; this distribution is a delta function for this (as for any) deterministic model with isolated fixed points.

Deterministic models with multiple equilibria such as those discussed above may mimic the occurrence of various possible atmospheric blocking states, but they predict that such states once achieved will persist forever. The finite lifetimes of such states in geophysical systems motivated Egger (1982) to introduce phenomenological fluctuations in the truncated equations, thereby making the lifetime of these states finite. He reasoned that the ignored modes act as a source of fluctuations that can induce transitions between the basins of the stable fixed points. Egger augmented Eqs. (5) by adding zero-centered, mutually independent, Gaussian delta-correlated fluctuations to each equation:

$$\dot{\psi}_j = F_j + f_j(t), \quad j = A, K, L \quad . \tag{6}$$

The strengths D_j of the fluctuations in the averages $< f_i(t) f_j(\tau) > = 2 D_j \, \delta(t-\tau)\delta_{ij}$ were determined by heuristic reasoning. Our analogic equation of motion is now replaced by $\dot{Q} = U'(Q) + f(t)$ and the asymptotic distributions appropriate to different initial conditions (Figs. 2b, 2c) now merge and broaden as shown in Fig. 3. The width of each peak is proportional to the strength D of the fluctuations $f(t)$ and is a measure of the rate at which the system makes a transition between stable states. In the blocking model these effects generalize to multivariable probability distributions peaked at the stable states 1 and 3 and of widths determined by the D_j's. The distribution has a global minimum at the unstable state 2.

Variants of Egger's treatment have been considered by other authors. For example, Benzi, Hansen and Sutera (1984) and also Moritz (1984) consider the case where only the flow parallel to the channel is subject to fluctuations, i.e. $f_K(t) = f_L(t) = 0$. This choice is made for computational convenience. In addition to the delta-correlated fluctuations mentioned above, they consider the effects of colored noise as embodied in a finite correlation time τ_A, $< f_A(t) f_A(\tau) > = (D_A/\tau_A) \exp[-|t-\tau|/\tau_A]$.

We note that the above procedure has a distinguished lineage. It is in the spirit of the phenomenological approach of Landau and Lifshitz (1959) whereby they introduced fluctuations in the stress tensor into the Navier-Stokes equation and fluctuations in the heat flux in the internal energy equation. This tradition was carried on by Thompson (1972) in his use of fluctuations in the modeling of two-dimensional vorticity in the atmosphere.

3. FLUCTUATIONS AND DISSIPATION IN REDUCED DESCRIPTIONS

It is worthwhile to recall that there exist a variety of simple model systems in the physical and chemical literature in which the origins, nature, and role of fluctuations and dissipation and the relations between them are well understood. We mention here two such examples with a view towards anticipating these effects in more complicated geophysical flow fields.

The first example is that of a "Brownian" oscillator in a thermal environment at temperature T (Ford, Kac and Mazur, 1965). Let $Q(t)$ denote the displacement and $P(t)$ the momentum of the oscillator. When isolated, such an oscillator is described by the equation $\ddot{Q}(t) = -\omega_o^2 Q(t)$ where ω_o is its natural frequency. In a thermal environment the dynamical equation becomes

$$\ddot{Q}(t) = -\omega^2 Q(t) - \int_0^t d\tau\, K(t-\tau)\, \dot{Q}(\tau) + f(t) \quad . \tag{7}$$

Here ω is a frequency shifted from ω_o by environmental effects. The integral term is a dissipative force representing the irreversible extraction of energy from the oscillator by the environment. If the environment is a fluid then the dissipative memory kernel can be associated with the fluid viscosity (Reichl, 1980). The state-independent or additive force $f(t)$ represents the rapid thermal fluctuations whereby the oscillator acquires energy from the environment. If the system is to be thermodynamically closed, then the fluctuations and the dissipation must be related to one another by the generalized fluctuation-dissipation relation (FDR) (Callen and Welton, 1951; Ford, Kac and Mazur, 1965; Lax, 1966; Reichl, 1980) $k_B T K(t-\tau) = \langle f(t) f(\tau) \rangle$ that represents the balance between the influx and the efflux of energy. Here k_B is Boltzmann's constant. This balance of energy fluxes insures that asymptotically the phase space distribution attains equilibrium, $W(Q,P) \propto \exp[-(P^2 + \omega^2 Q^2)/ 2 k_B T]$. If the frequency spectrum of the environment is sufficiently broad band then the dissipation is instantaneous, $K(t-\tau) = 2\lambda\, \delta(t-\tau)$, and (7) reduces to the more familiar form

$$\ddot{Q}(t) = \omega^2 Q(t) - \lambda \dot{Q}(t) + f(t) \tag{8}$$

together with the FDR $\langle f(t) f(\tau) \rangle = 2 k_B T \lambda\, \delta(t-\tau)$, i.e. the fluctuations are then delta-correlated in time.

A more complicated situation arises when the environment induces rapid fluctuations in the oscillator frequency. For a thermodynamically closed system it has been shown that the equation of motion replacing (7) now is (Zwanzig, 1973; Lindenberg and Seshadri, 1981)

$$\ddot{Q}(t) = -\omega^2(t)\, Q(t) - \int_0^t d\tau\, K(t-\tau) Q(t) Q(t-\tau) \dot{Q}(\tau) \quad . \tag{9}$$

The time-dependent squared frequency can be decomposed as $\omega^2(t) = \omega^2 + \gamma(t)$ where the parametric or multiplicative fluctuations $\gamma(t)$ are zero-centered. The fluctuations in (9) are state-dependent, being amplified or suppressed depending on the oscillator displacement. The dissipative term is now nonlinear, and the memory kernel is again related to the fluctuations by the generalized FDR $\langle \gamma(t) \gamma(\tau) \rangle = k_B T K(t-\tau)$. We note that phenomenological equations with such multiplicative fluctuations, but with linear dissipation, may be energetically unstable in some parameter regimes (Bourret, Frisch and Pouquet, 1973;

Lindenberg, Seshadri and West, 1980; Seshadri, West and Lindenberg, 1981). A third example we mention to establish a clear distinction from the other two is that of a system with a statistical but time-independent parameter (van Kampen, 1981). For instance, consider an ensemble of oscillators each of which has a constant frequency whose value varies randomly from member to member of the ensemble. The equation of motion for the *ith* member of the ensemble is

$$\ddot{Q}(t) = -\omega_i^2 Q(t) \quad . \tag{10}$$

This deterministic equation contains no dissipative contributions because it represents an oscillator in a frozen (i.e. unresponsive) random medium.

The examples sketched above have been derived starting from complete dynamical Hamiltonian descriptions of the system *and* the environment (Ford, Kac and Mazur, 1965; Zwanzig, 1973; Lindenberg and Seshadri, 1981). The reduced descriptions (7), (9) and (10) are a result of the elimination procedure alluded to in the last section. In the same way, we propose to arrive at reduced descriptions for geophysical flow fields starting from primitive equations in which the modes to be eliminated constitute the environment. The following questions immediately present themselves:

Are the fluctuations additive or multiplicative, rapid or frozen or combinations of these?

What is the form of the dissipative contributions to be associated with the fluctuations? Are there FDR's?

These questions are answered if one starts from the full set of hydrodynamic equations (4) and carries out the elimination procedure. Rather than carrying out this analysis here, we choose to illustrate its salient features using a simple linear set.

Let $X(t)$ represent a set of observables and $Y(t)$ a set of unobservables, e.g. unresolved hydrodynamic modes. Suppose that these variables satisfy the linear rate equations

$$\dot{X} = AX + BY, \qquad \dot{Y} = CX + EY \tag{11}$$

where A, B, C and E are constant matrices. A direct truncation of this system would yield the equation of motion $\dot{X} = AX$, whereas the phenomenological stochastic model would yield $\dot{X} = A'X + f(t)$. In the former the effect of BY has simply been excluded while in the latter it has been parameterized by modifying the elements of A (e.g. to include a dissipation) and adding the fluctuating vector $f(t)$.

The systematic procedure begins with the explicit solution of $\dot{Y}$:

$$Y(t) = e^{Ct}\, Y(0) + \int_0^t d\tau\, e^{C(t-\tau)} EX(\tau) \quad , \tag{12}$$

where $Y(0)$ is the initial vector of unobservables. To arrive at a consistent interpretation of the terms in the reduced description it is often necessary to write Eq. (12) in the equivalent form obtained by an integration by parts (Lindenberg and Seshadri, 1981; Lindenberg and West, 1984b),

$$Y(t)+C^{-1}EX(t) = e^{Ct}[Y(0)+C^{-1}EX(0)]+\int_0^t d\tau\, e^{C(t-\tau)}C^{-1}E\dot{X}(\tau)\,. \quad (13)$$

Substituting this latter equation into $\dot{X}$ yields

$$\dot{X}(t) = A'\; X(t) + f(t) - \int_0^t d\tau\, K(t-\tau)\,\dot{X}(\tau) \quad (14)$$

where $A' = A - BC^{-1}E$, $f(t) = Be^{Ct}[Y(0)+C^{-1}EX(0)]$, and $K(t-\tau) = -Be^{Ct(t-\tau)}C^{-1}E$. Equation (14) clearly has the *form* of a generalized Langevin equation [cf. Eq. (7)] provided that one can interpret the various terms appropriately.

The matrix A' differs from A by the shift $BC^{-1}E$ and thus represents a modified potential function due to the interaction with the unobserved degrees of freedom. The function $f(t)$ depends on the initial values of both observed and unobserved variables. Whereas the former can be specified, the latter by definition can not. Said differently, for any specified initial state of the observed system, there is a large number of possible initial states for the environment that are consistent with the dynamic equations. This uncertainty in the value of $Y(0)$ is made manifest through the specification of an ensemble distribution function $W(Y(0) = y)$. Consequently $f(t)$ is also uncertain and hence specified only via this distribution. This is precisely the sense in which a function is said to be stochastic in statistical mechanics: its value at each time varies randomly from one member of the ensemble to another. We note that in addition it may happen that each realization $f(t)$ varies erratically with time (e.g. $f(t)$ may be ergodic). Whether or not it does so depends on the spectrum of C and the weights contained in B.

Knowledge of $W(Y(0) = y)$ is sufficient (but not always necessary) to determine the statistical properties of $f(t)$. In particular, if the elements of $Y(0)$ are independent of one another and if B couples many of them to X then the Central Limit Theorem can be invoked to deduce that $f(t)$ is Gaussian. Its mean value is determined by that of $Y(0)$ and its correlation properties by the second moments of $W(Y(0) = y)$. If we write $<[Y(0) + C^{-1}EX(0)][Y(0) + C^{-1}EX(0)]^{+}> = Q^{-1}$ where Q is the hermetian correlation matrix then $<f(t)f^{+}(\tau)> = Be^{Ct}Q^{-1}e^{C^{+}\tau}B^{+}$. If furthermore C is antihermetian so that $Q^{-1}C^{+} = -CQ^{-1}$ then $<f(t)f^{+}(\tau)> = -Be^{C(t-\tau)}Q^{-1}B^{+}$ and therefore the fluctuations are stationary. We also note that the matrix of correlation functions is related to memory kernel matrix by $<f(t)f^{+}(\tau)> = K(t-\tau)M$ where $M = E^{-1}CQ^{-1}B^{+}$. If $K(t-\tau)$ is dissipative, then this is a *generalized FDR* that emerges naturally from our elimination procedure. If $M = k_B T\, 1$ (where 1 is the unit matrix) then this reduces to the familiar relation in thermodynamically closed systems as discussed in the previous section. We note that the elements of the matrix M in general provide a relative measure of the excitation of the unobserved degrees of freedom. Finally, the conditions on the elements of B, C, and E necessary to insure that $K(t-\tau)$ is indeed dissipative (i.e. positive definite) have been studied by a number of investigators (Ford, Kac and Mazur, 1965; Zwanzig, 1973; Haken, 1975, 1978; Lindenberg and Seshadri, 1981).

It is important to recall the integration by parts that lead to the form (14) in which the dissipation is proportional to $\dot{X}$ rather than X. Had we not performed this integration, the "dissipative" term would have been proportional to $X(\tau)$ and to a kernel $\overline{K}(t-\tau) = -Be^{C(t-\tau)}E$. We have elsewhere discussed the reasons why this latter form is inappropriate and, in particular, why $\overline{K}(t-\tau)$ and $< f(t) f^{+}(\tau)>$ are not related to one another by a proper FDR (Lindenberg and West, 1984b). In fact, we have shown that the convergence of one often implies the divergence of the other.

Finally, we note that an unresponsive (frozen) environment corresponds to the case $C = E = 0$. In this case $Y(t) = Y(0)$ and $\dot{X}(t) = AX + BY(0)$. Although $BY(0)$ is a random matrix, it is constant in time and has no associated dissipation.

In summary, our elimination procedure consists of the following steps:

1. Separate the dynamical equations into those of the system and those of the environment using separation of scales where possible.
2. Eliminate the degrees of freedom of the environment by explicitly solving their equations of motion. This could be done exactly for the simple linear example but must in general be done perturbatively for nonlinear systems.
3. Choose an ensemble of initial states for the environment.
4. Identify fluctuations by their dependence on the initial state of the environment. Identify the corresponding dissipation (if appropriate).
5. Verify the existence of a generalized FDR. Note that this relation implies that any perturbative solution to the equations of motion of the environment cannot be uniform: if fluctuations are retained to a given order in the expansion parameter ϵ, e.g. $0(\epsilon)$, then the dissipation must be retained to twice that order, e.g. $0(\epsilon^2)$ (Lindenberg and West, 1984a; Kottalam, West and Lindenberg, 1985; Cortés, West and Lindenberg, 1985).

The result of these steps is then the desired generalized Langevin equation for the physical observables.

4. APPLICATION TO ATMOSPHERIC BLOCKING

Let us now return to the geophysical model introduced in Section 2 and let us make the identifications $X \to \psi_A, \psi_K, \psi_L$, $Y \to$ all other modes. The equations of motion for the modes to be eliminated are, of course, highly nonlinear in this case. These modes describe the dynamics of the smaller spatial scales and oscillate rapidly when a mean zonal flow is present (Pedlosky, 1982). The results of our analysis lead to the following equations (Kottalam, West and Lindenberg, 1985):

$$\dot{\psi}_A = F_A(\psi_A, \psi_K, \psi_L) + f_A(t) - \int_0^t d\tau\, K_{AA}(t-\tau)\, \dot{\psi}_A(\tau) + e^{-\gamma t} u_A^o \,, \quad (15a)$$

$$\dot{\psi}_K = F_K(\psi_A, \psi_K, \psi_L) + f_K(t) - \int_0^t d\tau\, K_{KK}(t-\tau)\, \dot{\psi}_K(\tau) \,, \quad (15b)$$

$$\dot{\psi}_L = F_L(\psi_A, \psi_K, \psi_L) + f_L(t) - \int_0^t d\tau\, K_{LL}(t-\tau)\, \dot{\psi}_L(\tau) + e^{-\gamma t} u_L^o \,. \quad (15c)$$

Explicit forms for all the functions appearing in these equations in terms of the coefficients in the primitive equation (4) are available but will not be reproduced here. Rather, we describe the general features of these results.

1. The functions $f_j(t)$, j = A, K, L are rapid fluctuations that arise from the modes that have been eliminated. The statistical properties of these fluctuations are quite similar to those assumed in phenomenological models. In particular, we have been able to prove that the rapid fluctuations in the zonal flow, $f_A(t)$, are indeed stronger than those in the other components, $f_K(t)$ and $f_L(t)$, as assumed by Benzi *et al.* (1984) and by Moritz (1984).

2. The integral terms containing the memory kernels $K_{jj}(t-\tau)$ are the "dissipative" contributions that balance the rapid fluctuations. The quotation marks are used to indicate that we have not established that the kernels are positive definite. We have, however, established that the kernels are related to the correlation functions $< f_j(t) f_j(\tau) >$ by generalized FDR's. None of the phenomenological models recognize the necessity of including these dissipative terms to balance the corresponding additive fluctuating terms.

3. The quantities u_j^o are time-independent random quantities that arise from the orography. Their effects decay slowly and were not included in previous analyses. These frozen fluctuations have no associated dissipation as discussed in section 3.

We have thus arrived at a new model equation for the study of atmospheric blocking. The equation has the same multiple equilibria as has the model of Charney and DeVore (1979). However, the transition between these states may exhibit different characteristics than those predicted by Egger (1982) and others (Benzi, Hansen and Sutera, 1984; Moritz, 1984). One is immediately faced with the problem of constructing new measures of persistence of blocking events. In the past it has been adequate to apply existing first passage time ideas to the calculations of the transition between steady states (Benzi *et al.* 1984; Moritz, 1984). All of these ideas rely on the stochastic differential equations being time-local, ergodic and stationary. Thus, the nonstationarity, nonergodicity and non-locality in time in our stochastic blocking model requires the development of new ancilliary techniques.

5. REFERENCES

Baur, F., P. Hess and H. Nagel, 1944: 'Kalendar der Grosswetterlagen Europas 1881-1069. Bad Homburg v.d.H.

______, 1951: 'Extended range weather forecasting. *Compendium of Meteorology, Amer. Meteor. Soc.,* 814-833.

Benzi, R., A. R. Hansen and A. Sutera, 1984: 'On the stochastic perturbation of simple blocking models'. *Q. J. Roy. Meteor. Soc.,* **110**, 393-409.

Bourret, R. C., U. Frisch and A. Pouquet, 1973: 'Brownian motion of harmonic oscillator with stochastic frequency', *Physica* **65**, 303-320.

Callen, H. B., and T. A. Welton, 1951: 'Irreversibility and generalized noise'. *Phys. Rev.,* **83**, 34-40.

Charney, J. G., and J. G. De Vore, 1979: 'Multiple flow equilibria in the atmosphere and blocking'. *J. Atmos. Sci.,* **36**, 1205-1216.

Cortés, E., B. J. West and K. Lindenberg, 1985: 'On the generalized Langevin equation: Classical and quantum mechanical'. *J. Chem. Phys.,* **82**, 2708-2717.

Egger, J., 1982: 'Stochastically driven large scale circulations with multiple equilibria'. *J. Atmos Sci.*, **38**, 2606-2618.

Ford, G. W., M. Kac and P. Mazur, 1965: 'Statistical mechanics of assemblies of coupled oscillators'. *J. Math. Phys.*, **6**, 504-515.

Haken, H., 1975: 'Cooperative phenomena in systems far from thermal equilibrium and in nonphysical systems'. *Rev. Mod. Phys.*, **47**, 67-121.

_____, 1978: *Synergetics: An Introduction.* Springer-Verlag, 147-158.

Hasselmann, K., 1976: 'Stochastic climate models. Part I. Theory'. *Tellus*, **28**, 473-484.

Holloway, G., and M. C. Hendershott, 1977: 'Stochastic closure for nonlinear Rossby waves'. *J. Fluid Mech.*, **83**, 747-765.

_____, and B. J. West, Eds., 1984: *Predictability of Fluid Motions.* American Institute of Physics Conference Proceedings, Vol. **106**, 612 pp.

Kottalam, J., B. J. West and K. Lindenberg, 1985: 'Fluctuations and dissipation in multiple flow equilibria'. (Preprint).

Kraichnan, R. H., and D. Montgomery, 1980: 'Two-dimensional turbulence'. *Rep. Prog. Phys.*, **43**, 547-619.

Landau, L. D., and E. M. Lifshitz, 1959: *Fluid Mechanics.* Pergamon, 523-529.

Lax, M., 1966: 'Classical noise IV: Langevin methods'. *Rev. Mod. Phys.*, **38**, 541-566.

Leith, C. E., 1971: 'Atmospheric predictability and two-dimensional turbulence'. *J. Atmos. Sci.*, **28**, 145-161.

Lindenberg, K., V. Seshadri and B. J. West, 1980: 'Brownian motion of harmonic systems with fluctuating parameters. II. Relation between moment instabilities and parametric resonance'. *Phys. Rev. A* **22**, 2171-2179.

_____, and V. Seshadri, 1981: 'Dissipative contributions of internal multiplicative noise. I. Mechanical oscillator'. *Physica*, **109A**, 483-499.

_____, and B. J. West, 1984a: 'Fluctuations and dissipation in a barotropic flow field, *J. Atmos. Sci.*, **41**, 3021-3031.

_____, 1984b: 'Statistical properties of quantum systems: The linear oscillator. *Phys. Rev. A*, **30**, 568-582.

Lorenz, E. N., 1963: 'Deterministic nonperiodic flow'. *J. Atmos. Sci.*, **20**, 130-141.

_____, 1969: 'The predictability of a flow which possesses many scales of motion'. *Tellus*, **21**, 289-307.

Moritz, R. E., 1984: 'Predictability and almost intransitivity in a barotropic blocking model'. *Predictability of Fluid Motions.* G. Holloway and B. J. West, Eds., American Institute of Physics Conference Proceedings, Vol. **106**, 419-439.

Pedlosky, J., 1982: *Geophysical Fluid Dynamics.* Springer-Verlag, 108.

Reichl, L. E., 1980: *A Modern Course in Statistical Physics.* University of Texas Press, 545-595.

Seshadri, V., B. J. West and K. Lindenberg, 1981: 'Stability properties of nonlinear systems with fluctuating parameters'. *Physica*, **107A**, 219-240.

Thompson, P. D., 1957: 'Uncertainty of initial state as a factor in the predictability of large scale atmospheric flow patterns'. *Tellus,* **9**, 275-295.

_____, 1972: 'Some exact statistics of two-dimensional viscous flow with random forcing'. *J. Fluid Mech.,* **55**, 711-717.

van Kampen, N. G., 1981: *Stochastic Processes in Physics and Chemistry.* North Holland, 419 pp.

West, B. J., and K. Lindenberg, 1984a: 'Comments on statistical measures of predictability'. *Predictability of Fluid Motions.* G. Holloway and B. J. West, Eds. American Institute of Physics Conference Proceedings, Vol. **106**, 45-53.

_____, 1984b: 'Nonlinear fluctuation-dissipation relations'. *Fluctuations and Sensitivity in Nonequilibrium Systems.* W. Horsthemke and D. K. Kondepudi, Eds., Springer Proceedings in Physics 1, Springer-Verlag, 233-241.

Zwanzig, R., 1973: 'Nonlinear generalized Langevin equations'. *J. Stat. Phys.,* **9**, 215-220.

PART III

ATMOSPHERIC DYNAMICS

DETERMINISTIC AND STOCHASTIC ASPECTS OF ATMOSPHERIC DYNAMICS

Edward N. Lorenz
Center for Meteorology and Physical Oceanography
Massachusetts Institute of Technology
Cambridge, Massachusetts, U.S.A.

ABSTRACT. We raise the question as to whether the atmosphere should be treated as deterministic or stochastic, for the purpose of investigating atmospheric dynamics most effectively. Because the atmospheric equations are nonlinear, all but special solutions must be sought numerically. The range of scales which numerical models can handle explicitly is limited, and the influence of smaller scales must be introduced through parameterization. The most realistic parameterizations contain stochastic terms in addition to the deterministic ones. However, since realistic atmospheric models ordinarily possess aperiodic general solutions with or without their stochastic terms, they tend to yield similar results in either event. The choice between a deterministic and a stochastic formulation of the equations can therefore be dictated by convenience.

1. INTRODUCTION

Among the many questions which have inspired considerable debate among meteorologists, one in particular has also attracted some prominent mathematicians: Should the weather be treated as a deterministic or a stochastic process, for the purpose of making the best attainable weather forecasts? The differences of opinion have led to the development of two rather different objective methods of weather forecasting, popularly known as numerical and statistical weather prediction. In the former method one attempts to predict future atmospheric states by integrating formally deterministic systems of differential or integro-differential equations which represent the governing physical laws, using observed values of atmospheric variables as initial conditions. In the latter one attempts to establish formulas which minimize the expected mean-square error in prediction, using observations of past weather to determine the numerical values of the coefficients in the formulas. Former champions of the two methods include John von Neumann and Norbert Wiener [1].

One should not conclude that practitioners of numerical weather prediction believe that the atmosphere is deterministic. The assumption

C. Nicolis and G. Nicolis (eds.), Irreversible Phenomena and Dynamical Systems Analysis in Geosciences, 159–179.

is simply that, despite any possible randomness, a formally deterministic approach will produce acceptable forecasts. Likewise, the use of statistical weather prediction does not presuppose any randomness. The assumption is simply that the laws, even if deterministic, may not be perfectly known or may be too difficult to apply, and that empirical procedures offer an acceptable alternative.

It might be added that the equations generally used in numerical weather prediction, even though formally deterministic, are not derived exclusively from the physical laws, but contain some empirically determined functions and coefficients. Likewise, the selection of predictors to be used in a statistical forecasting scheme is often guided by a knowledge of the physical laws.

Since we shall be dealing with the general topic of atmospheric dynamics, let us pose the following more general question: Should the atmosphere be treated as deterministic or stochastic, for the purpose of investigating atmospheric dynamics most effectively? Our ensuing discussion will be directed toward reaching a suitable answer to this question.

We must immediately note that the system in whose deterministic or stochastic nature we are interested is actually not restricted to the atmosphere, but includes also those portions of the underlying oceans and continents which influence the atmosphere, and which in turn are significantly influenced by the atmosphere. It therefore includes at least the upper layers of the ocean and land, and the sea ice and continental snow and ice cover and soil moisture. We shall nevertheless find it convenient to refer to this system as the atmosphere.

We should also point out that neither question is equivalent to asking whether the atmosphere actually is deterministic or stochastic. We shall not dwell at length on the determinism of the atmosphere, and simply note that it is influenced to some extent by human activity, particularly when that activity consists of clearing large forests or building dams to create large lakes. Even on rather short time scales, intentional or inadvertent weather modification through cloud seeding or setting large fires sometimes occurs. Any claim that the future of the atmosphere is predetermined would therefore imply a claim that human activity is predetermined. However, our concern in this discussion is not whether the behavior of the atmosphere involves some randomness, but whether it is important to take any such randomness into account.

2. OBSERVATIONS AND PHYSICAL LAWS

To deal effectively with the dynamics of any time-dependent system, whether it is a spiral galaxy, a planetary atmosphere, a glacier, or a small waterfall, we need a set of observations and a set of governing physical laws. Observations are needed first of all to make us aware of the system's existence, and subsequently with some degree of precision to reveal the system's typical structure and behavior. The goal of dynamical studies is to explain the observed features in terms of the physical laws, and sometimes to anticipate or predict additional features which have not yet been observed.

At this point we may ask why the roles of observations and laws should not be reversed, i.e., why the goal of dynamical studies should not be to deduce the physical laws from the observations. The answer is that this might well be the goal in dealing with certain systems. In historical times the discovery of the laws of motion was facilitated by observations of the motions of the planets. In the case of present-day studies of systems like the atmosphere, we can still deduce rules, some of which may be useful for weather forecasting or other practical tasks, but we tend to think of the physical laws are something more basic than specialized rules; possibly an extensive set of rules could be analyzed into basic laws. We also assume that the most basic laws--the laws of motion and thermodynamics--are already known with sufficient precision.

Turning to the atmosphere, let us enumerate a few observed features. Some of the qualitative features are revealed by casual observation; quantitative measurements may require sophisticated instrumentation.

First, the atmosphere consists mostly of a gas, with small amounts of liquid and solid matter. Readily noticed properties are the wind and the temperature; these vary from one location to another, and at any location they vary from one time to another. The pressure and density vary similarly, although the changes at one elevation might go unnoticed in the absence of instruments. Most of the gaseous constituents occur in nearly constant proportions, the most notable exception being water vapor. At high levels, variations in ozone content are significant, while over long periods the carbon dioxide content appears to undergo progressive changes. Liquid and solid water occur in the form of droplets and small crystals which are suspended as clouds, and larger drops and flakes which fall out as precipitation. Dust and other solid matter also occur in variable concentrations. The state of the atmosphere may be expressed in terms of the spatial and temporal distributions of wind components, temperature, pressure, density, mixing ratios of the various phases of water, and concentrations of other substances such as dust.

Observations also show that the atmospheric variables are not randomly distributed, but that certain spatial and temporal distributions are highly favored over others, so that the atmosphere tends to be organized into identifiable structures, each having a typical size and shape, and life span and life history. A partial listing of these structures, arranged in order of decreasing size, could include circumpolar westerly wind belts, migratory extratropical cyclones, tropical cyclones, squall lines, thunderstorms, fair-weather cumulus clouds, tornado funnels, individual wind gusts, hailstones, snow crystals, and cloud droplets. Separate occurrences of a particular structure, other than a cloud droplet, are generally not exact repetitions, but they tend to have much in common.

The laws governing the atmosphere include the basic laws of motion and thermodynamics, and some more specialized laws involving such processes as the absorption, emission, and scattering of radiation by atmospheric constituents and the changes of phase of water. The latter are complicated by the occurrence of water in the form of cloud droplets and ice crystals, and the presence of hygroscopic particles. The laws are commonly expressed as a system of mathematical equations. A typical equation may be written

$$dx/dt = F, \tag{1}$$

where t represents time, x represents the value of an atmospheric variable, such as temperature or a wind component, and F represents the sum of the physical processes which change x.

As originally formulated the laws apply to a fixed mass. Because the atmosphere is a fluid whose different parts move at different velocities, an initially concentrated mass tends to be stretched and twisted, so that tracing a particular mass may prove difficult. It is thus more convenient to introduce a coordinate system which is fixed in the atmosphere, and to rewrite eq. (1) to apply to fixed locations. The result is

$$\partial x/\partial t = - \underline{v} \cdot \nabla x + F, \tag{2}$$

where $\underline{v}$ represents the three-dimensional wind vector.

3. APPLICATION OF THE PHYSICAL LAWS

The difficulties encountered in applying eq. (2) directly to the atmosphere become apparent when we ask what we mean by a point, at which eq. (2) is to be applied. Certainly we do not mean a geometrical point, which would be smaller than a molecule, whence the velocity and temperature at the point would not even be defined. The use of the gradient operator in eq. (2) implies that we are treating the atmosphere as a continuum, and any "point" must actually be large enough to contain many molecules. Similarly, except in clear air, any point must be large enough to include many cloud droplets. Values of the variables at such a point must actually be averages over a region with a diameter of at least a centimeter.

Further difficulties appear when we note that we do not have observations spaced at one-centimenter intervals, and, except in data sets gathered for special studies, we do not even have observations at ten-kilometer intervals. To apply the equations to globally distributed observational data, we must therefore interpret "values at a point" as meaning averages over regions with horizontal extents comparable to 100 kilometers. For dealing with the internal dynamics of individual structures, such as thunderstorms, the regions may be considerably smaller.

These requirements might seem to disappear when we apply the equations to idealized rather than observed distributions of the variables, but here another practical difficulty arises. The advective term $-\underline{v} \cdot \nabla x$ in eq. (2) contains the product of one variable, $\underline{v}$, with the gradient of another variable, x, and is therefore inherently nonlinear. Of course, the function F might also be nonlinear, as in the case when it represents the effect on temperature of radiative heating and cooling. Analytic solutions of nonlinear equations generally represent specialized cases, with different properties from the general solutions, and, particularly since the advent of high-speed computers, it has been customary to seek approximate solutions using numerical methods. Even the largest computers have their limitations, and at present it is not practical to solve the equations when the state of the atmosphere, or of the individual atmospheric structure which is being studied, is represented by more than about one million numbers. Even restriction of the data to four physical variables and ten elevations would allow only 25000 regions at each elevation, and for global coverage the diameter of a region would still have to exceed 100 kilometers.

It appears, then, that eq. (2) should be replaced by an equation governing changes of the average values of the variables over rather extensive regions--regions which may actually contain many of the smaller individual structures. One way to do this is to use the equation of continuity of mass,

$$\partial\rho/\partial t = -\underline{\nabla} \cdot (\rho\underline{v}), \tag{3}$$

where ρ represents density. When combined with eq. (3), eq. (2) becomes

$$\partial(\rho x)/\partial t = -\underline{\nabla} \cdot (\rho x\underline{v}) + \rho F\ ; \tag{4}$$

the quantities ρx and ρF are values per unit volume when x and F are values per unit mass. When averaged over a region, eq. (4) becomes

$$\partial\ \overline{\rho x}/\partial t = -\underline{\nabla} \cdot (\overline{\rho x}\ \overline{\underline{v}}) - \nabla \cdot \overline{(\rho x)'\underline{v}'} + \overline{\rho F}, \tag{5}$$

where a bar over a quantity denotes a regional average, and a prime denotes a local departure from a regional average. In addition to terms obtained simply by replacing quantities in eq. (4) by their averages, we find a new term which depends on variations within the region.

The averaging process which produces eq. (5) entails two new practical difficulties. First, the regional averages which may be computed from observational data are generally averages of rather small statistical samples, and may therefore contain sampling errors. If the observing stations in a particular region had been established at slightly different locations, the computed averages at any time would presumably be somewhat different. A region might contain a single thunderstorm, and the computed average temperature, wind, and water content will depend upon whether the thunderstorm coincides with an observing station. Thus, for example, a realistic processing of a data set, instead of concluding that the average temperature over a region is 18°C, might more realistically conclude that it is 18°C plus an error, whose expected absolute value is 1.2°C.

The second difficulty involves the term $-\underline{\nabla} \cdot \overline{(\rho x)'\underline{v}'}$ in eq. (5). This term includes the transport of the property represented by x across the boundary of the averaging region by the circulations associated with structures of smaller spatial scale than the region itself. If, for example, the regions are 200 kilometers square and 2 kilometers deep, the term includes the exchange, between vertically adjacent regions, of heat and water and possibly momentum by cumulus-cloud circulations. If the region is considerably smaller, it may still include transports by individual wind gusts.

Since these transports often account for important fractions of the total change of a variable, it is important that the equation which we finally use in our investigations should not disregard them altogether. It is presently standard practice to include the effects through parameterization [2]. That is, we assume that the effects can be expressed reasonably well as functions of the averaged variables which now appear as dependent variables in our equations.

For example, we might assume that the number and size distribution and typical structure of cumulus clouds in a region is fairly well determined by the average temperature, moisture, and wind velocity in the region, and the manner in which these averages vary from this region to the regions immediately above and below. From the assumed cumulus-cloud statistics we could evaluate the amounts of heat, moisture, and momentum carried upward or downward by the cloud circulations. We would then include expressions for these amounts, in terms of the averaged quantities, as additional terms in our equations.

It has been found through experience that systems of equations which parameterize the effects of unresolved processes can perform considerably better than those which merely disregard the effects. Nevertheless, the distribution of cumulus clouds or other small-scale structures contained in any region or influencing the region at any instant constitutes at best a statistical sample drawn from the set of distributions which could conceivably have been present. A cloud which occupies a region may move out of the region within a few minutes, or it may alter its shape considerably, during which time the average properties of the region may not have detectably changed. We therefore face a sampling problem again; in addition to a deterministic term, which represents the "expected" or most probable effect of the unresolved structures, the equations should contain a stochastic term, representing the distribution of departures from the expected effect. It appears that this uncertainty in the equations far outweighs any possible uncertainty due to unpredictable human behavior.

We thus find that averaging produces two types of uncertainty, one in estimating the initial state, and one in formulating the governing laws. The effects of these uncertainties on operational weather forecasts appear to be far from negligible.

Instead of introducing regional averages we may expand the fields of the variables in series of orthogonal functions, such as spherical harmonics if the investigation is global, or multiple Fourier series if it is local. The coefficients in the series then become the new dependent variables. However, in order to retain a finite system, we must discard all but a finite number of coefficients in each series; ordinarily these represent features of small spatial scale. We find that the difficulties introduced by averaging, although perhaps slightly alleviated, are by no means eliminated. After all, a Fourier coefficient is nothing more than an average of the product of a variable and a trigonometric function, and statistical sampling problems remain.

It would thus seem that the atmosphere might best be treated as a stochastic system. That this is not necessarily the case will become evident after we consider in detail the phenomenon of chaos.

4. CHAOS

The term "chaos" has been used in mathematical and physical works with a number of meanings [3]. Often it is used as a synonym for randomness or lack of complete determinism, so that, in this sense, any stochastic process would be chaotic. More recently the term has been used to describe any system which varies aperiodically, or perhaps more often any system of equations where, in some sense, almost all solutions are aperiodic [4]. Under the category of periodic solutions we include not only those which exactly repeat themselves, but also those which eventually acquire a state arbitrarily close to some previous state, provided that the evolution following the near repetition remains arbitrarily close to the evolution following the original occurrence. We also include any other solutions which asymptotically approach those solutions which we have already included as periodic.

The distinction between the concepts of chaos more or less disappears if we confine our attention to finite systems of linear ordinary differential equations, since the solutions of these equations are generally periodic if the equations are deterministic, and not exactly periodic if stochastic terms are added. The concepts differ when we turn to nonlinear equations, which often have aperiodic general solutions even if they are deterministically formulated.

Some investigators prefer to reserve the term "chaos" for those aperiodically varying systems which are governed by formally deterministic equations. Others liberalize the definition to include stochastic systems, provided that it appears that the system would remain aperiodic even if the stochastic part of the governing equations were eliminated. This modification makes it possible to include real physical systems, whose actual determinism is likely to be in doubt.

The feature of aperiodically varying systems which has earned them the designation of "chaos" is their sensitive dependence on initial conditions [5]. If a system possesses a finite number of variables, and if each variable continues to oscillate between fixed upper and lower bounds, the system will in due time necessarily assume a state arbitrarily close to some previously encountered state. By definition, if the system is aperiodic, the evolution following the near repetition cannot forever remain arbitrarily close to the evolution following the original occurrence. If there is no semblance of periodicity, the evolutions following the two occurrences will ultimately go their own ways. Thus two states which are nearly alike will ultimately evolve into two states which lack any resemblance. If, for example, the system is a chaotic atmosphere and the observations are anything but exact, there will be no basis for choosing among a number of possible evolutions, and weather forecasting at some sufficiently distant range will be impossible.

The idea that deterministic equations may have aperiodic general solutions is not particularly new. For a long time, investigators of fluid turbulence have worked with deterministically formulated systems, and have assumed that the turbulent motion which satisfies these equations is non-repetitive. What is relatively new is the general realization that systems consisting of very few simple nonlinear ordinary differential equations may have aperiodic solutions.

The possibility of studying chaos with small systems, together with the general availability of high-speed computers, has made it feasible to examine the attractors of chaotic systems [4]. An attractor is actually a kind of multidimensional graph, and it is most easily described in terms of the phase space of a system. This is a Euclidean space with as many dimensions as the number of variables, and these varibles serve as coordinates. An instantaneous state is thus represented by a point in the phase space, while a time-variable solution is represented by an orbit.

A particular point which is approached arbitrarily closely, arbitrarily often, by a point traversing a given orbit is an attracting point for that orbit. A point which has a greater-than-zero probability of being an attracting point for a randomly selected orbit is a point of the attractor set. This set may be connected, or it may consist of a number of disjoint connected sets, in which case each of these is an attractor.

If almost all solutions of a system of equations approach a single repeating solution asymptotically, the attractor is simply the closed orbit representing this solution. If the system is chaotic, the attractor set is generally more complicated. When the system in question is the atmosphere, points on the attractor set represent states which are likely to be approximated again and again as the weather continues to evolve, i.e., states which are compatible with the climate. For example, hypothetical states where the poles are warm and the equator is cold, where the surface winds are everywhere of hurricane strength, or where the winds blow the wrong way around most of the high and low pressure centers are represented by points which are _not_ on the attractor set.

To obtain an approximate picture of an attractor, we may select an arbitrary initial state and perform an extended numerical integration. We discard the leading part of the solution as possibly representing transient conditions, and assume that the remaining part lies as close to the attractor as the resolution will allow [6]. Unless the system consists of only two equations, an actual picture is likely to be the projection of the attractor on a plane, or the intersection with a plane.

A procedure which is equally good in concept although more difficult to approximate in practice consists of taking a small sphere centered at an arbitrary point, and finding the successive shapes into which the interior of the sphere is deformed as each point in the interior moves along its orbit. Ultimately the deformed sphere should look like an attractor, or perhaps several attractors connected by infinitesimal threads. If the system is dissipative, the volume of the deformed sphere will shrink toward zero. If it is also chaotic, with sensitive dependence on initial conditions, the maximum diameter will grow. If the sphere is initially very small, it will for a while be deformed into an approximate ellipsoid. The long-term average rates of stretching or compression of the axes of the ellipsoid are called the Lyapunov exponents, and the condition for chaos is that at least one exponent should be positive [4]. If the equations are differential rather than difference equations, one exponent will also be zero, indicating that two points within the sphere moving along the same orbit will tend to maintain their initial separation. If the system is dissipative, the sum of the exponents will be negative.

As the deformation continues, the ellipsoidal shape will be lost. In the case of a dissipative chaotic system of three ordinary differential equations, the deformed sphere will come to resemble a strip of paper, which is continually becoming thinner but increasing in area. As the strip is stretched, it is bent and twisted so that it continues to fit within a reasonably confined volume. In due time different parts of the strip will be brought close to one another, so that locally two and then several sheets of paper will appear to be pressed together, although they will never actually merge. In the limit there will be an infinite number of sheets, which an ordinary picture might resolve into several. A transverse line will intersect these sheets in a Cantor set. An attractor with such a Cantor-set structure is called a strange attractor [7].

The above arguments may be generalized to systems of more than three equations. It would be difficult to draw a picture of a chaotic attractor, or even visualize its shape, in a high-dimensional system, so mathematicians who are principally interested in the topology of attractors have tended to use small systems as illustrative examples. Obtaining a picture of an obviously strange attractor is often an effective way of convincing oneself that a given system is chaotic.

5. EXAMPLES OF CHAOS

For a first example we shall choose one of the simplest possible nonlinear systems--the quadratic difference equation

$$x_{n+1} = (x_n - 2)^2 \tag{6}$$

in the single variable x. Starting with an initial value x_0 of x, with $0 \leqq x_0 \leqq 4$, we let x_n be the value of x after n applications of eq. (6). It is evident that $0 \leqq x_n \leqq 4$ for all values of n.

If x_0 is an even integer, the sequence $x_0, x_1, \ldots$ soon becomes a repetition of 4's, while if x_0 is an odd integer, it becomes a repetion of 1's, but, for most non-integer values of x_0, the sequence is aperiodic. Special values of x_0 where the sequence is periodic but not steady include $x_0 = (3 + \sqrt{5})/2 = 2.618$, when x alternates between $(3 + \sqrt{5})/2$ and $(3 - \sqrt{5})/2$.

Table I shows values of x when $x_0 = 0.5$. The lack of periodicity is apparent. These values are compared with values when $x_0 = 0.5001$, and the differences are also tabulated. The continual although non-uniform increase in the difference, until it becomes large, is typical of the sensitive dependence on initial conditions exhibited by aperiodically varying systems.

To demonstrate that the solutions are truly aperiodic, we let y_n be one of the solutions of the equation

$$x_n = 2(1 + \cos (2\pi y_n)), \tag{7}$$

with $0 \leqq y_n < 1$. It follows from eq. (6) that

$$x_{n+1} = 2(1 + \cos (4\pi y_n)), \tag{8}$$

so that eqs. (6) and (7) can be satisfied for all values of n, with $0 \leqq y_n < 1$, if

$$y_{n+1} = 2y_n (\text{mod } 1) . \tag{9}$$

Table I. Values x_n' and x_n'' of x_n determined by successive iterations of eq. (6) from initial values x_0' and x_0'', and difference $\varepsilon_n = x_n'' - x_n'$.

n	x_n'	x_n''	ε_n
0	0.5000	0.5001	0.0001
1	2.2500	2.2497	-0.0003
2	0.0625	0.0624	-0.0001
3	3.7539	3.7545	0.0006
4	3.0762	3.0782	0.0020
5	1.1582	1.1626	0.0044
6	0.7087	0.7013	-0.0074
7	1.6676	1.6866	0.0191
8	0.1105	0.0982	-0.0123
9	3.5701	3.6169	0.0468
10	2.4653	2.6143	0.1490
11	0.2165	0.3774	0.1609
12	3.1809	2.6329	-0.5480
13	1.3945	0.4006	-0.9939
14	0.3667	2.5582	2.1915
15	2.6677	0.3116	-2.3561
16	0.4458	2.8506	2.4048
17	2.4155	0.7236	-1.6919
18	0.1726	1.6292	1.4566
19	3.3394	0.1375	-3.2019
20	1.7940	3.4689	1.6750

If y_n is expressed in binary notation, eq. (9) simply shifts the bits of y_n one place to the left and drops the leading bit. The sequence $y_0, y_1, \ldots$, and hence the related sequence $x_0, x_1, \ldots$, is therefore periodic or aperiodic according to whether the bits of y_0 are arranged periodically or aperiodically, i.e., according to whether or not y_0 is a rational fraction. Thus almost all choices of y_0, and hence of x_0, lead to aperiodicity. For example, the value $x_0 = 1/2$ of Table I corresponds to $y_0 = (1/2\pi) \cos^{-1}(-3/4)$, which is not a rational fraction.

It will not surprise anyone to learn that one can obtain aperiodic solutions of a discontinuous equation simply by choosing an aperiodic infinite sequence of 0's and 1's and continually shifting left and removing the leading member. What is not so obvious until the above analysis is performed is that the possibility of doing so implies also that one can find aperiodic solutions of simple continuous equations, with simple rational numbers as initial values.

Eq. (6) does not possess an attractor with a Cantor-set structure, because it is not dissipative. In fact, the attractor is the entire interval $0 \leqq x \leqq 4$, and it is only because two distinct values of x_n can produce the same value of x_{n+1} that a small one-dimensional sphere, i.e., a segment, is not stretched to infinite length. The single Lyapunov exponent is log 2, i.e., on the average, the length of a small segment doubles with each iteration.

The more general equation

$$x_{n+1} = (x_n - c)^2 \tag{10}$$

possesses aperiodic solutions for some values of c between 1.4 and 2.0, although in most cases it cannot be converted to an equation like eq. (9) by a trigonometric transformation. Identification of the transitions from periodic to chaotic and from chaotic to periodic behavior which occur as c continually increases constitutes an interesting problem in dynamical-systems theory.

For examples of chaos more closely related to the atmosphere we turn first to a system of 12 ordinary differential equations which we introduced about 25 years ago for the specific purpose of obtaining a meteorological model with an aperiodic general solution [8]. As we have noted, even a million numbers governed by a million equations would give a somewhat incomplete picture of the atmosphere, so a 12-variable model must be very crude indeed. The model was obtained by representing the horizontal wind components by a stream function, expanding the fields of the vertically averaged stream function and the vertically averaged temperature in orthogonal functions, and then truncating each series to six terms. Vertical variations of the wind were identified with horizontal temperature gradients through the geostrophic relation. Other variables were inferred implicitly or disregarded altogether. The equations assumed the general form

$$dx_i/dt = \Sigma\, a_{ijk} x_j x_k + \Sigma\, b_{ij} x_j + c_i \ , \tag{11}$$

with x_i representing a stream function if $0 < i \leq 6$ and a temperature if $6 < i \leq 12$. In each equation at most four coefficients a_{ijk} and at most two coefficients b_{ij} differed from zero.

We integrated the equations numerically, using six-hour time steps, for a total of about 20 years. Fig. 1 shows the variations of x_1, representing the strength of the globally averaged westerly wind current, during a typical 18-month interval. Although no true periodicity is apparent, we see a succession of episodes, each lasting a month or somewhat longer, and each bearing a fair resemblance to the others. Each episode is marked by a rapid rise from very weak to very strong westerlies, followed by a somewhat less rapid fall to weak westerlies. The episode is completed by oscillations with periods of a week or two, generally about low values of x_1, but occasionally, as from days 140 to 170 and 380 to 410, about rather high values. The appearance of pronounced regularities, which, however, fall short of exact repetitions, is a typical feature of chaos which is deterministically generated. The succession of episodes may be regarded as a model of the atmosphere's index cycle, although the true index cycle is less regular [9].

Some of the properties of the 12-variable model are more easily illustrated by turning to a 3-variable model, which may be derived from

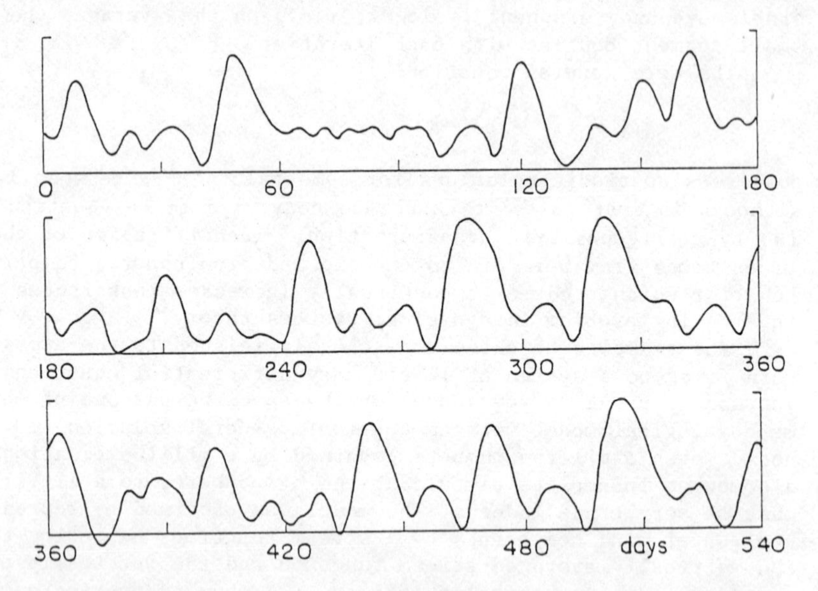

Fig. 1. Variations of the variables x_1 in the 12-variable model governed by eq. (11), representing the strength of the globally averaged westerly wind, during a particular 18-month interval.

the larger model by replacing the 12 variables by 12 linear combinations, and then performing additional truncations [10]. As might be expected, the new model is an even cruder approximation to the real atmosphere than the old one, yet it retains some of the real atmosphere's properties.

The equations of the new model are

$$dx/dt = -y^2 - z^2 - ax + aF \, , \tag{12}$$

$$dy/dt = xy - bxz - y + G \, , \tag{13}$$

$$dz/dt = bxy + xz - z \, . \tag{14}$$

Here x denotes the strength of the globally averaged westerly current, which is identified through the geostrophic relation with the cross-latitude temperature contrast, while y and z denote the cosine and sine phases of a chain of superposed waves, whose troughs and ridges are constrained to tilt westward with increasing elevation. The waves transport heat poleward, thus reducing the temperature contrast, as indicated by the $-y^2$ and $-z^2$ terms in eq. (12). The energy thus removed from the zonal current is added to the waves, as indicated by the xy and xz terms in eqs. (13) and (14). The waves are also carried along by the current, as indicated by the $-bxz$ and bxy terms. The linear terms represent mechanical and thermal damping, while the constant terms

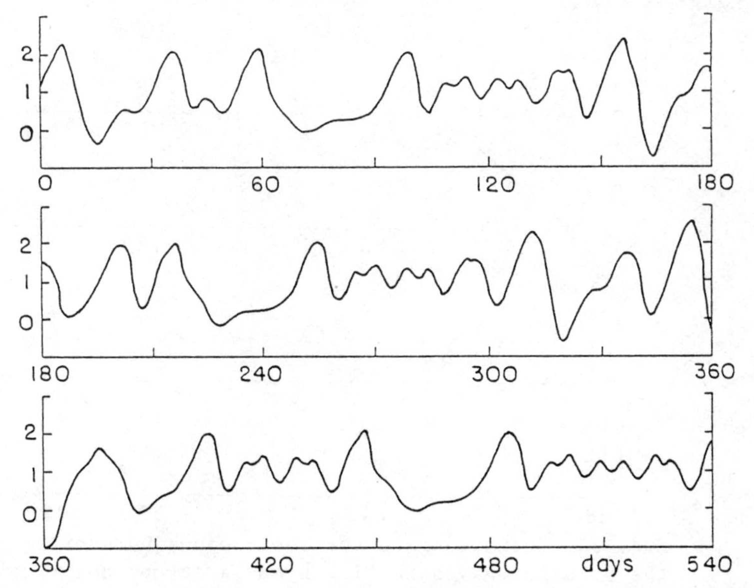

Fig. 2. Variations of the variable x in the 3-variable model governed by eqs. (12)-(14) with a = 1/4, b = 4, F = 8, and G = 5/4, representing the strength of the globally averaged westerly wind, during a particular 18-month interval

represent thermal forcing. The time unit equals the damping time for the waves, which is assumed to be five days.

For suitable choices of a, b, F, and G, eqs. (12)-(14) produce chaos. Such choices include $a = 1/4$, $b = 4$, and $F = 8$, with G some number between 0.85 and 1.3. Fig. 2, which is like Fig. 1, shows the variations of x for a typical 18-month interval, when $G = 1.25$. Again we see aperiodic variations, but with certain preferred types of behavior. The oscillations may again be regarded as modeling the atmospheric index cycle, but a six-month sequence produced by eq. (11) would probably not be mistaken for one produced by eqs. (12)-(14), nor, presumably, would a sequence produced by either model be mistaken for a real atmospheric index-cycle sequence.

A simple measure of the difference between two states is the distance in phase space. Fig. 3 shows the growth of such a difference, during a 12-month interval. The two initial states are the initial state of Fig. 2 and the same state with a small perturbation added. Eventually the difference becomes large, but the significant increases seem to be confined to the phase of the index cycle when the westerlies are approaching a maximum, and there are intervals as long as three months with no growth at all. The Lyapunov exponents prove to be 0.18, 0.00, and -0.52; the first exponent indicates that, on the average, small differences double in about 3.6 time units, or 18 days. We might add that by real atmospheric standards this growth is unreasonably slow.

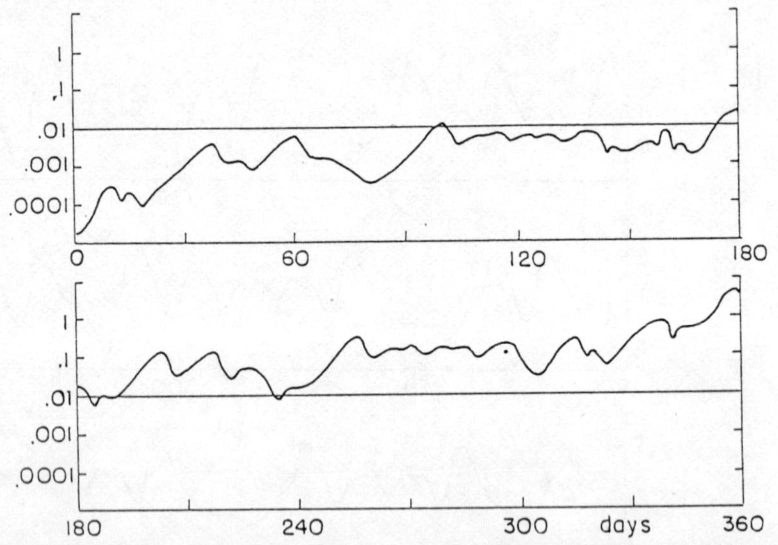

Fig. 3. Variations of the root-mean-square difference between the solution given in Fig. 2 and a second solution obtained by adding 0.00001 to the initial value of each variable, during the first 12 months of the 18 month interval of Fig. 2.

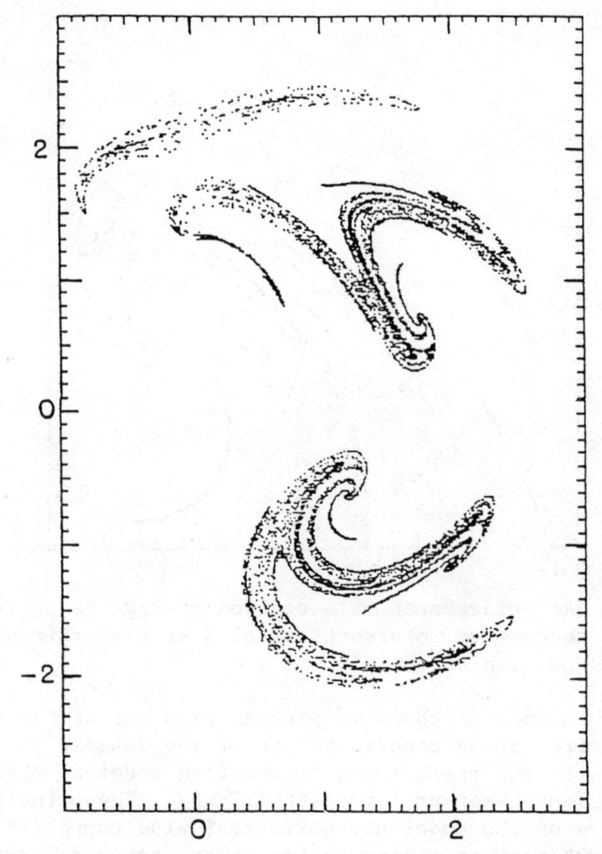

Fig. 4. The intersection of the attractor of eqs. (12)-(14) with the plane z = 0, when a = 1/4, b = 4, F = 8, and G = 5/4, as represented by 15000 successive intersections of a single orbit with the plane z = 0.

One feature of the 3-variable model which is easily examined is its attractor set. Fig. 4 shows the intersection of the single attractor with the plane z = 0, as approximated by 15000 successive intersections of a single orbit with the plane; these took place during about 160 years. The points appear to be concentrated on a few dozen curves, and nearby curves are approximately parallel. Between the curves are large areas which are avoided. In particular, the line y = 0 is avoided, indicating that states where y = z = 0, i.e., where the flow is independent of longitude, are never approached.

Fig. 5 shows an enlargement of a portion of Fig. 4, while Fig. 6 is an enlargement of part of Fig. 5. Additional curves are resolved. Further enlargements, not shown, reveal further curves, and it seems evident that a line cutting across all the curves would intersect them in a Cantor set, i.e., that the attractor is strange.

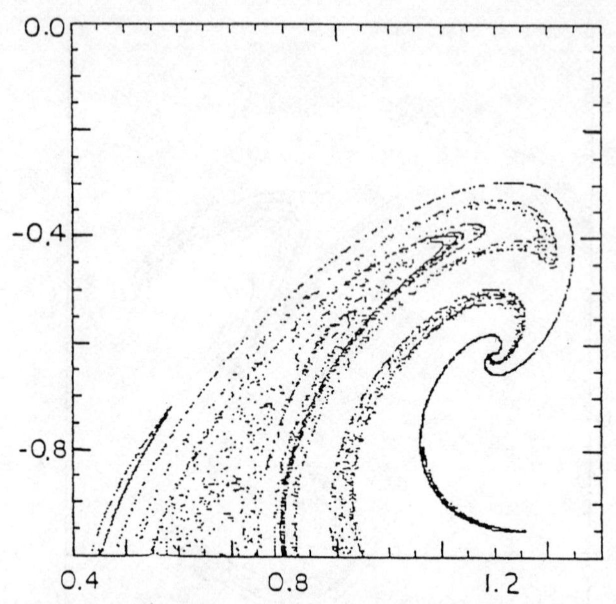

Fig. 5. An enlargement of a portion of Fig. 4, as represented by 8000 successive intersections of a single orbit with the included portion of the plane z = 0.

For our next example of chaos we proceed from one of the smallest possible "global circulation models" to one of the largest yet constructed. This is the operational forecasting model of the European Centre for Medium Range Weather Forecasts (ECMWF). The principal dependent variables of the model are horizontal wind components, temperature, and water-vapor mixing ratio; other variables are determined from these by auxiliary diagnostic formulas. The variables are independently defined at 15 elevations, and, in a recent version of the model, each horizontal field is represented by more than 10000 spherical-harmonic coefficients. The model thus consists effectively of more than 600000 ordinary differential equations in as many variables.

The model contains such physical features as orography. The effects of structures which are unresolved by the model, such as cumulus clouds, are included via parameterization. The intent is to make the model as good an approximation to the real atmosphere as is practical, in view of today's observation and computation systems. Diagnostic studies are regularly performed to determine how closely the climate produced by the model resembles the real atmosphere's climate, and significant differences generally lead to further research aimed at eliminating the discrepancies.

As the name of the Centre might imply, the principal purpose of the model is to produce weather forecasts at the "medium range" extending from a few days to a week or two. The present operational routine

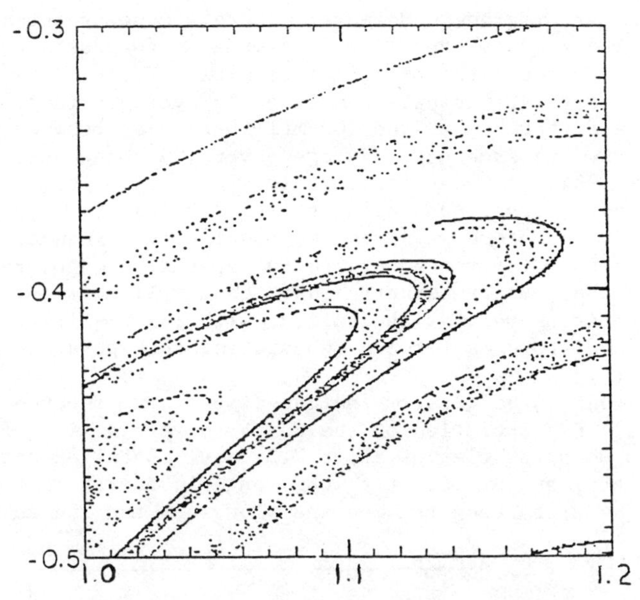

Fig. 6. An enlargement of a portion of Fig. 5, as represented by 6000 successive intersections of a single orbit with the included portion of the plane z = 0.

involves preparing, every day, a ten-day forecast of the global atmospheric state, using the present day's state as initial conditions. Since the equations are solved by stepwise integration, forecasts for intermediate ranges are automatically produced, and one-day, two-day, . . ., ten-day forecasts are routinely archived and made available for further research. However, forecasts more than ten days in advance are not generally prepared, and anything like an 18-month time series, comparable to Fig. 1 or 2, is unavailable.

Since the climate of the model differs from that of the real atmosphere, initial states determined from the real atmosphere need not lie on the model's attractor, and, since transient effects may well take more than ten days to die out, not even one point on the model's attractor set is known, let alone an entire attractor. That the model behaves chaotically rather than periodically is best determined by examining it for sensitive dependence on initial conditions.

We have performed a detailed examination of this sort. It would have been computationally expensive to perform many additional runs, in which the operationally used initial states were slightly modified. Instead we have capitalized on the fact that the model produces rather good one-day forecasts, so that the state predicted for a given day, one day in advance, may be regarded as equal to the state subsequently observed on the given day, plus a relatively small error. By comparing the one-day and two-forecasts for the following day, the two-day and three-day forecasts for the day after that, etc., we can determine how rapidly the error grows. Moreover, there are no practical barriers to averaging the results over a large sample of forecasts.

Fig. 7 presents the principal results. Points labeled i,j, where i and j are integers, indicate the globally averaged root-mean-square temperature difference at the 500-millibar level between i-day and j-day forecasts for the same day, averaged over 100 consecutive days beginning 1 December 1984. A 0-day forecast is simply an initial analysis.

The upper curve, connecting points labeled 0,j, for different values of j, therefore measures the model's performance, and indicates how rapidly the difference between two states, one governed by the model and one by the real atmospheric equations, will amplify. The lower curve, connecting points labeled i,j, with j - i = 1, indicates how rapidly the difference between two states, both governed by the model, will amplify.

The lower curve clearly indicates sensitive dependence on initial conditions. Extrapolation of the curve to very small differences suggests a doubling time of about 2.5 days. Detailed forecasting of weather states at sufficiently long range is therefore impractical. However, the difference between the slopes of the two curves indicates

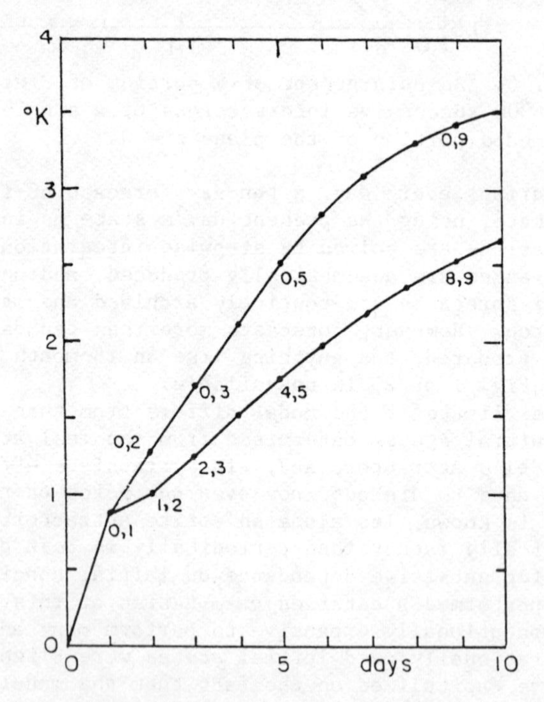

Fig. 7. Root-mean-square differences between i-day and j-day forecasts of the 500-millibar temperature for the same day, made by the ECMWF operational model, averaged over 100 days beginning 1 December 1984. Numbers i, j appear beside selected difference values, which are plotted against values of j.

that there is still considerable room for improvement in forecasting, and implies that we may, for example, some day produce one-week forecasts as good as today's three-day forecasts. Fig. 7 closely resembles a figure constructed from an earlier version of the ECMWF model [11], and both studies tend to confirm the results of earlier studies performed with less elaborate models [12].

Our final example of chaos is the weather itself. In contrast to the case of large atmospheric models, our evidence for chaotic behavior is mainly the absence of any tendency for exact repetitions, and the accompanying presence of continua in the many available variance spectra. We cannot perturb the atmosphere and observe what happens, and at the same time know what would have happened if we had not introduced the perturbation. In principle we could wait for an atmospheric state which closely resembles a previous state, and regard the new state as equal to the old state plus a small perturbation, but in practice we would have to wait too long. We recently estimated that we would have to wait 140 years to obtain one pair of states with a difference of one half of the difference between randomly chosen states [13].

Frequently we observe atmospheric states which closely resemble one another over limited regions; for example, two extratropical cyclones may look very much alike. After a few days the local resemblance will be much weaker, but it is not certain whether this is so because of local amplification or because of the influence of more distant regions where the states are quite different.

Probably our confidence in the chaotic nature of the atmosphere is fortified by the fact that the various large global models exhibit behavior resembling that of the real atmosphere fairly closely, and all of these models show sensitive dependence on initial conditions and agree fairly well as to the rate of error growth. We may also be influenced by our familiarity with baroclinic instability, where perturbed states will depart from unperturbed states.

6. CONCLUSIONS

We may now return to our question as to whether, in investigating atmospheric dynamics, we ought to treat the atmosphere as a deterministic or a chaotic system. The possibly surprising answer is that for most investigations it does not matter. The system of equations which we will be using to study the atmosphere will necessarily involve some approximations, and it may be regarded as a model. Provided that the model is realistic enough to produce a chaotic atmosphere with essentially correct gross features, its behavior will be about the same whether or not it contains some stochastic terms. Here we are assuming that the magnitude of these terms is not completely out of proportion with the actual randomness present in the laws governing the atmosphere.

Our choice between a formally deterministic and a stochastic model will therefore be one of convenience. If our reasoning can be facilitated by the knowledge that our equations contain no randomness, we should use a deterministic formulation. If explicit randomness will aid our investigation, we should introduce it.

As with most general conclusions, there are particular exceptions. If we are studying the growth of the difference between two atmospheric states, using a model in which the smaller scales have been parameterized, and if the initial difference is very small, it will grow quasi-exponentially and require a number of days to become appreciable, if the parameterization is deterministic. With a stochastic parameterization the difference, even if it is initially zero, will quickly become appreciable, possibly during the first day. The latter type of behavior seems more realistic, since it appears that if the small scales could be carried explicitly, uncertainties in these scales would rapidly spread to the larger scales [14], [15]. Once the differences in the resolved scales have become appreciable, it matters little whether the parameterization is deterministic or stochastic.

We are not maintaining that a system of equations with no random terms, and the same system with random terms added, can produce quantitatively identical results. Qualitatively the results may be nearly indistinguishable, or they may be quite different if some of the constants in the system are close to their bifurcation values. In the latter event, the addition of small random terms may still be nearly equivalent to making small alterations in the numerical values of the constants.

ACKNOWLEDGMENT. This work has been supported by the GARP Program of the Atmospheric Sciences Section, National Science Foundation, under Grant 82-14582 ATM.

REFERENCES

1. Wiener, N., 1956: I am a Mathematician. New York, Doubleday, 380 pp (see p. 259).
2. Haltiner, G., 1971: Numerical Weather Prediction. New York, Wiley, 317 pp.
3. Prigogine, I. and Stengers, I., 1983: Order out of Chaos. New York Bantam, 349 pp.
4. Guckenheimer, J. and Holmes, P., 1983: Nonlinear Oscillations, Dynamical Systems, and Bifurcations of Vector Fields. New York, Springer-Verlag, 453 pp.
5. Lorenz, E. N., 1963: 'Deterministic nonperiodic flow'. J. Atmos. Sci., **20** , 130-141.
6. Ruelle, D., 1981: 'Small random perturbations of dynamical systems and the definition of attractors' Comm. Math. Phys., **82** , 137-151.
7. Ruelle, D. and Takens, F., 1971: 'On the nature of turbulence' Comm. Math. Phys., **20** ,167-192; **23** , 343-344.
8. Lorenz, E. N., 1962: 'The statistical prediction of solutions of dynamic equations' Proc. Internat. Sympos. Numerical Weather Prediction, Meteorol. Soc. Japan, 629-635.

9. Namias, J., 1950: 'The index cycle and its role in the general circulation' J. Meteorol., **7** , 130-139.
10. Lorenz, E. N., 1984: 'Irregularity: a fundamental property of the atmosphere' Tellus, **36A** , 98-110.
11. Lorenz, E. N., 1982: 'Atmospheric predictability experiments with a large numerical model' Tellus, **34** , 505-513.
12. Smagorinsky, J., 1969: 'Problems and promises of deterministic extended range forecasting' Bull. Amer. Meteorol. Soc., **50** , 286-311.
13. Lorenz, E. N., 1969: 'Atmospheric predictability as revealed by naturally occurring analogues' J. Atmos. Sci., **26** , 636-646.
14. Lorenz, E. N., 1969: 'The predictability of a flow which possesses many scales of motion' Tellus, **21** , 289-307.
15. Leith, C. E., 1971: 'Atmospheric predictability and two-dimensional turbulence' J. Atmos. Sci., **28** , 148-161.

9. [illegible], 1952: "The [illegible] and its [illegible] in the general circulation", J. Meteorol., 7, [illegible].
10. [illegible], 1964: "[illegible] property of the [illegible]", Tellus, 16, 98-110.
11. Lorenz, E. N., 1965: "A [illegible] predictability [illegible] with a [illegible] model", Tellus, 17, 321-333.
12. [illegible], 1969: "[illegible] and [illegible]", Bull. Amer. Meteor. Soc., 50, [illegible].
13. Lorenz, E. N., 1969: "Atmospheric predictability as revealed by naturally occurring analogues", J. Atmos. Sci., 26, 636-646.
14. Lorenz, E. N., 1969: "The predictability of a flow which possesses many scales of motion", Tellus, 21, 289-307.
15. Leith, C. E., 1971: "Atmospheric predictability and two-dimensional turbulence", J. Atmos. Sci., 28, 145-161.

THE BLOCKING TRANSITION

J. Egger
Meteorologisches Institut
Universität München
FRG

ABSTRACT. The atmosphere tends to reside in relatively persistent circulation patterns called grosswetterlagen. Blocking is one of these. It has been suggested by Charney and DeVore (1979) that these grosswetterlagen are linked to the existence of multiple equilibria of the dynamic equations underlying the large-scale flow. Transitions between grosswetterlagen can be seen as a consequence of instabilities of these equilibria. Work on this topic is reviewed. It is demonstrated that the dynamic mechanism for establishing these equilibria as proposed by Charney and DeVore is not effective in the atmosphere.

A different approach is discussed as well where grosswetterlagen and their transitions are seen as the result of the interaction of large-scale flow with flows at smaller scale. One obtains quite realistic simulations of the blocking process if this interaction is prescribed according to data. This suggests that models of just the large-scale flow will not be sufficient to clarify the blocking problem.

1. GROSSWETTERLAGEN AND TRANSITIONS

The term grosswetterlage has been coined (Baur et.al. 1944) in order to characterize persistent circulation patterns of large scale in Europe and over the northern Atlantic. By definition a pattern must persist for at least three days to be acceptable as a grosswetterlage (correspondingly, there are days when there is no grosswetterlage). The original classification contained as many as 21 grosswetterlagen. However, Hess and Brezowsky (1969), when applying a somewhat modified scheme, proposed also a classification where all these patterns are lumped together to form three categories: zonal, meridional and mixed. In Fig. 1 we show a grosswetterlage of the meridional type. A enormous ridge dominates the eastern Atlantic with well

C. Nicolis and G. Nicolis (eds.), Irreversible Phenomena and Dynamical Systems Analysis in Geosciences, 181–197.

developed troughs to the east and west. Correspondingly one has meridional flow over central Europe. The category of meridional grosswetterlagen contains blocking situations as a subset. The grosswetterlage depicted in Fig. 1 is a blocking situation. Flow patterns like those shown in Fig. 1 have been called "blocks" since low pressure systems of smaller scale which travel towards the blocking anticyclone from the west are hindered by the block to move on towards the east. Thirty years ago quite a number of criteria had to be satisfied for a flow pattern to qualify as a block. These criteria were derived from synoptic experience and were adapted to visual map analysis (e.g. Rex, 1950). More modern selection criteria are geared to computerized map analysis. Typically one searches for anomalies (deviations from long-term means) of the stream function at 500 mb, say, which exceed a threshold value for more than five consecutive days. It turned out that both methods yield about the same results.

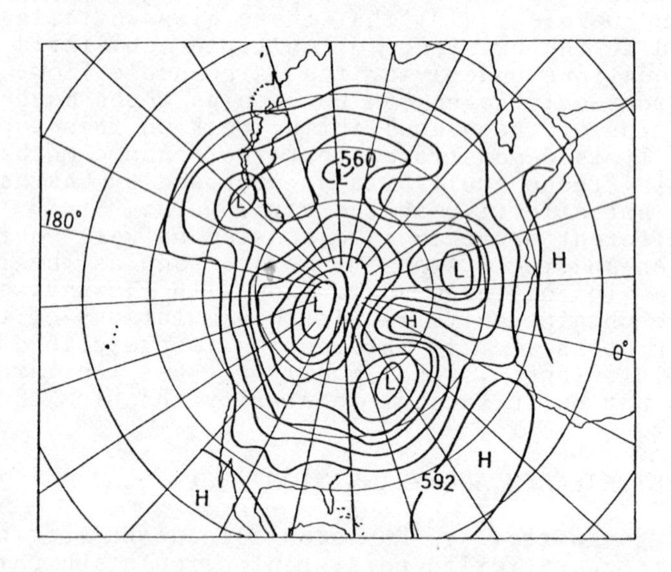

Figure 1. Example of an Atlantic blocking situation. 500 mb height in dam on 24 June 1949. Adapted from Rex (1950). Distance of isolines is 8 dam.

A zonal grosswetterlage is characterized by strong westerly flow over the Atlantic. Correspondingly the isolines of geopotential height (stream function) are roughly parallel to latitude circles.

Strictly speaking grosswetterlagen are defined for Europe only. However there is no reason why a similar classification could not by be worked out for other parts of the globe. In particular, blocking is known to occur frequently over the Pacific (e.g. Treidl et.al., 1981). Nevertheless it should be kept in mind that all these patterns do have a relatively local character. It may be difficult if not impossible to define hemispheric grosswetterlagen.

The synoptician and the dynamic climatologist will not be content to have a complete list of grosswetterlagen at hand. They would also like to know the mean residence time for each type and, perhaps even more urgently, the probability of transition from one type to the other. Spekat et.al. (1983) have evaluated these statistical characteristics for the three types of grosswetterlagen mentioned above. They found a mean residence time of about seven days for the meridional type, of six days for the zonal grosswetterlagen and of five days for the mixed situation. In Fig. 2 we show transition probabilities on a day to day basis. The transition probabilities are of the order 0.08. Transitions from meridional to zonal are seen to be relatively unlikely.

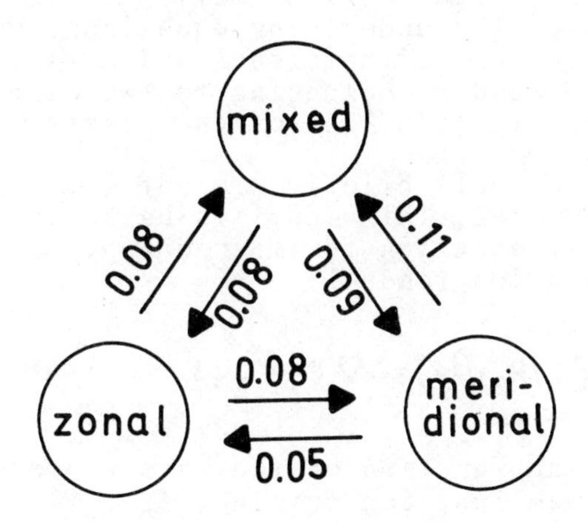

Figure 2. Transition probabilities between three categories of grosswetterlagen on a day to day basis. After Spekat et al. (1983).

These findings call for an explanation. Why is it that we observe just these types of grosswetterlagen? Why do they persist for such a long time? What processes cause the transitions?

In this paper we want to discuss these problems. In

the following section the dynamic systems approach of Charney DeVore (1979) to the problem is described. Problems with this approach are discussed in Sect. 3 and extensions and modifications of the original research programme of Charney and DeVore are presented in Sec. 4. A rather different approach to the problem, which relies on observed vorticity transfers, is presented in Sect. 5.

2. THE CDV-MODEL

Since the forties, theoretical work on the blocking problem was mainly concerned with elucidating the basic mechanism of blocking development. Almost all conceivable mechanisms have been proposed: hydraulic jump (Rossby, 1950), linear resonance (Tung and Lindzen, 1979), non-linearly modified resonance (Egger; 1978; Malguzzi and Speranza, 1981) and baroclinic instability (Schilling, 1982). Charney and DeVore (1979; hereafter referred to as CdV) took a more global view partly inspired by the work of Lorenz (1963) on different circulations regimes in a differentially heated annulus. They suggested that grosswetterlagen can be associated with stationary or oscillating states of the underlying equations. More specifically, they suggested that zonal and meridional flow types can be interpreted as belonging to two distinct, quasi-stable stationary solutions of the barotropic vorticity equation.

To fix ideas we shall briefly present a simplified version of the CdV-model. This model is based on the barotropic vorticity equation for forced ß-plane flow over topography. The equation reads

$$\frac{\partial}{\partial t}\nabla^2\psi + J(\psi, \nabla^2\psi + f_0 h/H_0 + \beta y) = C\nabla^2(\psi^* - \psi) \qquad (2.1)$$

where ψ is the stream function of the barotropic atmospheric flow, ψ^* is a stream function forcing, f_0 a midlatitude value of the Coriolis parameter with ß as the derivative with respect to latitude. With H_0 we introduce a scale height ($\sim 10^4$ m), h is the orographic profile and C is a damping parameter. The symbol ∇^2 denotes the Laplacian and the Jacobian

$$J(a, b) = -\frac{\partial a}{\partial y}\frac{\partial b}{\partial x} + \frac{\partial a}{\partial x}\frac{\partial b}{\partial y}$$

in (2.1) describes the advection of absolute potential vorticity $q = \nabla^2\psi + f_0 h/H_0 + \beta y$ by the flow. With respect to the atmosphere it is best think of ψ as the vertically averaged flow in the troposphere.

CdV used a channel of width L/2 and length L as flow domain. It is a main point of the approach of CdV that it is sufficient to resolve only the largest scales of motion in the model. Grosswetterlagen are by definition of relatively large scale. Smaller scale low pressure systems travelling within this coarse grain flow are seen as superimposed perturbations. Such a coarse grain representation of the flow is best carried out by projecting the stream function on Fourier modes

$$\psi = \sum_{m=0}^{M} \sum_{n=1}^{N} \psi_{mn} \exp(ik_m x) \sin k_n y - u_o y, \qquad (2.2)$$

where $k = 2\pi\, m/L$, and ψ_{mn} are complex Fourier coefficients. The zonal mean flow when averaged over the flow domain is denoted by u_o with positive values of u_o denoting westerly flow. The coarsest resolution possible is to restrict the flow to one wave mode and the zonal mean flow u_o. The stream function of the coarse grain flow is

$$\psi = (v_r k_1^{-1} \cos k_1 x + v_i k_1^{-1} \sin k_1 x) \sin k_1 y - u_o y \qquad (2.3)$$

where $v_r = \mathrm{Re}\,(\psi_1)\, k_1$, $v_i = \mathrm{Im}\,(\psi_1)\, k_1$.

The subscript 1 will be dropped in the following. Inserting (2.3) in (2.1) yields after standard manipulations the three prognostic equations of the simplified CdV model:

$$\frac{dv_r}{dt} = k v_i (u_o - \beta/2k^2) - C v_r + F_1 \qquad (2.4)$$

$$\frac{dv_i}{dt} = -k v_r (u_o - \beta/2k^2) + f_o u_o \tilde{h}/2H - C v_i + F_2 \qquad (2.5)$$

$$\frac{du_o}{dt} = -f_o v_i \tilde{h}/H - C(u_o - u_o^*) + F_3 \qquad (2.6)$$

where u_o^* is the forcing of the mean flow corresponding to the stream function forcing in (2.1). The orography has been chosen such that $h = \tilde{h} \cos kx \sin ky$. An hitherto unspecified forcing is provided by the terms F_i.

The first terms on the right hand side of (2.4), (2.5)

represent advective processes. Linear resonance is said to occur if $u_o = ß/2K^2$. Due to the special choice of the orographic profile it is only v_i which can be affected directly by the orography. The third equation can be seen as a momentum equation for the mean flow. The mean momentum can be transferred to the earth through interaction of the wave mode with the orography or by friction.

Imposing steady-state conditions it is straightforward to derive from (2.4)-(2.6) a third-order polynomial for the equilibrium zonal flow $\hat{u}_o$. This equation may have one or three real roots $\hat{u}_{oi}$. Correspondingly we have one or three equilibria. In Fig. 3 we give the steady state solution $\hat{u}_o$ and the corresponding values of $\hat{v}_r$, $\hat{v}_i$ as a function of the parameter L. For $L \sim 2 \times 10^6$m there is one equilibrium E_1 with $\hat{u}_{o1} \sim u_o^*$. According to (2.6) the mountain torque term must be small in that case

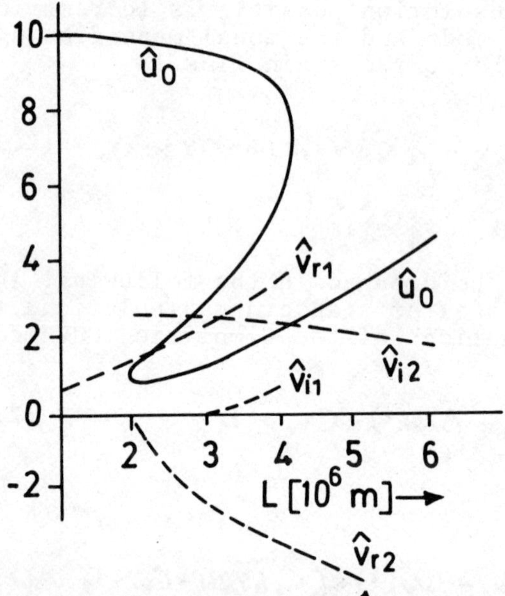

Figure 3. Equilibrium solution $\hat{u}_o$ as a function of wave length L. Also given are the velocity components of the stable equilibria. Velocities in ms^{-1}; $u_o^x = 10\ ms^{-1}$,

and, indeed, $v_{i1} \sim 0$. With increasing L two more equilibria appear, an intermediate equilibrium E_3 and an equilibrium E_2 with a rather small value of u_o. At E_2 the mountain torque term must be large and negative to balance the forcing term $C\ (u_o^* - u_o)$. Strong zonal flow prevails at E_1 and a pronounced wave is seen at E_2. The equilibrium E_1 can be associated with a grosswetterlage of the zonal type whereas E_2 corresponds with a grosswetterlage of the meridional type. It is customary to call E_2 a blocking equilibrium although the corresponding flow pattern is only vaguely reminiscent of blocking. A conventional stability analysis shows that E_1 and E_2 are stable with respect to perturbations in the model. E_3 is unstable.

The situation described above constitues the starting point of the "research programme" (Lakatos, 1978) proposed by CdV: grosswetterlagen are associated with multiple equilibria of the underlying equation. In particular the meridional equilibrium is characterized by a strong mountain torque and low values of the zonal flow. At the zonal equilibrium we have strong zonal flow and a small mountain torque. As is typical of research programmes at an early stage the CdV-model should not be expected to answer all the basic questions posed above. The basic model must be improved and extended step by step in order to overcome inconsistencies and digest conflicting evidence. Part of this work has been done by Charney himself and coworkers. Major problems are as follows:

1. Relevance of the physical mechanisms incorporated in the CdV model: it is by no means obvious that the mountain torque plays such a central role in atmospheric dynamics.

2. Transition: in the CdV-model there are no transitions between stable equilibria even in a time dependent formulation of the model.

3. Stability with respect to motions and physical processes excluded in the model.

In what follows we shall comment on these problems.

3. VALIDATION OF THE CDV-MODEL

The CdV model predicts a high index state characterized by strong zonal flow $u_0 \sim u_0^*$ and weak mountain torque and a low index state with a strong mountain torque and relatively weak zonal flow. This prediction can be tested by looking at data. One has to estimate the mountain torque and the zonal mean flow for blocking and nonblocking situations. This has to be done for a hemispheric domain since boundary fluxes become extremely important if just a certain longitude sector is chosen for the data analysis.

Metz (1985) has carried out this analysis. Some of his findings are presented in Figs. 4, 5 where the deviations of the zonal flow and the mountain torque from the respective time mean are shown for blocking situations in winter. For a strict comparison with CdV the curves have to be integrated over latitude. The zonal flow has a negative anomaly before the onset of blocking throughout the northern hemisphere. For mature blocks, however, negative deviations are found to the north of 40°N only. But even there typical deviations of the zonal wind are smaller than 0.75 ms^{-1}. There is nothing like the dramatic drop of u_o required by the CdV-model for blocking situations. Quite to the contrary. The hemispheric mean u_o may have even anomalously positive values for mature blocks. The deviations of the mountain torque from the time mean are given in units of ms^{-1} day^{-1}. This corresponds with (2.6). The curves in Fig. 5, when integrated over latitude, give the change of u_o per day as enforced by the "anomalous" mountain drag during blocking. The mountain torque term is "anomalously" positive in the onset of blocking period but negative later on.

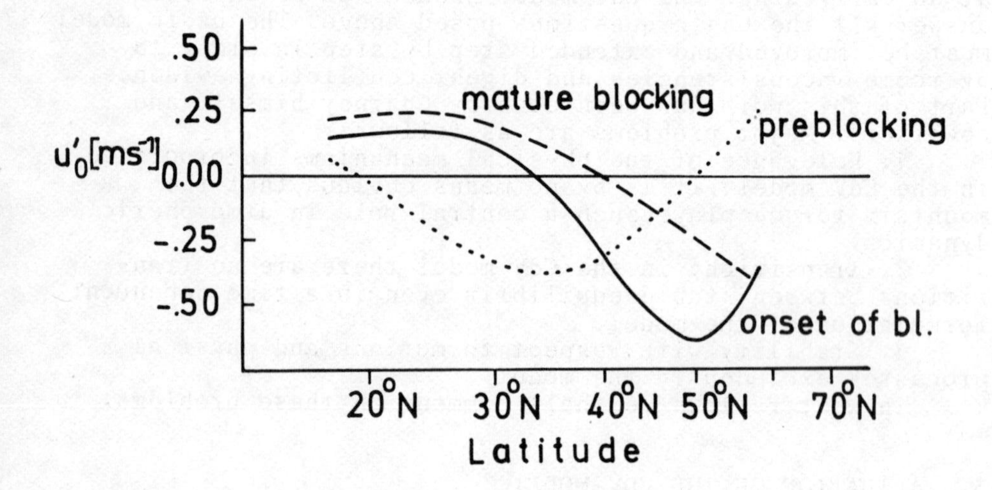

Figure 4. Deviation of the zonal mean wind from climatological mean for three stages of the blocking process. After Metz (1985).

Again the deviations are quite small. Even if the negative values of the onset period would stay on throughout the whole blocking period the mean flow would be reduced by not more than 1 ms^{-1} within ten days. One has to conclude that the blocking mechanism as envisaged by CdV does not exist. The CdV-model must be seen as an illustrative model. Using this model helps to formulate a description of large-scale weather phenomena in terms of attractor sets, stochastic forcing etc. However, to eventually resolve the problem of blocking and the corresponding transition problem this way one has to turn to different mechanisms and models thereof.

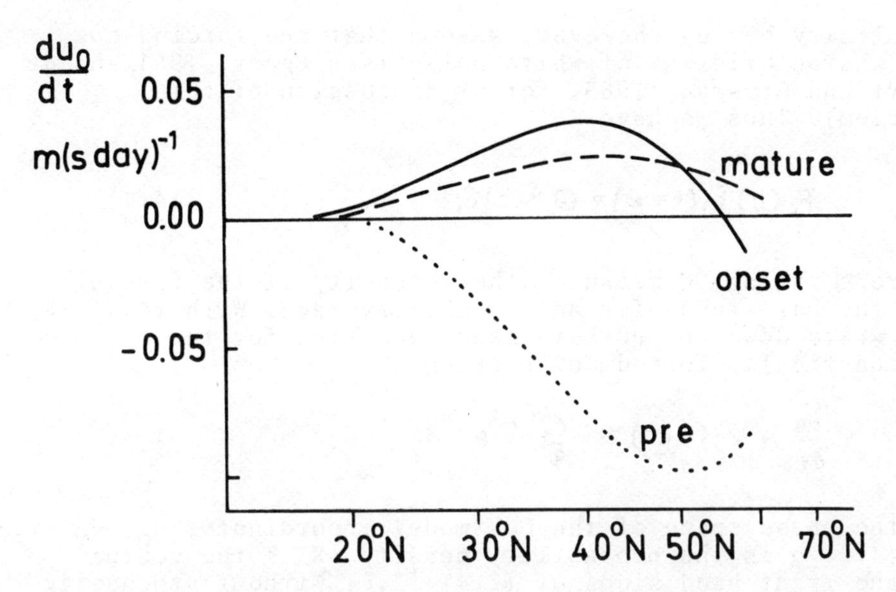

Figure 5. Contribution of mountain-torque to the tendency of the zonal wind. Deviations from climatological mean for three stages of the blocking process. After Metz (1985)

4. THE BLOCKING TRANSITION IN MODIFIED CDV-MODELS

As we have seen the CdV -model cannot serve as a basic model of the dynamics of grosswetterlagen. Nevertheless some of the principal problems can be studied using this model. In this section we wish to adress the transition problem in the light of extended and modified versions of the CdV model. Of course, CdV were aware that a more realistic model of grosswetterlagen would have to include the possibility of transitions between the attractor basins. They proposed that each of the equilibria has its own class of smaller scale instabilities and that each of them would be capable of undergoing transitions with the aid of these instabilities. To activate the instabilities one has to include more modes in the model (e.g. Legras and Ghil, 1985). However, there is a somewhat simpler way to induce transitions. In a high-resolution model many modes would interact with the modes resolved in the CdV-model. We can take this interaction into account in the CdV-model by prescribing it as a random forcing. In other words we have to specify the F_i in (2.4)-(2.6). We choose a forcing which resembles the interactive forcing found in the atmosphere. Egger and Schilling (1983) determined this forcing using atmospheric data and found that the forcing has the characteristics of red noise. For the sake of simplicity let us, however, assume that the forcing has the characteristics of white noise (see Egger, 1981, and de Swart and Grasman, 1985, for a discussion of this problem). Thus we have

$$\overline{F_i(t)F_j(t+\tau)} = Q\,\delta(\tau)\,\delta_i^j \tag{4.1}$$

where τ is lag, Q measures the intensity of the forcing and the bar stands for an ensemble average. With (4.1) we can write down the Fokker-Planck equation for the stochastically forced CdV-problem:

$$\frac{\partial p}{\partial t} + \nabla\cdot(p\underset{\sim}{K}) = \frac{Q^2}{2}\nabla^2 p \tag{4.2}$$

in the phase space of the CdV model (coordinates u_o, v_v, v_i). In (4.2) p is the probability density, $\underset{\sim}{K}$ is the vector of the right hand sides of (2.4)-(2.6) without stochastic forcing (see, e.g. Egger, 1981, for details).

Steady-state solutions of (4.2) give the climatological probability to find the system in a certain part of the phase space. The Fokker-Planck equation (4.2) has been solved numerically and it has been found that both the zonal and the meridional equilibrium are enclosed by surfaces of constant probability density. This means that transitions occur between the attractor basins of both equilibria. One can also calculate residence and exit times for neighbourhoods of the equilibria. De Swart and Grasman showed that the residence times are $\sim\exp(D_i/Q^2)$ for small forcing (D_i constant). They found that the residence times are much longer than those given by Spekat et al. (1983) if realistic forcing intensities are chosen (see also Benzi et al. 1984).

Legras and Ghil (1985) essentially took up the original idea of CdV and extended the CdV-model to include up to 25 modes. Orography is still restriced to one large-scale mode. Legras and Ghil were able to determine the equilibria even for this rather large system. The equilibria exhibit blocking and zonal flow patterns and indeed most of these equilibria are unstable with respect to perturbations in the model. Therefore, transitions between the various neighbourhoods of the equilibria must be expected.

Long-term integrations of the model have been carried out. Transitions and residence times can be monitored by computing the distance

$$C(t) = \|\underset{\approx}{\psi}(t+\tau) - \underset{\approx}{\psi}(t)\|\,\tau^{-1}$$

between two states Ψ of the model with lag τ in the phase space of the variables. For small C we have a persistent behaviour. States drift rapidly apart if C is large. As an example we show in Fig. 6 the function C as obtained in a time integration.

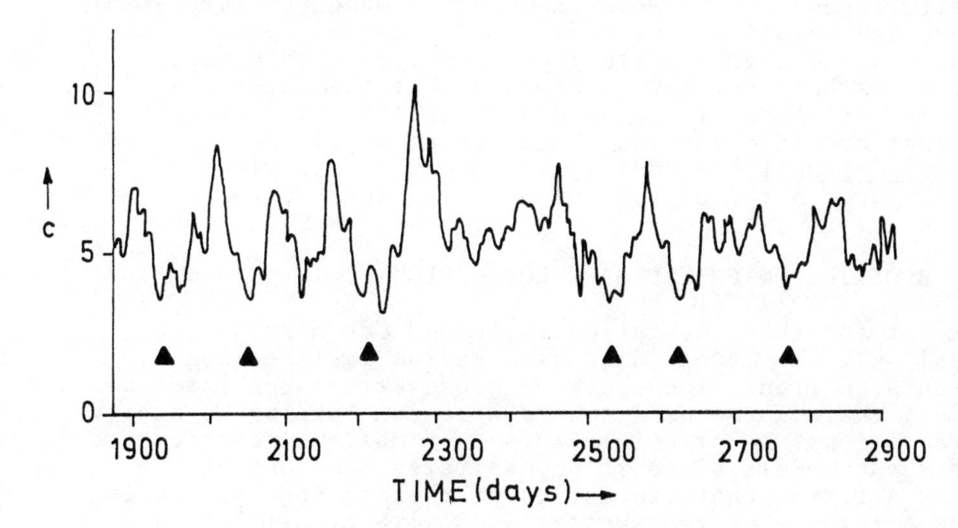

Figure 6. Evolution of the distance function C in an integration of the extended CdV-model according to Legras and Ghil (1986). The lag τ equals two days. Units of C are normalized. The triangles mark blocking periods with high persistency.

It is seen that relatively short periods where C is small, interrupt longer periods with relatively large C values. By computing composite flow patterns for the low-C periods Legras and Ghil where able to show that these are periods of blocking where the composite pattern show a good resemblance to the blocking equilibrium pattern. It is however, not possible to detect all the other equilibria in the time dependent solution. We can conclude that multiple equilibria have a clearly visible influence on the transient behaviour of an CdV model with many modes. Transition between flow regimes occur and are caused internally.

So far we have considered barotropic models. In 1980 Charney and Straus extended the CdV model to include baroclinic effect. Reinhold and Pierrehumbert (1982) used the same highly truncated representation of large-scale flow as Charney and Straus but added a wave of smaller scale. This wave was chosen such that the equilibria of the large-scale flow were baroclinically unstable with respect to this wave mode. Integrating this model in time Reinhold and Pierrehumber found that the state of the model aperiodically vacillates between two distinct grosswetterlagen which are not located near any of the stationary equilibria of the large-scale flow. Both grosswetterlagen are characterized by a pronounced large-scale wave and belong to the meridional category. Transitions between these grosswetterlagen are caused internally and occur rather rapidly. It appears that the CdV-model's flow characteristics are changed drastically once baroclinic effect are incorporated. More recent developments of the theory of multiple equilibria in baroclinic flows are described in the article of Benzi in this volume.

5. BLOCKING IN FORCED BAROTROPIC FLOW

Except for the stochastically forced CdV model (Egger, 1981) all the models discussed in the last section generated transitions between grosswetterlagen internally, i.e., there is no need to add a random forcing to achieve transitions. On principle this is a most welcome feature of these models since it demonstrates that one of the most striking characteristics of the atmosphere, namely the existence of grosswetterlagen with transitions, are also found in low-order models. On the other hand it is hard to believe that the dynamics of grosswetterlagen in the atmosphere are governed by just a few modes. Instead it is quite likely that many modes interact to create and destroy grosswetterlagen. This suggests to extend the approach chosen by Egger (1981) where both the large-scale flow and flow at smaller scales were represented.

We use the barotropic vorticity equation (2.1) to describe the large-scale flow. With $M = 5$, $N = 5$ the large-scale flow patterns are described by 25 wave modes. Atmospheric flow modes with higher wave numbers in either zonal ($m > 5$) or meridional direction ($n > 5$) are called synoptic-scale modes and are not included explicitly in the model. Orographic forcing is not included. Correspondingly there is no mountain torque in the model and we obtain from the mean flow equation in analogy to (2.6) $u_o = u_o^*$. To simplify matters as far as possible we discard even all nonlinear terms in (2.1). The resulting system of equations has an extremely simple attractor: $u_o = u_o^*$; $\psi_{mn} = 0$. Nevertheless we can create blocking patterns and other grosswetterlagen in this model if the interaction of the large-scale modes with the synoptic-scale modes is prescribed as forcing. In analogy to (2.4)-(2.6) we have to specify Fi for all wave modes of the large-scale flow. To obtain realistic forcing functions it is best to turn to observations. The corresponding procedures are described in Egger and Schilling (1983; 1984). To compute of the modes included in the model with the motion at smaller scales on a day-by-day basis using atmospheric data. If this is done one obtains a time series of the forcing functions F_i for each mode of the model.

A typical numerical experiment is carried out as follows (Metz, 1986). We start a numerical integration of the large-scale flow equation from the initial state $u_o = u_o^*$, $\psi_{mn} = 0$ and prescribe the forcing terms Fi at each day of the model run according to the observations. This way we induce wave motions in the model and a large-scale flow forms. Metz (1986) has analyzed this flow with particular emphasis on the blocking problem. He found that blocks form in the model with surprisingly realistic characteristics. In Fig. 7 we show the mean stream function anomaly averaged over all Atlantic blocking cases in the model and the corresponding mean as obtained from the data for the same period.

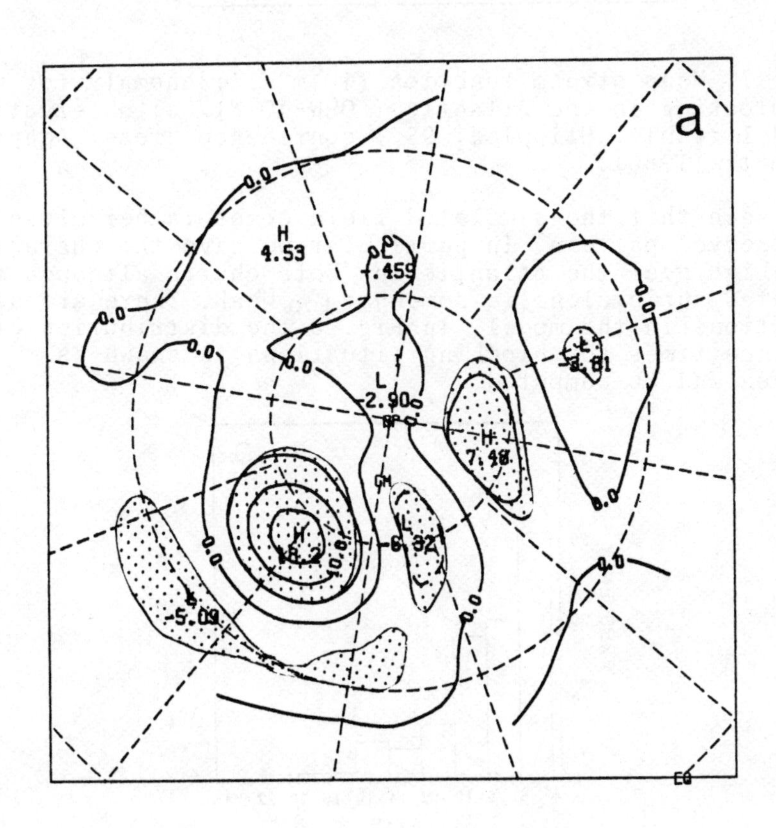

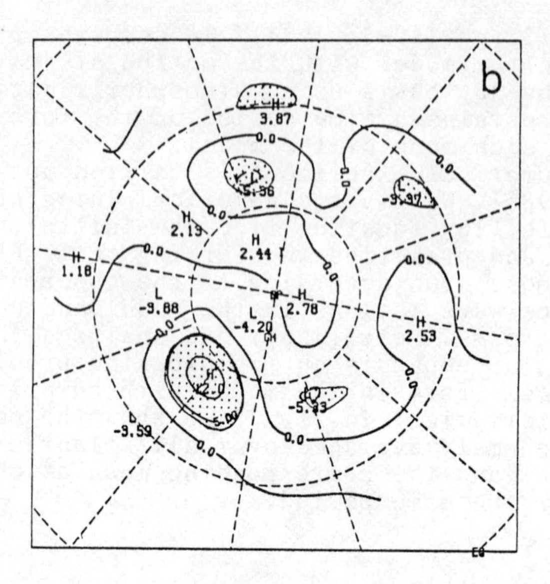

Figure 7. Mean stream function ($10^6 m^2 s^{-1}$) anomaly for winter time blocking in the Atlantic (40°W-10°W). a) observations b) model result. Stippled: 95 % confidence areas. Adapted from Metz (1986).

It is seen that the simulated field comes rather close to the observed pattern. In particular we have the characteristic high over the Atlantic in both charts although the simulated anticyclone is somewhat too weak. There are also transitions in the model. In Fig. 8 the distribution of residence times for blocking situations is shown as observed and as computed.

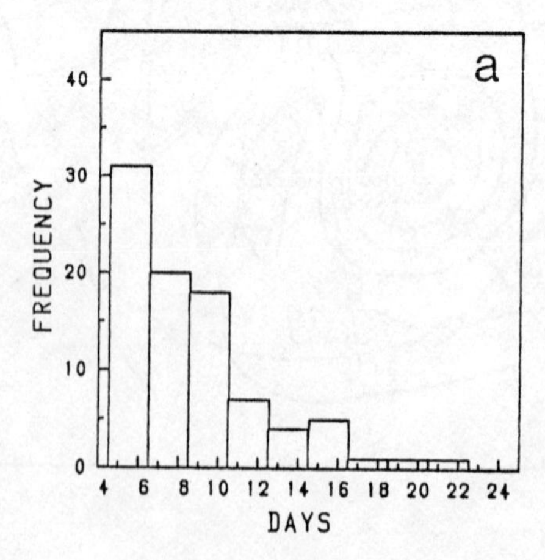

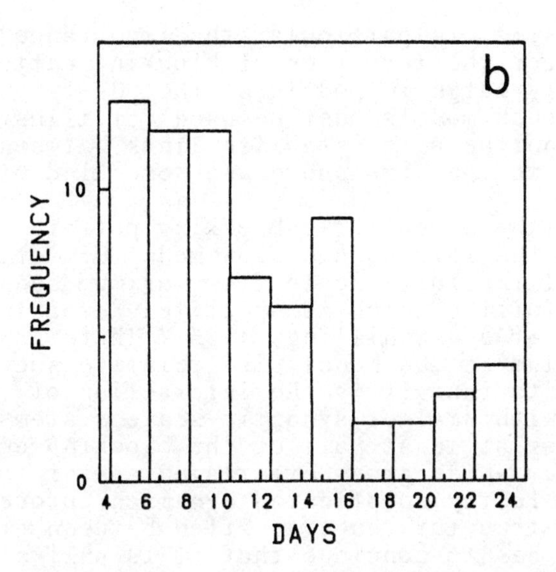

Figure 8. Frequency distribution of blocking highs with duration as observed (a) and as obtained in the forced model (b). Adapted from Metz (1986).

The blocks in the model are too persistent as compared to the atmospheric blocks but again we have a reasonably good agreement. This means that grosswetterlagen of the meridional type can be created even in linear large-scale flow if the forcing is chosen as observed. Since the flow is linear transitions are not due to instabilities. It is the forcing and the linear dynamics of large-scale Rossby waves which are responsible for the buildup and decay of blocking situations in the model.

6. CONCLUSION

Let us try to answer the questions raised in the introduction. Charney and DeVore (1979) suggested that the existence of grosswetterlagen in the atmosphere and their respective residence times and transition probabilities are linked to the existence of multiple equilibria and their attractor basins. Low-order dynamical models of large-scale flow with multiple equilibria have been developed and links between grosswetterlagen in the model and these equilibria have been established. Transitions can be enforced (Egger, 1981) or occur due to instabilities (Legras and Ghil, 1986; Reinhold and Pierrehumbert, 1982). However, the model used so far are rather simple and somewhat unrealistic. In particular the importance of the mountain torque for the formation of blocking patterns is unrealistically large in models of the CdV-type. Correspondingly such models must be used for illustrative purposes but cannot help to establish links between grosswetterlagen in the atmosphere and some kind of equilibria.

On the other hand realistic blocking patterns are seen to occur in the large-scale flow model of Metz (1985). In this model all the large-scale flow must be seen as a response to the forcing which is specified according to observations (Egger and Schilling, 1983). There is no non-trivial equilibrium in the model. The relative success of this model suggests that it is the interaction of large planetary modes with smaller synoptic-scale systems which is responsible for at least part of the blocking activity observed. In other words grosswetterlagen appear to be partly caused by fairly complicated dynamical processes. The same must be true for the transition between grosswetterlagen. One has to conclude that it is unlikely that a simple concept like that of multiple equilibria will be able to solve the problem. It should be kept in mind, however, that forcing experiments like those carried out by Egger and Schilling (1983, 1984) and Metz (1985) cannot solve the problem either. In these experiments the forcing turns out to be of crucial importance. This forcing is taken from observations and it remains, therefore, to explain the structure of these forcing terms. What appears to be necessary is to combine both approaches.

REFERENCES

Benzi, R., A. Hansen and A. Sutera, 1984: On stochastic perturbation of simple blocking models. Quart.J.Roy.Met. Soc. 110, 293-409.

Charney, J. and J. DeVore, 1979: Multiple flow equilibria in the atmosphere and blocking. J.Atm.Sci. 36, 1205-1216.

Baur, F. P. Hess und H. Nagel, 1944: Kalender der Großwetterlagen Europas, 1881-1939 ,Bad Homburg v.d.H.

Charney, J. and D. Straus, 1980: Form-drag instability, multiple equilibria and propagating planetary waves in baroclinic, orographically forced, planetary wave systems. J.Atm.Sci, 37, 1157-1176.

DeSwart, H. and J. Grasman, 1985: Effect of stochastic perturbations on an atmospheric spectral model: the interest of unstable equilibria. Unpub. Manusc.

Egger, J., 1978: Dynamics of blocking highs. J.Atm. Sci. 35, 1788-1801.

Egger, J. 1981: Stochastically driven large-scale circulations with multiple equilibria. J.Atm.Sci. 38, 2608-2618.

Egger, J. and H.-D. Schilling, 1983: On the theory of the long-term variability of the atmosphere. J.Atm.Sci. 40, 1073-1085.

Egger, J. and H.-D. Schilling, 1984: Stochastic forcing of planetary scale flow. J.Atm.Sci. 41, 779-788.

Hess, P. und H. Brezowsky, 1969: Katalog der Großwetterlagen Europas. Ber.Deutsch.Wetterd. 113.

Lakatos, I. 1978: Falsification and the methodology of scientific research programmes. In I. Lakatos and A. Musgrave (edts): Criticism and the growth of knowledge, Cambridge University press.

Legras, B. and M. Ghil, 1986: Persistent anomalies blocking and variations in atmospheric predictability. J.Atm.Sci., 42, to appear.

Lorenz, E. 1963: The mechanics of vacillation. J. Atmos.Sci. 20, 448-464.

Malguzzi, P. and A. Speranza, 1981: Local multiple equilibria and regional atmospheric blocking. J.Atm. Sci., 38, 1939-1948.

Metz, W. 1986: Transient cyclone-scale vorticity forcing of blocking highs. Subm. to J.Atm.Sci.

Metz, W. 1985: Wintertime blocking and the mountain forcing of the zonally averaged flow: a cross-spectral time series analysis of observed data. J.Atm.Sci. 42, to appear.

Reinhold, B. and R. Pierrehumbert, 1982: Dynamics of weather regimes: quasi-stationary waves and blocking. Mon.Weath.Rev. 110, 1105-1145.

Rex, D., 1950: Blocking action in the midde troposphere and its effect upon regional climate. I. An aerological study of blocking action. Tellus 2, 196-211.

Rossby, C.-G. 1950: On the dynamics of certain types of blocking waves. J.Chin.Geophys. 2, 1-13.

Schilling, H.-D., 1982: A numerical investigation of the dynamics of blocking waves in a simple two-level model. J.Atmos.Sci. 39, 998-1017.

Spekat, A., B. Heller-Schulze und M. Lutz, 1983: Über Großwetterlagen und Markov-Ketten. Met. Rdsch. 36, 243-248.

Treidl, R., E. Birch and P. Sajecki, 1981: Blocking action in the northern hemisphere: a climatological study. Atmos. Oc., 19, 1-23.

Tung, K.-K. and R. Lindzen, 1979: A theory of stationary long waves. Part I: A simple theory of blocking. Mon.Weath.Rev. 107, 714-734.

DYNAMICS AND STATISTICS OF ATMOSPHERIC LOW FREQUENCY VARIABILITY

R. Benzi
IBM ECSEC Rome.
A. Speranza
CNR-FISBAT, Bologna.

1. INTRODUCTION.

One of the basic problems in meteorology, if not the basic one, is to explain the general circulation of the atmospheric fluid, namely its average flow and statistical properties of its time fluctuations. Even if many fundamental steps have been taken in the process of solving this problem, the meteorological community is still waiting for a unified solution in which theoretical and observational results so far obtained are assembled.

As noted by Lorenz in 1967, the difficulty we face in trying to resolve this problem is a measure of our limited ability deducing the observed circulation from the equations governing the motion of the relevant physical quantities such as the wind velocity, temperature, pressure, humidity and so on. For the sake of comparison, we may say that this problem proposes the same sort of difficulty as that of deducing the thermodynamic behavior of a real gas starting from Newton's law of mechanics and the atomic hypothesis concerning the structure of the matter. In this latter case, the major breakthrough followed the introduction of the idea that there are well defined average quantities whose dynamic evolution is much slower than that of their atomic constituents. The presence of such a time-gap separation allowed the establishment of a statistical mechanical approach which solved the problem. Pushing further this comparison, we may also say that what is missing in our problematic context is an "atomic hypothesis", physically connected with the existence of a time-gap separation. In meteorology we are used to define a number of scales of motion, but no time separation scale spectral gap has been observed up to know. Still it is a common feeling that the slow time evolution of the general circulation is governed by a limited number of degrees of freedom. If this is the case, the first question to answer is: what are the slow component of the physical system and how we can describe their dynamics without explicitly resolving the dynamics of the "fast" component? Such question is the main object of these lectures. We shall try to give a systematic approach to this problem with the aim of clarifying the nature of the physical problem rather than solving it.

2. THE EQUATIONS OF MOTION.

Let us consider the quasi geostrophic equation of motions on a β plane:

C. Nicolis and G. Nicolis (eds.), Irreversible Phenomena and Dynamical Systems Analysis in Geosciences, 199–239.

$$(\partial_t + \vec{U} \bullet grad)(\Delta\psi + \beta\psi) - \frac{f_0}{\rho_s}\partial_z(\rho_s w) = 0 \tag{2.1}$$

$$(\partial_t + \vec{U} \bullet grad)\psi_z + \frac{N^2}{f_0} w = \frac{gQ}{f_0 c_p T_s} \tag{2.2}$$

where the notation used is the same as in Charney (1973). The boundary conditions of equation (2.1,2.2) are:

$$w|_{z=0} = \frac{1}{2} D\Delta\psi|_{z=0} + \vec{U} \bullet gradh|_{z=0} \tag{2.3}$$

where h is the bottom topography (note that the boundary condition are applied in z=0 and not in z=h). A systematic derivation of equations (2.1,2.2,2.3) from the basic fluid dynamic equations can be found in Charney (1973). In these lectures all the dynamical interpretation of long time, large space scale behavior of the atmosphere is discussed in the framework of the quasigeostrophic approximation.

Integrating equation (2.1) in z and using bottom boundary conditions we obtain:

$$\partial_t \Delta\psi + J(\psi, \Delta\psi + \beta y + f_0 \frac{h}{H}) = forc. + diss. \tag{2.4}$$

that is the so-called barotropic version of the quasi geostrophic equations. Equation (2.4) describes the evolution of a flow which is homogeneous in the vertical. In several respects this approximation fails as we shall see later, for the real atmosphere. A flow described by equation (2.4) cannot perform any conversion of potential energy into kinetic energy.

3. OBSERVATIONAL PRELIMINARIES

It is known that tropospheric variability at time scales exceeding 10 days is dominated by planetary flow patterns (see for example Sawyer 1970, Blackmon 1976, Dole 1982). The maximum variance is localized in fixed geographical areas. The meteorological features that contribute to such variance are usually stationary or slowly migrating. Such properties are clearly evidenced by space-time spectral analysis. In the space-time spectral domain the low-frequency component of the variance appears as a distinct peak in the region corresponding to periods exceeding 10 days and to zonal wavenumbers less than 5 (see for example Freidrich and Bottger 1978). In so far as propagating and non-propagating wave-components can be separated (for summary and discussion of this problem see Hayashy, 1982), the low-frequency variability is almost totally non-propagating on average. Analysis of specific segments of the time evolution of single wave component seems to show only during limited extents of time a tendency for wavenumber 1-2 to propagate to the West and for wavenumbers above 4 to propagate to the East (in agreement to Rossby dispersion formula). At this point, it is worth mentioning that the results concerning the partition of the variance into stationary and propagating somewhat depend on the nature of the eigenfunctions chosen for projection. Lindzen et.al. (1984), for example, chose as an expansion base the eigenfunctions of the Laplace tidal equations. Such functions (Hough functions) are fully global. They average therefore the Northern Hemisphere variance where, as mentioned, the low-frequency variability is almost non-propagating and the southern hemisphere variance where the low-frequency variability is essentially of the propagating type (Mechoso,Hartman 1982; Fraedrich,Kietzig 1983). However, even in this projection, identifiable wave trains are characterized by variances that are generally small (5%-20%) compared to the total low-frequency variance (Straus, 1984). It therefore seems impossible to explain the low-frequency variability in terms of traveling neutral waves produced by some readjustment process on the global scale. This impression is confirmed by the analysis of the spectral energetics and its time variations (Chen 1982; Chen and Marshall, 1984) which reveals the presence of a well-marked peak in all the cross-correlations in a band of periods between 10 and 20 days. Care must be taken here not to interpret the presence of a distinct process in a specific spectral band as a sign of periodicity. In fact, in our case, the process is distinctly intermittent: it is the transformation operation that projects the signal on a periodic support. As mentioned, the results of spectral analysis depend to a certain extent on the projection operations. The analysis we refer to above are all based on transformation in longitude. This may cast doubts on the robustness of the properties of the Fourier spectrum with respect to different latitudinal decompositions. However, preliminary analysis of two dimensional spectral distributions in horizontal domain (Lambert 1984) confirm the results obtained with spectral decomposition only in the zonal direction.

In a sequence of papers Lindzen and Tung (Tung and Lindzen, 1979a,b; Tung, 1979) give a full account of the linear theory of three dimensional, neutral planetary waves in the middle latitude circulation. They discuss the possible interpretation large amplitude stationary waves and blocking anomalies in terms of neutral Rossby waves forced by zonal non-homogeneities in the boundary conditions (topography) or in the external forcing (heat sources and sinks). Low frequency variability is interpreted as caused by the linear "adjustment" of the circulation to changes in the basic zonal wind, the evolution of which is conceived as determined by independent dynamics. In the language of dynamical systems, the picture that emerges is that of an atmospheric circulation performing "small" (linear) oscillations around an equilibrium point essentially determined by the dynamics of the zonal flow (not discussed in the above mentioned papers).

A detailed analysis of the linear theory for three dimensional planetary waves forced by orography and heat sources can be found in Charney (1973) and Held (1983). The two kind of forced waves have different vertical structures: waves forced by orography are vertically coherent while heat sources produce westward vertical shift. The observed vertical structure of stationary planetary waves seems to be near the one predicted by orography forcing at middle latitude. In the following we shall concentrate on the orographic forcing as main source of forced stationary waves.

4. BAROTROPIC THEORIES

Let us consider the barotropic equation (2.4). We start discussing a quite simple picture of orographically modulated flow by means of a unidimensional version of Charney DeVore model (1979, hereafter CDV). Our preference for unidimensional formalism is due not only to its intrinsic simplicity, but also to the fact that it provides us with a field theory characterized by some useful mathematical properties. In particular, the stationary solution of the unidimensional problem we intend to discuss, are solutions of the barotropic equation. The same is not true for the stationary solution of the original CDV system. As a matter of fact, this difficulty is encountered each time, following a classical procedure, the equations of motions are projected onto the eigenfunctions of the Laplace operator. Such eigenfunctions form a complete set, but their linear combinations are not necessarily solutions of the stationary nonlinear problem: the stationary solutions of any truncated version of the projected equation is not, in general, a solution of the original stationary problem.

In the case of unidimensional formalism, on the other hand, we shall deal with stationary solutions that are solutions of the full barotropic equations. In future work this property may prove to be crucial to an appropriate representation of localized (soliton type) solutions.

Orography is assumed to be of the form:

$$h(x) = h_0 \cos k_m x \tag{4.1}$$

The streamfunction is composed of a uniform zonal flow plus a wave of the same shape as the topography:

$$\psi(x, t) = -U(t)y + A(t)\cos k_m x + B(t)\sin k_m x \tag{4.2}$$

Inserting (4.2) into (2.4) it can be seen that the wave is maintained by the zonal flow over the topography slope and dissipated by friction:

$$\dot{A} + \frac{\nu}{k_m} A + \left(U - \frac{\beta}{k_m^2}\right) B = 0 \tag{4.3a}$$

$$\dot{B} + \frac{\nu}{k_m} B - \left(U - \frac{\beta}{k_m^2}\right) A + U \frac{f_0 h_0}{H k_m^2} = 0 \tag{4.3b}$$

and the zonal flow is forced by some momentum convergence mechanism U^*, dissipated by linear friction and accelerated or decelerated by mountain drag:

$$\dot{U} = f_0 k_m \frac{h_0}{4H} B - \nu'(U - U^*) \tag{4.4}$$

Equation (4.4) is called the form-drag equation and cannot be derived from the barotropic vorticity equation (2.4) by simple substitution of the form of the solution (4.2) (see Buzzi et al. 1984). Indeed equation (4.4) is a statement of angular momentum conservation for the model atmosphere. Note that for $\nu \to 0$ equations (4.3) and (4.4) conserve the total kinetic energy. Equilibrium solutions of eqs.(4.3) and (4.4) can easily be found by solving the linear system (4.3) for B (the amplitude of the wave component out of phase with respect to topography), as a function of U, and by searching for the intersection of the resulting curve B(U) with linear relationship corresponding to the stationary version of (4), as shown in Fig 1. Once the appropriate points in the plane B,U are selected, A is also determined along the curve A(U). By means of stability analysis, equilibrium E_1 was identified by CDV as "zonal", E_2, as blocked, and the third, E_3, as an "intermediate" equilibrium, unstable with respect to a new kind of instability, called "orographic", because of the role played in its growth by the orographic drag action appearing in (4.4).

It is clear that the presence of multiple equilibria is due to a simple folding of the state surface in the direction U^* in parameter space. Although the physics is simple and interesting, one already notices at this early stage some of the theory's inadequacies. For example, the locations of the equilibria along the U axis are not realistic: the corresponding value of the zonal flow are never observed in the real atmosphere (we shall return to this point later). This sort of distortion of the observed dynamics can perhaps be tolerated in such a simple model. However, even more serious inadequacies are found in the statistical properties of the CDV model atmosphere. Given the equilibrium structure of phase space illustrated in Fig. 1, it is obvious that the asymmetric time behavior of the time-dependent system (4.3,4.4) is characterized by only two possibilities: orbits in phase-space can,in a certain span of time, die either in E_1 or in E_2. In order to produce transitions between zonal and blocked states, we have to superimpose on the stationary solution some sort of perturbation. This can be done by increasing the dimensionality of the model or, more simply, by taking into account the effect of small random noise on the deterministic CDV dynamics. This last approach was first taken by Egger (1981), who integrated numerically the Fokker-Plank equation for the evolution of probability density in phase space of the stochastically perturbed CDV system and obtained the intuitive result that the asymptotic statistics of occupation of states in the CDV system are characterized by maxima centered at the stable equilibria. The same problem was also studied by Benzi,Hansen and Sutera (1984) (from now on, BHS), with an analytic technique for estimating the expectation values of exit times, From the potential wells of the CDV theory, use of this analytical technique, instead of numerical integration, makes the overall picture of the model's statistical properties more controllable. The estimate is based on the assumption

$$\frac{1}{\nu} < < \frac{1}{\varepsilon} \tag{4.5}$$

where $\frac{1}{\nu}$ is the time typical of dissipative processes and $\frac{1}{\varepsilon}$ is the time-scale of random noise. In the limit (4.5) it can be proved that the behavior of the entire system is approximated by the single stochastic differential equation for the zonal wind:

$$dU = -\left(\frac{\partial V}{\partial U}\right)dt + \sqrt{\varepsilon}\, dW \tag{4.6}$$

where V(U) is the equivalent of a potential for the motion of perturbed CDV system in the phase-space, and W is a Wiener process with amplitude $\sqrt{\varepsilon}$. The corresponding stationary probability density is

$$P(U) = N\exp\left(-\frac{2V(U)}{\varepsilon}\right) \tag{4.7}$$

where N is a normalization constant. in the limit $\varepsilon \to 0$, the expectation values ($<..>$) of the exit times $\tau(\bullet)$ can be estimated as:

$$<\tau(E_2, E_1)> \leq \pi\left|\frac{dV}{dx^2}(U_2)\frac{dV}{dx^2}(U_3)\right|^{-\frac{1}{2}}\exp\left(\frac{2\Delta_1 V}{\varepsilon}\right) \tag{4.8a}$$

$$<\tau(E_1, E_2)> \leq \pi\left|\frac{dV}{dx^2}(U_1)\frac{dV}{dx^2}(U_3)\exp\left(\frac{2\Delta_2 V}{\varepsilon}\right)\right. \tag{4.8b}$$

where $\Delta_{1,2} = V_3 - V_{1,2}$ represent the potential barriers to be overcome in order to escape from the blocked a nd zonal potential wells. Exit times are shown in Table I. It is clear that dramatic changes take place when the external forcing U^* changes. The structural dependence on U^* can be stabilized if we include red noise in the zonal momentum equation (see BHS for details),

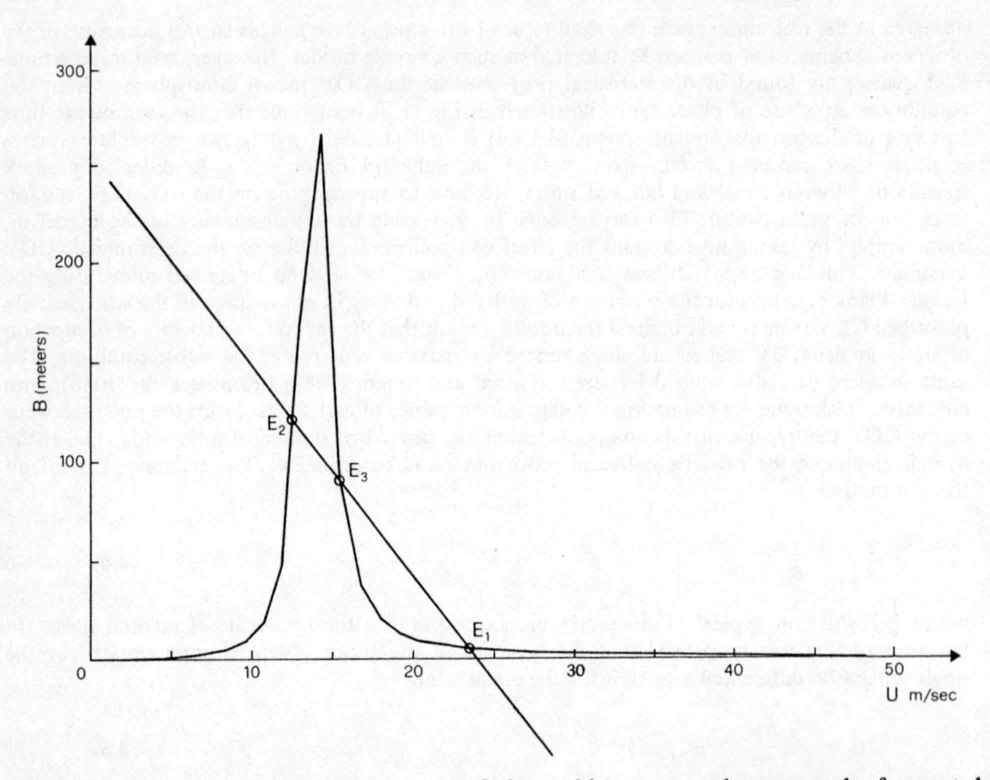

Fig 1. Absolute value of the component out of phase with respect to the topography fcomputed from the unidimensional model equation (2.3). The amplitude is expressed in meters and the resonant wind is about 13 m/sec. The straight line represents the form-drag relation as in CDV. E_1¢, ¢E_2¢, ¢E_3 indicate respectively the "zonal, blocked and intermidiate" equilibria of CDV.

Zonal forcing	T (block)	T (zonal)
10 meters/sec.	days	days
.64	43.2	101.6
.77	10.2	24.6
.89	6.1	28.1
1.02	4.3	44.7
1.15	3.3	90.0
1.28	2.6	223.0
1.41	2.2	669.6
1.54	1.9	2400.
1.66	1.7	10000
1.79	1.6	54000

Table I. Exit times for the Charney-Devore model as a function of the zonal forcing U^*. T(block) and T(zonal) are the times for the blocking-to-zonal and zonal-to-blocking transitions, respectively.

however the problem of fitting the observed statistics remains unsolved because the remarked separation in intensity of the zonal wind of the two stable equilibrium states makes the statistics too dispersed in U itself.

To give a physical picture of the nature of the problem, we have represented in Fig. 2 the distribution of instantaneous states in the CDV parameter space B-U computed from observations of 500 mb height during two winters (1977-78 and 1978-1979). It is clear that the distribution of observed values of zonal wind is much less dispersed than in CDV. No similar disagreement with the range of observed variability can be noticed in the wave amplitude.

It is clear that, if we accept the observed zonal momentum distribution as a phenomenological law and still insist on the idea of multiple equilibria, we need a mechanism capable of stabilizing, at finite amplitude, the wave in a limited range of zonal wind values. Going back to the barotropic equation we see that there are two main contribution to the modification of the wave amplitude: wave-wave interaction $J(\psi, \Delta\psi)$ and wave-orography interaction $J(\psi, f_0 \frac{h}{H})$. In CDV theory the only effect taken into account is $J(\psi, f_0 \frac{h}{H})$. Hence the linear structure of the wave equation does not allow the existence of different wave amplitude for the same zonal wind U. On the other hand insertion of non-linearity in the wave equation, by means of $J(\psi, \Delta\psi)$, can easily "bend" the resonance, thus creating a folding with three branches of stationary solutions superimposed in the wave amplitude direction of parameter space.

The easiest way of producing resonance bending consists of introducing weak non-linearity into the solutions of the stationary, inviscid equation (Malguzzi and Speranza, 1981, from now on MS):

$$J(\psi, \Delta\psi + \beta y + f_0 \frac{h}{H}) = 0 \tag{4.10}$$

A solution of this equation is:

$$\Delta\psi + \beta y + f_0 \frac{h}{H} = F(\psi)$$

for any function F. If F is a linear function then equation (4.11) is identical to the inviscid version of CDV theory. We are interested in the case of nonlinear F. For simplicity we shall assume $\partial_y \cong 0$, i.e.

$$\psi_{xx} + \beta y + f_0 \frac{h}{H} = -k^2\psi - c\psi^3 \tag{4.11}$$

Equation (4.11) can be studied using multiple scale analysis with $h \sim c \sim \varepsilon << 1$ and assuming:

$$\psi = \psi_0 + \varepsilon\psi_1 + \varepsilon^2\psi_2 + ...$$

At the zeroth order we obtain:

$$\psi_{0xx} + k^2\psi_0 + \beta y = 0$$

the solution of which is

$$\psi_0 = -Uy + Ae^{ikx} + *$$

with $k^2 = \frac{\beta}{U}$. Introducing the "long" scale $X = \varepsilon x$ and writing the orography as

$$h(x, X) = h_0 e^{ikx + i\Delta X} + (*)$$

$$\Delta k = k_m - k$$

we obtain at the first order:

$$\psi_{1xx} + k^2\psi_1 = -2ikA_X e^{ikx} - f_0 \frac{h_0}{H} e^{ikx + i\Delta k} +$$

$$-c(3A|A|^2 e^{ikx} + 3U^2 y^2 A e^{ikx}) + *$$

All terms on the r.h.s. must sum to zero in order to avoid resonances for ψ_1. Assuming:

$$A(X) = ae^{i\Delta kX}$$

we obtain

$$2k\Delta ka + f_0 \frac{h_0}{H} + 3ca|a|^2 + 3U^2 y^2 ca = 0$$

Let us consider the simple case y = 0:

$$3ca^3 + 2k\Delta ka + f_0 \frac{h_0}{H} = 0$$

Assuming $c > 0$ we can easily check that multiple equilibria exist for $\Delta k < 0$, i.e. for $U < \frac{\beta}{k_m^2}$.

What is new and important for this model respect to the CDV theory is that multiple equilibria do exist for the same value of U.

In order to produce resonance bending in the context of the forced-dissipative dynamics of CDV, we shall introduce a nonlinearity associated with wave selfinteraction. Before proceeding with the necessary modification of CDV, it is useful to consider briefly the nature of the different non-linearities appearing in the quasi-geostrophic equations. Non linearity of the Jacobian operator has two distinct parts: wave-wave interaction and wave-zonal flow interaction. The latter has very frequently been considered in theoretical studies (see, for example, Pedlosky, 1981), while the former has usually been neglected. This is because, in the usual one-wave representation of the wave field near neutral stability, direct wave self-interaction vanishes. In the following theoretical treatment of equilibrium solutions of the barotropic equation (BMSS), we prefer to consider as basic non-linearity (that bends the resonance in the B,U plane) the self-interaction of waves. Here again, of crucial importance is the choice of the eigenfunction in terms of which the streamfunction field is written: our choice is that of separated modes of the form A(x,t)g(y) with a latitudinal structure such as to guarantee wave self-interaction. In mathematical terms, we expand the barotropic wave equation with dissipation in terms of functions of the form

$$\psi(x,y,t) = \Sigma_n A_n(x,t) g_n(y) \tag{4.12}$$

$$h(x,y) = \Sigma_n h_n(x) g_n(y) \tag{4.13}$$

and, multiplying by $g_n(y)$, integrating in y and keeping only one mode, we obtain:

$$(\partial_t + U\partial_x)(A_{1xx} + \alpha^2 A_1) + \frac{f_0}{H} U h_x + \beta A_{1x} + \frac{3}{2}\delta A_1 A_{1x} + \tag{4.14}$$

$$-\nu(A_{1xx} + \alpha^2 A_1) = \gamma[A_1(A_{1xx} + \frac{f_0}{H} h) - A_1(A_{1xxx} + \frac{f_0}{H} h_x)]$$

where the non-linear self-interaction coefficients have been defined as ($\{....\} = \int ....dy$):

$$\alpha^2 \equiv \{g_1 g_{1yy}\} \tag{4.15}$$

$$\gamma \equiv \{g_1 g_1 g_{1y}\}$$

$$\delta \equiv \{g_1 g_{1y} g_{1yy}\}$$

The nature of the coefficients defined in (4.15) clarifies the meaning of the previous remarks concerning non-linearity in our model. This non-linearity depends on the latitudinal structure of the modes: it would vanish for an eigenfunction of Laplace operator of the kind considered by Pedlosky (1981) and many others. It can be proved (Speranza et.al., 1984) that local topography, for example, is able to produce the latitudinal structure needed to produce self-interaction. In the following we shall assume $\gamma = 0$ because of lateral b.c. and concentrate on the role of the nonlinearity associated with δ. The ordering of parameters that turns out to be necessary for resonance bending in CDV parameter space is (ε is a small parameter):

$$\delta \equiv O(\varepsilon) \tag{4.16}$$

$$h_0 \equiv O(\varepsilon^2)$$

$$\nu u \equiv O(\varepsilon^2)$$

which, in physical terms, means that Jacobian interaction is stronger than topographic action and dissipation. Introducing long time and space scales

$$T = \varepsilon^2 t \tag{4.17}$$

$$X = \varepsilon^2 x \tag{4.18}$$

and assuming the amplitude of the wave mode is of the form

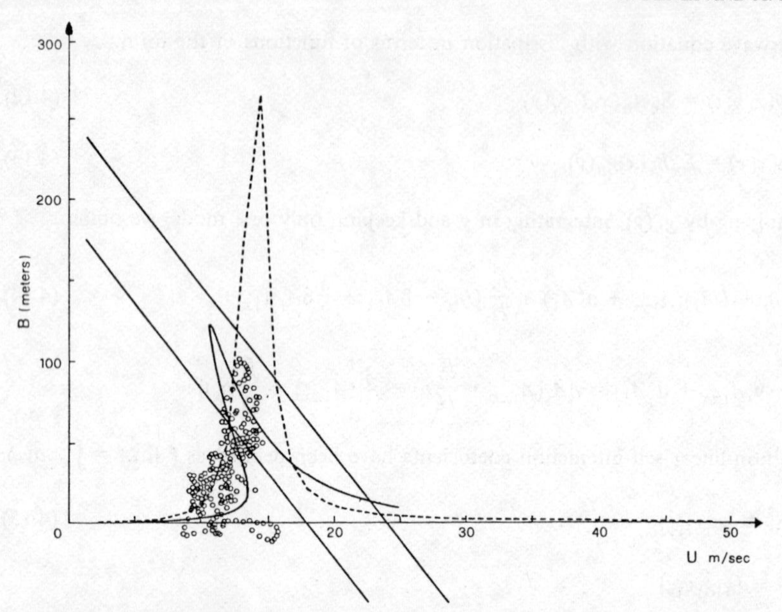

Fig 2. Plot of the nonlinear resonance, linear resonance and formdrag relationship. The circles represent the amplitude of the out of phase component with respect to the topography of wavenumber 3 (data from winters 66/67 and 77/78).

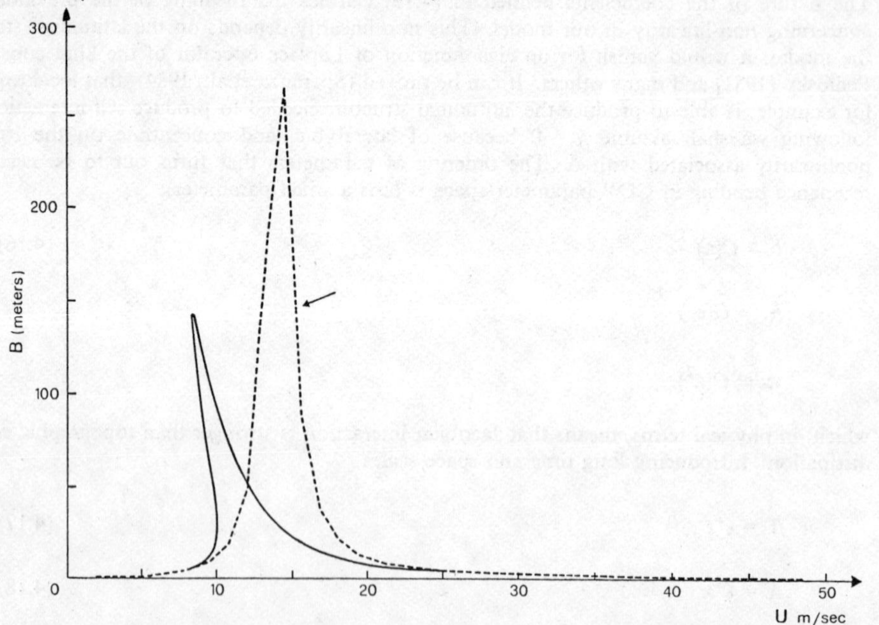

Fig 3. Amplitude of the wave component out of phase with topography computed from the stationary version of (2.18) as a function of the zonal wind U. The resonant wind is the same as in fig. 1 and $\alpha = 0$, $k_m = 10^{-7} m^{-1}$ and $\delta = \dfrac{1}{1000^3} km$. The dashed line represents the linear case ($\delta = 0.$).

$$A_1(x, X, t, T) = A^{(0)}(x, X, T) + \varepsilon A^{(1)} + \varepsilon^3 \ldots\ldots\ldots \tag{4.19}$$

we obtain from (4.12-13), at zero order in ε

$$(U\alpha + \beta)A_x^{(0)} + UA_{xxx}^{(0)} = 0 \tag{4.20}$$

The solution of (4.20) is:

$$A^{(0)} = B(X, T)\exp(ik_s x) + (*) \tag{4.21}$$

where

$$k_s{}^2 = \frac{U\alpha^2 + \beta}{U} \tag{4.22}$$

that is, a modified wavenumber of the stationary Rossby wave (4.21). At first order in ε, we obtain:

$$(\beta + U\alpha^2)A_x^{(1)} + UA_{xx}^{(1)} + \delta A^{(0)}A_x^{(0)} = 0 \tag{4.23}$$

A particular solution of (4.23) is

$$A^{(1)} = -\frac{\delta|B|^2}{\beta + U\alpha^2} - \frac{\delta}{2}B^2\frac{\exp(2ik_s x)}{\beta + U\alpha^2 - k_s^2 U} \tag{4.24}$$

At second order in ε, secular terms appear and the solvability condition reads:

$$\partial_T(\alpha^2 - k_s^2)B + ik_s U f_0 \frac{h}{H} - 2(\beta + \alpha^2 U)B_X \tag{4.25}$$

$$-\frac{7}{6}\delta^2 ik_s|B|^2\frac{B}{(\beta + \alpha^2 U)} = -\nu(\alpha^2 - k_s^2)B$$

The introduction of slight detuning $\Delta k = k_m - k_s$, allows us to write

$$h(X) = h_0(i\Delta k X) + (*) \tag{4.26}$$

which, introduced into (4.25), gives the stabilization equation

$$\partial_T(\alpha^2 - k_s^2)B + ik_s U f_0 \frac{h}{H} - 2i\Delta k(\beta + \alpha^2 U)B + \tag{4.27}$$

$$-\frac{7}{6}\delta^2 ik_s|B|^2\frac{B}{(\beta + \alpha^2 U)} = -\nu(\alpha^2 - k_s^2)B$$

The stationary version of this equation is a cubic condition which produces resonance bending in the CDV parameter space (as shown in fig. 3). Although the expansion outlined above is based on ad hoc assumptions, the phenomenology of resonance bending is very robust and can easily be observed in many physical systems. Moreover in the context of barotropic dynamics such phenomenology is confirmed by numerical experiments with both finite difference (Rambaldi and Mo 1984) and spectral (Legras and Ghil 1983) models of high spatial resolution in both of which, however, the dominant non-linear interaction is associated with wave-zonal flow, rather than wave-wave interaction as in our case.

The structure of the wave amplitude equilibrium solutions that we have drawn above is obviously apt to produce a time dependent behavior with statistical properties in essential agreement with observations. However, in order to write a consistent dynamical system, we must close (4.27) with a prognostic equation for zonal wind. Here, we face the essential difficulty of all barotropic theories: in order to produce orographic waves of the observed amplitudes, we must postulate an orographic drag much smaller than the observed one (see BMSS for a discussion). The problem is most easily expressed in terms of the energetics of the barotropic models. The form-drag equation is exactly what we energetically require in order to transfer energy from the zonal flow to the wave: in absence of potential energy sources or direct wave forcing (i.e. in the CDV type of energetics) the only way of amplifying planetary waves is through the action of mountain drag. However, observational analysis of this term, both for seasonal average waves (Holopainen 1970) and their time fluctuations (Schilling, 1982), shows that its role is consistently negligible ($\sim 10^{-2} \frac{Watts}{m^2}$). More recent analysis of winter data in general by Chen (1982), Chen and Marshall (1984), and more specifically, of blocking versus non-blocking cases by Hansen and Shilling (1985) confirm the above results, together with the general observation that the mountain drag term does not play an important role in the real atmosphere (see Speranza 1984 for a discussion). As we shall see in the next section, form-drag can operate as an essential catalyst for baroclinic conversion, although it directly forces only irrelevant amounts of energy in the ultra-long waves.

5. BAROCLINIC ENERGETICS AND NON-LINEAR RESONANCE

The barotropic model has served well the purpose of isolating the mechanism of non-linear stabilization required in order to produce theoretical multiple equilibrium states in the observed range of variability of the Northern Hemisphere circulation. However, as we have seen, the energetics of the observed variability is dominated by baroclinic conversion. We shall now discuss here how to extend the concepts concerning non-linear stabilization to a baroclinic model atmosphere. For mathematical convenience we make use of the quasi-geostrophic, two-level model in the standard formulation (Pedlosky, 1964):

$$\partial_t q_1 + J(\psi_1, q_1) = F_1 \tag{5.1a}$$

$$\partial_t q_2 + J(\psi_2, q_2) = F_2 \tag{5.1b}$$

where

$$q_1 = \Delta\psi_1 + F(\psi_2 - \psi_1) + \beta y$$

$$q_2 = \Delta\psi_2 + F(\psi_1 - \psi_2) + \beta y + h$$

In (5.1) $F_{1,2}$ represent any kind of forcing and/or dissipation.

Equation (5.1) can be derived by (2.1),(2.2) and (2,3) by a descretization in the vertical axis z (see Pedlosky 1964). Separation of symmetric and non symmetric flow components is obtained, in analogy with the procedure used in setting the barotropic model, by assuming that :

$$\psi_1(x,y,t) = -U_1(t)y + \varphi_1(x,y,t) \tag{5.2a}$$

$$\psi_2(x,y,t) = -U_2(t)y + \varphi_2(x,y,t) \tag{5.2b}$$

and wave-wave nonlinearity by defining:

$$\varphi_1(x,y,t) = \Sigma_{n=1,N}\, g_n(y) A_{1,n}(x,t) \tag{5.3a}$$

$$\varphi_2(x,y,t) = \Sigma_{n=1,N}\, g_n(y) A_{2,n}(x,t) \tag{5.3b}$$

where the $g_n(y)$ satisfy the same conditions as in the barotropic case. We will consider one wave approximation (N= 1) with topography h(x,y) projected onto the same g function. The forcing on the wave field is pure laplacian dissipation:

$$F_1 = -\nu_1 \Delta\psi_1 \tag{5.4a}$$

$$F_2 = -\nu_2 \Delta\psi_2. \tag{5.4b}$$

Substituting (5.2-3) into (5.1) and projecting onto g(y), we obtain the quasi unidimensional wave equation:

$$\partial_t[A_{1xx} - \alpha^2 A_1 + F(A_2 - A_1)] + U_1\partial_x(A_{1xx} - \alpha^2 A_1) + \beta\partial_x A_1 + \delta A_1 A_{1x} \tag{5.5a}$$

$$+F(U_1 A_{2x} - U_2 A_{1x}) = -\nu_1(A_{1xx} - \alpha^2 A_1)$$

$$\partial_t[A_{2xx} - \alpha^2 A_2 + F(A_1 - A_2)] + U_2\partial_x(A_{2xx} - \alpha^2 A_2) + \beta\partial_x A_2 + \delta A_2 A_{2x} \tag{5.5b}$$

$$+F(U_2 A_{1x} - U_1 A_{2x}) + U_2 h_x = -\nu_2(A_{2xx} - \alpha^2 A_2)$$

where h has only projection on g(y), i.e. $h(x,y) \to h(x)g(y)$ and:

$$\alpha^2 = -\int g g_{yy}\, dy \tag{5.6}$$

$$\delta = \int g g_y g_{yy}\, dy$$

are the nonlinearity coefficients of our system. Another coefficient $\gamma = \int g^2 g_y\, dy$ originally appears in the projected equations (2.11), but vanishes because of the boundary conditions As in the barotropic case, the choice of non-trigonometric g(y) allows non linear interaction of waves to take place in the midlatitude channel.

In order to close the wave dynamics described by (5.5), we need to write additional equations for the time evolution of the zonal flow. As already mentioned for the barotropic equation (see Buzzi et al. 1984 for a discussion), since the symmetric circulation defined in (5.2) has no horizontal vorticity, the derivation from the potential vorticity equation is not straightforward, involving a limit to uniform latitudinal structure (Hart 1979). An alternative is to derive the quasi unidimensional approximation for the zonal flow directly from the momentum equations. Since neither procedure is particularly illuminating in the context of the problem in question, we do not give a account here, the results being in any case very familiar.

Energetic closure of (5.5) is guaranteed by two equations. The first is the form-drag equation for the average zonal momentum $U = 0.5(U_1 + U_2)$:

$$\partial_t U = -\overline{A_2 h_x} - \nu_U (U - U^*) \tag{5.7}$$

where the overbar represents zonal average and ν_U is the inverse of a typical relaxation time of the system. As usual U^* is the external momentum forcing, provided in the real atmosphere mostly by convergence of zonal momentum fluxes associated with small and fast eddies which are not explicitly represented in our minimal model. Such forcing is often omitted in the baroclinic case (as, for example, in Charney and Straus 1980) because the average baroclinicity associated with planetary latitudinal thermal contrast is available as a primary driving of the circulation, as we shall see in the discussion of the next closure equation. However, this procedure leads to the consequence that, in order to mantain a stationary balance in (5.7), dissipation must be, against any observational evidence, compensated by formdrag acceleration. We therefore maintain explicit momentum forcing U^* as in the barotropic case. The second closure equation describes the time evolution of the average baroclinicity $m = 0.5(U_1 - U_2)$:

$$C\partial_t m = 2F\overline{A_2 A_{1x}} + \overline{A_2 h_x} - \nu_m (m - m^*) \tag{5.8}$$

where, again, ν_m is the inverse of a relaxation time and m^* is external baroclinicity generation associated with global latitudinal thermal contrast. C is a constant representing the inertia of the average baroclinicity field m. This constant is determined by two contribution: the first, associated with the horizontal vorticity of the zonal flow, is kinematic, while the second, associated with vertical shear, is thermodynamic. The thermodynamic inertia dominates in the earth atmosphere.

Equations (5.5,5.7,5.8) form a closed system of partial differential equation in the zonal shape of the wave field $A_{1,2}(x, t)$, the total zonal momentum U(t) and the average baroclinicity m(t). The purpose of the next section is to show that this system of equations admits multiple wave amplitude stationary solution in a limited range of values of U and m, as required by observations.

In order to set up an approximation suitable for the search of multiple equilibrium solutions of (5.5,7,8), we observe that, in the presence of orography, the symmetric circulation $U_2 = 0$ produces rather realistic instability patterns (see Buzzi et.al. 1984 for a discussion of orographic baroclinic in the unidimensional formulation). Of particular importance are the baroclinic character of the instability, as well as its vertical coherence and its stationary phase. Moreover, we are interested in limited deviations from symmetric circulation. We shall therefore consider states with most of the momentum concentrated in the upper level. The rest of our scaling consists in the classical quasi resonant approximation (see BMSS) which provides the nonlinear resonance folding.

Specifically, having defined a small parameter ε (usually the wave detuning with respect to resonance), we assume:

$$\delta = O(\varepsilon) \rightarrow \delta\varepsilon \tag{5.9}$$

$$U_2 = O(\varepsilon^2) \rightarrow \varepsilon^2 U_2$$

$$\nu = O(\varepsilon^2) \rightarrow \varepsilon^2 \nu$$

$$X = \varepsilon^2 x$$

$$\tau = \varepsilon^2 t$$

where all the dissipations have been assumed equal to ν and slow space-time modulations have been introduced. The arrows indicate substitution of the scaled variables to the original ones. We expand the wave field:

$$A_{1,2}(x,t,X,\tau) = A^{(0)}_{1,2}(x,t,X,\tau) + \varepsilon A^{(1)}_{1,2}(x,t,X,\tau) + \varepsilon^2 A^{(2)}_{1,2}(x,t,X,\tau) + \ldots.. \tag{5.10}$$

and consequently expand equation (2.11). At the lowest (zero) order we obtain:

$$U_1(A^{(0)}_{1xxx} - \alpha^2 A^{(0)}_{1x}) + \beta A^{(0)}_{1x} + F A^{(0)}_{2x} U_1 = 0 \tag{5.11a}$$

$$(\beta - FU_1) A^{(0)}_{2x} = 0 \tag{5.11b}$$

which define a nondissipative , free baroclinic Rossby wave :

$$A^{(0)}_1(X,\tau,x) = a_1(X,\tau) \exp ik_s x + (*) \tag{5.12a}$$

$$A^{(0)}_2(X,\tau,x) = 0 \tag{5.12b}$$

The wave is stationary, with resonant wave number $k_s = \sqrt{(\frac{\beta}{U_1} - \alpha^2)}$ except for slow modulation in space and time. The vanishing of $A^{(0)}_2$ reveals that the wave is essentially confined to the upper level as the zonal flow: only at second order in ε corrections in the wave field of the lower level will appear.

At the first order in the small parameter ε:

$$U_1(A^{(1)}_{1xxx} - \alpha^2 A^{(1)}_{1x}) + \beta A^{(1)}_{1x} - \delta A^{(0)} A^{(0)}_{1x} + F A^{(1)}_{2x} U_1 = 0 \tag{5.13a}$$

$$(\beta - FU_1) A^{(1)}_{2x} = 0 \tag{5.13b}$$

which describe a non-linearly modified, nondissipative, free Rossby wave:

$$A^{(1)}_1(X,\tau,x) = \delta \frac{[a_1(X,\tau)]^2}{2(\beta - \alpha^2 U_1)} + \delta \frac{[a_1(X,\tau)]^2}{2(\beta - \alpha^2 U_1 - 4k_s^2 U_1)} e^{2ik_s x} + (*) \tag{5.14a}$$

$$A_2^{(1)} = 0 \tag{5.14b}$$

modulated on the zonal wavenumbers 0 and $2k_s$. The slow space time modulation $a_1(X, \tau)$ is determined by the balance with frictional dissipation and topographic action at the second order in ε:

$$\partial_\tau [A_{1xx}^{(0)} - (F + \alpha^2) A_1^{(0)}] \tag{5.15a}$$

$$+ U_1(A_{1xxx}^{(2)} - \alpha^2 A_{1x}^{(2)}) + \beta A_{1x}^{(2)} + (\beta - \alpha^2 U_1) A_{1X}^{(0)} + 3U_1 A_{1xxX}^{(0)}$$

$$- \delta A^{(1)} A_{1x}^{(0)} - \delta A^{(0)} A_{1x}^{(1)} + F(A_{2x}^{(2)} U_1 - A_{1x}^{(0)} U_2) = -\nu(A_{1xx}^{(0)} - \alpha^2 A_1^{(0)})$$

$$\partial_\tau F A_1^{(0)} + \beta A_{2x}^{(2)} + F(U_2 A_{1x}^{(0)} - U_1 A_{2x}^{(2)}) + U_2 h_x = 0 \tag{5.15b}$$

Equation (5.15b) is linear in $A_2^{(2)}$:

$$A_{2x}^{(2)} = -\frac{U_2 h_x + F U_2 A_{1x}^{(0)} + F \dot{A}_1^{(0)}}{(\beta - F U_1)} \tag{5.16}$$

which, substituted into (5.15a), gives:

$$-\partial_\tau [(\alpha^2 + k_s^2) a_1 + F a_1] - 2k_s^2 U_1 a_{1X} - \frac{\delta^2 i k_s a_1 |a_1|^2}{3k_s^2 U_1} \tag{5.17}$$

$$- F U_2 i k_s a_1 - \nu(\alpha^2 + k_s^2) a_1 - \frac{F U_1}{\beta - F U_1} [U_2 i k_m h_0 \exp i\Delta k X + F U_2 i k_s a_1 + F \dot{a}_1] = 0$$

where we have assumed:

$$h(x) = 2h_0 \cos k_m x = h_0 \exp i(k_s + \Delta k)x + (*) \tag{5.18}$$

and $\Delta k = k_m - k_s$ is the detuning with respect to resonance. Choice of $a_1(X, \tau) = a(\tau) \exp i\Delta k X$ finally gives the nonlinear equation for near resonant wave amplitude:

$$-\partial_\tau [(\frac{F^2 U_1}{\beta - F U_1} + F + \alpha^2 + k_s^2) a] - 2ik_s^2 U_1 \Delta k a + - F U_2 i k_s a - \frac{F^2 U_1 U_2}{\beta - F U_1} \tag{5.19}$$

$$- \nu(\alpha^2 + k_s^2) a - \frac{\delta^2 i k_s a |a|^2}{3k_s^2 U_1} = \frac{F U_1 U_2 h_0 i k_m}{\beta - F U_1}$$

Substituting (5.16) and (5.19) into the closure equations (5.7) and (5.8) we obtain:

$$\partial_\tau U = \frac{-1}{\beta - FU_1}[FU_2 h_0 ik_s(a - a^*) + Fh_0(\dot{a} + \dot{a}^*)] - \nu(U - U^*) \tag{5.20}$$

$$C\partial_\tau m = \frac{1}{\beta - FU_1}[FU_2 h_0 ik_s(a - a^*) + Fh_0(\dot{a} + \dot{a}^*)] \tag{5.21}$$

$$\frac{+2F}{\beta - FU_1}[F(a\dot{a}^* - a^*\dot{a}) - ik_m U_2 h_0(a - a^*)] - 2\nu(m - m^*).$$

Equations (5.19),(5.20),(5.21) form a closed dynamical system describing the dynamics of the nonlinear (cubic) folding of resonant response to orographic modulation in an almost zonally symmetric baroclinic flow. Here our major concern is to find the stationary states and their stability properties, because of their impact on the statistics. We find stationary solutions by solving for a, as a function of U and m, the cubic algebraic condition resulting from the stationary version of (5.19), and consequently choosing U^* and m^*, satisfying the stationary version of (5.20-21) for those values of U and m that allow multiple wave amplitude solutions with a choice of a corresponding to its maximum equilibrium value. Having fixed U^* and m^*, we can then compute the multiple equilibria for all other states by means of a Newton iteration procedure. Convergence is quite fast.

An illustration of our results is given in Table II. The first set of values (Table IIa) corresponds to multiple equilibria near the symmetric circulation ($U_2 = 0$): the two states of maximum and minimum amplitude are stable and separated by about 100 meters of geopotential. The intermediate state is unstable with respect to nontravelling baroclinic distrubances of orographic nature. The difference in zonal flow and average baroclinicity between the large and small wave amplitude stable states is confined in the range of meters per seconds as required by the observed statistics. The maintenance of the stable equilibria is, again in agreement with observations (Hansen 1985a), essentially baroclinic (see Benzi, Speranza and Sutera 1986).

It is interesting at this point to contrast the above results with those obtained by studying a barotropic nonlinear system computed from the same equations, setting $m = 0$. To this purpose we show in Table IIb a second set of equilibrium values computed for the above-defined barotropic dynamics. One can immediately see the large differences in the zonal wind of the two extreme (in wave amplitude) equilibria. This difference is due to the balance between zonal and wave kinetic energy, emphasized before as a necessary consequence of barotropy.

In conclusion, our results support the interpretation of the discrepancy between the prediction of barotropic theories and the observed statistics: different wave amplitude states are maintained by baroclinic conversion, with form-drag exerting only a catalytic role. This explains how the wave amplitude can display widely varying values and at the same time bimodal statistics, while the zonal flow has almost coincident equilibrium values and, consequently, essentially unimodal statistics.

As regards the theoretical manipulation of minimal models of atmospheric circulation, it is also worth noting that the asymptotic stability of both extreme amplitude equilibria is here critically dependent on the process of resonance bending. It is, in fact, this bending which permits the coexistence of different equilibria on the same side of the linear resonance, therefore displaying similar stability properties (Speranza 1985b). We thus escape the embarassement of finding globally attractive large amplitude states like other authors (Charney and Straus 1980, Yoden 1983, Rambaldi and Salustri 1984, Itho 1985) dealing with the same system, but without resonance bending.

	U	m	A1	A2
E1	2.1	1.7	0.73+i0.78	0.12+i0.11
E2	2.15	1.72	-0.76+i0.61	-0.1+i0.1
E3	2.37	1.83	-0.15+i0.01	-0.01+i0.002

Table IIa. Stationary solutions for equations (5.19), (5.20) and (5.21) computed for the following set of parameter: $C = 5$, $\alpha = 0.2$, $\delta = 0.5$, $F = 1.$, $\beta = 1.$, $U^* = 2.38$ and $m^* = 1.84$. The rescaling of variables has been performed using a wind scale of 10 meters per second and a characteristic length of 1000 km. In this way the amplitude of the waves is expressed in units of hundreds of meters.

	U	m	A1	A2
E1	2.1	0.	2.08+i0.63	4.16+i1.21
E2	4.72	0.	-1.53+i0.13	-1.83+i0.17
E3	5.1	0.	-0.58+i0.02	-0.61+i0.02

Table IIb. Stationary solutions of equations (5.19), (5.20) and (5.21) for the case $m = m^* = 0$. The values of the parameters are the same of Table Ia.

6. OBSERVATIONAL EVIDENCE OF MULTIPLE EQUILIBRIA

It is very important, although apparently obvious, that any theoretical explanation of low frequency variability in the atmosphere must be subjected to observational verification using the same physical quantities for which the theory has been developed. In the preceding sections we have constructed a self consistent theory of low frequency variability which predicts multiple equilibria of ultralong planetary wave amplitude for nearly the same value of the zonal flow and the zonal vertical shear. In this section we shall verify whether there is any observational evidence of such statistical behaviour in the real atmosphere.

The specific mechanism of resonance folding described in the preceding sections is robust with respect to generalization to more spectral components (Benzi,Iarlori, Lippolis and Sutera 1986). In the light of such extension of the theory, we have to expect a complex region of folding, more spread in the spectral domain than in the simple process of fig. 3. For this reason, in the data analysis described below, we use as an indicator of ultra-long wave amplitude the quantity:

$$A(t) = [A_2(t)]^2 + [A_3(t)]^2 + [A_4(t)]^2 \qquad (6.1)$$

that represents the total power present in the near resonant band.

Our data consists of twice-daily (0000GMT and 1200 GMT) National Meteorological Center final analysis of the Northern Hemisphere 500 mb heights from 1966 to 1977 as provided by Roy Jenne. As is well known, the original data is interpolated on the NMC octagonal grid (51x47 points). Following the quasi unidimensional approach discussed in the preceding section, after subtracting the zonal average, we integrated the geopotential field between 30 and 75 degree north. The resulting instantaneous fields were then Fourier decomposed in longitude. The wave amplitude indicator was then constructed by means of formula (6.1) and the time series of the zonal wind by averaging in latitude between 30 and 75N and in longitude. Because of the quasi-geostrophic balance, the zonal field is simple proportional to the difference in geopotential height averaged along the 30N and the 75N latitude circles. The physical process we want to study is concentrated between 10 and 40 days. This range is intermediate between the 2-5 days range typical of ordinary baroclinic instability and slow modulations on seasonal scale or longer (probably associated with variations in the macroscopic parameters of the system). Previous studies (Dole 1982,1983) quote an order of magnitude of 20% for season to season variability of long lasting anomalies in the extratropical circulation. A certain amount of filtering and detrending of the data is therefore required if we want to be able to superimpose different winters. The filtering is performed with a spectral procedure: periods below 5 days and above 1 year are eliminated directly from the Fourier spectrum of the time series. The results of applying this procedure are shown in fig.4 where filtered and unfiltered series are compared for each winter. The filtered time series still contain some trends. This is shown in fig. 5 where the results of filtering with a different choice of periods on which the Fourier spectrum is computed are shown for the winter 1975-76. Application of a procedure of digital detrending at 45 days removes the residual trend as shown in fig. 6. At this point we are left with two time series of instantaneous values of the wave-amplitude indicator and the average zonal wind that presumably describe some projections of the fluctuations of the atmospheric circulation in the low-frequency range (10-30 days). These constitute the basis of our analysis.

Having constructed an suitable data set, we proceed now to the analysis of statistical properties of the observed circulation in the parameter space of quasi unidimensional theory. The first

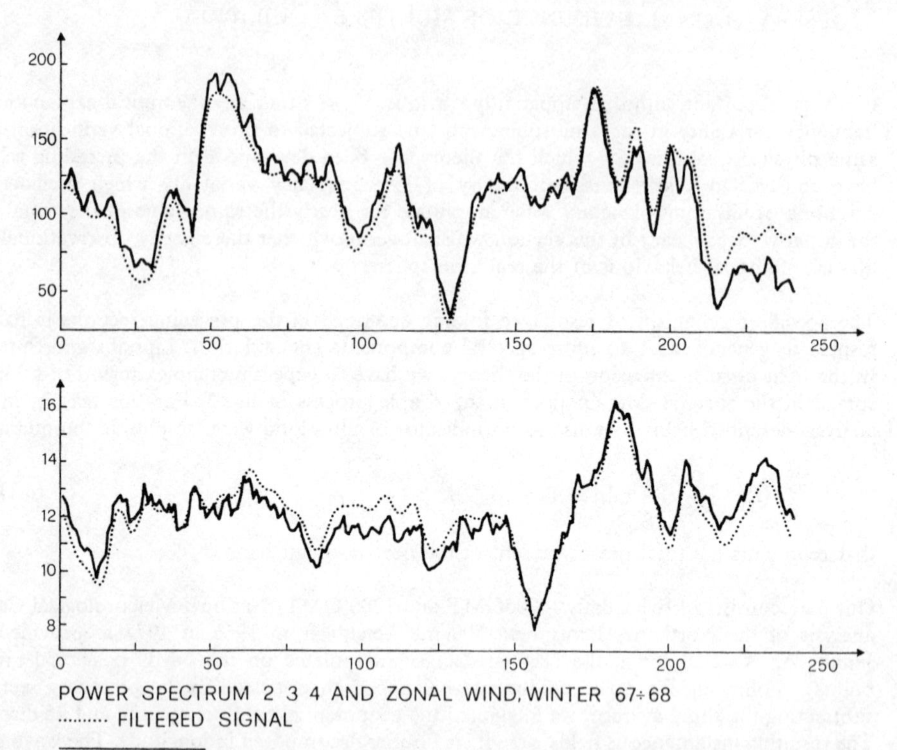

Fig. 4 $I = \sqrt{A_2^2 + A_3^2 + A_3^2}$ and U versus time for the winters analyzed in the paper. A_n is the amplitude of Fourier mode n of $\frac{1}{L}\int_{30N}^{75N} \eta(x,y)dy$ where η is the geopotential at 500 mb and L is the width of the channel. U is the average zonal flow: $U = \frac{1}{L}\int(\eta(x, 75^o) - \eta(x, 30^o))dx$ The dashed lines represent filtered signals, as discussed in the text.

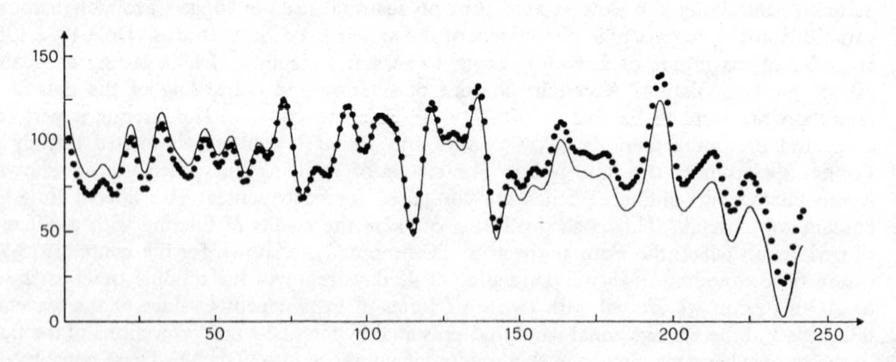

Fig. 5.Filtered signal of I for winter 75-76 as computed from the Fourier spectrum of the period 65-77 and 69-79. We note the appearance of some trend, possibly due to the seasonal cycles.

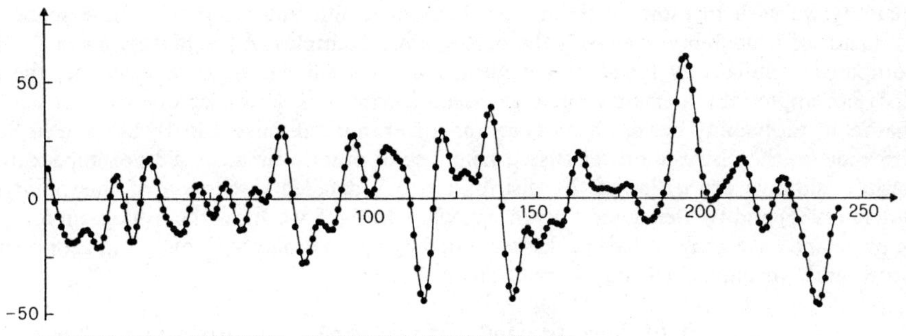

Fig 6. Same as in figure 5 after a digital detrending at 45 days.

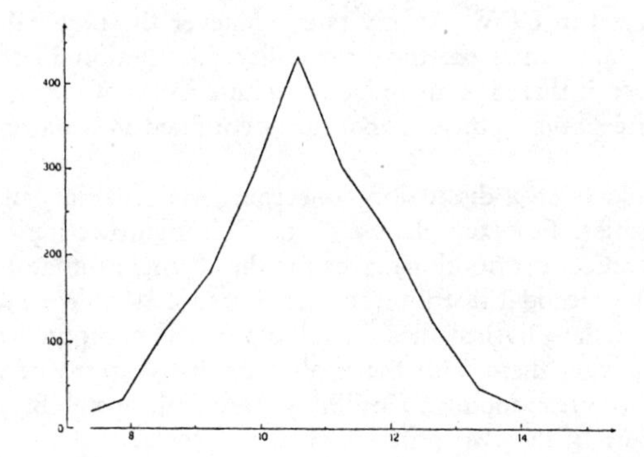

Fig 7a. Histogram 11 bins for the zonal wind U for the winters 67-78.

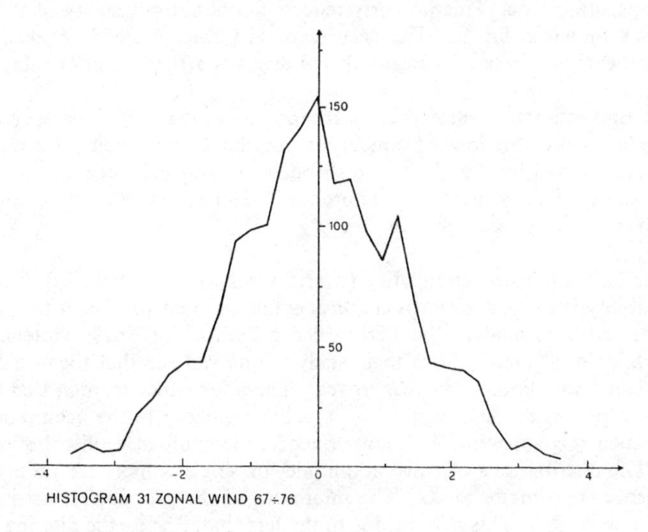

Fig 7b. Histogram 31 bins for the zonal wind U for the winters 67-78.

quantity we wish to estimate is time occupation in different regions of phase space. A classical estimator of probability density is the histogram. Examples of the histogram of the zonal wind, computed cumulatively for several winters, are shown if fig. 7. As we can see, the distribution does not appreciably deviate from a gaussian distribution. However one can consider other estimators of probability besides histogram (as for example discussed in BMSS and in Sutera 1984). For most of the distributions discussed here, use of other estimators did not appear to change the basic qualitative properties of the distributions. A detailed discussion of this point is given by Sutera (1985) and by Benzi and Speranza (1986). In fig. 8 we show the histogram for our indicator of planetary wave activity: two peaks, separated by approximately 30 meters in geopotential, appear consistently in almost all the distributions.

Fig. 9 and 10 show the combined two-dimensional histogram in the U-A parameter space of quasi unidimensional theory: two maxima in the projection onto the wave amplitude axis correspond to a single peak in the projection onto the zonal flow axis. This result seems to be more in agreement with a process of folding than with the stabilization due to the wave-zonal flow interaction envisaged in CDV. At any rate, whatever the physical mechanism of generation of low-frequency variability may be, the probability distribution in the wave amplitude seems rather spread. Moreover, if there is a quantized structure in the wave amplitude, the different wave equilibrium values correspond to mean zonal flows confined in a narrow range.

At this point, we should open a discussion concerning the reliability of the computed estimators for the probability densities. For example, use of the Kolmogorov criterion (based on the maximum distance from the theoretical distribution) gives for the Kernel estimators a reliability consistently larger than 95% for the bimodal distribution (see Sutera 1984). We prefer, however, to give the reader an impression of the physical meaningfulness of the computed distributions of ultra-long wave activity by contrasting them with the analogous distributions of wave-activity in the range of wavenumbers 7-18 (corresponding to ordinary baroclinic instability) displayed in fig. 11: the different statistical nature of the two processes is quite clear.

As already mentioned in the introduction, most of the variance of the process we are studying is of the non propagating type. This property reflects the nonhomogeneity of the probability distribution of phases shown in fig. 12. The occurrence of phases is clearly peaked in a limited range of values. Moreover the phase of wavenumber 3 displays a marked bimodality.

However, it is important to realize that in the presence of the various spectral components of orographic modulation at the lower boundary it may be very difficult get a good correspondence between the days selected by the bimodal distribution of our indicator and those obtained by the bimodal distribution of the phase of wavenumber 3. Hence, we are not certain as to the specific physical identification of phase behavior.

Up to now, we have discussed cumulative (over several winters) statistical distributions. The low frequency variability is characterized by a considerable amount of season to season variability. It is therefore interesting to analyze the statistical distribution for single winters. This is discussed by Benzi and Speranza (1986). From their analysis one can see that the zonal wind distribution shows very unequal variations from year to year. The appearance is often that there is a superposition of different processes. At any rate, no specific regularity in the occurrences of peaks in the occupation frequency is apparent. As a consequence, the cumulative distribution is unimodal (see again fig. 7). The distributions of wave amplitude for single winters are more regular: they quite consistently display two major peaks. The minimum of the probability distribution shows some variation from year to year. This may be due to the fact that, despite the filtering and the detrending procedure used, some year to year variability is still present in our data.

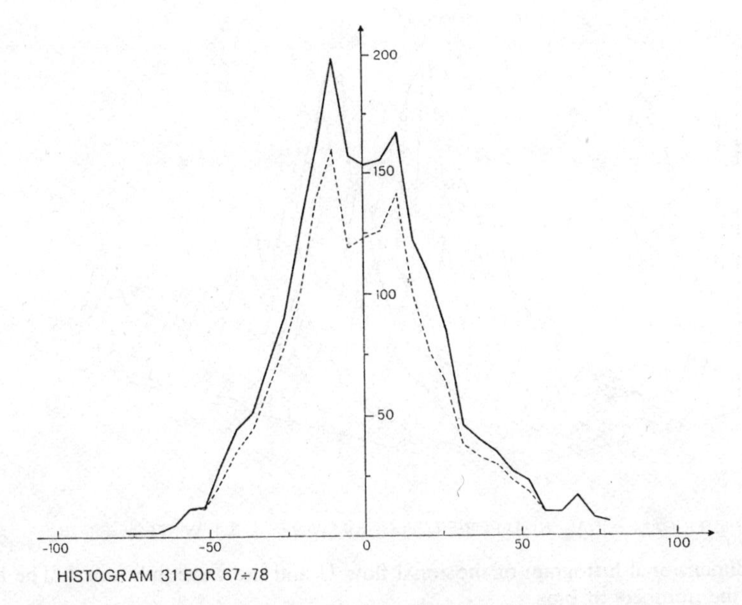

Fig 8a. Histogram 31 bins for the cumulative probability distribution of the detrended indicator I for winters 67-78. The dashed line refers to the period 67-76.

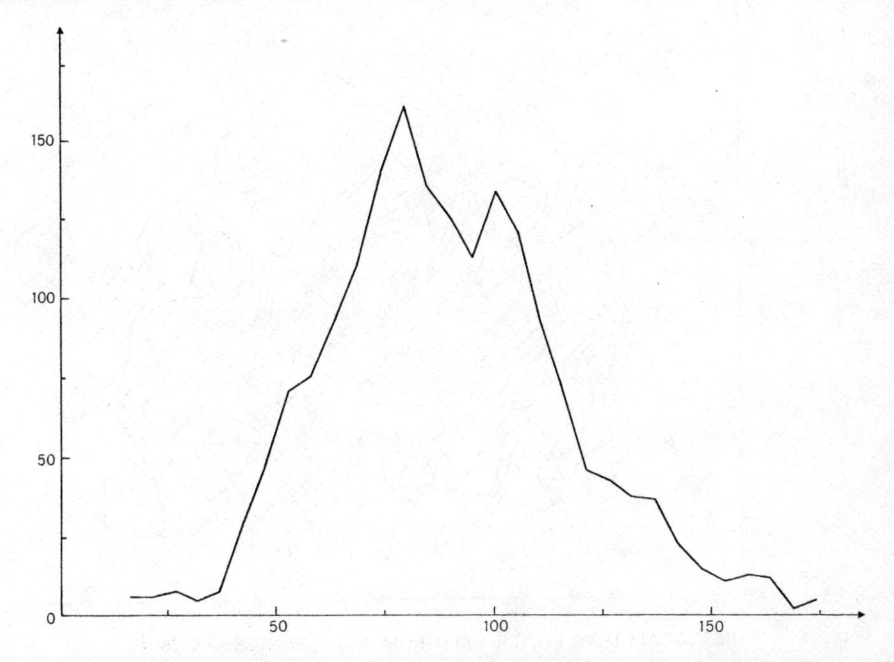

Fig. 8b Histogram 31 bins for the filtered but not detrended indicator for the period 67-76.

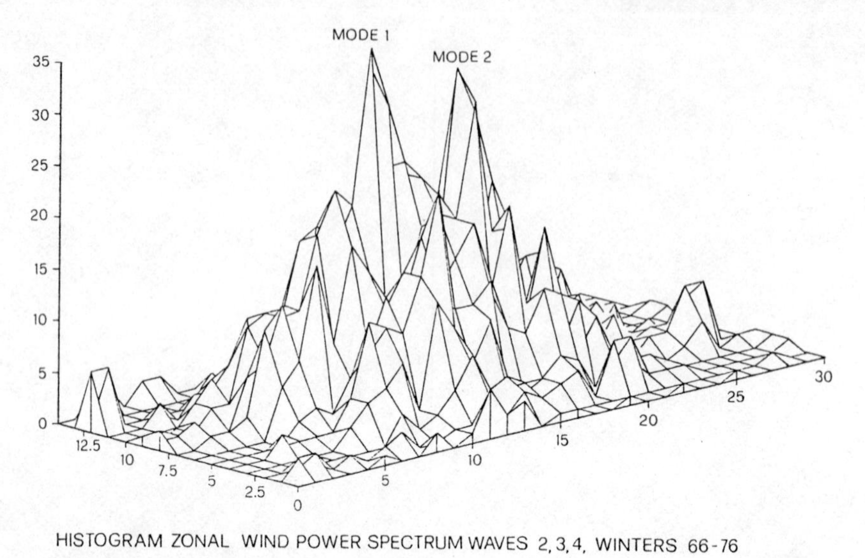

Fig. 9. Two dimensional histogram of the zonal flow U and the wave indicator I. The horizontal axis refers to the numbers of bins.

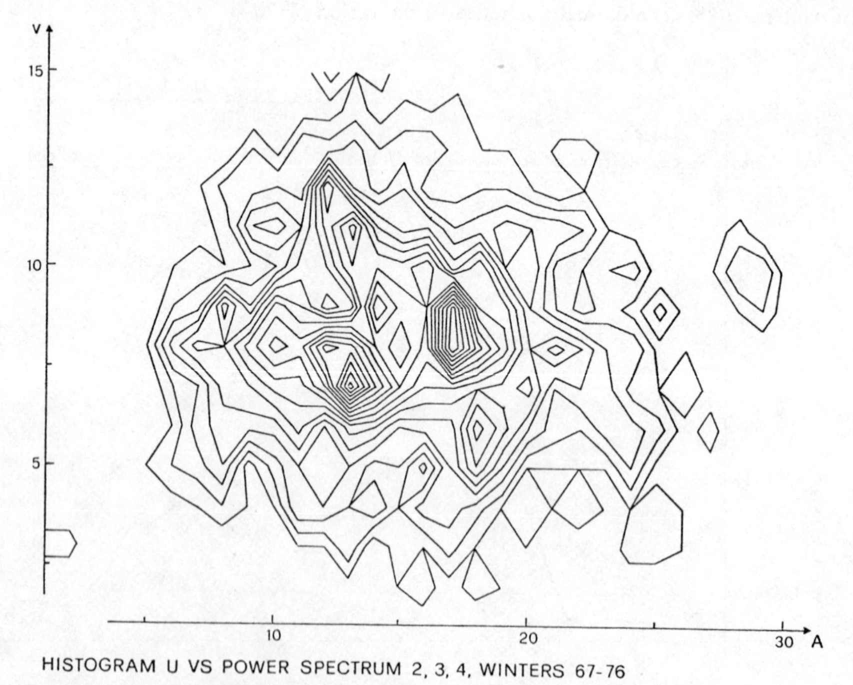

Fig. 10. Same as in figure 9 using contour interval.

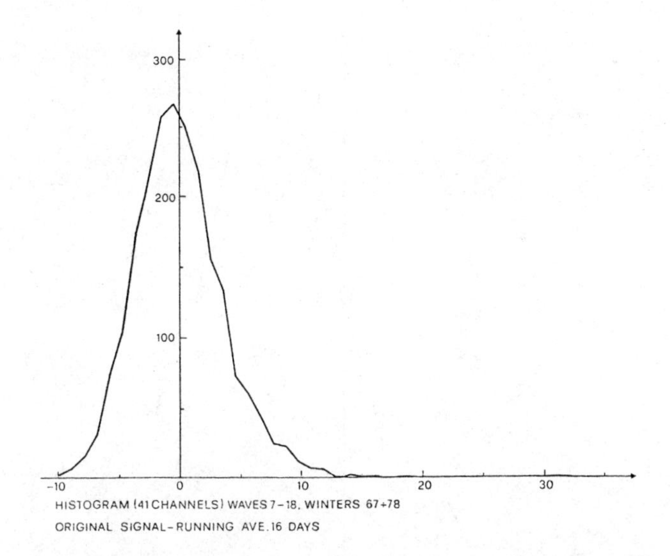

Fig. 11. Histograms of the power spectrum in wavenumber 7-18 for the period 67-78. From the original signal we have subtracted a running average of 16 days.

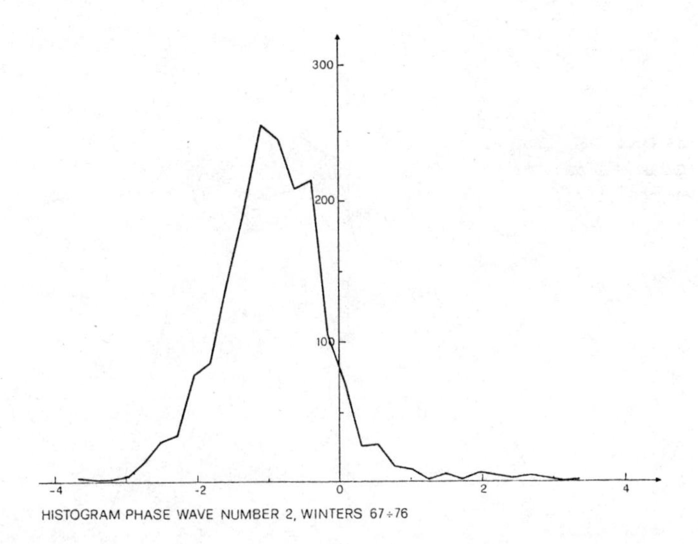

Fig. 12.a

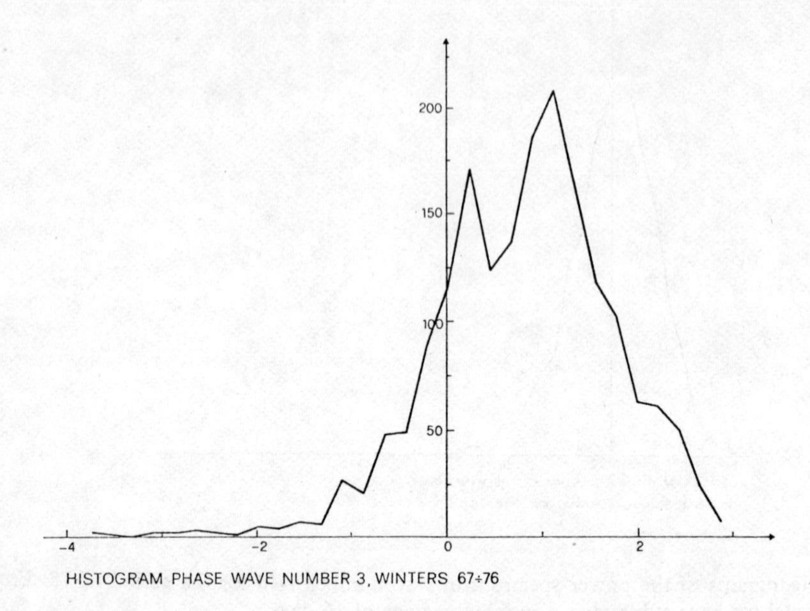

Fig. 12.b

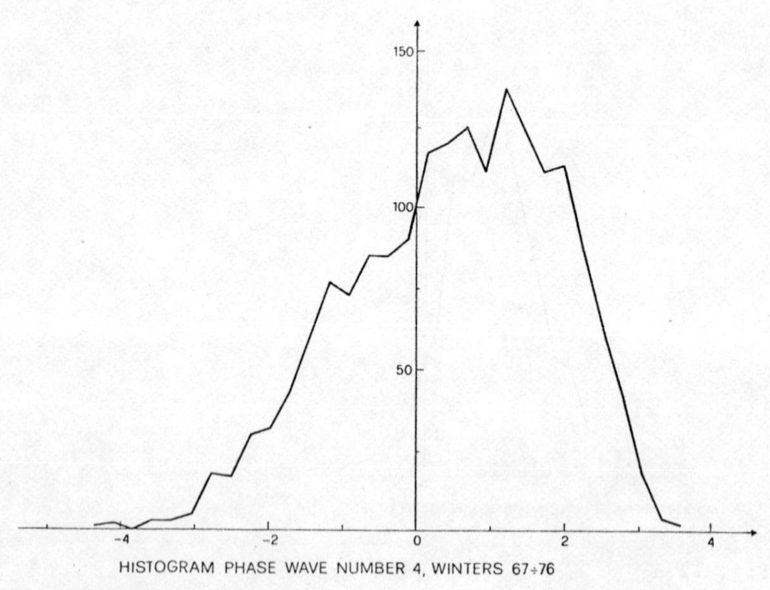

Fig. 12.c

Fig. 12. Histograms of the phase of wavenumber 2(12a), 3(12b), 4(12c) for the period 67-76. The horizontal scale is expressed in radiants.

It is interesting to show what one can be found when analyzing the statistical distribution of values of meteorological quantities measured at specific locations (White 1980, Dole 1982) instead of spectral projections. Fig. 13 shows the histograms of the wave amplitude indicator computed at selected longitudes. It can be clearly seen that the bimodal character appears only at some locations (d,e,g of Fig. 13) which coincide perfectly with the regions of persistent anomalies individuated by Dole (1982, 1984). Obviously, the total fields analyzed by White and Dole also contain all the other spectral components that, presumably, mask the bimodal character of the distribution.

Up to now we have considered pure statistical properties of the low-frequency variability. In such a range of variability we find atmospheric features of great meteorological interest, in particular blocking. It has been known for some time that synoptic definition of blocking, in any of its different formulations, cannot be simply reduced to the amplification of ultralong waves., although many blocking patterns project quite substantially onto the first few wavenumbers. This, as we shall see, has also been our own experience in the course of the present research. Therefore, the problem of explaining low frequency variability must, for the time being, be maintained distinct from that of understanding the physical mechanism of generation and maintenance of blocking anticyclones.

A first impression regarding the synoptic structure of the meteorological fields associated with low-frequency variability can be gleaned from the composite of all the maps falling near the two peaks of the statistical distribution of wave amplitude that we indicate with MODE 1 (small wave amplitude) and MODE 2 (large wave amplitude). An example of such composite is shown in Fig. 14. As expected the main difference between the two "modes" consists in the amplification of a wave number 3 pattern in MODE 2. The main amplification takes place during periods of major blocking activity although, as mentioned, this cannot be taken to signify the identity of the two phenomena in question. Hansen (1984) shows the Hovmuller diagram of the wave amplitude indicator for the winter 1981-1982. From such a diagram it can be seen the ridging activity over the Pacific and the Atlantic (it is worth noting that there is no sign of phase propagation in the Hovmuller for any period); the wave amplitude indicator essentially captures Pacific and large scale Atlantic blocking, but misses very localized Atlantic "dipoles". Such blocks, presumably maintained by "synoptic" forcing, rather than baroclinic conversion (Hansen-Sutera 1983, Shutts, 1984, Egger et.al. 1984), seem to be captured by other indicators associated with wave-wave interaction as opposed to the amplitude of ultra-long waves. An example is given in fig.s 15,16 and 17 where we discuss the meridional transport of enstrophy $Q = [\Delta\psi]^2$ average on the channel for the winter 1967-68. Following the theoretical considerations of section 4, we can easily see that:

$$\int v_y \partial_y Q dx dy = \int A_x g (A_{xx}^2 g^2 + 2 A A_{xx} g g_{yy} + A^2 g_{yy}^2) dx dy =$$

$$2 \int A A_x A_{xx} dx \int (g g_y g_{yy} - g^2 g_{yyy}) dy \sim \delta \int A A_x A_{xx} dx$$

We considered the power spectrum of $\int A A_x A_{xx} dx$ analyzing the contributions on wavenumbers 2,3,4 decomposing it into two parts: one part induced by wavenumbers 2,3,4 themselves and the other one by all wavenumbers greater than 4 The dashed lines in Fig.s 15,16 and 17 refer to the days selected by the our indicator for wave amplitude activity in the MODE 2. We see that large and small scale meridional advection of enstrophy are active during MODE 2 events except for in cases (pointed out by the arrow in fig. 17) where small scale advection is active but no large scale activity is present. The 10 days composite 500 mb geopotential map for this event shows a well defined synoptic blocking present on the Atlantic region.

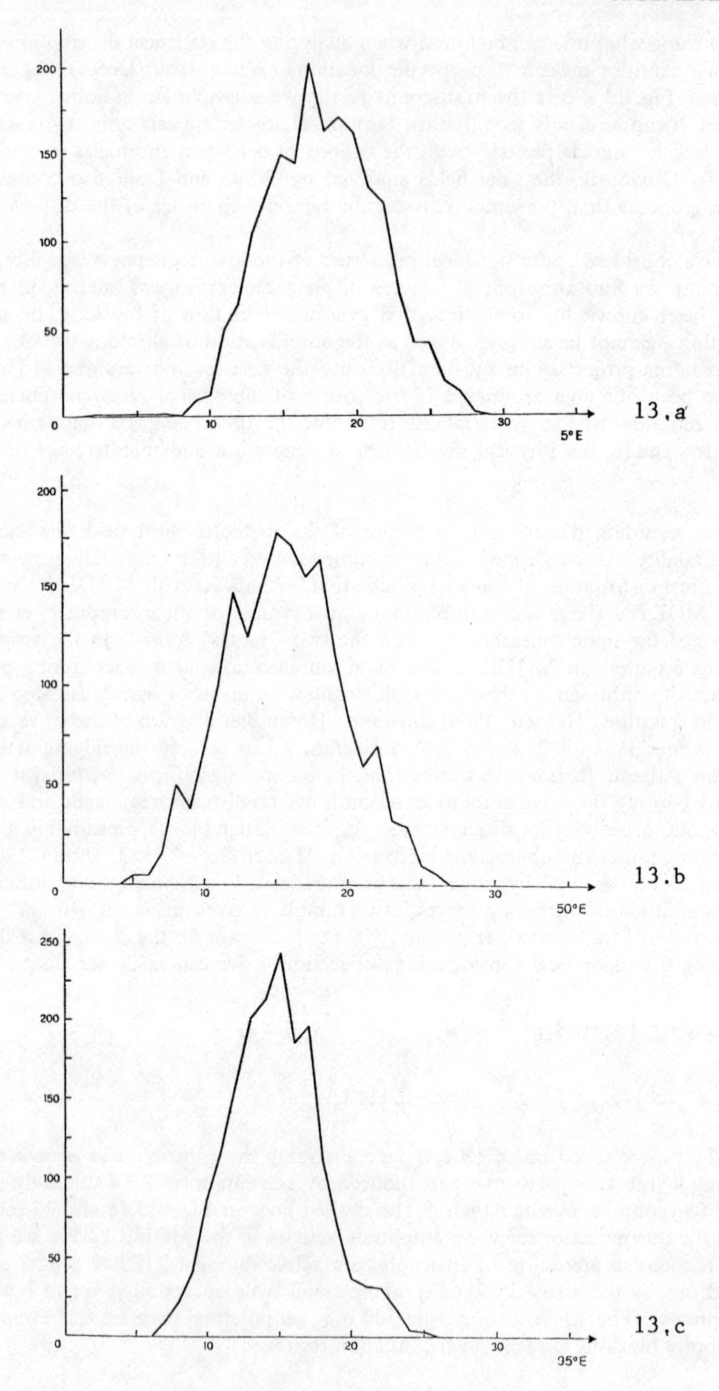
200
150
100
50
0
10
20
30
5°E
13, a
200
150
100
50
0
10
20
30
50°E
13, b
250
200
150
100
50
0
10
20
30
95°E
13, c

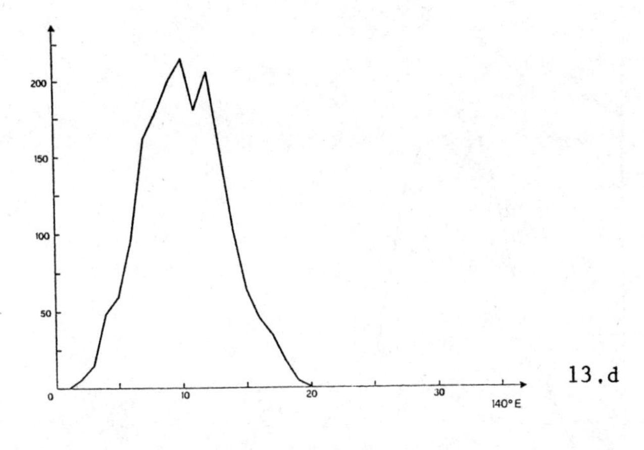
200
150
100
50
0
10
20
30
140° E

13.d

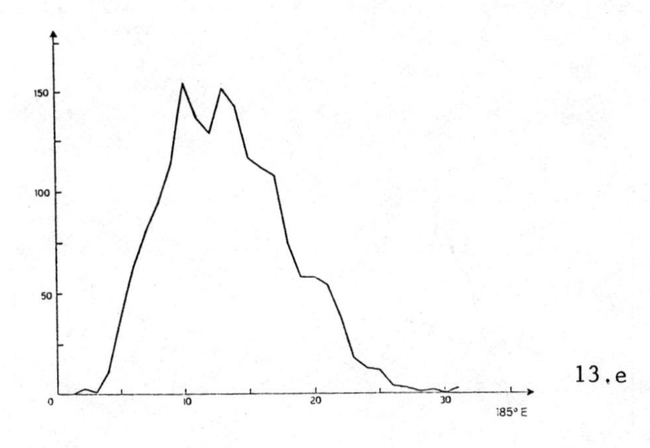
150
100
50
0
10
20
30
185° E

13.e

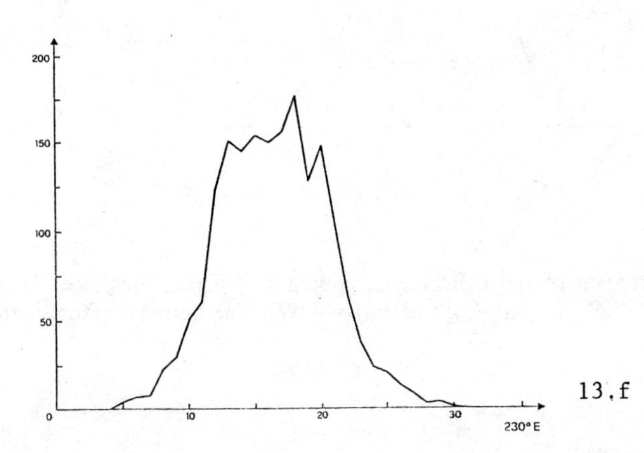
200
150
100
50
0
10
20
30
230° E

13.f

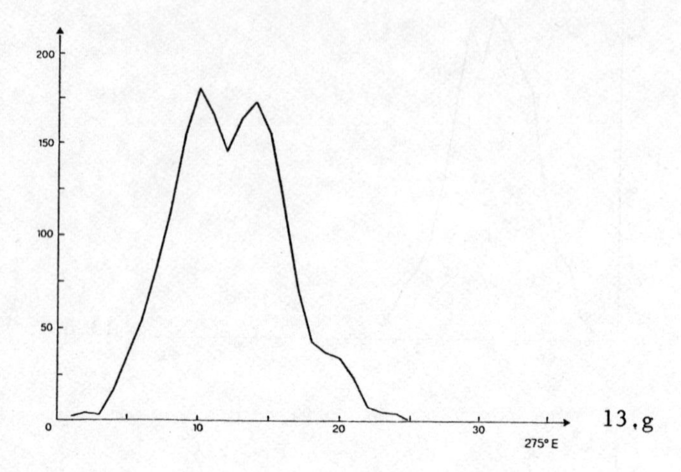

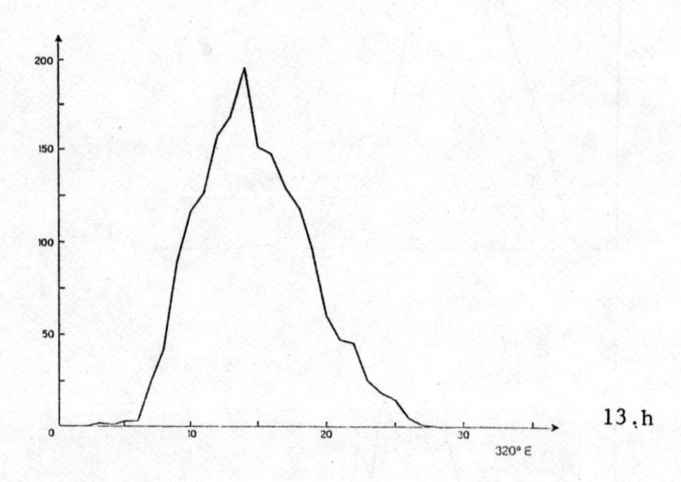

Fig. 13. Histogram of the indicator computed at different longitudes (a = 5E, b = 50E, c = 95E, d = 140E, e = 175W, f = 130W, g = 85W, h = 40W). The bimodal nature corresponds to cases d,e,g.

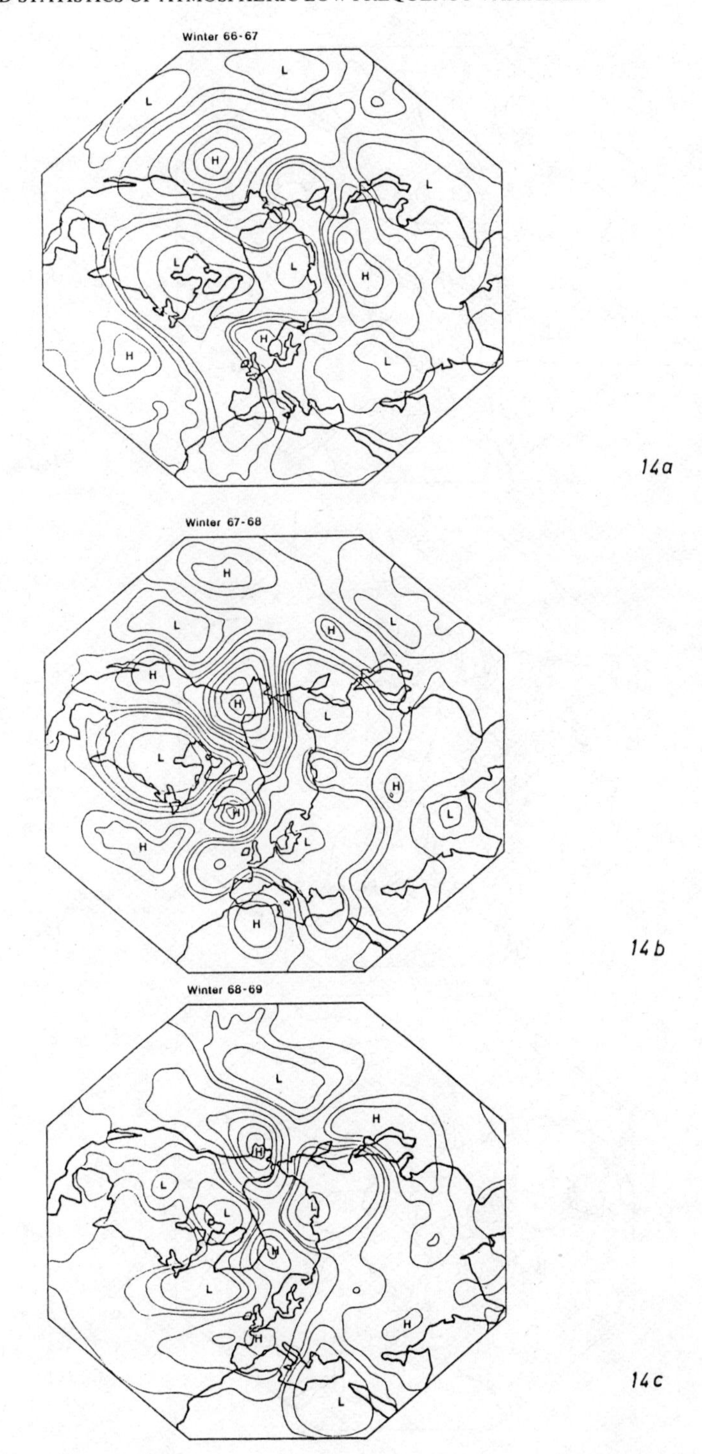
Winter 66-67
14a
Winter 67-68
14b
Winter 68-69
14c

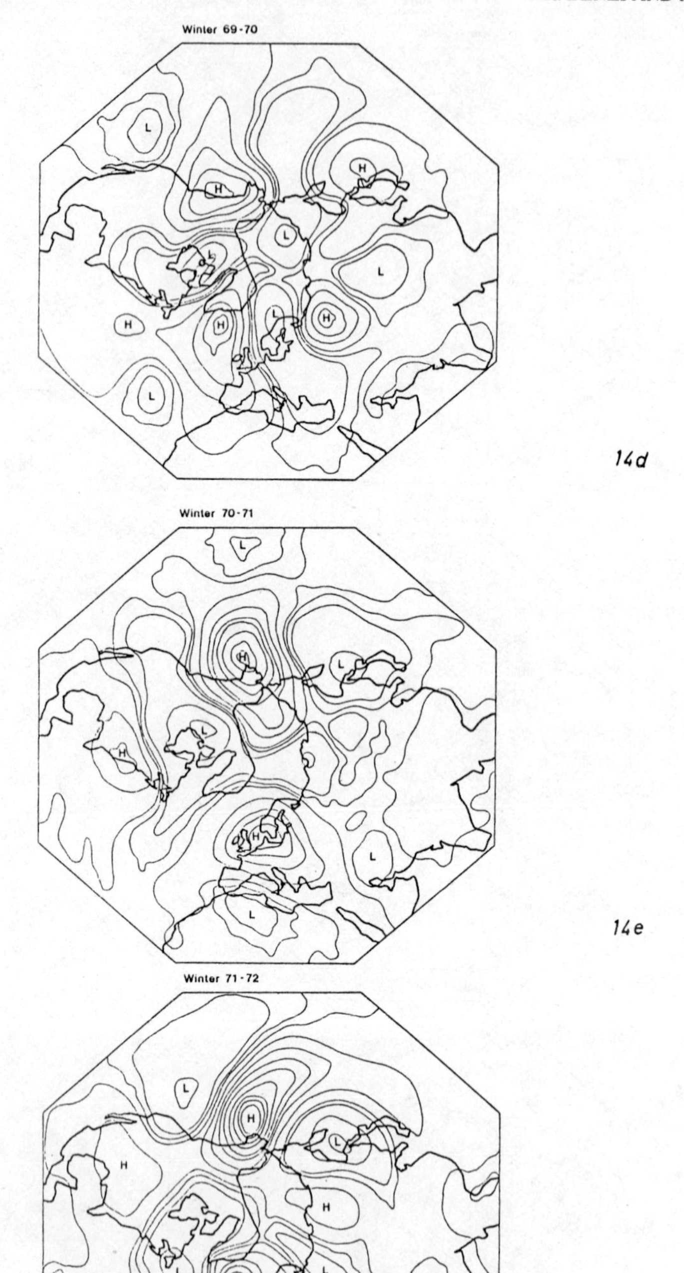
Winter 69-70
14d
Winter 70-71
14e
Winter 71-72
14f

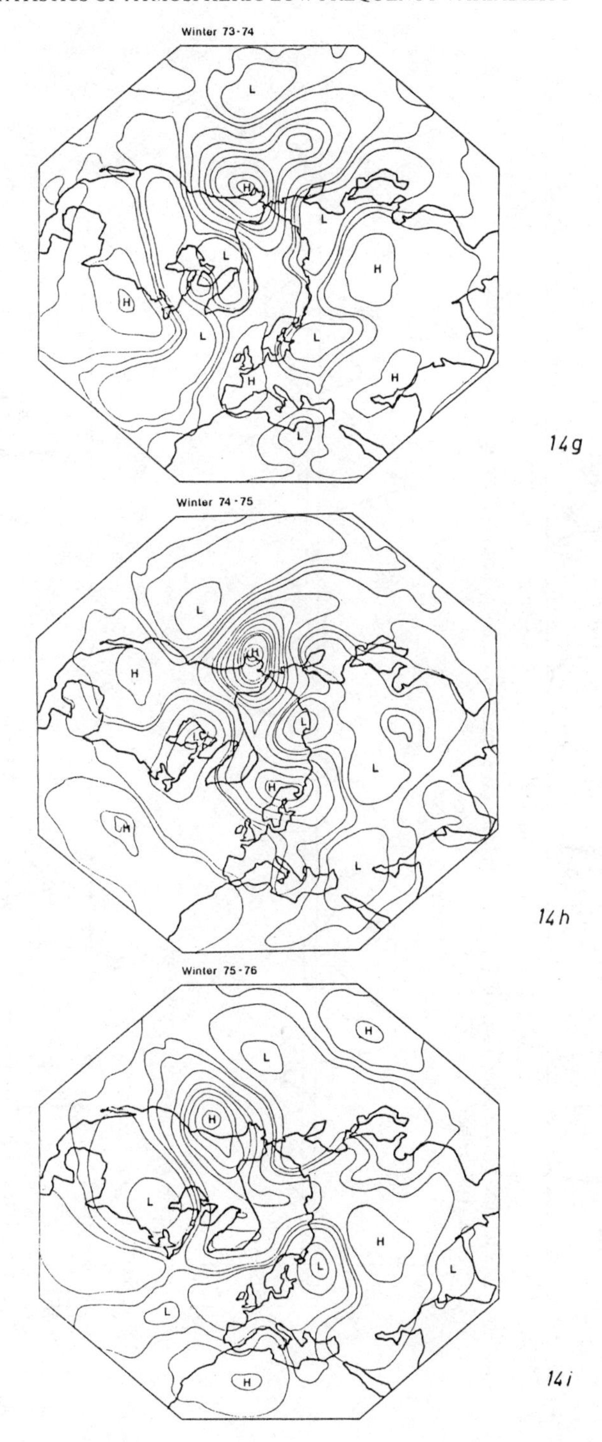

Fig. 14. Geopotential difference MODE 2 - MODE 1 year by year for the period considered in the paper.

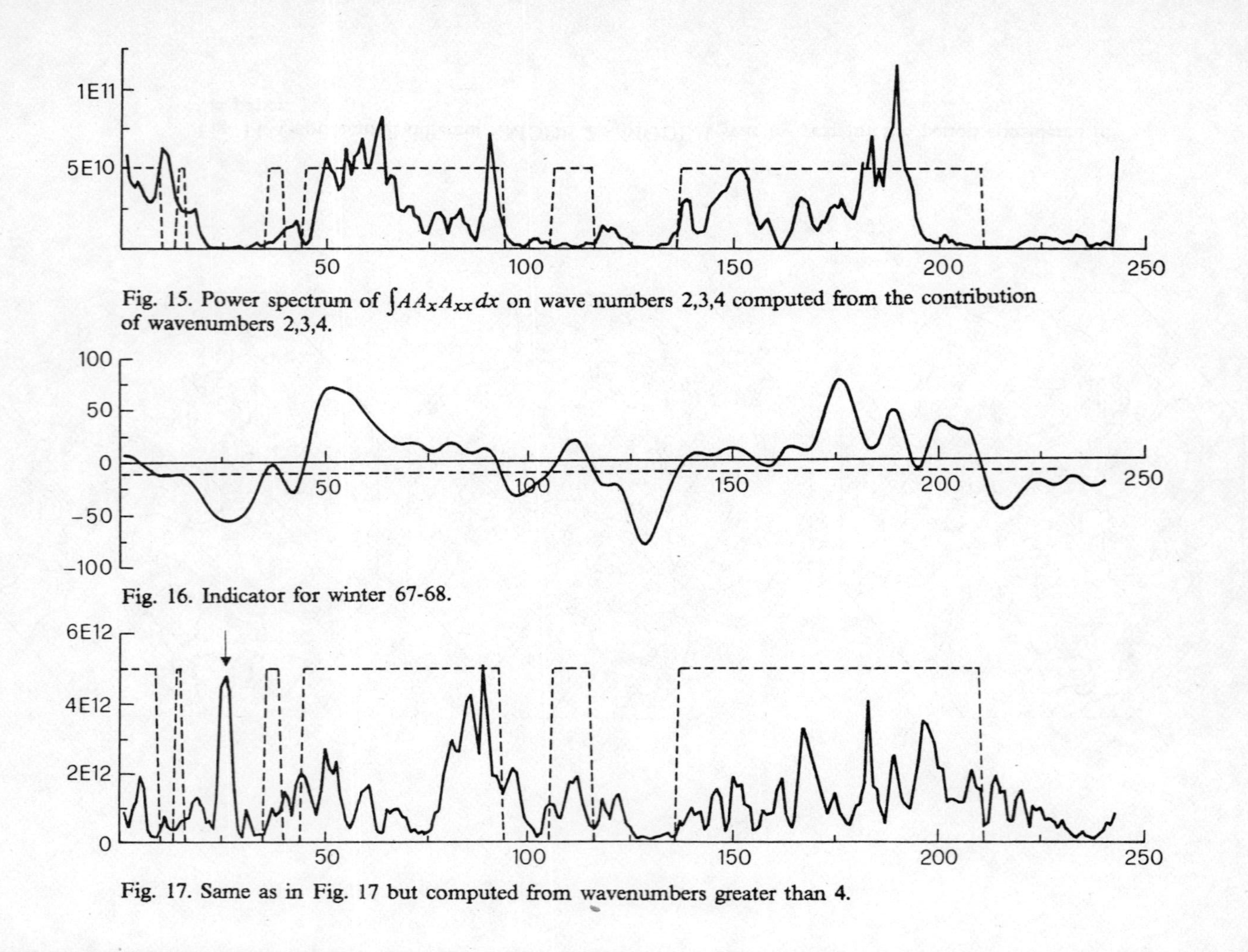

Fig. 15. Power spectrum of $\int AA_x A_{xx}\, dx$ on wave numbers 2,3,4 computed from the contribution of wavenumbers 2,3,4.

Fig. 16. Indicator for winter 67-68.

Fig. 17. Same as in Fig. 17 but computed from wavenumbers greater than 4.

If the qualitative properties of the statistical indicators are associated to the presence of a different "equilibrium flow configuration" of the general circulation, a fundamental dynamical role must be played by the instability causing the divergence in phase space associated with the minimum in the occupation frequency of wave amplitude. It is therefore important to study not only "equilibria", but also the transitions between them. Such transitions occur typically as sudden amplifications of the ultra long waves.

Cases of strong ultra long wave amplifications have been analyzed by Itho (1983). The cases selected by Itho are those exceeding one standard deviation of the amplitude over the period 1965-1974 (only winter season). Fig. 4 of Itho's paper shows the time evolution of amplitudes and phases of wavenumber 3 at 55N for for 500 mb height and 850mb temperature. At least two aspects are striking. First of all, the phases are locked at a specific value (around -30 degrees) for all the events. Second, the amplitude pattern grows with an e-folding time around 7 days with little dispersion around the average over the ensemble of cases.

The evolution suggests of the presence of a tube in phase space in which the system rapidly moves along the dilatation axis, while the other components essentially contract. The energetics of this process is predominantly baroclinic although not of ordinary type. That this is the case is quite clearly shown by the synoptic analysis of transients performed by Dole (1984) for the onset of Pacific anomalies: a large scale, vertically coherent, baroclinic conversion cell develops with no sign of phase propagation leading to the formation of a mature block in less than one week. We can conclude that available information points in the direction of a process non-propagating, vertically coherent with a strong baroclinic component. At the present there seems to be not known instability displaying such properties except "orographic baroclinic instability". This instability requires not only a source of available potential energy, but also a mountain induced feedback onto the zonal flow for its existence (see Buzzi, Trevisan and Speranza 1984 for a discussion of the theory of orographic instability in various conditions), although topography does not apply a direct energetic role in the unstable wave generation.

We therefore expect the time variation of zonal wind to be small althoug correlated with simultaneous variations of the ultra long waves. In fig. 18,19 and 20, we show the cross correlation between the time derivative of the zonal flow and the amplitude of different waves. A striking correlation appears in the range 10-20 days. This is in agreement with previous findings by Chen and Miller (1984). This result does not entail that the zonal flow is forcing the wave. As already mentioned, the energy comes from available potential energy and the zonal flow somehow enters (catalizing action of the form-drag ?) the process of ultra-long wave amplification. Our investigation in the spectral domain confirms those of the extensive spectral analysis of Chen and Miller (1984): all the cross correlations of different physical quantities (different forms of energy, conversions, etc....) show a distinct peak in the range of low-frequency variability (10-20 days). Inspection of the time evolution of the meteorological features contributing to this portion of the variance shows that the process is intermittent, with rapid entrance into and exit from the "equilibrium" configurations. An idea of the meteorological patterns associated with transitions is given by the difference between geopotential fields in MODE 1 and MODE 2 shown in fig. 14. The structure of this flip-flop pattern is characterized by a wavelike pattern centered over the Rocky mountains and moving north of the Hymalayas.

Returning now to the statistical approach, we can try to outline some other physical properties of the transition process. Fig. 21 shows the probability of the persistence of events corresponding to the two "equilibrium" states. As we can see, they are similar and are characterized an expected value of decay time of about 7 days; this is in good agreement with the previous estimates based on the analysis of selected cases. It is important to note here that this is the result which is to be

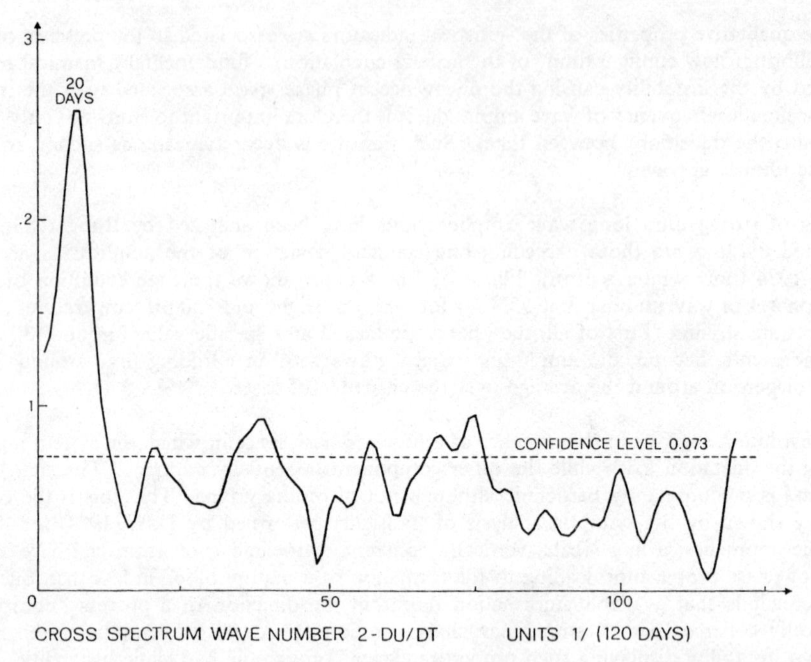

Fig. 18. Cross spectrum dU/dt and amplitude of wavenumber 2 for the period 67-76. Time lag for the computation 120 days.

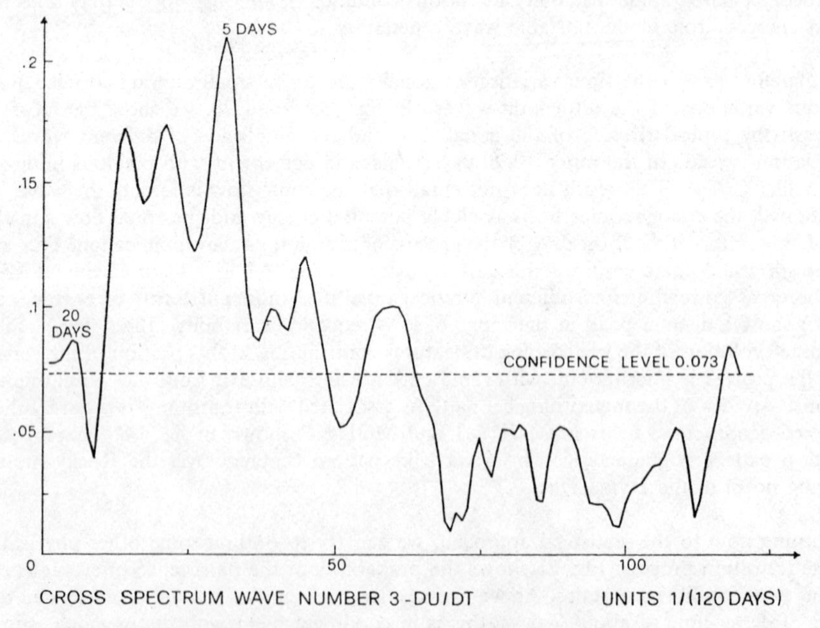

Fig. 19. As in figure 20 for the amplitude of wave number 3.

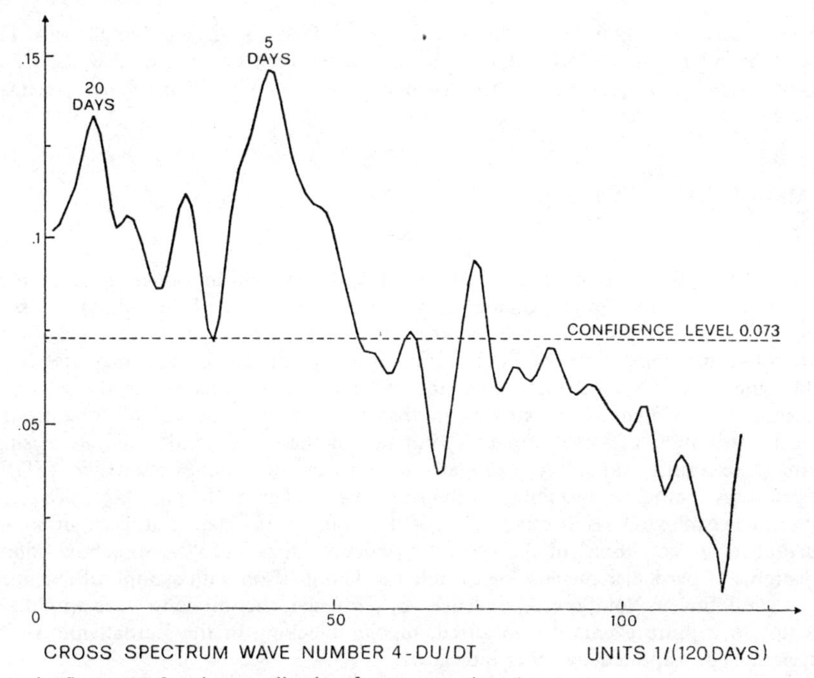

Fig. 20. As in figure 20 for the amplitude of wave number 4.

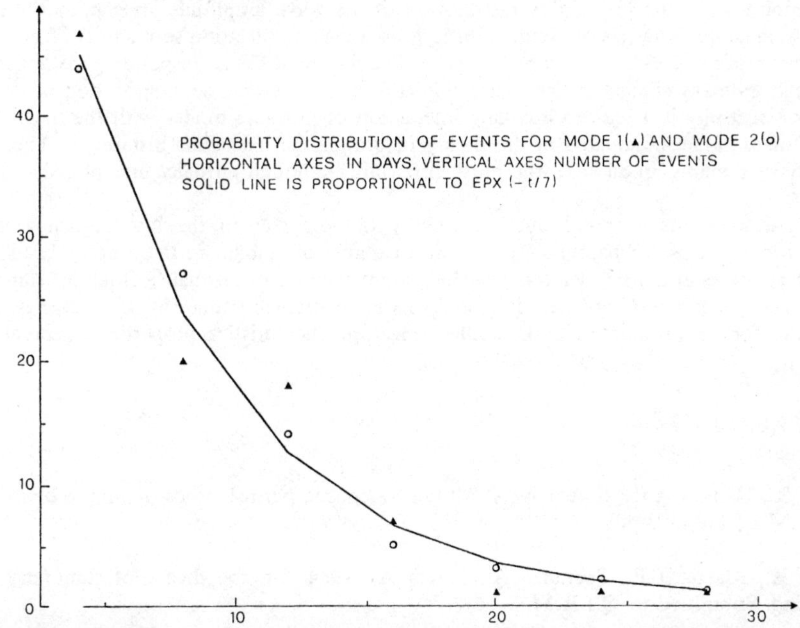

Fig. 21. Probability distribution of persistence on MODE 1 (triangle) and MODE 2 (dots). MODE 2 (MODE 1) is defined to be all the days for which the indicator is larger (smaller) than the minimum of the probability distribution.

expected on theoretical grounds for the statistics of exit from localized potential wells. The physical picture of an intermittent process due to transitions between two potential wells, separated by a region of divergence in phase space associated with an instability of the system, seems not to be contradicted by data analysis.

7. SUMMARY AND CONCLUSION

In these lectures, after a reexamination of the CDV theory, and introduction of the modifications (essentially the introduction of nonlinearities different from that of form-drag) necessary to limit the variability of the zonal flow within the observed range, we have analyzed the occupation statistics of an ultralong wave amplitude indicator and of the zonal wind integrated in the latitude over to band 30N-75N as deduced from the 500 mb observations during the period 1966-1978. The results of this analysis are striking in that they show some sort of "quantization" in the statistical distribution of wave amplitude, but not of the zonal wind. Analysis of single winters confirms these results. variability. Comparison with other techniques of analysis (EOF) and with other processes (baroclinic instability in the short wave range of the planetary spectrum) confirms the physical meaningfulness of our results. If the peaks in the statistical distribution are taken as representative of "equilibria" in the general circulation, they must be somehow connected with flow patterns of particular meteorological interest. Comparison with synoptically defined blocking shows a good, but not total correspondence. In particular, the ultralong wave amplitude indicator seems not to capture events of localized, dipolar blocking in the Euroatlantic region (which, however, can be captured by other indicators).

The minimum in the probability distribution of the wave amplitude appears as the most stable statistical feature. Analysis of events of ultra-long wave amplification seems to confirm the presence of a region of rapid diffluence in phase space. The nature of the corresponding instability is unclear although its main characteristics (baroclinic conversion, vertical coherence, non-propagation) are in favor of a physical process invoking interaction of unstable modes with the mean flow of the kind, for example, postulated in the theory of orographic baroclinic instability. The problem of transitions remains, in all case, the most important, although virtually unexplored.

Our analysis seems to point out consistently the presence in the low-frequency range of an intermittent process of baroclinic conversion capable of stabilizing the amplitude of the longest planetary waves at discrete values. To what extent theories of resonance "folding" can account for such phenomenology is not yet clear: progress in theoretical studies in this sector is, a necessary condition for the continuation of the studies concerning the statistical properties of general circulation.

REFERENCES.

Benzi R., Hansen A.R., Sutera A., 1984: On Stochastic perturbations of simple blocking models. Q.J.R.M.S., 110, 393-409.

Benzi R., Malguzzi P., Speranza A., Sutera A., 1984: On the theory of stationary waves and blocking. Submitted to Q.J.R.M.S.

Benzi R., Speranza A., Sutera A., 1985: A minimal baroclinic model for the statistical properties of low frequency atmospheric variability. Submitted to J. Atmos. Sci.

Benzi R., Speranza A.,1985: Statistical properties of Low Frequency Variability in the Northern Hemisphere. Submitted to Mon. Wea. Rev.

Benzi R., Iarlori S., Lippolis G., Sutera A., 1986: Work in preparation

Blackmon M. L., 1976: A climatological spectral study of the 500 mb geopotential height of the Northern Hemisphere. J. Atmos. Sci., 33, 1607-1623.

Buzzi A., Trevisan A., Speranza A, 1984: Instability of a baroclinic flow related to topography forcing. J. Atmos. Sci., 41, 637-650.

Charney J.G. 1973: Planetary Fluid Dynamics. In "Dynamic Meteorology" Ed. E. Morel.

Charney J.G., DeVore J.G., 1979: Multiple flow equilibria in the atmosphere and blocking. J. Atmos. Sci., 36, 1205-1216.

Charney J.G.,Straus D.M.,1980: Form-drag instability , multiple equilibria and propagating planetary waves in the baroclinic , orographically forced , planetary wave system.J.Atmos.Sci.,37,1157-1176.

Chen T.C., 1982: A further study of spectral energetics in the winter atmosphere. Mon. Wea. Rev., 110, 947-961.

Chen T.C., Marshall H.G., 1984: Time variation of atmospheric energetics in the FGGE winter. Tellus, 36A, 251-268.

Dole R.H., 1982: Persistent anomalies of the extratropical Northern Hemisphere wintertime circulation. Ph.D. thesis, M.I.T., Cambridge, Ma.

Dole R.H., 1983: Persistent anomalies of the extratropical Northern Hemisphere wintertime circulation. In "Large Scale Dynamical processes in the Atmosphere" edited by B.J. Hoskins and R.P. Pierce, Academic Press.

Dole R.H., 1984: The life cycles of persistent anomalies and blocking over the North Pacific. In "Large scale anomalies and blocking", edited by R. Benzi, A. Wiin-Nielsen, B. Saltzman, Academic Press.

Egger J., 1981: Stochastically driven large-scale circulation with multiple equilibria. J. Atmos. Sci., 38, 2606-2618.

Egger J., W. Metz and G. Muller, 1984: Forcing of planetary-scale blocking anticyclones by synoptic-scale eddies. In "Large scale anomalies and blocking", edited by R. Benzi, A. Wiin-Nielsen, B. Saltzman, Academic Press.

Fraederich K., Bottger H., 1978: A wavenumber - frequency analysis of the 500 mb geopotential at 50N. J. Atmos. Sci., 35, 745-750.

Fraederich K., Kietzig, 1983: Statistical analysis and wavenumber spectra of the 500 mb geopotential along 50 S. J. Atmos. Sci.,

Hansen A. R., 1985a: Observation characteristics of atmospheric planetary waves with bimodal amplitude distribution. In "Large scale anomalies and blocking", loc. cit.

Hansen A. R., 1985b: Transition between modes of a bimodal planetary wave amplitude distribution. Mon. Wea. Rev. submitted.

Hansen A.R., Sutera A., 1983: A comparison of the spectral energy and enstrophy budgets of blocking versus non-blocking periods. Tellus, 36A, 52-63.

Hart J.E., 1979: Barotropic, quasi-geostrophic flow over anisotropic mountains. J. Atmos. Sci., 38, 2606-2618.

Hayashy Y., 1982: Space-time spectral analysis and its applications to atmospheric waves. J. Meteor. Soc. Jap., 60, 156-171.

Held I. 1983: The theory of stationary eddies. In "Large Scale Dynamical processes in the Atmosphere" edited by B.J. Hoskins and R.P. Pierce, Academic Press.

Holopainen E.D., 1970: An observational study of the energy balance of the stationary disturbances in the atmosphere. Q. J. R. M. S., 96, 626-644.

Itoh H., 1983: An observational study of the amplification of planetary waves in the troposphere. J. Meteor. Soc. Jap., 61, 568-589.

Lambert, 1984: A global APE-KE budget for the FGGE year in terms of the twodimensional wavenumber. In "Reports of the seminar on progress in diagnostic studies of the global atmospheric circulation as a result of the global weather experiment" GARP special report No 42, WMO Geneva.

Legras B., Ghil M., 1983: Blocking and variations in atmospheric predictability In "Predictability of Fluid Motion" edited by G. Holloway and B.J. West, American Inst. of Physics, N. 106.

Lindzen R.S., Straus D.M., Katz B., 1984: An observational study of large scale atmospheric Rossby waves during FGGE. J. Atmos. Sci., 41, 1320-1335.

Malguzzi P., Speranza A., 1981: Local multiple equilibria and regional blocking. J. Atmos. Sci., 38,1939-1948.

Mechoso C.R., Hartman D.L., 1982: An observational study of travelling planetary waves in the southern hemisphere. J. Atmos. Sci., 39, 1935.

Molteni F., Sutera A., Turci N., 1986: Personal Comunication.

Pedlosky J., 1981: Resonant topographic waves in barotropic and baroclinic flows. J. Atmos. Sci, 38, 2626-2641.

Rambaldi S., Mo K.C., 1984: Forced stationary solutions in a barotropic channel: multiple equilibria and theory of non-linear resonance. J. Atmos. Sci., Dec. 1984.

Sawyer J.S., 1970: Observational characteristics of fluctuations with a time scale of a month. Quart. J. Roy. Met. Soc., 96, 610-625.

Schilling H.D., 1982: personal comunication.

Schilling H.D., 1985: On Atmospheric Blocking Types and Blocking Numbers. In "Large Scale Anomalies and Blocking" edited by R. Benzi, B. Saltzman and A. Wiin-Nielsen, Academic Press.

Shutts G., 1984: A case study of eddy forcing during an atlantic blocking episode. In "Large Scale Anomalies and Blocking" edited by R. Benzi, B. Saltzman and A. Wiin-Nielsen, Academic Press.

Speranza A., 1984: Deterministic and statistical properties of northern hemisphere, middle-latitude circulation: minimal theoretical models. In "Large Scale Anomalies and Blocking" edited by R. Benzi, B. Saltzman and A. Wiin-Nielsen, Academic Press.

Sutera A., 1984: Probabilistic distributions of ultra long planetary waves in winter circulations. In "Large scale Anomalies and Blocking", edited by R. Benzi, B. Saltzman and A. Wiin-Nielsen, Academic Press.

Straus D.M., 1984: The behaviour of global Rossby waves during the FGGE year. In "Reports of the seminar on progress in diagnostic studies of the global atmospheric circulation as a result of the global weather experiment" GARP special report No 42, WMO Geneva.

Tung K.K. and R.S. Lindzen, 1979: A theory of stationary long waves. Part I: A simple theory of blocking. Mon. Wea. Rev., 107, 714-734.

Tung K.K. and R.S. Lindzen, 1979: A theory of stationary long waves. Part II: Resonant Rossby waves in the presence of realistic vertical shears. Mon. Wea. Rev., 107, 735-750.

Tung K.K. ,1979: A theory of stationary long waves. Part III: Quasi-normal modes in a singular waveguide. Mon. Wea. Rev., 107, 751-774.

Yoden S. 1983: Nonlinear interaction in a two layer, quasi geostrophic, low order model with topography. J. Metor. Soc. Japan, 61, 1-18.

Wallace J.M., Blackmon M.L., 1983: Observations of low-frequency atmospheric variability. In "Large Scale Dynamical processes in the Atmos phere" edited by B.J. Hoskins and R.P. Pierce, Academic Press.

White G.H., 1983: Skewness, Kurtosis and extreme values of Northern Hemisphere geopotential heights. Mon. Wea. Rev., 108, 1446-1455.

DYNAMICS, STATISTICS AND PREDICTABILITY OF PLANETARY FLOW REGIMES

Michael Ghil
Department of Atmospheric Sciences and
Institute of Geophysics and Planetary Physics
University of California
Los Angeles, CA 90024, U.S.A.

ABSTRACT. We consider regimes of low-frequency variability in large-scale atmospheric dynamics. The model used for the study of these regimes is the fully-nonlinear, equivalent-barotropic vorticity equation on the sphere, with simplified forcing, dissipation and topography. It is found that certain limited regions in the system's phase space are visited recurrently and for extended periods by model solutions. Flow patterns associated in physical space with these regions correspond to synoptically defined zonal and blocked Northern Hemisphere midlatitude flows.

Based on recent ideas from dynamical system theory, it is shown that the system's macrodynamics can be described by two or more planetary flow regimes, the expected residence time in each regime, and the transition probabilities from one regime to another. Connections are made with the classical concepts of statistical-synoptic long-range forecasting (LRF). In particular, transitions between regimes in the model's nonlinear, deterministic dynamics turn out to occur along the directions of maximum variance, i.e., along the first few principal components.

These model-derived ideas are further tested by applying them to a time series of atmospheric data from the Southern Hemisphere. The effect of anomalous surface conditions on planetary flow regimes and their predictability is discussed and perspectives for practical LRF are foreshadowed.

1. INTRODUCTION

After explaining the words in the title, I shall try to give you some feeling for the extent to which climate on the time scale of months to years might be predictable. The lecture is not about actually forecasting climate, for that is still in the experimental stage.

C. Nicolis and G. Nicolis (eds.), Irreversible Phenomena and Dynamical Systems Analysis in Geosciences, 241–283.

1.1. Motivation

The motivation of our work is first of all to understand the *low-frequency* variability of the atmosphere. At this point I should say what "low frequency" is and in what sense one talks about climate as opposed to weather. For operational forecasting purposes, anything that has periods longer than one-to-two weeks has low frequency. One to two weeks is roughly the life time of a cyclone or a family of cyclones, which is definitely what one understands, at least in middle latitudes, by "weather". *Predictability* of atmospheric motions is limited to about one or two weeks precisely by the rather wild behavior of these cyclones -- the fact that they evolve and move in unstable and hence unpredictable patterns. Beyond two weeks the detailed predictability of, say, rainfall or temperature at a point at a given instant is lost and we start talking about low-frequency variability and climate, rather than weather.

Three days are virtually the limit of practical predictability for anyone who listens to radio weather forecasts, and ten days or perhaps two weeks is a limit of theoretical predictability (Lorenz, 1985). In other words, no matter how good our models will ever get, how fine a resolution they will have, how well they will represent physical processes, we shall still not be able to tell whether a hurricane will hit New Orleans two years from now on a certain day of the month of May. But if we understood low-frequency variability better, we might also be able to extend the practical range of numerical weather prediction, in a sense that should become apparent forthwith. So much for the explanation of one word in the title, at least for now.

I emphasize that we first have to understand, since there is really no hope of confident prediction without understanding. A further motivation for our undertaking is that there are new, more extensive, and more accurate data sets on atmospheric flows and fine statistical tools which give us more information about atmospheric behavior as it really is. Next, certain advances in nonlinear fluid dynamics, as well as in computing power, give us new tools for both understanding and prediction.

Finally, the phrase "order in chaos" is making the rounds of the scientific community these days. There is a feeling, not only in the geosciences, but in many areas of continuum physics and even biology, that very disorderly behavior, which in the past might have seemed chaotic or random, is actually subject to certain deterministic laws. Previously we had only statistical tools to describe this chaotic, irregular behavior, like the fluctuations of temperature in Heraklion from day to day. Now we are trying to bring dynamics to bear on it so that the statistics and the dynamics are combined. Basically this lecture is an illustration of how this combination looks and works.

First of all, what is the object of our study? Fig. 1a is a stereographic projection of the Northern Hemisphere with contours of a particular height of the 500mb pressure surface, namely the contour of 5520 geopotential meters which is a crude indicator for the polar front (Reinhold, 1981). It is not just a map of the 500mb surface on a given day, but rather of this particular contour on successive days -

- not a snapshot, but a movie which concentrates on one particular part of the picture. Across the Atlantic the contour only undergoes small changes from one day to the next and remains nearly zonal, i.e., nearly parallel to circles of latitude; over the Eastern Pacific, the contour positions vary widely in time. This picture covers the 18 days from January 26 to February 15, 1980. The agitation over the Eastern Pacific is characteristic of cyclone disturbances and explains basically why you will not be able to predict the weather over the United States beyond a few days.

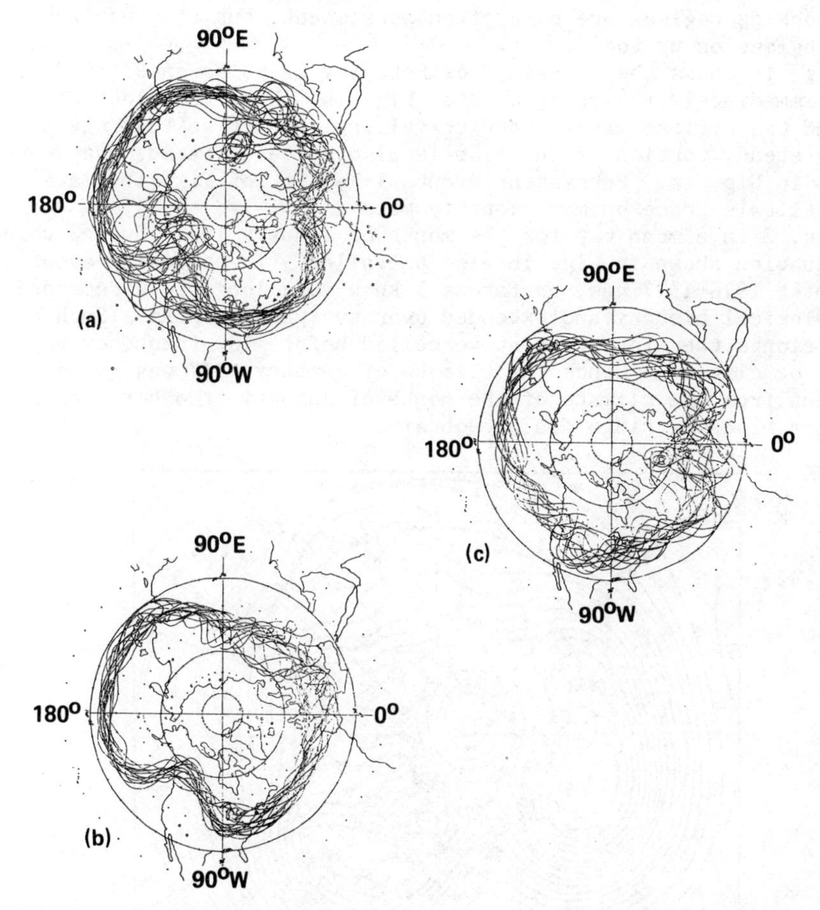

Fig. 1. Successive positions of the 552 dekameter contour on the 500 mb surface. (a) January 26 to February 15, 1980. The flow is aperiodic, or weakly turbulent. The disturbances over the North Atlantic arise from cyclones with characteristic life times of 3 to 5 days. (b) February 5 to February 20, 1977. The flow regime is persistent with a blocking high over the American West Coast. (c) February 22 to March 8, 1977. The flow is weakly turbulent everywhere (after Reinhold, 1982).

Fig. 1b shows a different situation, the 16 day period from February 5 to 20, 1977. The flow is much better behaved and furthermore that nice behavior is associated not with zonal flow like in Fig. 1a but with a *blocking* ridge which deflects the flow. There appears to be hope in a situation where this flow regime prevails that one could predict U.S. weather more accurately for a longer period of time. This is not to say that blocked flows are always, or necessarily, more persistent and therefore more predictable: in Fig. 1a it is the zonal flow which is persistent, and thus predictable. We shall see later that blocking regimes are more often persistent, but very persistent zonal regimes occur too.

Fig. 1c shows the 15 days from February 22 to March 8, 1977, the period immediately following that of Fig. 1b. Here the block breaks down and the cyclone waves start travelling through so that the previously steady portion of the flow is also quite agitated, even more so than in Fig. 1a. Persistent events like that of Fig. 1b leave their telltale trace on mean monthly maps.

Fig. 2 is a mean map for the month of January 1977, during which the situation shown in Fig. 1b also prevailed. The blocking event of the winter 1976-1977 was, as far as I know, the longest in recorded meteorological history and extended over two whole months. Such an event belongs therefore to what we called before low-frequency variability, or climate. Hence the climate of January 1977 was rather different from the climate of the month of January of other years, when this blocking ridge did not obtain.

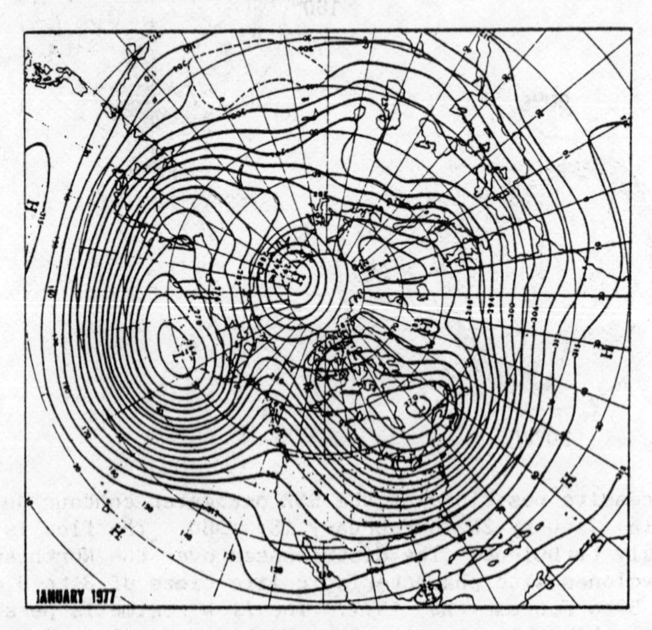

Fig. 2. Mean 700 mb height contours for January 1977; contour interval is 3 dekameters (from Wagner, 1977).

1.2. Multiple Regimes

Next the question arises of what creates these particularly persistent sequences? The major topic of dynamic meteorology over the past 30 or more years has been to study weather, the cyclone disturbances. These have been studied very successfully, and in particular they have been studied precisely as small disturbances of zonal flow. The current generation of meteorologists was raised on the basic idea that everything that can be known about the atmosphere will be known by doing some perturbation analysis about zonal flow. J.G. Charney and E.T. Eady were most instrumental in establishing this point of view, with two now classical papers in the late 1940s (Charney, 1947; Eady, 1949).

While Eady did not live to see the cascade of variations on his theme (Drazin, 1978), Charney, 30 years later, following up on some work of J.Egger (1978), turned around and said: well it doesn't really look very likely that such a persistent ridge as the one in Figs. 1b and 2 will only be a small perturbation of zonal flow, it just might be something entirely different. So he studied a very simple dynamical model, the well-known quasi-geostrophic equation of equivalent barotropic potential vorticity conservation in rectangular geometry (Pedlosky, 1979), using a very crude, low-resolution representation of the atmospheric fields which supposedly satisfy that equation.

With this model, Charney and DeVore (1979) obtained the following picture: as a forcing parameter in the problem was changed (Fig. 3a) there was not just one stable stationary solution, namely the zonal flow which everybody expected until then, but a second solution which looked like blocked flow. Figs. 3b and 3c show the contours of these two stationary solutions for a very simplified Northern Hemisphere topography in which there are basically two oceans and two continents of equal size. The zonal flow pattern (Fig. 3b) is not perfectly zonal of course, but it has only small amplitude waves and these waves are in phase with the topography. The other solution pattern (Fig. 3c) has first of all much stronger meandering, and furthermore is out of phase with the topography. The high can be thought of as representing the West Coast ridge in the previous figures.

Of course, it is very thrilling that you can solve a nonlinear problem and obtain two stable equilibrium solutions instead of one, but when you look at the data, the atmosphere is not quite in equilibrium. It is not in one equilibrium, but it is not really in two equilibria either. This can be demonstrated with the diagrams in Fig. 4 prepared by R.M. Dole (1982) who looked in detail at the statistics of such persistent weather anomalies.

In Dole's observational work the Atlantic and Pacific do not have equal width and equal properties as in the simple barotropic channel model of Charney and DeVore. By the time of this work, the question of multiple regimes and blocking had become somewhat of a bone of contention among various groups of meteorologists; so Dole tried to avoid those words and only talked about positive and negative geopotential anomalies. An anomaly is a deviation from the climatological mean pattern, defined in operational use as the 30 year average pattern.

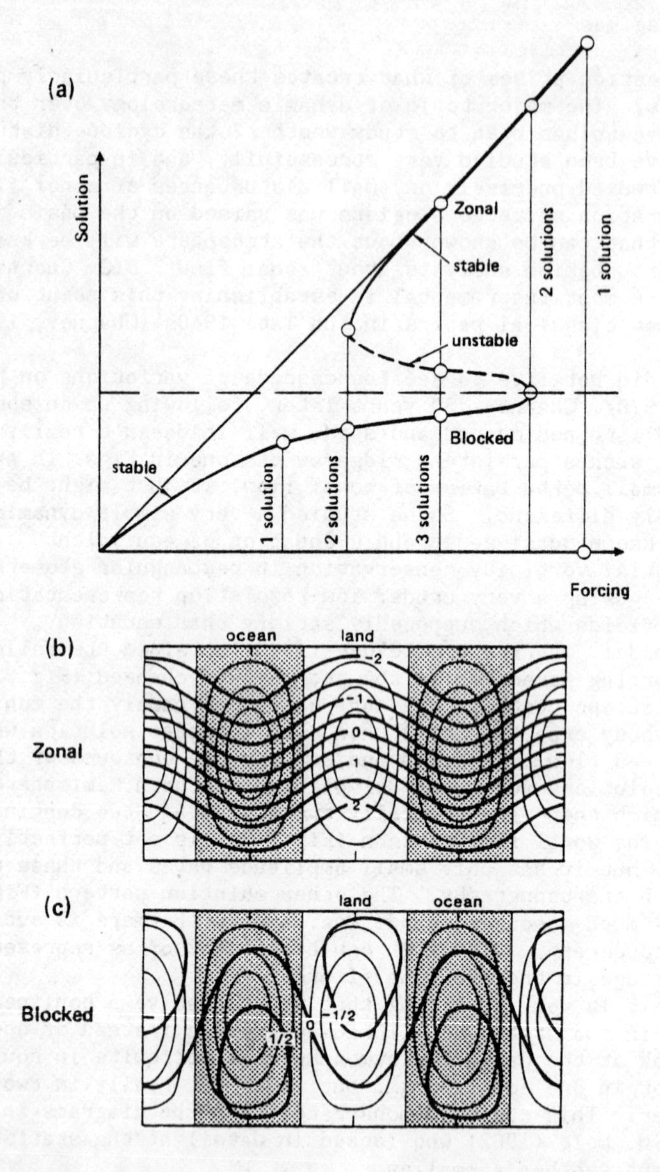

Fig. 3. Multiple flow regimes of a simple equivalent barotropic model. (a) Change in number and nature of solutions as the forcing parameter changes (after Ghil and Childress, 1986). (b) and (c) Flow patterns of two stable stationary solutions: Zonal (b) and Blocked (c) (after Charney and DeVore, 1979).

Since the climatological mean pattern contains both blocked and zonal events, a negative anomaly is more straight and more zonal than the mean (Fig. 4b). By contrast, Fig. 4a shows for the Atlantic a positive anomaly which corresponds to a blocking ridge and to a splitting of the westerly jet around that block; sometimes such a pattern is called an Ω-block, after the shape of the Greek letter.

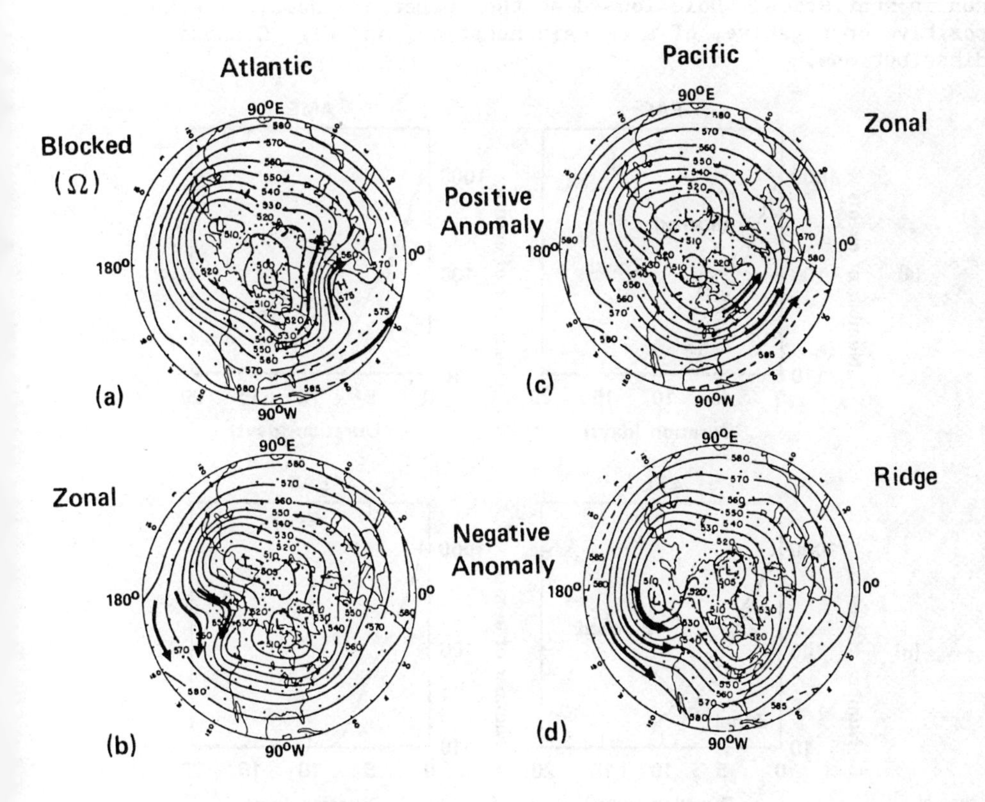

Fig. 4. Spatial patterns of observed persistent anomalies: (a, b) North Atlantic positive and negative; (c, d) North Pacific positive and negative, respectively (after Dole, 1982).

In the Pacific the situation is somewhat different, because the Pacific is wider and most of the points selected by Dole to define anomalies lie in the mid-Pacific. A positive anomaly there (Fig. 4c) corresponds in fact to zonal flow across the West Coast, and a negative anomaly at these points (Fig. 4d) often goes with blocked flow over the West Coast.

What Dole and many of us were interested in was whether these anomalies are in any way different from a first-order auto regressive process, or Markov process, or "red noise". In other words, if you have a purely random motion, with fluctuations decaying towards the mean, there still is a certain chance of the fluctuation staying above or below zero for a certain extent of time. That is called a *run* in statistics. Dole looked at the number of anomalous events, positive or negative, of a certain duration, and Fig. 5 shows their distributions.

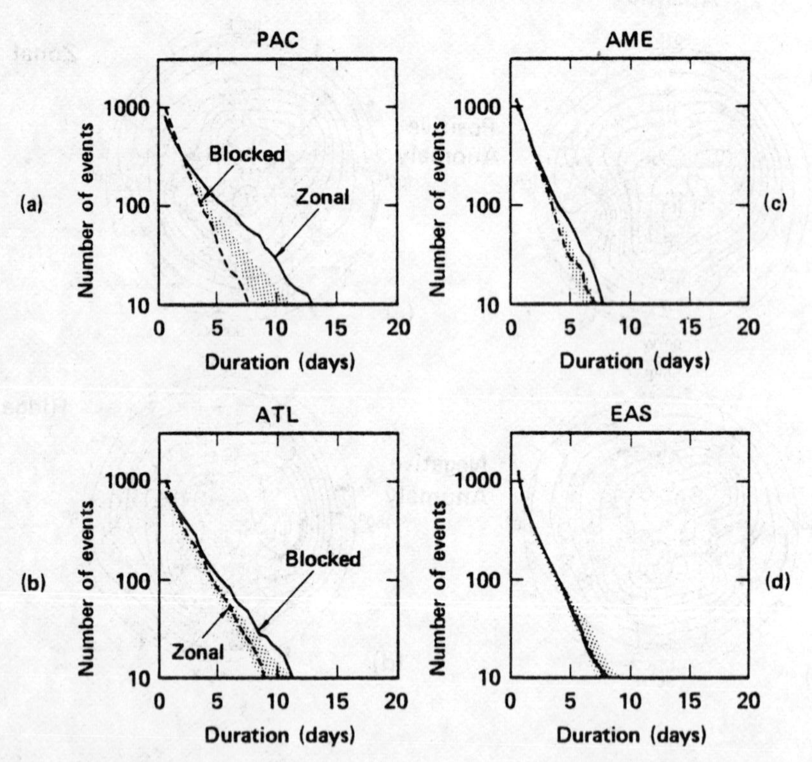

Fig. 5. Persistence properties of circulation anomalies (normalized by standard deviation). Note the differences between regions: positive (solid) and negative (dashed) anomalies differ significantly from a red-noise process (dotted) for the Pacific (PAC) and Atlantic (ATL) regions, but not for Eastern North America (AME) or East Asia (EAS). Regimes tend to be of shorter duration over the latter than over the former (after Dole and Gordon, 1983).

Positive anomalies again imply blocking flow in mid-Pacific, with zonal flow across the West Coast, and negative anomalies the reverse pattern. The areas in which the corresponding curves for a red-noise process would be expected to lie with 90% probability have been shaded. This shows that not only are the positive and negative anomalies distinct from one another in pattern (Fig. 4), but that they are distinct also in a well-defined statistical sense as to persistence. Depending on actual duration (notice cross-over in Figs. 5a and 5b between two and three days), positive anomalies can have more often a given persistence than negative ones, or *vice versa*, in either case, the two types of flow patterns, zonal and blocked, have distinct persistence properties.

The conclusion from these observations is that in the atmosphere we have "multiple" - yes, but "equilibria" - no. In other words, we have multiple regimes which are more complicated than simple equilibria. That explains two more words out of the lecture's title, *planetary flow regimes*. Atmospheric flows have several preferred regimes. The word of mouth around the community was that among these, blocking is characterized by strong persistence. Blocking contours certainly have weak vacillations; in other words they stay pretty constant. Against that mean zonal flow was regarded as weakly persistent, with strong vacillations.

1.3. Dynamics and Statistics

At this point, we have to see how *statistics* comes into our business. I shall again illustrate with a very simple model, formulated by E. N. Lorenz (1963). It has three components, and is a model of thermal convection, rather than of atmospheric circulation. It was the first deterministic model to exhibit clearly chaotic behavior. That behavior is displayed schematically in Fig. 6 which shows a solution projected onto the coordinate planes of three-dimensional *phase space* defined by the variables X, Y, Z of the model. The solution moves around the points marked C and C', which correspond to two states of steady convection, in an entirely random fashion. You could think about one lobe as being zonal flow and the other lobe as being blocked flow, except for the fact that of course the symmetry in the Lorenz model does not have a counterpart in the atmospheric flow regimes that we are interested in.

Now let us adopt the point of view of some observer who knows nothing about the underlying dynamics, and suppose that such a person would try to describe statistically what is going on. Especially in models with many more than three components, one then tries to determine the directions of *maximum variability*. Those directions of maximum variability are called the "principal components" or the *empirical orthogonal functions* (EOFs) of the second-order statistics or "moments" of the phenomenon. More precisely, EOFs are just the eigenvectors of the covariance matrix, i.e., of the matrix of second-order moments.

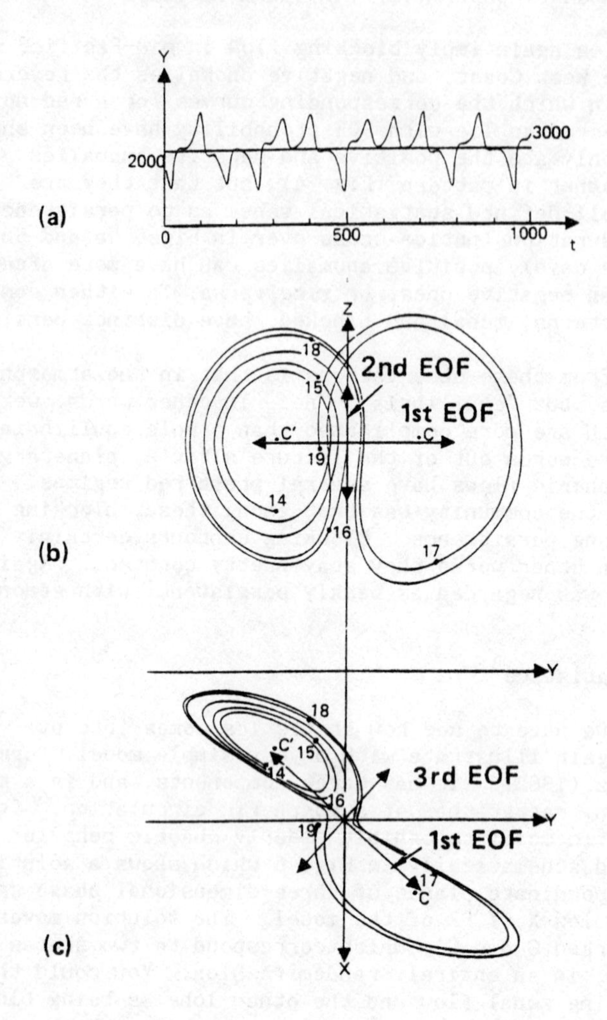

Fig. 6. Numerical solution of the simplified convection equations. Panel (a) shows the evolution in time of one variable, Y; panels (b) and (c) show the projections of the trajectory in phase space onto the (Y,Z)-plane and the (X,Y)-plane, respectively. Numerals in (b) and (c) indicate successive epochs, 1,000 time units apart along the solution (after Lorenz, 1963). In contrast to large-scale circulations, both convection regimes in this model have an extremely simple flow pattern. Moreover they both have the same persistence characteristics, whereas the persistence of zonal vs. blocked flow is asymmetric.

But physically speaking, these vectors determine the directions of maximum variability. For the Lorenz model, a statistician who would not know where his time series came from and just computed its first EOF will get something that pointed from one lobe to the other; EOFs are only defined up to sign: they indicate directions.

To make clearer what is meant by the connection between *dynamics* and statistics let us look more generally at a system with a small number of variables. If the system is subject to forcing and dissipation, it can only behave in one of the following ways in the long term: 1) In time all variables can approach a constant value (Fig. 7a), corresponding to a fixed point in phase space (Fig. 7b). 2) Alternatively the system can approach a limit cycle in phase space (Fig. 7d), corresponding to a periodic solution in time (Fig. 7c). 3) The case of the Lorenz model is more complex and described by a strange attractor (Ruelle, 1980, 1985).

What is a *strange attractor?* First of all, the term "strange" has to be taken as limited in meaning to its mathematical definition, just as in the case of "irrational" or "complex" or "imaginary" numbers, which are inseparable from modern physics. In the same way I am confident that you will not be able to conceive of physics, or even biology, in the next two decades without strange attractors. The term describes long-time behavior in a system with three or more variables: Because of dissipation, the flow along trajectories in phase space converges to a subset of zero volume, but of high complexity.

This subset is made of an infinity of sheets of dimension two or higher, according to whether the number of variables is three or higher. The sheets are closely packed together, like folded linen or onion skin, but still leave very small gaps between them. Thus two trajectories lying in two adjacent sheets can be arbitrarily close at one moment and still separate soon thereafter to visit very different portions of the attractor for the foreseeable future. Every point on the attractor is stable in the direction or directions perpendicular to the sheet it lies in; but it is unstable in all directions tangent to the attractor. Consequently two trajectories which come together no matter how closely will diverge again, exactly like two sequences of atmospheric flow or two solutions of general circulation models (Lorenz, 1985).

To illustrate the concept of strange attractors and to give you an idea of what an EOF would do for such an attractor, let us look first at the simpler case of the limit cycle or periodic motion. In Fig. 7e we see the periodic motion of a planet around the Sun. The speed of the planet at each point of its orbit is given by Kepler's Law of Areas, according to which equal areas are swept out by the radius vector in equal periods of time. In other words, the planet slows down considerably as it comes to aphelion and it speeds up as it nears perihelion.

By symmetry, the mean (the first-order statistic) has to lie on the major axis of the elliptic orbit and it will be closer to the part where the concentration of points lying equal intervals of time apart is bigger. Again if you did not know what law was governing the motion of the planet, you would just take successive observations and see that the cloud of observations is much denser around the aphelion. So the time mean is closer to this point; furthermore the direction of maximum

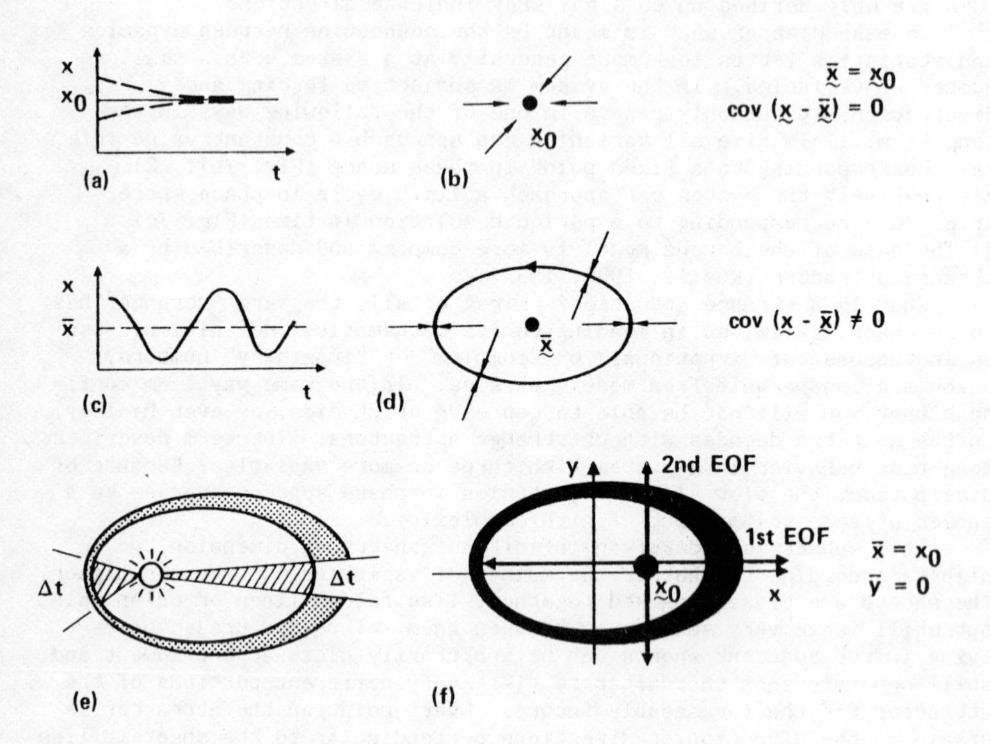

Fig. 7. Long-time behavior of solutions to forced dissipative models of nonlinear dynamics. Panels (a) and (c) indicate how one variable, x, behaves in time t, with $\bar{x}$ indicating the long-time mean. Panels (b) and (d) indicate the phase-portrait of the attractor set: fixed point (b) or limit cycle (d), and the directions of stability and motion. Panels (e) and (f) show in further detail the illustrative case of elliptic motion of a planet around the Sun, with time spent along different portions of the orbit (e), and the EOFs which obtain (f). The dark shading in (e) and (f) is proportional in thickness to the local density ρ of successive points equally spaced in time, with $\rho \sim 1/v$, v being the local tangent velocity.

variability, or of the first EOF, just points from this cloud where it is thickest to where the fewest points are found near the perihelion. Similarly in the case of Fig. 6 the direction of the first EOF simply joins the centers of the two concentration maxima in phase space.

Having now explained the various words in the title we are ready to get into the substance of the lecture. In order to make the ideas more precise, I shall formulate a model which is slightly more complicated than the ones discussed up to now, with 25 variables rather than two or three. We shall pass in review stationary solutions, their multiplicity and stability, and the flow patterns associated with them. More realistic nonstationary flow regimes occur in this model, resembling the ones that appear in the atmosphere and in particular we shall consider the persistence and predictability of their flow patterns. In other words, we shall try to answer the question, "What can really be predicted beyond two weeks, aside from just means and variances?"

We shall come back to the EOFs, and shall also discuss some Southern Hemisphere data. In addition, something will be said about 30-to-50 day oscillations, an important phenomenon in the low-frequency variability of the global atmosphere. Much, but not all of what I have to say is contained in greater detail in two books. One of these (Ghil *et al.*, 1985) has just come out and contains the contributions by Lorenz and Ruelle already cited; the other (Ghil and Childress, 1986) will appear shortly.

2. MODEL DESCRIPTION AND DYNAMICS

The model, like those of Egger (1978) and Charney and DeVore (1979), is barotropic. The atmosphere's low-frequency variability is characterized by rather *barotropic*, nearly two-dimensional structures: the mean monthly ridge at the 700 mb level (Fig. 2), which is the one operational long-range forecasters like to look at (e.g., Namias, 1968), would appear pretty much the same at 500 mb and 850 mb, and perhaps even at 300 mb. That such a ridge is not significantly distorted or displaced from one level to another is one of the intriguing differences between the low-frequency variability we discuss here and synoptic-frequency variability: cyclone waves are definitely *baroclinic*, tilting with height, and hence truly three-dimensional, while mean monthly features are predominantly barotropic. Baroclinic phenomena are important in transitions between persistent anomalies, and even in their maintenance, but once an anomaly is established and persists it tends to have a barotropic *structure*.

2.1. Model equations

The model is governed by the equivalent-barotropic form of the equation for the conservation of potential vorticity on the sphere (Ghil and Childress, 1986, Chapters 3 and 6; Legras and Ghil, 1985):

$$\frac{\partial}{\partial t}(\Delta - L_R^{-2})\psi + J\left\{\psi, \Delta\psi + f\left(1 + \frac{h}{H}\right)\right\} = \alpha\Delta(\psi^* - \psi). \quad (1)$$

Here ψ is the streamfunction, Δ is the horizontal Laplacian, J is the Jacobian operator, f is the Coriolis parameter, h is the topographic height, H is the scale height of the atmosphere, ψ^* is the forcing streamfunction and α^{-1} is the relaxation time. Horizontal coordinates are the longitude ϕ and the sine of the latitude $\mu = \sin\theta$.

All variables are scaled by the radius of the Earth a, its angular velocity Ω and a characteristic speed U, yielding the nondimensional variables

$$L_R = a\lambda,\ h = Hh',\ t = t'/2\Omega; \qquad (2a,b,c)$$

$$(\psi,\psi^*) = aU(\psi',\psi^{*'}),\ \alpha = 2\,\Omega\,\alpha',\ f = 2\Omega\,\mu. \qquad (2d,3,f,g)$$

The nondimensional form of the equation, dropping the primes, is thus given by:

$$\frac{\partial}{\partial t}(\Delta - \lambda^{-2})\psi + \rho J(\psi,\Delta\psi) + J\{\psi,\ \mu(1+h)\} = \alpha\Delta(\psi^* - \psi)\ , \qquad (3)$$

where the nondimensional number $\rho = U/2\Omega a$ measures the intensity of the forcing.

Equation (3) is discretized through a truncated expansion in *spherical harmonics* $Y_n^m(\phi,\mu) = P_n^m(\mu)e^{im\phi}$,

$$\psi(\phi,\mu,t) = \sum_{n=0}^{N}\ \sum_{m=-n}^{n}\ \psi_n^m(t)\ Y_n^m(\phi,\mu)\ , \qquad (4)$$

where $P_n^m(\mu)$ are associated Legendre functions, and N = 9. Equatorial symmetry of the flow, as well as a sectorial periodicity (mod π) in longitude are assumed. The resulting $M = (N+1)^2/4 = 25$ modes allow 132 triadic nonlinear interactions.

The topography represents two equal continental masses separated by two equal oceans, i.e., a crude approximation of Northern Hemisphere topography,

$$h = 4h_0\mu^2(1 - \mu^2)\cos 2\phi\ , \qquad (5a)$$

where $h_0 = 0.1$. The mean forcing is a zonal jet, expanded in the first two zonal components.

$$\psi^* = -\kappa\mu^3 = aY_1^0(\mu) + bY_3^0(\mu)\ ; \qquad (5b)$$

κ is a nondimensional constant chosen so that the maximum forcing speed, which occurs at 50 N, has a dimensional value of 60 ms^{-1} for $\rho = 0.20$. The forcing ψ^* models, in the absence of explicit baroclinic effects, the mean thermal wind which would be observed in the case of an idealized purely zonal circulation, with no meridional mass transfer. Higher ρ corresponds therewith to the winter circulation, lower ρ to the summer.

The right-hand side of (1) models, at the same time as the forcing by ψ^*, the dissipation across a hypothetical Ekman boundary layer.

The characteristic relaxation time α^{-1} in mid-troposphere, at the equivalent barotropic level, is of the order of 10 days.

Figure 8a shows the wavenumber combinations, or "modes" which are allowed, giving a reasonable representation of the largest scales of the flow; at the same time they permit a rather detailed analysis of what is going on dynamically. Fig. 8b is a map showing the same simplified model topography already used by Charney and DeVore in a periodic midlatitude channel (Fig. 3). The plot uses stereographic

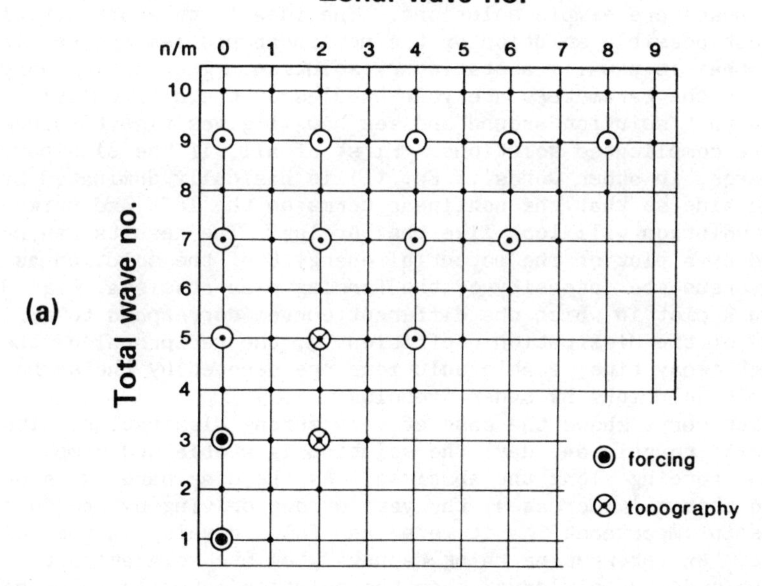

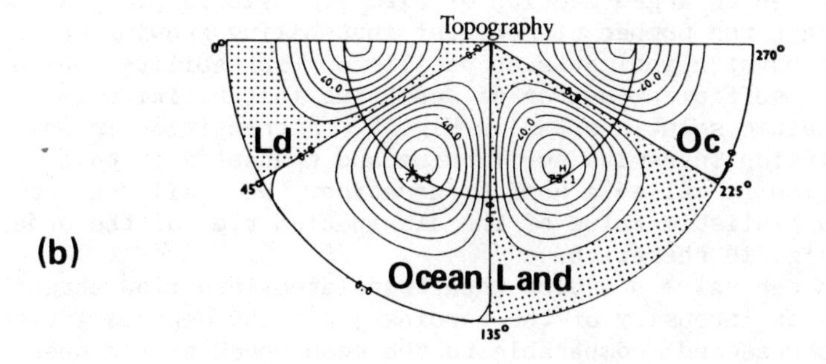

Fig. 8. Model truncation and topography. (a) Truncation of model. Active "modes" are circled. Wave-wave interactions are allowed. (b) Simplified relative topography of model (after Legras and Ghil, 1985).

projection in which 270 degrees have been mapped onto a half disc; we have one full ocean and one full continent, together with half an ocean on one side, and half a continent on the other side, showing the repetition of the pattern.

2.2. Stationary Solutions

The general approach in nonlinear dynamics toward searching for order in chaos requires some guide posts on its difficult route, and the guide posts are simple solutions. The idea is to start out with the simplest possible solution of the nonlinear problem one can lay hands on; that is usually a stationary solution. Then as you vary parameters-- the parameters are your handles on the difficulty-- you can follow that solution around and see how it gives rise to successively more complicated solutions. First of all, if the dissipation is very large, in other words if Eq. (1) is basically dominated by its right-hand side so that the nonlinear terms on the left are unimportant, the solution will look like the forcing. The results can be summarized by a plot of the potential energy E of the solution as ordinate versus the intensity of the forcing ρ as abscissa. Fig. 9 shows such a plot in which the different curves correspond to different values of the dissipation coefficient α, the reciprocal of the "e-folding" decay time; stable solutions are denoted by the symbol x and unstable solutions by other symbols.

The top curve shows the case of very strong dissipation, with a decay time of roughly one day; the solution is stable and simply follows the forcing along the abscissa. As the dissipaton α is decreased so that α^{-1} increases, the very strong driving by the forcing diminishes in importance, as it were, and folds develop in the solution curve. An interesting thing happens when the relaxation time reaches the order of six days; then the solution actually folds back on itself, in a saddle-node bifurcation, so that more than one solution exists for a single value of the forcing parameter. Furthermore, at least one of the coexisting solutions is unstable.

Fig. 10 is an enlarged section of Fig. 9. Symbols along solution branches indicate the number of modes of instability growing on that solution: a plus sign indicates an oscillatory instability, while a dot indicates a sufficient number of competing instabilities in order to generate chaotic solutions, as in Fig. 6. A transition in which a stationary solution loses its equilibrium and transmits it to a *periodic* solution is called a *Hopf bifurcation*. We shall concentrate hereafter on a realistic value of the dissipation time of the order of 20 days, and Fig. 10 therewith.

The parameter value $\rho = 0.20$ when translated into flow strength corresponds to an intensity of the forcing jet at 50 degrees latitude of 60 meters per second, comparable to the mean speed of the observed westerly winds. Near this value of the forcing there are plenty of stationary solutions; however all of them except those marked by x are actually unstable. So what good are these solutions if they are

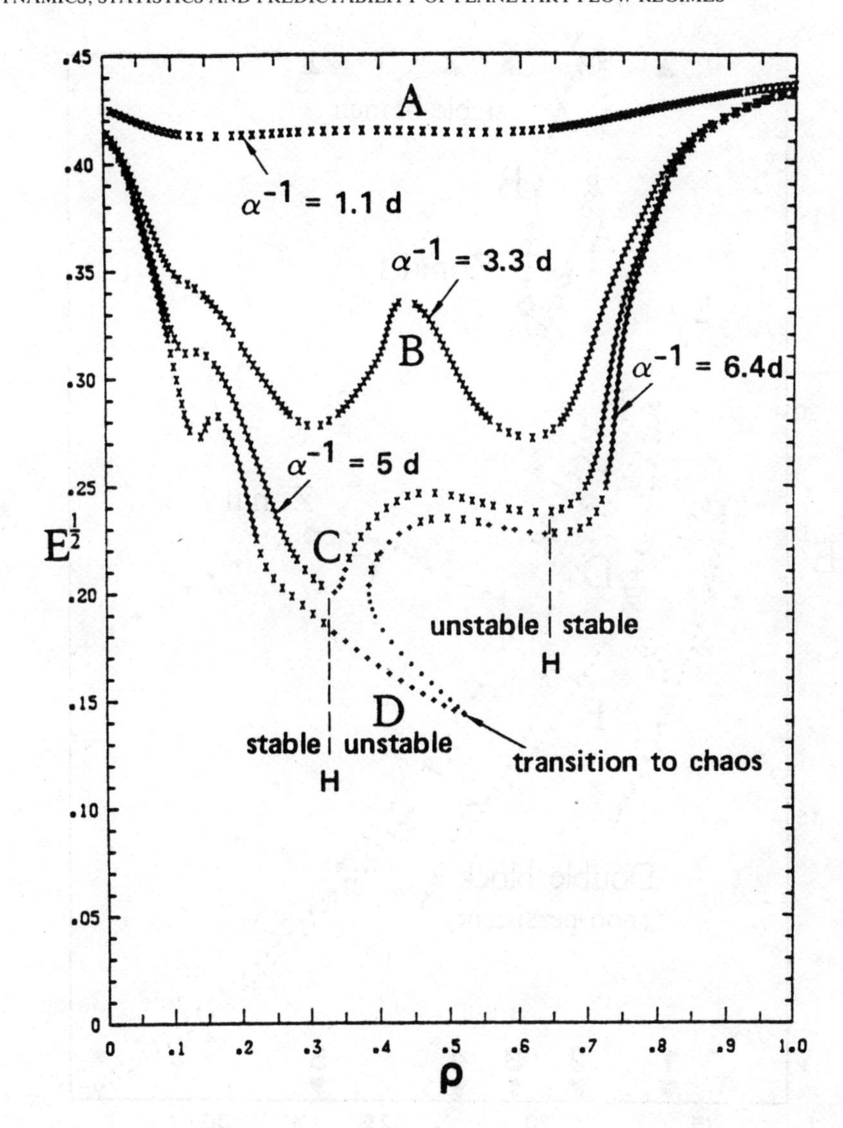

Fig. 9. Stationary solutions of the model as a function of the forcing parameter ρ and the dissipation parameter α: x indicates stability, H stands for Hopf bifurcation (after Legras and Ghil, 1985).

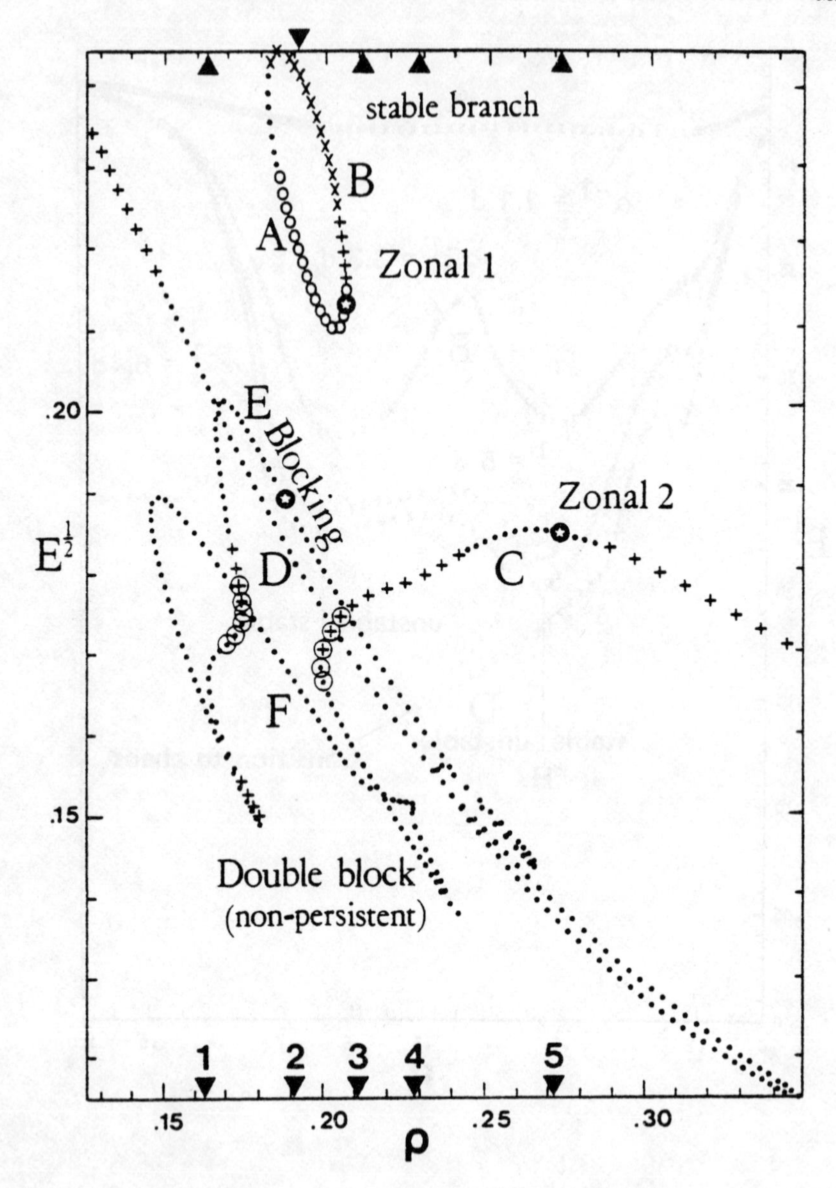

Fig. 10. Enlargement of part of $(\sqrt{\bar{E}},\rho)$ cross-section in Fig. 9 for α^{-1}= 20 days. Numbered pointers indicated ρ-values for which time-dependent model solutions were studied in detail (from Legras and Ghil, 1985).

unstable? We have been raised to think that only stable solutions can be observed in an experiment, whether in a laboratory experiment or in a numerical experiment or in nature itself. Still, it will soon become apparent that these stationary solutions, although unstable, play their role of guide posts very well. To prove this and to justify the descriptive names which are attached to the branches in Fig. 10 we turn now to the flow patterns represented by solutions.

Figs. 11 a, b show the flow patterns of the solutions marked Zonal 1 and Zonal 2 in Fig. 10. The flow in both cases is largely zonal, with the suggestion of a climatological trough over the eastern part of the continent. The flow pattern of the solution branch marked Blocking in Fig. 10 is shown in Fig. 11c and exhibits a closed high over the west coast with splitting of the jet into two branches passing north and south of the high.

These patterns prove that the resolution used here provides realistic-looking solutions. A summary of their dependence on the parameters is shown in Fig. 12. The dissipation parameter α is the abscissa and the forcing parameter ρ is the ordinate. For larger dissipation (smaller relaxation time) the solutions will look like the forcing; everywhere to the right and above the hatched area stationary solutions look a lot like zonal jets and are stable. However as one moves in towards more realistic values of dissipation and forcing the hatching indicates that no stationary solutions are stable, and that their place is taken by either periodic or even more interestingly aperiodic solutions. In other words, solutions appear which behave somewhat like the atmosphere.

2.3. Time-Dependent Solutions

It is really this region of aperiodic or chaotic flow that we are interested in when searching for order in the atmosphere's chaos. How do we inspect, describe and understand this aperiodic behavior? Since we are interested in persistence, one way of looking for persistent sequences is simply to measure the root-mean-square difference, of "RMS distance" between successive maps. Fig. 13 shows time sequences of such successive distances for three different values of the forcing parameter ρ, increasing from top to bottom. When the successive distances are large the flow is non-persistent; when the change in distance per unit time is small, then the solution is persistent. The tick marks on the abscissa in Fig. 13 are 50 days apart so that every minimum in the plotted curve corresponds to a few tens of days. Of course we can look at such a long record only because the model is relatively simple and solutions can be computed numerically for a very long time.

The minima in the top curve (Fig. 13a), marked by arrows, are persistent blocking episodes. As the value of ρ increases, persistent blocking alternates with persistent zonal episodes (Fig. 13b). Finally in the bottom curve (Fig. 13c) zonal episodes have the lowest minima and are most persistent, while blocking episodes are very transient. Note that the range of the ordinate in the top curve is only from 0 to 0.011, while the ordinate in the bottom curve ranges up to

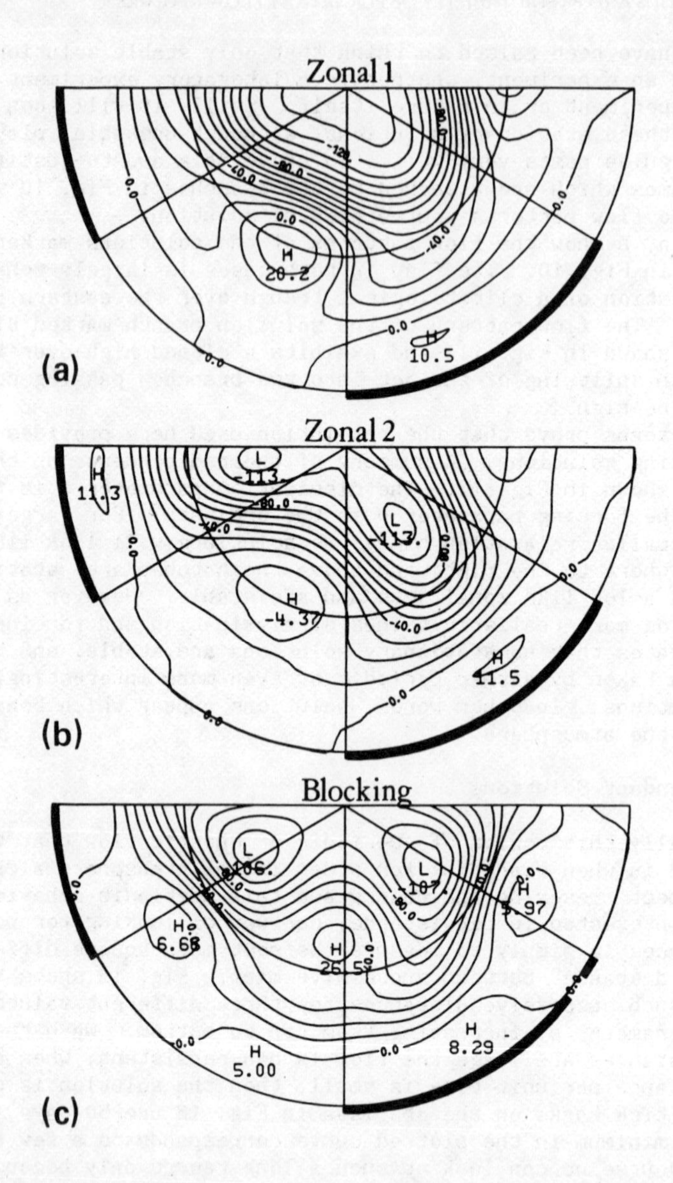

Fig. 11. Flow patterns of solutions indicated by stars in Fig. 10. Panels (a) and (b) are two types of zonal solution; panel (c) is a blocked solution. The position of the model's continents is shown by a heavy dark segment on the outside of the disc (after Legras and Ghil, 1985).

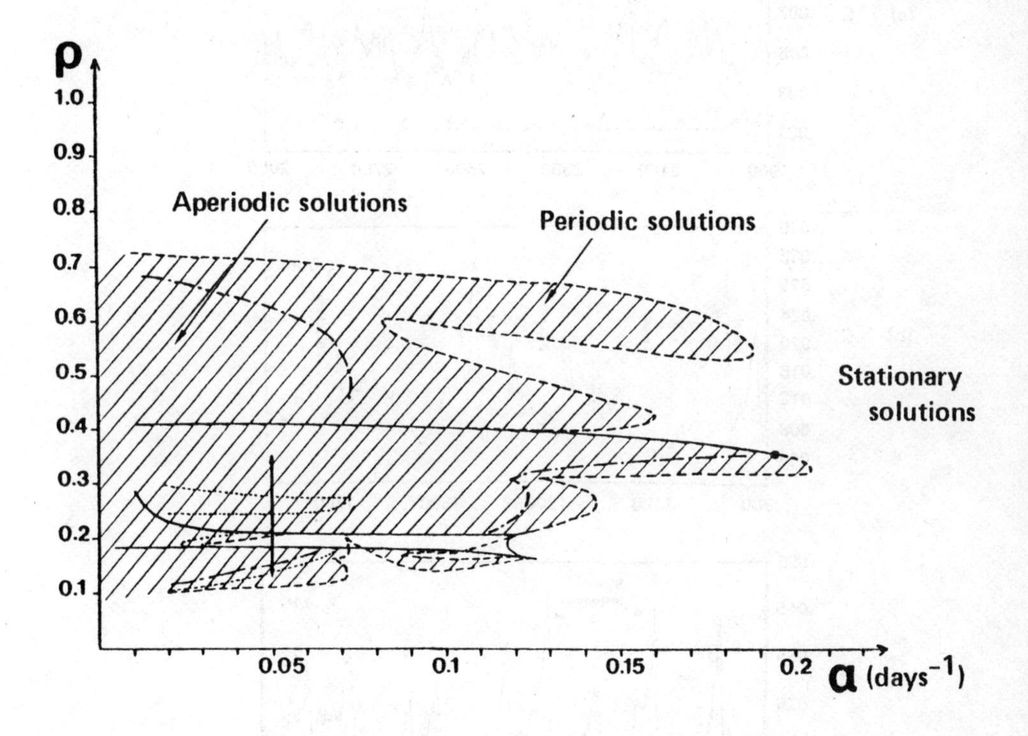

Fig. 12. Parameter dependence of solutions and their stability. Stationary solutions (one or more) are stable outside the hatched area. Hopf bifurcation is indicated by dashed lines, transition to aperiodic solutions by dash-dotted ones. Double-pointed arrow at $\alpha = 0.05$ indicates position of (E,ρ)-cross section in Fig. 10 (after Legras and Ghil, 1985).

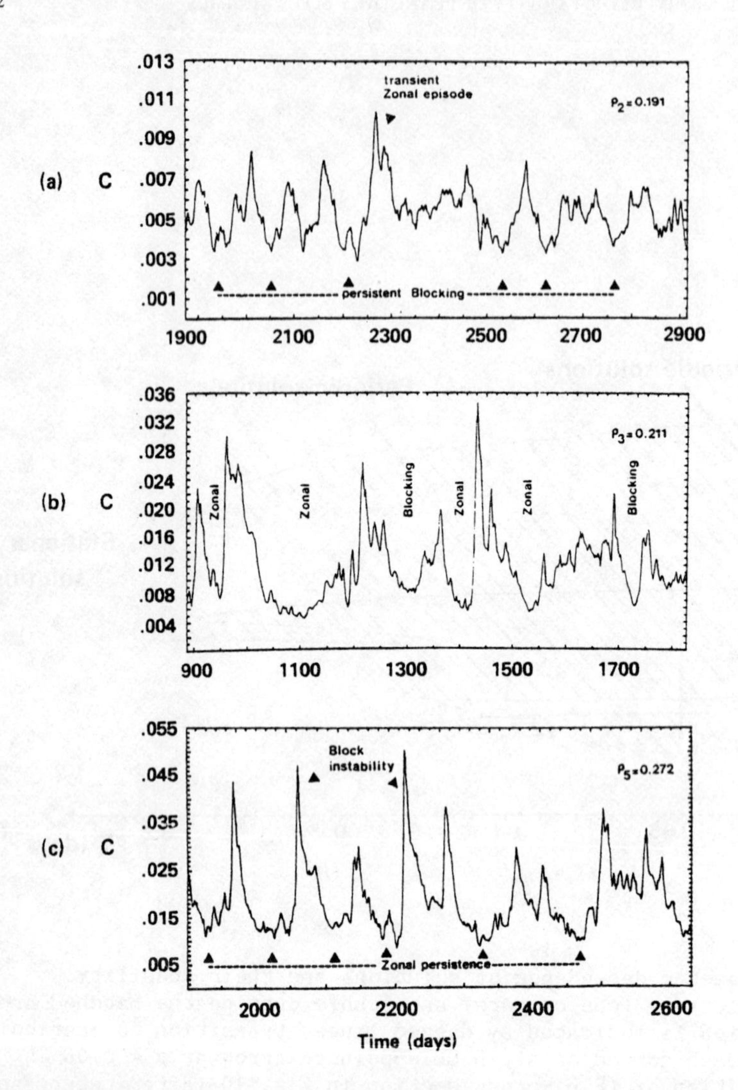

Fig. 13. Persistence of nonstationary solutions. The ordinate is the change C in RMS difference ("distance") between successive maps for unit time τ, $\tau \cong 2$ days,

$$C(t) = \{\Sigma_{m,n}[\psi_n^m(t+\tau)-\psi_m^n(t)]^2\}^{1/2}/\tau ,$$

ψ is the streamfunction. When C is small the solution is persistent. The three panels show successive map distances for increasing value of the forcing parameter ρ. Zonal flows in the model are agitated, while blocked flows are more quiescent, in agreement with observations (after Legras and Ghil, 1985).

0.050. Thus, when the flow is predominantly zonal you can have very persistent sequences of zonal flow; but the deviations from them are very large, while when the flow is dominantly blocked the deviations are relatively small.

To see how the persistent sequences look, it suffices to average all the maps which occur during a given sequence. The composite map of a blocking sequence is shown in Fig. 14a. The very large deflection of streamlines and some of the splitting which characterized the unstable stationary solution called Blocking are still visible in the composite. Hence persistent flow episodes occur close in phase space to the stationary solution they resemble visually. Transient episodes in the solution look much more zonal (Fig. 14b). In the opposite situation where zonal flow is dominant (Figs. 14c,d) the persistent composite again looks much like the unstable stationary solution designated in the same way, while the transient shows a well formed dipole structure, like the ones associated sometimes with Atlantic European blocking.

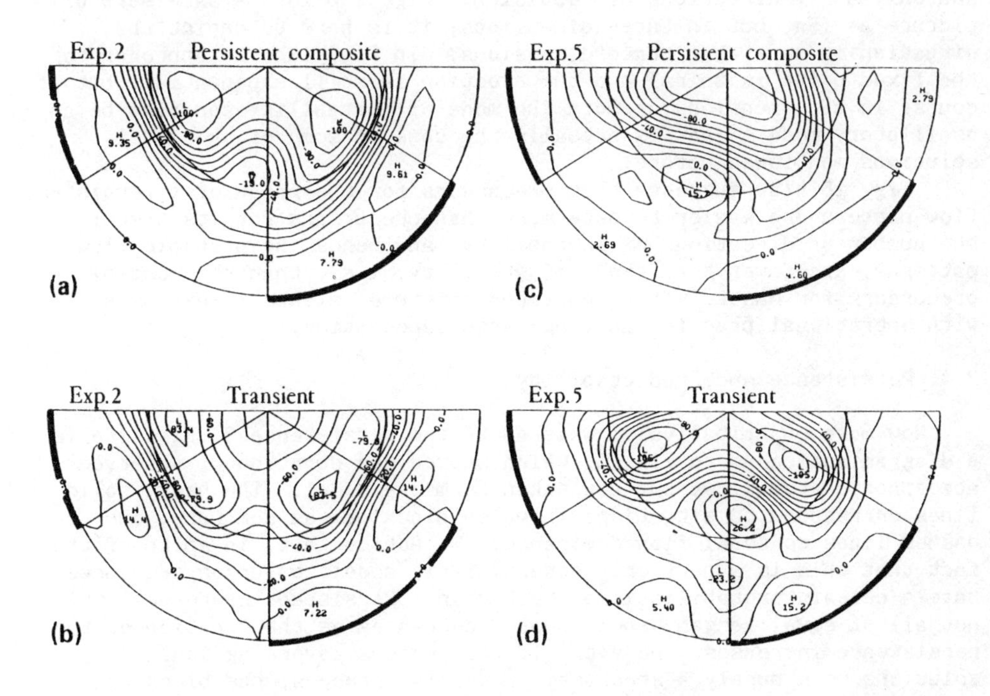

Fig. 14. Composite maps of (a) the blocking episode centered at t = 2640d in Fig. 13, for $\rho = \rho_2$; (b) the transitional period indicated by a pointer in Fig. 13a; (c) a zonal episode centered at t = 2330d in Fig. 13c, for $\rho = \rho_5$; and (d) a transitional period centered at t = 2220d in the same figure (from Legras and Ghil, 1985).

That this can be so is again known by synopticians: Ω-blocks are sometimes very transient, and can last as little as two days only.

To clarify the connection between unstable stationary solutions and persistent sequences we go again to very, very simple models in two or three variables. Fig. 15a shows a fixed point with one direction of stability and one direction of instability; such an unstable fixed point is called a saddle. What happens in the presence of such a saddle? Let us say that a solution is ejected because of the instability from the neighborhood of the fixed point and then wanders around, but eventually gets close to the nonlinear continuation of the stable direction,i. e., to the so-called *stable manifold* of the solution. If it actually moved precisely along the stable manifold it would take an infinite time to get into the fixed point; but even if it misses the manifold by just a little it will stay for a very long time. So, basically, persistent sequences in many variables can be generated by aperiodic solutions getting into the neighborhood of a generalized saddle point that has a lot of directions of attraction and only a few directions of repulsion. Fig. 15b is the same sort of picture as 15a, but in three dimensions; it is hard to depict the situation in more than three dimensions. In Fig. 15b the approach to the fixed point is spiral and the ejection is still exponential, but of course in 25 dimensions or more the mode of instability can also be an oscillatory mode. That is actually the case in some of the model solutions mentioned before.

Fig. 15 also suggests that precursors for the break of a persistent flow pattern are easier to determine than precursors for its onset: the number of directions of instability, and hence the unstable flow patterns, are fewer than those of stability, i.e., than the possible precursors for onset. This conceptual picture is thus in agreement with operational practice in long-range forecasting.

2.4. Persistence and Predictability

Now more quantitatively, what about the persistences? Fig. 16a is a diagram analogous to Fig. 5, which showed the duration of observed atmospheric anomalies, but now taken form the model. The heavy solid lines correspond to situations in which blocking was dominant, and dashed lines to zonal flow dominance. We notice that, in spite of the fact that this is a perfectly deterministic model, solution sequences have a certain probabilistic distribution. Persistent sequences are not all of equal length, their number decreases as the duration of the persistence increases. However, if the process governing model solutions were purely a stochastic red-noise process, the blocking distribution line and also that for zonal flow would have to be perfectly straight in these coordinates.

The behavior of the deterministic model has an aperiodic component, but its persistence curves are significantly different from a straight line. For a certain range of durations they fall off more rapidly; for another range they fall off more slowly. There are definite breaks in both the solid and the dashed lines. A similar plot based on Southern

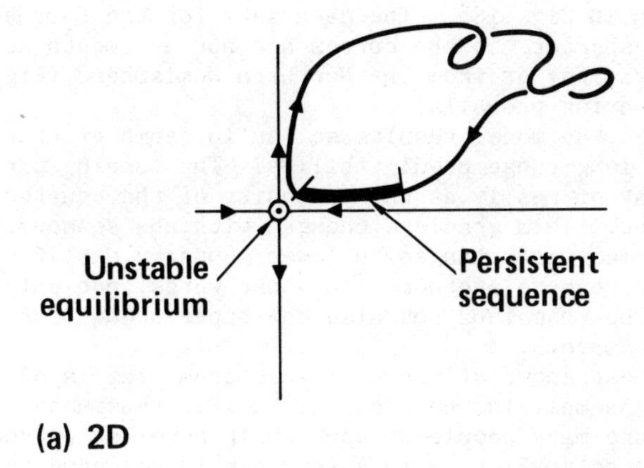

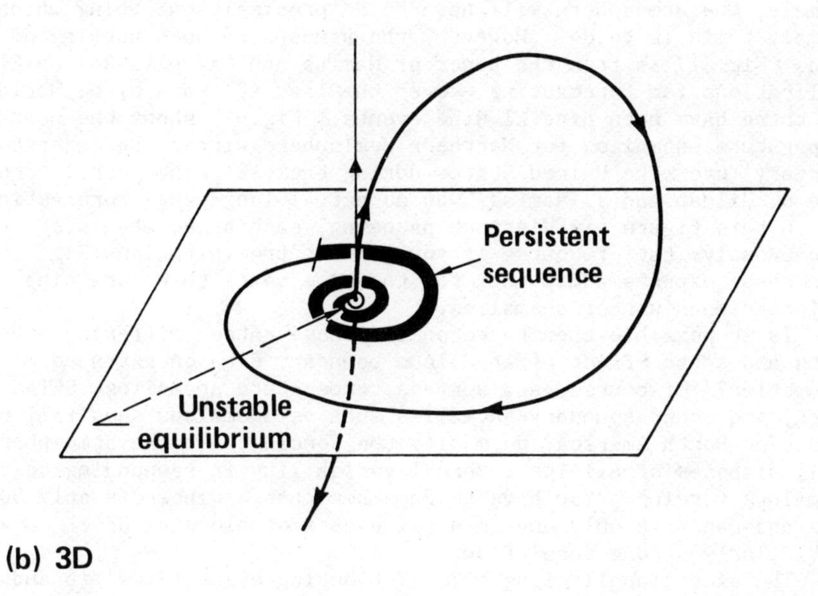

Fig. 15. Solution orbits in relation to an unstable fixed point in (a) two dimensions, and (b) three dimensions (after Ghil and Childress, 1986).

Hemisphere data is shown in Fig. 16b. The data sets for the Southern Hemisphere are a little shorter, so the curves are not as smooth as the ones from the model (Fig. 16a) or from the Northern Hemisphere (Fig. 5), but the same type of behavior prevails.

How can we interpret the model results so far in terms of the real atmosphere and its long-range predictability? The forcing parameter ρ has to be thought of really as the intensity of the equator-to-pole temperature gradient. This gradient changes with the seasons. We also know that it varies with changes in lower boundary data for the same season from one year to another. In other words, not only the atmosphere likes to be anomalous but also the upper ocean, for instance, likes to be anomalous.

The strongest and best known of these oceanic anomalies is of course the El Niño warm anomaly in the tropical Pacific (Rasmusson and Wallace, 1983). There are many people who put their hope for extended range prediction or climate prediction in a very simple response to such boundary forcing. The idea is that, if you have such a strong anomaly, the atmosphere will have to do precisely one thing which that anomaly tells it to do. However, the atmosphere does nothing of the kind. Fig. 17 is from the paper of Namias and Cayan (1984) on El Niño implications for forecasting. Over the last 40 years since World War II, there have been nine El Niño events. Fig. 17 shows the mean temperature anomalies for Northern Hemisphere winter (December to February) over the United States during these El Niño years. People like D. Gilman and J. Namias, who do actual long-range forecasting, see in this figure six distinct patterns, each associated with an El Niño anomaly; but of course if you look at precipitation (Fig. 17b) even these experts admit that for the nine years there are nine different mean winter anomalies.

Is it possible then to reconcile these rather differing statements about the effect of anomalous boundary data on extended prediction? Of course, sea surface temperature anomalies, SSTAs for short, and other boundary anomalies such as anomalous snow fall over Europe or North America, do modify the forcing, but the atmosphere still disposes of all its internal variability in responding to the anomalous forcing. You have to remember that a winter is only 90 days long and can have only one or a few events of blocking or of particularly strong zonal flow.

The exceptionally long 1976-77 blocking event (Figs. 1b and 2), for instance, was associated with an El Niño and the very strong zonal flow of 1982-83 (see last panels in Figs. 17 a,b) also occurred in an El Niño year. During those 90 days if you have an El Niño SSTA there may be a stronger likelihood of a certain type of atmospheric persistent anomaly to occur and to last longer. It doesn't mean that it will be the only response of the atmosphere in that winter, but there may be a better chance that in that winter you will have one or two events of the favored type. In other words, what the forcing does is to select one of the curves in Figs. 5a, 5b, or 16a; it does not select the actual event which will occur, it will only select the preference for a certain type of event being more prevalent and more persistent.

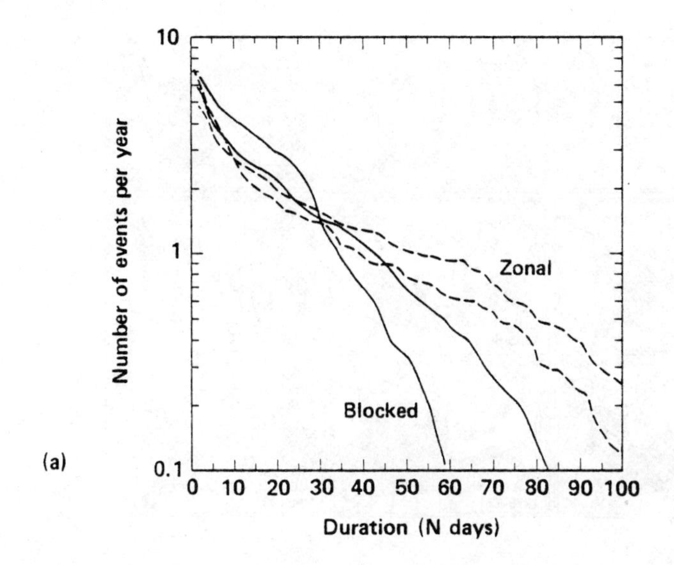

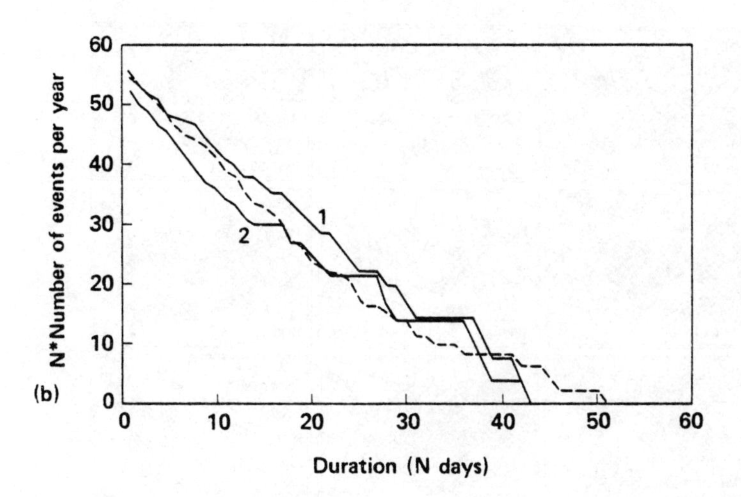

Fig. 16. (a) Persistence properties of model circulation anomalies for blocking episodes (solid: $\rho = \rho_1$ and $\rho = \rho_2$, respectively) and zonal episodes (dashed: $\rho = \rho_4$ and $\rho = \rho_5$, respectively); after Legras and Ghil, 1985. (b) Persistence properties of anomalies observed in the Southern Hemisphere winter circulation. Solid lines indicate persistence for EOFs 1 and 2, respectively, dashed lines for a simulated red-noise process with variance equal to the sum of that associated with the first three EOFs (from Mo and Ghil, 1986).

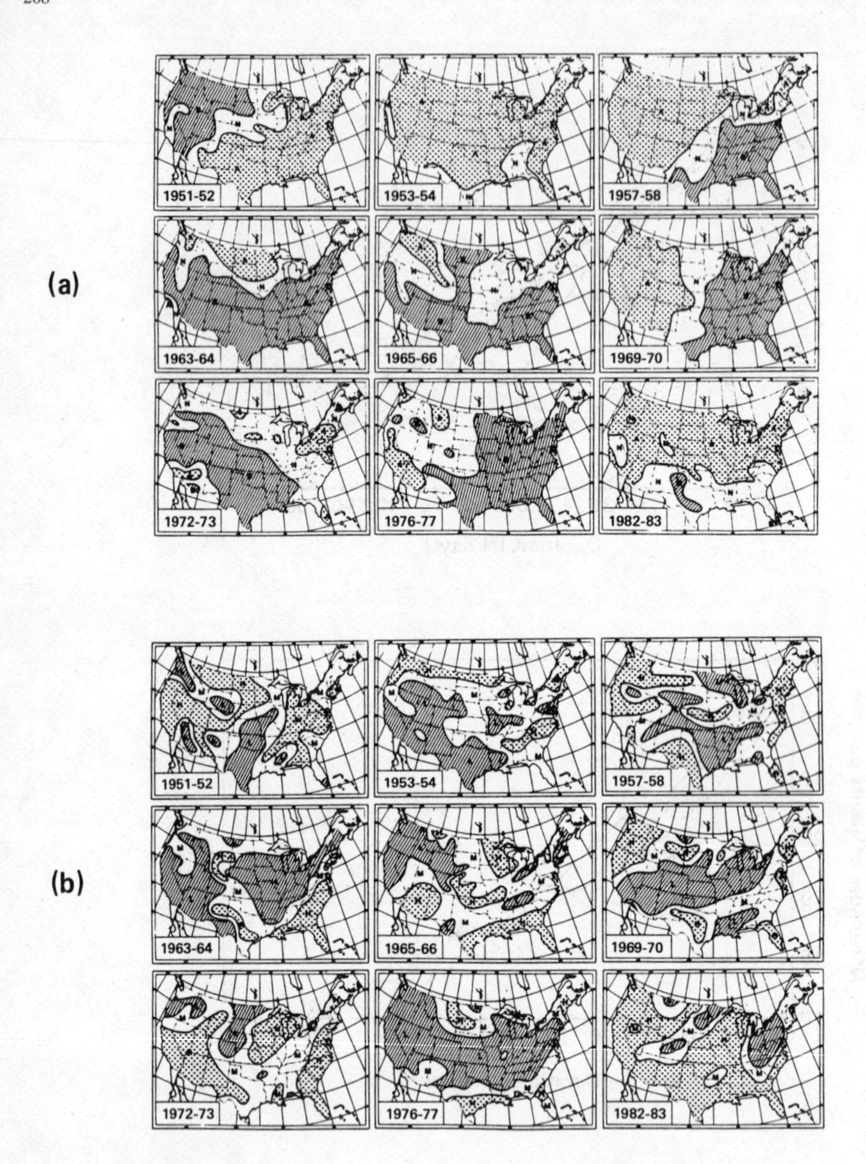

Fig. 17. Mean winter (December through February) anomalies during recent El-Niño years. (a) Temperature anomalies: N - normal, A - above normal, B - below normal. (b) Precipitation anomalies: M - moderate, H - heavy, L - light (from Namias and Cayan, 1984).

3. EOFs AND NONLINEAR DYNAMICS

For a further illustration of the connection between the statistics and the dynamics of persistent anomalies, I shall use model results which parallel some of the Southern Hemisphere data of K. Mo (1986). Fig. 18 repeats Fig. 10 with some changes of notation, for greater ease of reference. The subscripts *b*, *z*, and *m* stand for blocking-dominated, zonal-flow dominated and mean pattern, respectively.

For the sake of brevity, I shall only discuss results corresponding to the parameter value ρ_m, at which blocked and zonal flow are about equally likely. Circulation patterns of model solutions will be shown from now on plotted stereographically onto the full Northern Hemisphere (NH) disk, to be compared more easily with the observational Southern Hemisphere (SH) data discussed later. The position of the model continents is indicated by heavy dark lines on the outside of the disk, as before.

3.1. Persistent Sequences and EOFs

First of all, Fig. 19 is the grand mean of a solution segment 7000 time units τ long, with $\tau \cong 3$ days. Clearly a remnant of the blocking ridge of Fig. 11c or 14a is preserved in the mean. Fig. 19b gives the anomaly of the stationary blocking solution, i.e., the solution of branch E of Fig. 18 minus the mean in Fig. 19a. Here the anomaly's low over the ocean is more prominent than the anomalous high over the west coast. Fig. 19c shows the corresponding zonal flow anomaly; this is based on the solution marked Bm in Fig. 18, with a blocking ridge over the ocean and zonal flow across the west coast of the continent. Notice that the anomaly of Fig. 19c has almost exactly the same shape as, but sign opposite to that of Fig. 19b. Hence the mean of this time-dependent solution lies almost precisely in the middle between the model's blocked and zonal equilibria. Fig. 19d is yet another type of zonal flow, corresponding to the solution marked Cm in Fig. 18.

It turns out that the objective persistence criterion used in Section 2, Fig. 13, and based on distance between successive maps, does not yield all the persistent sequences a skilled synoptician would notice. A better objective criterion seems to be computing the pattern correlation between successive maps and requiring that this correlation be higher than 0.5 for five successive days. This was done for all solutions marked by pointers in Fig. 18. The nature of persistent sequences, their number and average duration is shown for each ρ-value in Table 1.

The corresponding composite anomaly maps, for ρ_m only, are given in Fig. 20. The blocking sequence map (Fig. 20a) shows the anomalous low over the ocean and the extended blocking ridge. In the zonal flow composite (Fig. 20b), the main anomaly is positive over the ocean. Finally, the composite map Fig. 20c shows a wave train similar to that found in recent observational studies by Blackmon, Lee and Wallace (184) as a feature of intermediate-frequency variability (10 to 30 days) in the Northern Hemisphere.

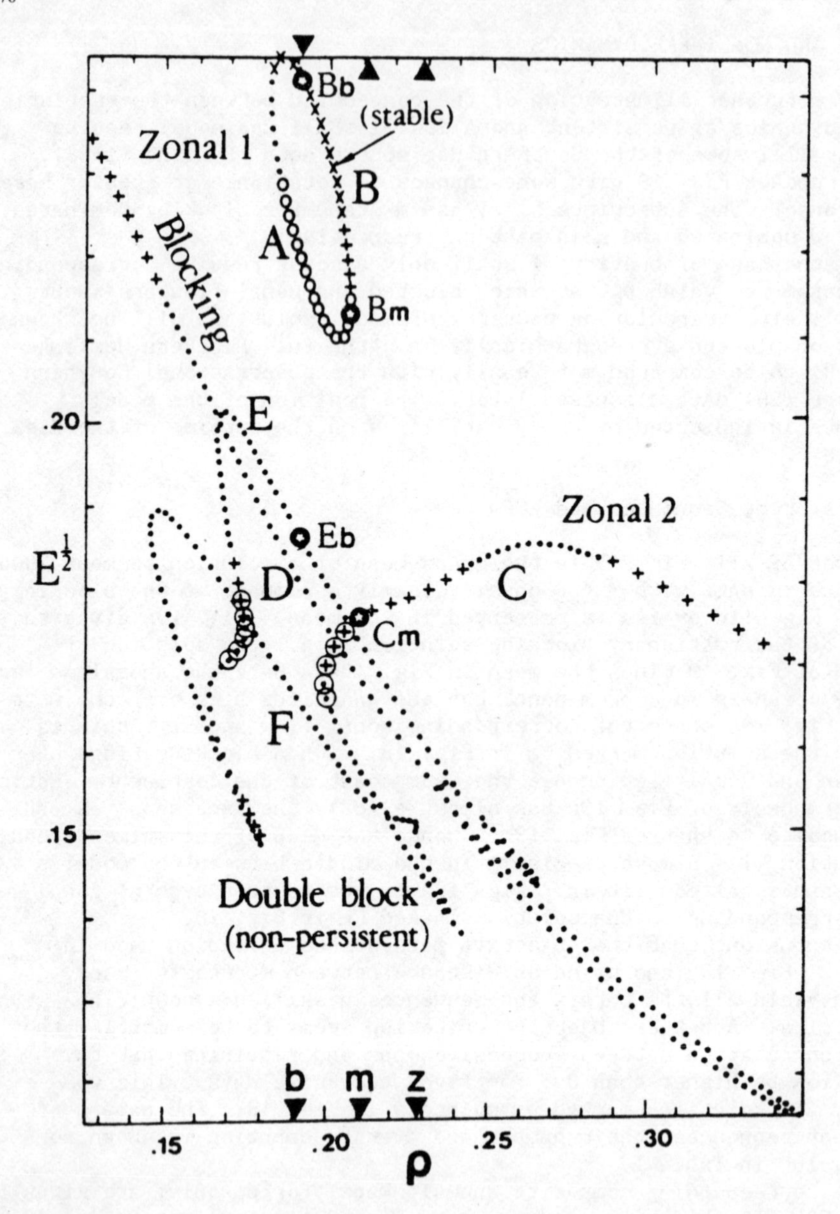

Fig. 18. Stationary model solutions for α^{-1} = 20d. Labeled pointers indicate ρ-values for which the dynamics and statistics of solutions were studied in further detail; the correspondence with Fig. 10 is given by $\rho_b = \rho_2$, $\rho_m = \rho_3$ and $\rho_z = p_4$ (from Mo and Ghil, 1986).

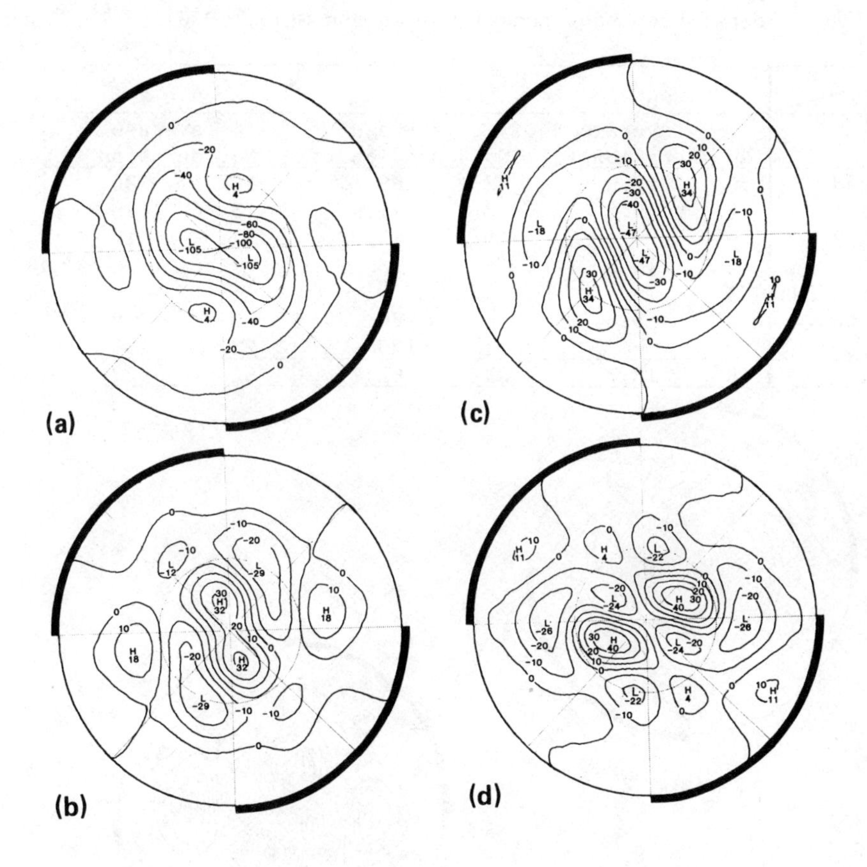

Fig. 19. Flow patterns of model solutions for the mean value of forcing, $\rho = \rho_m$: (a) time-mean pattern for 20,000 simulated days; (b) anomaly pattern of blocking equilibrium Em; (c) anomaly of zonal equilibrium Bm: (d) anomaly of zonal equilibrium Cm (from Mo and Ghil, 1986).

Table 1. Classification and average duration of the model's persistent sequences (from Mo and Ghil, 1986).

ρ / Events	ρ_b		ρ_m		ρ_z	
	No.	average duration (τ)	No.	average duration (τ)	No.	average duration (τ)
Blocked	6	13.8	8	89	16	30
Zonal 1	2	10	3	71.6	13	25
Zonal 2	–	–	1	13	7	12
PNA	4	14.5	–	–	–	–
Reverse PNA	4	9	–	–	–	–
Wave train	2	9.5	6	20	3	9
Time mean	–	–	11	120	23	36
Total	18	13.3	29	62.7	62	23.8

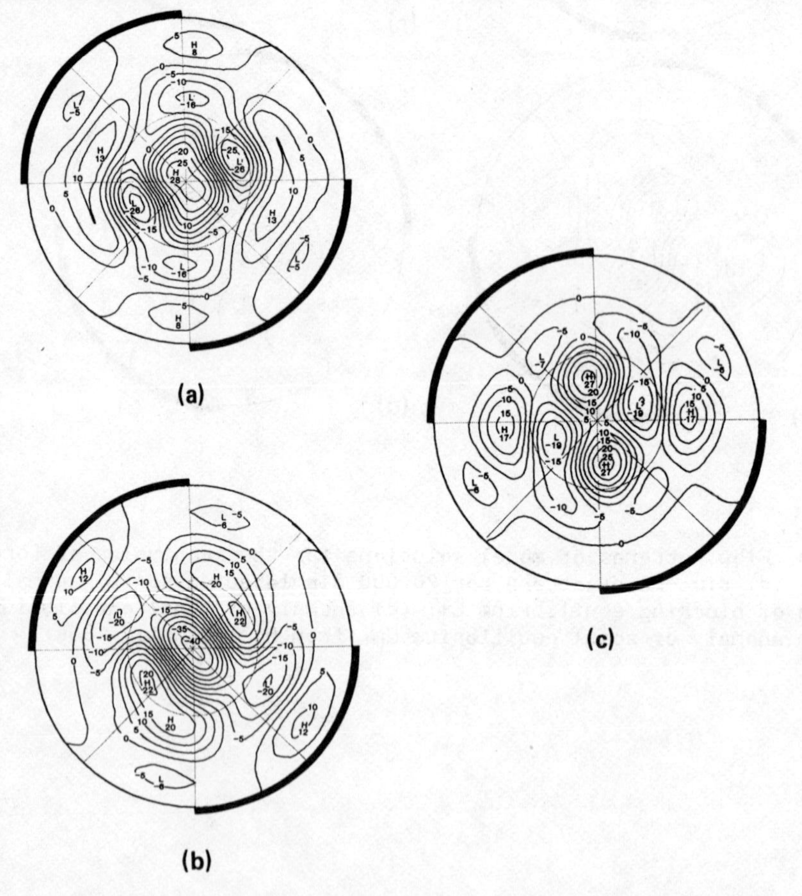

Fig. 20. Flow patterns of persistent model events: (a) similar to Em anomaly, (b) similar to Bm anomaly, (c) wave-train anomaly (from Mo and Ghil, 1986)

The EOFs computed for this time sequence and the variance fractions associated with them are shown in Fig. 21. The first EOF (Fig. 21a) really defines a line joining the blocked sequences (Fig. 20a) and the zonal sequences (Fig. 20b). This corresponds to the fact that the time mean (Fig. 19a) lies almost exactly between the blocked and zonal equilibria, as was the case for the Lorenz model (Fig. 6). Fig. 21b is the second EOF: it shows a wave-train pattern. The third EOF (Fig. 21c) resembles the alternative zonal flow of Fig. 19d; notice that EOFs are defined only up to sign, so that in Figs 21c and 19d highs and lows are interchanged.

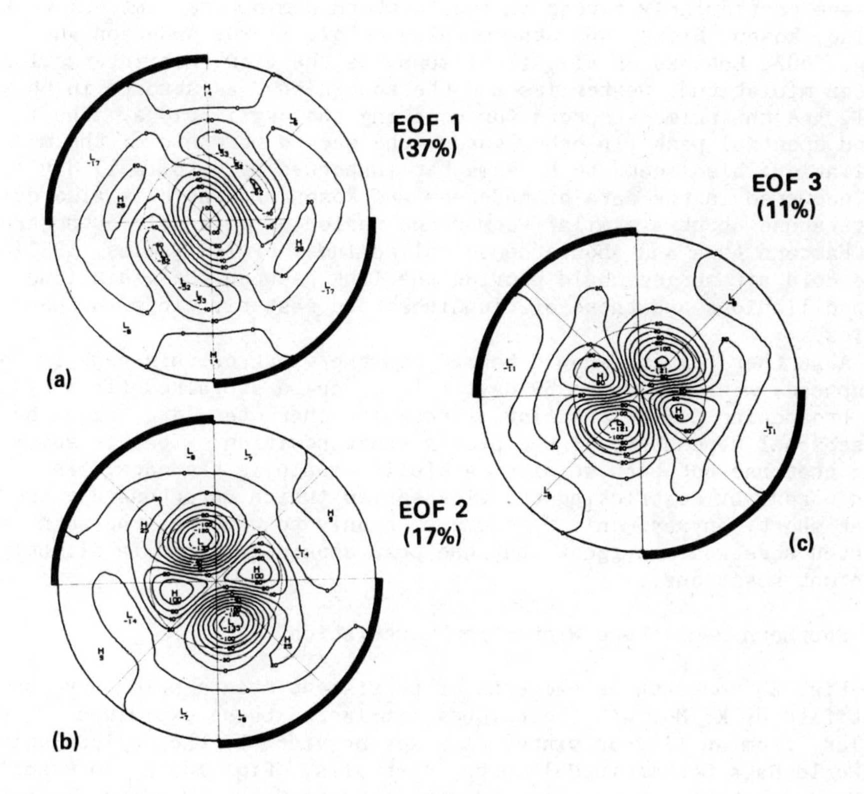

Fig. 21. Empirical orthogonal functions (EOFs) of a 20,000 day model sequence. Numbers give the percentages of associated variance (after Mo and Ghil, 1986).

3.2. The 30-50 day oscillation

Spectral analyses of the amplitudes of the EOF fluctuations in time are shown in Fig. 22. Note that there are peaks at 28 days and 14 days in these time series. The exact position of each peak depends on the intensity of the forcing; for the lower value of the forcing ρ_b the larger peak is at 36 days, for the higher value ρ_z it is at 37 days.

At first 30-to-50 day oscillations were discovered in the tropics (Madden and Julian, 1971). But recently people have realized that there are oscillations with this period and an amplitude of $0(10^{-1})$ of the mean in the global angular momentum of the atmosphere, and that they are particularly strong in the Northern Hemisphere. Recent work by Hide, Rosen, Dickey and others (Hide *et al.*, 1980; Anderson and Rosen, 1983; Eubanks *et al.*, 1985) supports the view that interactions between midlatitude westerlies and the topography, as studied in this model, are the prime suspects for creating the oscillations. The second spectral peak (in other words, the second harmonic of the model oscillation) also seems to be somewhat supported by data; a 17 day peak occurred in the data of Anderson and Rosen and there is also quite a literature about a similar recurrence period of polar air outbreaks over Eastern Asia and about the so-called *index cycle* (Namias, 1950). These cold air surges could provide the link between the midlatitude jet oscillations and those of cloudiness and easterly winds in the tropics.

A further comment should be made on the width of this peak in the atmosphere, between 30 and 50 days. This can be explained first by line broadening due to the flow's irregular character, and second by the seasonal dependence of the peak's exact position. That is something that has not been studied carefully enough in the data. In other words, by restricting the time series (which unfortunately are rather short, anyway) only to summers or only to winters, the work reported here would suggest that the peak should be found in slightly different positions.

3.3. Southern Hemisphere Wintertime Circulation

Fig. 23 shows three patterns of persistent events which have been identified by K. Mo, with techniques similar to those explained earlier, from an 11 year winter data set provided by the Meteorological World Data Center in Melbourne, Australia. Fig. 23a is an example of very interesting persistent events in these data which are clearly dominated by wavenumber three, and also contain a significant wavenumber-one component in the sense that features are much stronger in the Western Hemisphere than in the Eastern Hemisphere. Fig. 23b shows anomalies which are roughly the reverse of the previous ones, while the amplitude of wavenumber one here is clearly larger. Fig. 23c shows a somewhat less frequently occurring, but rather persistent event, dominated by an anomaly pattern with wavenumber four.

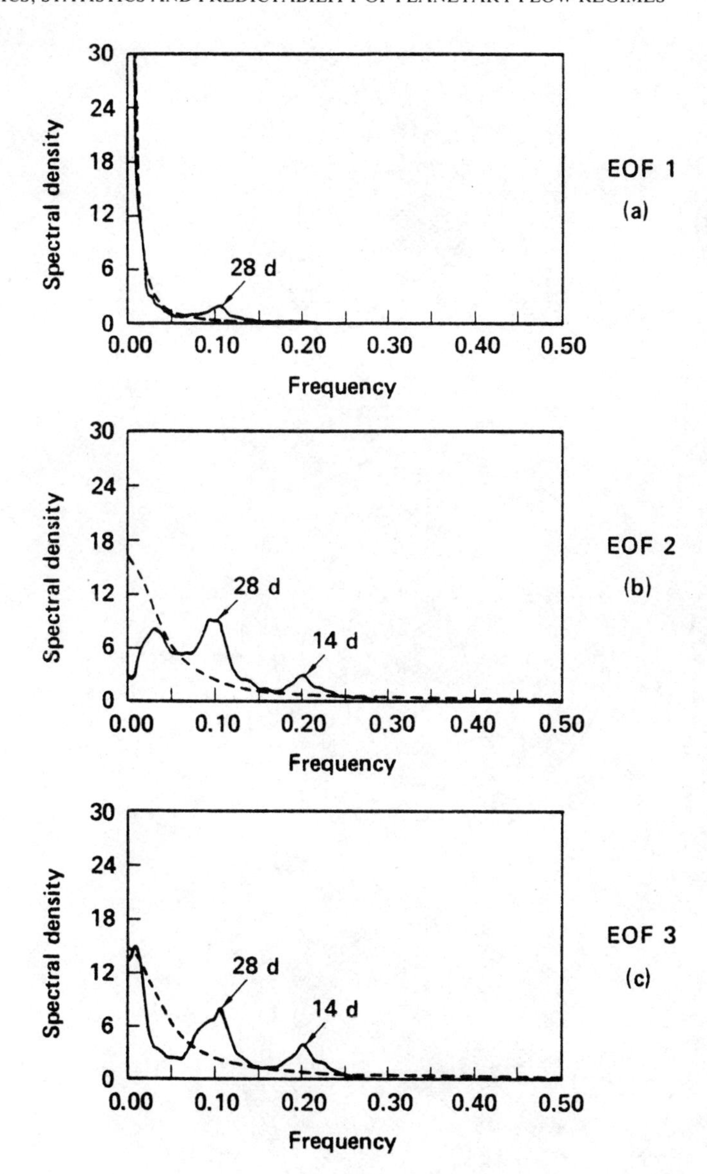

Fig. 22. Power spectra of EOF coefficient time series for the model solution depicted in Figs. 19-21 (solid) and for a red-noise process with the same mean relaxation time (dashed). The frequency is in non-dimensional units, with 0.1 corresponding to a period of 28.3 days (after Mo and Ghil, 1986).

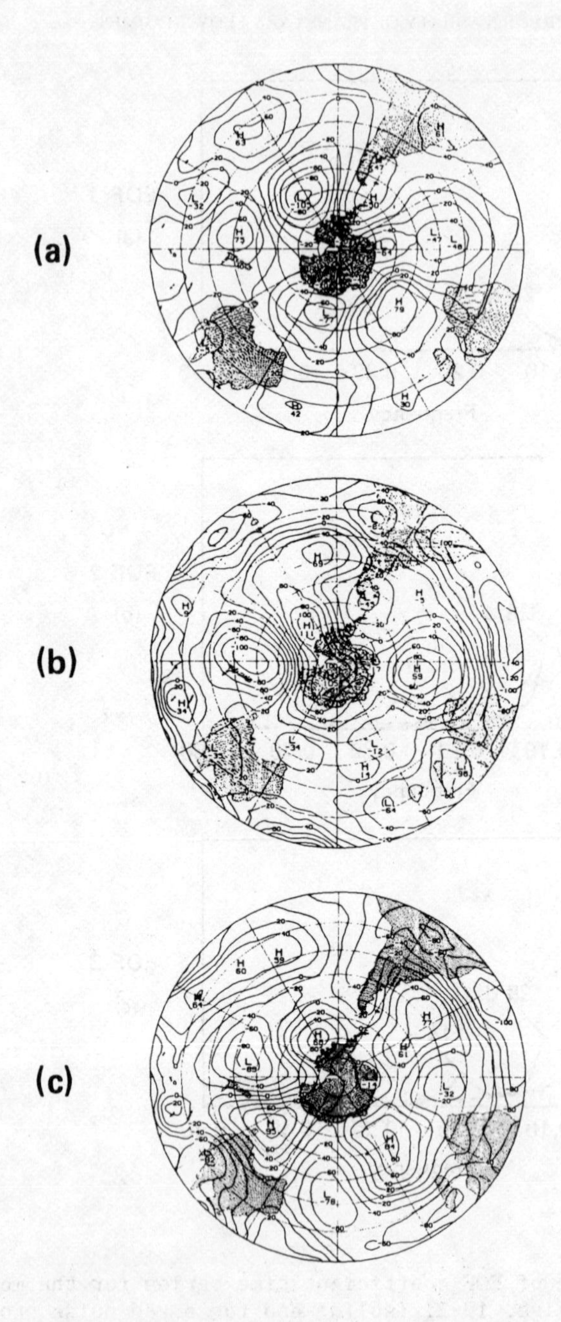

Fig. 23. Observed Southern Hemisphere persistent circulation anomalies. (a) wavenumber-three pattern (average duration 8 days); (b) reversed wavenumber-three pattern with stronger wavenumber-one component (duration 6 days); (c) wavenumber-four pattern (duration 8.5 days) (after Mo and Ghil, 1986).

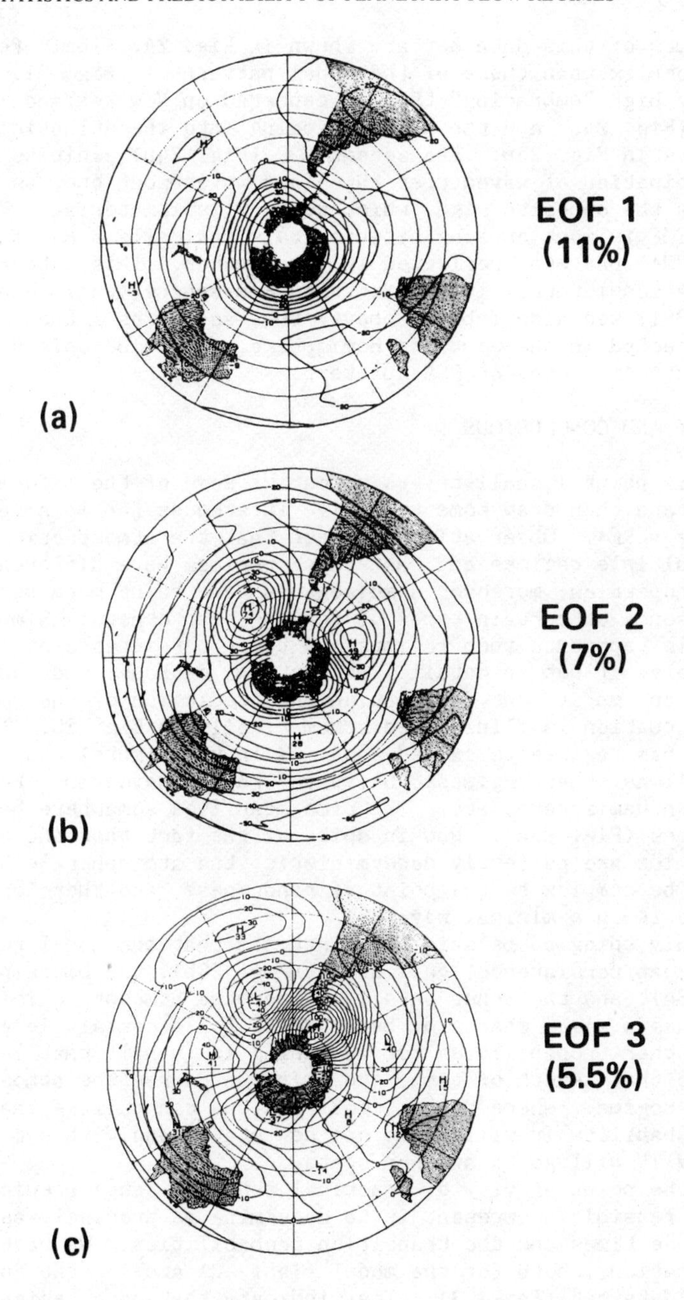

Fig. 24. EOFs of Southern Hemisphere data (after Mo and Ghil, 1986).

The EOFs of this data set are shown in Fig. 24. Their features are more complex than those of the model patterns in Fig. 21. However, the Pacific high "embracing" the low centered on New Zealand in the first EOF (Fig. 24a) and the high extending into the Atlantic have counterparts in Fig. 23b. The second EOF (Fig. 24b) exhibits again a strong combination of wavenumber three and wavenumber one, as does Fig. 23a; it has the opposite sign, which we know is immaterial. Finally, the third EOF is most amusing because it is basically a *Pacific-North American* (PNA) pattern, reflected in the Equator. This pattern, documented very consistently in the Northern Hemisphere (Wallace and Gutzler, 1981; see also Table 1 above) but, as far as I know, not previously detected in the Southern Hemisphere, should be called the *Pacific-South American*, or PSA, pattern.

4. SUMMARY AND CONCLUSIONS

At this point I shall try to summarize some of the information presented, and then draw some tentative inferences for long-range forecasting (LRF). Observations suggest that the atmospheric circulation has multiple regimes and that these regimes have different persistence properties; moreover some regimes tend to be more persistent in the presence of certain types of forcing than others. Simple dynamical models reproduce such regimes and their persistence at least qualitatively, if not in detail. Statistical methods, such as EOF analysis, can improve the description and prediction of the regimes.

The situation is illustrated schematically in Fig. 25. The atmosphere has regimes we called *zonal* (Z in the figure) and *blocked* (B), as well as other regimes: PNA, wave trains, wavenumber three in the Southern Hemisphere, etc. The time mean lies somewhere between these regimes (Fig. 25a). Now in spite of the fact that the underlying dynamics are perfectly deterministic, the atmosphere's behavior appears to be complex to the point of randomness, and therefore we want to describe it in a minimal way.

The only thing to be said, at first, is that the zonal regime has a certain mean persistence, call it τ_Z (Fig. 25b), the blocking regime τ_B (Fig. 25c), and the other ones τ_C. Once the atmosphere is in one regime it has a given chance of being in it for a certain length of time, and other probabilities for remaining in it half that length of time, twice that length of time, etc. In due course the atmosphere leaves the regime. Where does it go? All one can tell is that with a certain probability it will go to one new state, and with some other probability it will go to another state.

From the point of view of practical extended range prediction, what seems feasible at present is to determine as precisely as possible the residence times and the transition probabilities. We have done such computations, both for the model (Table 2) and for the Southern Hemisphere data set (Table 3). They indicate that each regime has nonzero probabilities of transition to two or more other regimes, that reentries into the same regime are possible, and that certain regimes are preferred way stations on the path between two other regimes.

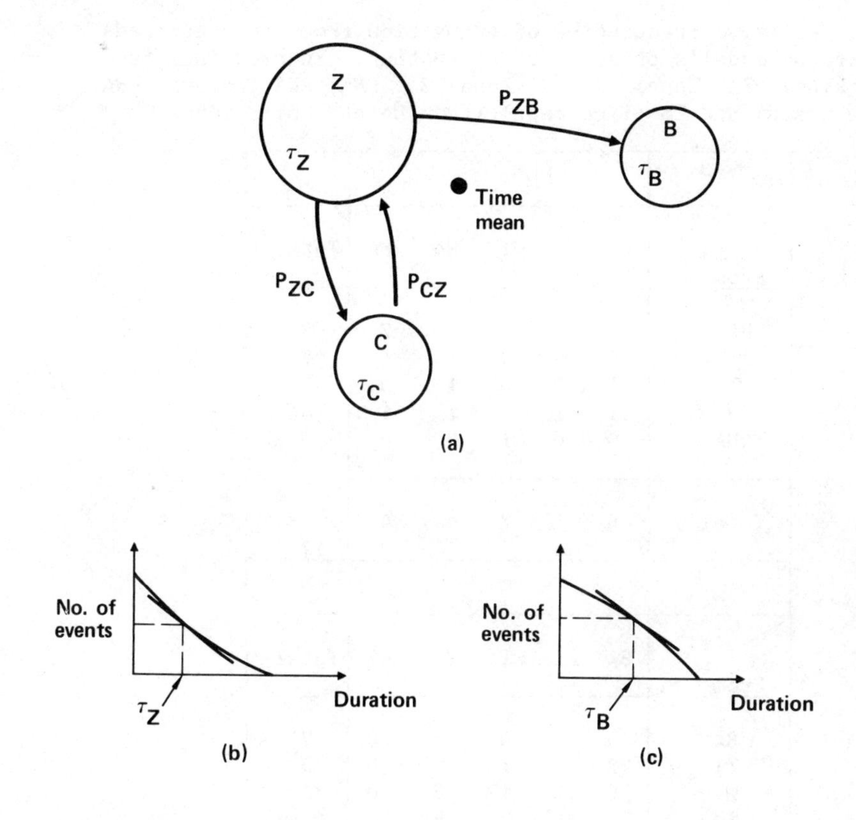

Fig. 25. "Markov chain" description of multiple regimes. Notation: p_{XY}- transition probability from regime X to regime Y; τ_X - average duration of regime X.

Since atmospheric data sets are limited in extent and the model described here falls considerably short of total verisimilitude, it is tempting to try using much more detailed general circulation models (GCMs) in this enterprise. The task is to see whether a GCM does have the same persistent flow patterns with the same frequencies of occurrence, and, to the extent to which that can be told from the data available, the same persistence characteristics and transition probabilities. If that is so then you can artificially extend your data set while you are running your model on ever bigger computers, improve the statistics, and perhaps determine preferred directions of instability

Table 2. Relative frequencies of transition from one preferred region of the model's phase space to another. Entries indicate Bℓ (Blocking), Z1 (Zonal 1), Z2 (Zonal 2), PNA, RNA (reverse PNA), WT (Wave train) and TM (Time mean) (from Mo and Ghil, 1986).

ρ	ρ_b					
From \ To	Bℓ	Z1	WT	PNA	RNA	Total
Bℓ	1	0	1	1	2	5
Z1	0	0	0	1	1	2
WT	1	0	0	1	0	2
PNA	1	1	0	1	1	4
RNA	3	0	1	0	0	4
Total	6	1	2	4	4	17 \ 17
	ρ_m					
From \ To	Bℓ	Z1	WT	TM	Z2	Total
Bℓ	4	0	0	3	0	7
Z1	2	0	1	0	0	3
WT	0	2	1	3	0	6
TM	2	1	3	5	0	11
Z2	0	0	1	0	0	1
Total	8	3	6	11	0	28 \ 28
	ρ_z					
From \ To	Bℓ	Z1	WT	TM	Z2	Total
Bℓ	2	2	0	10	2	16
Z1	4	3	1	3	2	13
WT	1	0	0	2	0	3
TM	5	7	1	7	3	23
Z2	3	1	1	1	0	6
Total	15	13	3	23	7	61 \ 61

Table 3. Relative frequencies of transition from one persistent type of event to another in the SH wintertime circulation. Entries indicate (+3) for the most frequently occurring wavenumber 3 pattern (Fig. 23a), (-3) for the similar pattern of opposite sign (Fig. 23b), (4) for the wavenumber 4 pattern (Fig. 23c), TM for the events with pattern resembling the time mean, and Other (from Mo and Ghil, 1986).

From \ To	+3	-3	4	TM	Other	Total
+3	3	1	1	1	2	8
-3	0	1	1	0	1	3
4	0	1	1	0	2	4
TM	1	0	0	0	0	1
Other	4	1	1	0	0	6
Total	8	4	4	1	5	22 \ 22

(recall Fig. 15). In other words, improve upon transition probabilities by saying, not only do I know that my model is in this particular regime, but I also think it approaches a particular direction of instability which will throw it into regime X, rather than into regime Y.

Here I come to the last conclusion: there is clearly a long way to go. At the same time as we improve our understanding of the existing atmospheric data sets and extend them by the use of GCMs, we also have to understand better yet the low-frequency variability that we are trying to predict; and we have to include into the simple dynamical models (which can be dissected at will) baroclinic processes, higher resolution, and various local features. In other words, the gap between the resolution you have seen here and the actual details of atmospheric events is to be bridged not so much by ever increasing the resolution of the simple models or decreasing that of GCMs, but rather by also having models which look in a more specialized way at the local nature of those features.

Acknowledgments. It is a pleasure to acknowledge the fruitful collaboration with Drs. B. Legras and K. Mo, on which much of what might be learned from this lecture is based. Discussions on the 30-50 day oscillation with Drs. R. Hide, J. Dickey, H. Itoh, M. Eubanks and J. Tribbia were also very useful, as were discussions on operational long-range forecasting with Drs. D. Gilman and J. Namias. The author's research was supported by the National Aeronautics and Space Administration under grants NSG-5130 and NAS-5,713, and by the National Science Foundation under grant ATM-8514731. The text of this lecture was prepared using the tape recording of a similar lecture delivered at the Cooperative Institute for Research in Environmental Sciences of the University of Colorado at Boulder. Dr. Uwe Radok's help in obtaining and editing the tape transcript are most gratefully acknowledged. Bianca Gola typed the manuscript and Kathy Martelli drew the figures.

REFERENCES

Anderson, J. R., and R. D. Rosen, 1983. 'The latitude-height structure of 40-50 day variations in atmospheric angular momentum'. *J. Atmos. Sci.*, **40**, 1584-1591.

Blackmon, M. L., Y. H. Lee and J. M. Wallace, 1984. 'Horizontal structure of 500 mb height fluctuations with long, intermediate and short time scales.' *J. Atmos. Sci.*, **41**, 961-979.

Charney, J. G., 1947. 'The dynamics of long waves in a baroclinic westerly current'. *J. Meteorol.*, **4**, 135-163.

Charney, J. G., and J. G. DeVore, 1979. 'Multiple flow equilibria in the atmosphere and blocking'. *J. Atmos. Sci.*, **36**, 1205-1216.

Dole, R. M., 1982. 'Persistent Anomalies of the Extratropical Northern Hemisphere Wintertime Circulation'. Ph.D. Thesis, M.I.T., 226 pp.

Dole, R. M., and N. D. Gordon, 1983. 'Persistent anomalies of the extratropical Northern Hemisphere wintertime circulation: geographical distribution and regional persistence characteristics'. *Mon. Wea. Rev.*, **111**, 1567-1586.

Drazin, P. G., 1978. 'Variations on a theme of Eady'. In *Rotating Fluids in Geophysics*, P. H. Roberts and A. M. Soward (eds.),

Academic Press, New York, pp. 139-169.
Eady, E. T., 1949. 'Long waves and cyclone waves'. *Tellus*, **1**, 33-52.
Egger, J., 1978. 'Dynamics of blocking highs'. *J. Atmos. Sci.*, **35**, 1788-1801.
Eubanks, T. M., J. A. Steppe, J. O. Dickey and S. P. Callahan, 1985. 'A spectral analysis of the Earth's angular momentum budget'. *J. Geophys. Res.*, **90**, 5385-5404.
Ghil, M., and S. Childress, 1986. *Topics in Geophysical Fluid Dynamics: Atmospheric Dynamics, Dynamo Theory and Climate Dynamics*. Springer-Verlag, New York, in press, 479 pp.
Ghil, M., R. Benzi and G. Parisi (eds.), 1985. *Turbulence and Predictability in Geophysical Fluid Dynamics and Climate Dynamics*. North-Holland,Amsterdam/Oxford/New York/Tokyo, 449 pp.
Hide, R., N. T. Birch, L. V. Morrison, D. J. Shea and A. A. White, 1980. 'Atmospheric angular momentum fluctuations and changes in the length of day'. *Nature*, **286**, 114-117.
Legras, B., and M. Ghil, 1985. 'Persistent anomalies, blocking, and variations in atmospheric predictability' *J. Atmos. Sci.*, **42**, 433-471.
Lorenz, E. N., 1963. 'Deterministic nonperiodic flow'. *J. Atmos. Sci.*, **20**, 130-141.
Lorenz, E. N., 1985. 'The growth of errors in prediction'. In Ghil *et al.* (1985), pp. 243-265.
Madden, R. A., and P. R. Julian, 1971. 'Detection of a 40-50 day oscillation in the zonal wind in the Tropical Pacific'. *J. Atmos. Sci.*, **28**, 702-708.
Mo, K. C., 1986. 'Quasi-stationary states in the Southern Hemisphere'. *Mon. Wea. Rev.*, in press.
Mo, K., and M. Ghil, 1986. 'Combined statistical and dynamical study of persistent anomalies'. *J. Atmos. Sci.*, submitted.
Namias, J., 1950. 'The index cycle and its role in the general circulation'. *J. Meteorol.*, **7**, 130-139.
Namias, J., 1968. 'Long range weather forecasting--history, current status and outlook'. *Bull. Amer. Meteorol. Soc.*, **49**, 438-470.
Namias, J., and D. Cayan, 1984. 'El Niño - implications for forecasting'. *Oceanus*, **27**, 40-45.
Pedlosky, J., 1979. *Geophysical Fluid Dynamics*. Springer-Verlag, New York/Heidelberg/Berlin, 624 pp.
Rasmusson, E. M., and J. M. Wallace, 1983. 'Meteorological aspects of the El Niño/Southern Oscillation'. *Science*, **222**, 1195-1202.
Reinhold, B. B., 1981. 'Dynamics of Weather Regimes: Quasi-Stationary Waves and Blocking'. Ph.D. Thesis, M.I.T.
Ruelle, D., 1980.'Strange attractors'. *Math. Intelligencer*, **3**, 126-137.
Ruelle, D., 1985. 'The onset of turbulence: A mathematical introduction'. In Ghil *et al.* (1985), pp. 3-16.
Wagner, A. J., 1977. 'Weather and circulation of January 1977: the coldest month on record in the Ohio Valley'. *Mon. Wea. Rev.*, **105**, 553-560.
Wallace, J. M., and D. S. Gutzler, 1981. 'Teleconnections in the geopotential height field during the Northern Hemisphere winter'. *Mon. Wea. Rev.*, **109**, 784-812.

[illegible]

[illegible] Press, New York, [illegible] pp.

[illegible], 1964: [illegible] waves and [illegible] water. [illegible]

[illegible], 1976: [illegible] of [illegible]. [illegible], 605-[illegible].

Johnson, [illegible], 1987: [illegible] simulation of the [illegible]. J. Geophys. Res., 90, [illegible].

[illegible] and [illegible], 1980: [illegible] Press, New York, [illegible] pp.

[illegible], 1982: [illegible] and [illegible]. [illegible]

[illegible], 1981: Atmospheric [illegible] the length of day. Nature, 289, [illegible].

[illegible], 1978: [illegible] J. Atmos. Sci., [illegible].

Lorenz, E. N., 1963: Deterministic nonperiodic flow. J. Atmos. Sci., 20, 130-141.

[illegible], 1984, [illegible].

[illegible], 1979: [illegible]. [illegible]

[illegible], 1969: [illegible] in the Northern [illegible].

[illegible], 1974: [illegible] and [illegible].

Namias, J., 1959: [illegible]. [illegible]

Namias, J., 1963: [illegible]. Bull. Amer. Meteor. Soc., [illegible].

[illegible], 1981: [illegible].

[illegible], 1977: [illegible]. [illegible]

[illegible], 1983: [illegible]. [illegible]

[illegible], 1985: [illegible] anomalies. [illegible]

[illegible], 1980: [illegible]. [illegible]

[illegible], 1973: [illegible]. [illegible]

[illegible], 1984: [illegible]. [illegible]

AN ANALYTIC APPROACH TO ATMOSPHERIC WAVES IN THREE DIMENSIONS

E. Rebhan*
Service de Chimie Physique II
CP-231, Campus Plaine
Free University of Brussels
1050 Brussels, Belgium

ABSTRACT. Assuming homogeneous incompressible flow and using the momentum equation of ideal hydrodynamics, the problem of linear three-dimensional atmospheric waves is treated in true spherical geometry. In order to avoid instabilities associated with a nonuniform equilibrium height of the atmosphere, it turns out to be necessary to take into account the geoid shape of the Earth or planet under consideration. It is demonstrated that spatial gradients of the vertical velocity are in general equally as important as gradients of the horizontal velocity components. Complete neglection of intrinsic vertical velocities proves to be in contradiction with momentum conservation. After employing a self-consistent approximation scheme, a linear second-order differential equation for the wave motion is derived and solved analytically for several special cases. This equation depends on the ratio $R = (2 \times$ gravity force $\times$ height of atmosphere)/(centrifugal force $\times$ planetary radius) as a parameter which is ≈ 0.7 for the Earth's atmosphere. There are strong indications that for $R \to 0$ a spectrum of unstable modes arises. Extrapolation to the parameter value of the Earth leads to typical growth times in the range of a few days.

INTRODUCTION

In the treatment of atmospheric problems usually a series of approximations is introduced. Prominent among these are the neglection of intrinsic vertical velocities and the so-called β-plane approximation (see, e.g., [1]) in which the influences coming from the spherical planetary shape are essentially dropped. On the other hand, in treating specific problems various other effects are frequently introduced, such as E kman pumping [2], orography [3], etc. This procedure implies, of course, that the neglected terms are small as compared to the terms which are kept on or taken up. The question arises whether or not some of the neglected terms could not sometimes lead to the same effects as some of

---

*On leave of absence from Institut für Theoretische Physik, Universität Düsseldorf, Universitätsstr. 1, 4000 Düsseldorf, FRG.

C. Nicolis and G. Nicolis (eds.), Irreversible Phenomena and Dynamical Systems Analysis in Geosciences, 285–309.

the additional terms considered. For example, E kman pumping involves originally vertical velocities, although technically it leaves the problem two-dimensional. Therefore, one wonders if Eckman pumping could not sometimes be replaced by frictionless vertical motions which are explicitly introduced into the equations.

The introduction of vertical velocity components and the consideration of all spherical effects will of course appreciably complicate the theory, and there are certainly a number of interesting problems for which one can well do without these complications. However, in this paper we are interested in very large scale perturbations of the atmosphere where this approach would be very doubtful. It can be shown that, on the contrary, some of the neglections commonly made are in fact prohibitive in this particular case. It turns out, e.g., that in general derivatives of the vertical velocity can by no means be neglected, even though the vertical velocity itself is small.

In this paper we take a new approach to the problem of large-scale atmospheric waves in which all spherical curvature effects are taken into account as well as vertical velocity perturbations. Starting with a basic state which corresponds to a homogeneous and rigidly rotating atmosphere, we study the linear stability of this problem. The influence of the thermal wind is not taken into account in this paper. It appears possible, however, that it can be included in the same kind of treatment. We do not want to exclude the applicability of our theory to planetary atmospheres with a larger ratio of centrifugal force over gravitational force from the very beginning, and shall therefore treat these forces as independent quantities. It will turn out that this kind of treatment necessitates taking into account the geoid shape of the Earth or planets under consideration. The complications that arise with curvature effects, geoid shape and vertical velocities can be handled by employing a self-consistent approximation scheme which takes advantage of the smallness of several parameters of the problem. In the end, the problem of linear wave motion can be brought into a tractable form.

The structure of the paper is as follows: In section 1, we introduce the basic equations of the problem in a suitable dimensionless form. In section 2, we introduce the equilibrium state of the atmosphere and discuss its implications for the height of the atmosphere. Section 3 deals with the boundary conditions of the problem. In section 4, an approximation scheme is introduced and applied to the linearized equations and boundary conditions of the problem. In section 5, we derive a second-order differential stability equation for one of the perturbation velocity components. Section 6 deals with uniqueness and regularity conditions for the solutions of this equation. Finally, in section 7 we solve this equation analytically for several special cases.

1. BASIC EQUATIONS

In order to exclude all complications connected with sound waves we shall deal only with incompressible flows of uniform density $\tilde{\rho}$. (All dimensional quantities are marked by a tilde.) Furthermore, we shall neglect all effects of fluid viscosity. These assumptions imply that in addition to the corresponding equation $\tilde{\nabla}\cdot\tilde{v} = 0$ of mass conservation, we have only momentum conservation to consider. We do this in an inertial frame and have the equation

$$\partial_{\tilde{t}}\, \underline{\tilde{v}} + \underline{\tilde{v}}\cdot\tilde{\nabla}\underline{\tilde{v}} + \frac{\tilde{\nabla}\tilde{p}}{\tilde{\rho}} = \tilde{\nabla}\, \frac{\tilde{g}\tilde{a}^2}{\tilde{r}} \quad ,$$

where $\tilde{t}$ = time, $\underline{\tilde{v}}$ = flow velocity, $\tilde{p}$ = pressure, $\tilde{a}$ = radius of the Earth (or planet) at the equator, and $\tilde{g}$ = gravity acceleration at $\tilde{r} = \tilde{a}$. The gravitational force which is employed in this equation is due to a spherical mass and does not take into account the contribution which comes from the small pad filling the space between a sphere and a geoid. As will be seen, this neglection is of no relevance for our problem. The transition from our inertial frame to the frame of the rotating Earth (or planet) will be made at a later stage, where it is more convenient (section 5). Defining $\tilde{T}$ = 1 day/2π, $\tilde{V} = \tilde{a}/\tilde{T}$ = rotational velocity of the equator ($\approx$ 464 m/sec for the Earth), we convert to dimensionless quantities by setting $t = \tilde{t}/\tilde{T}$, $r = \tilde{r}/\tilde{a}$, $\underline{v} = \underline{\tilde{v}}/\tilde{V}$, and $p = \tilde{p}/(\tilde{\rho}\tilde{V}^2)$. In these, our basic equations read:

$$\nabla\cdot\underline{v} = 0 \quad , \tag{1.1}$$

$$\partial_t\, \underline{v} - \underline{v} \times \underline{\omega} + \nabla\left(p + \frac{v^2}{2}\right) = \nabla\left(\frac{1}{\varepsilon r}\right) \quad ; \tag{1.2}$$

where

$$\varepsilon = \left(\frac{\tilde{V}^2}{\tilde{a}}\right)/\tilde{g} \tag{1.3}$$

is the ratio of centrifugal force and gravity force at the equator,

$$\underline{\omega} = \mathrm{curl}\,\underline{v} \tag{1.4}$$

is the vorticity and where use has been made of the identity $\underline{v}\cdot\nabla\underline{v} = \nabla v^2/2 - \underline{v} \times \underline{\omega}$.

In our dimensionless quantities we have of course $r = 1$ and rotational velocity $v = 1$ at the equator for all planets under consideration. The ratio ε = centrifugal force/gravity force is $\approx 3.4 \cdot 10^{-3}$ for the Earth, ≈ 0.09 for Jupiter, and ≈ 0.16 for Saturn. Considering the height of the atmosphere we employ the model of a homogeneous atmosphere and have (in dimensionless units) $H \approx 1.25 \cdot 10^{-3}$ for the Earth (corresponding to $\tilde{H} \approx 8$ km, $\tilde{a} \approx 6378$ km), $H \approx 2 \cdot 10^{-3}$ for Jupiter (corresponding to $\tilde{H} \approx 150$ km, $\tilde{a} \approx 71422$ km) and $H \approx 1.7 \cdot 10^{-2}$ for Saturn (corresponding to $\tilde{H} \approx 10^3$ km and $\tilde{a} \approx 59800$ km). The numbers for Jupiter and Saturn are very rough estimates and have been deduced from atmospheric data given in [4] and [5].

Taking the curl of Eq.(1.2), we get rid of the two gradient terms and obtain the integrability condition

$$\partial_t \underline{\omega} - \operatorname{curl} \underline{v} \times \underline{\omega} = 0 \quad . \tag{1.5}$$

Equation (1.5) with (1.4) suffices for calculating the velocity field. For calculating the pressure, one has to return to Eq.(1.2).

The steady state solution of Eqs.(1.1) - (1.2) presented in section 2 will be denoted by $\underline{V}$ for velocity, $\underline{\Omega}$ for vorticity, and P for pressure. For studying its linear stability, we replace $\underline{v} \to \underline{V} + \underline{v}$, $\underline{\omega} \to \underline{\Omega} + \underline{\omega}$ and $p \to P + p$, where after the replacement $\underline{v}$, $\underline{\omega}$ and p are time-dependent perturbations. Linearization of the basic equations leads to the following set of linear stability equations:

$$\nabla \cdot \underline{v} = 0 \quad , \tag{1.6}$$

$$\underline{\omega} = \operatorname{curl} \underline{v} \quad , \tag{1.7}$$

$$\partial_t \underline{\omega} - \operatorname{curl}(\underline{v} \times \underline{\Omega}) - \operatorname{curl}(\underline{V} \times \underline{\omega}) = 0 \quad , \tag{1.8}$$

$$\nabla p = \underline{v} \times \underline{\Omega} + \underline{V} \times \underline{\omega} - \nabla(\underline{v} \cdot \underline{V}) - \partial_t \underline{v} \quad . \tag{1.9}$$

2. THE BASIC SOLUTION

We shall consider a uniformly rotating atmosphere which is at rest in the reference frame of the rotating Earth (planet) as the basic solution for our stability analysis. Although we shall be dealing with a geoid-shaped Earth (planet), it is still useful to employ spherical coordinates r, θ, ϕ (see Fig. 1). It is easily seen that

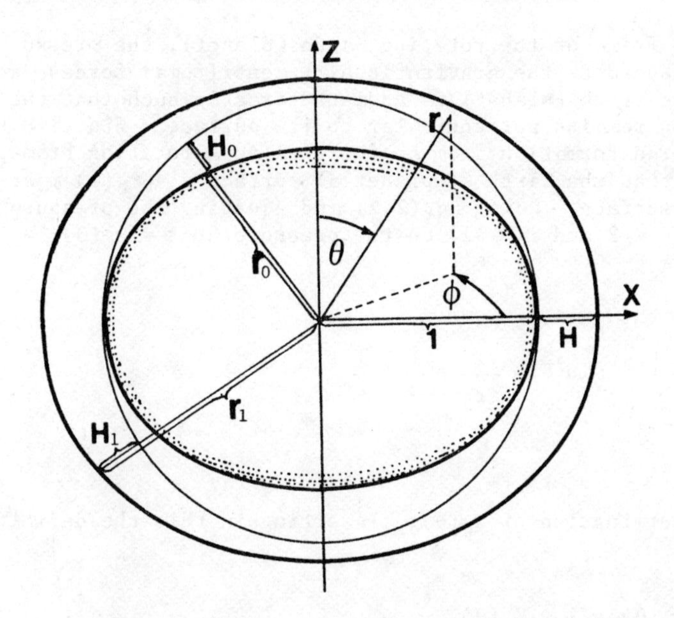

Fig. 1. Coordinate system and graphical representation of the quantities r_0, H_0, r_1, H_1 and H.

$$\underline{V} = r\sin\theta\ \underline{e}_{\phi} \ , \tag{2.1}$$

and

$$P = P_0 + \frac{1}{2} r^2 \sin^2\theta + \frac{1}{\varepsilon r} \tag{2.2}$$

is a solution of Eqs.(1.1) - (1.2). The vorticity $\underline{\Omega}$ = curl $\underline{V}$ corresponding to this solution is

$$\underline{\Omega} = 2(\cos\theta\ \underline{e}_r - \sin\theta\ \underline{e}_{\theta}) \ . \tag{2.3}$$

Since $\cos\theta\ \underline{e}_r - \sin\theta\ \underline{e}_{\theta} = \underline{e}_z$ is a unit vector in the direction from pole to pole and since V = 1 at the equator, it is obvious that our solution does in fact represent a uniformly rotating atmosphere.

In the frame of the rotating Earth (planet), the pressure force ∇P must be balanced by the gravitational + centrifugal forces, and the geoid shape of the Earth (planet) is adjusted exactly such that the sum of these forces remains perpendicular to its surface. Since ∇P is invariant under the transformation from an inertial to a rotating frame, we have the result that the Earth's (planet's) surface $r = r_0(\theta)$ must be a P = const. surface. Using Eq.(2.2) and equating the pressure at the equator ($\theta = \pi/2$ and $r_0 = 1$) to the pressure at $r = r_0(\theta)$, we obtain the equation

$$\frac{1}{2} r_0^2 \sin^2\theta + \frac{1}{\varepsilon r_0} = \frac{1}{2} + \frac{1}{\varepsilon} \tag{2.4}$$

for the determination of $r_0(\theta)$. Inserting in this the definition

$$r_0(\theta) = 1 - H_0(\theta) \tag{2.5}$$

we get the implicit equation

$$H_0(\theta) = \frac{\varepsilon}{2}(1 - H_0)[1 - (1 - H_0)^2 \sin^2\theta] \tag{2.6}$$

for determining $H_0(\theta)$. According to section 1, ε is a very small number for the Earth and still a relatively small number for Jupiter and Saturn. Expanding Eq.(2.6) with respect to ε, we get to the lowest significant order

$$H_0(\theta) = \frac{\varepsilon}{2}(1 - \sin^2\theta) \quad . \tag{2.7}$$

Even for Saturn, the next order term adds only a correction of $\approx 2.5\%$ to this, so that we may consider (2.7) as a reasonably good approximation.

The upper boundary $r_1(\theta)$ of the atmosphere is determined by the condition that the pressure P should vanish there. Introducing

$$r_1(\theta) = 1 + H_1(\theta) \quad , \tag{2.8}$$

and denoting the height $H_1(\pi/2)$ of the atmosphere at the equator by H (see Fig. 1), we get from this condition

$$P_0 = - \left[\frac{1}{2}(1 + H)^2 + \frac{1}{\varepsilon(1 + H)}\right] \tag{2.9}$$

at the equator and

$$P_0 + \frac{1}{2} r_1{}^2(\theta)\sin^2\theta + \frac{1}{\varepsilon r_1(\theta)} = 0 \tag{2.10}$$

elsewhere. Combining Eqs.(2.8) - (2.10) we obtain the implicit equation

$$H_1(\theta) = H - \frac{\varepsilon}{2}(1 + H)(1 + H_1)[(1 + H)^2 - (1 + H_1)^2\sin^2\theta] \tag{2.11}$$

for determining H_1. This time, we make a simultaneous expansion with respect to ε and H to the lowest significant order, obtaining

$$H_1(\theta) = H - \frac{\varepsilon}{2}(1 - \sin^2\theta) \quad . \tag{2.12}$$

Again, this is a very good approximation for the Earth and a reasonably good one for other planets, since H also is in general a rather small number.

Using the approximations (2.7) and (2.12) and the defining equations (2.5) and (2.8), for the local height $r_1(\theta) - r_0(\theta)$ of the atmosphere we get the result

$$r_1(\theta) - r_0(\theta) = H \quad , \tag{2.13}$$

i.e., the height of the atmosphere is constant for this order of approximation.

It can now be explained why it is in fact inevitable to account for the geoid shape of the Earth or the planet under consideration. For spherical shape we would have $r_1(\theta) \equiv 1$, and the height of the atmosphere would be given by Eq.(2.12). First of all, it would become impossible to use the concept of a homogeneous atmosphere, since for the numbers given in section 1 we would obtain negative values of H_1 at the poles. Even assuming an equator height of $H \approx \varepsilon$, one would obtain a height $\tilde{H}_1$ of ≈ 22 km at the equator and of only ≈ 11 km at the poles in the case of the Earth. The artificial hump created in this way might be expected to cause unrealistic instabilities, and a corresponding calculation confirmed this expectation.

3. BOUNDARY CONDITIONS

The total velocity $\underline{V} + \underline{v}$ must be tangential to the planetary surface at $r = r_0(\theta)$, and since this is true for $\underline{V}$ the same must hold for $\underline{v}$. Using the fact that ∇P stands perpendicular there, we obtain the boundary condition

$$\underline{v}\cdot\nabla P\Big|_{r=r_0(\theta)} = 0 \quad .$$

Since this condition is already linear it is not changed by linearization. Inserting Eqs.(2.1) and (2.5) we get

$$\left(\varepsilon r_0(\theta)\sin^2\theta - \frac{1}{r_0^{\,2}(\theta)}\right)v_r\Big|_{r=1-H_0} + \varepsilon r_0(\theta)\sin\theta\cos\theta\, v_\theta\Big|_{r=1-H_0} = 0 \tag{3.1}$$

from this. If the upper boundary of the atmosphere is perturbed from its equilibrium value (2.8) into

$$r_1(\theta,\phi,t) = 1 + H_1(\theta) + h(\theta,\phi,t) \quad , \tag{3.2}$$

h being a small perturbation, the relative fluid motion must be tangential to this surface. This means that inserting the trajectory r(t), $\theta(t)$, $\phi(t)$ of a fluid element into Eq.(3.2) must yield an identity. The time derivative of this is

$$\dot{r}(t) = \dot{\theta}(t) d_\theta H_1(\theta) + \dot{\theta}(t)\partial_\theta h + \dot{\phi}(t)\partial_\phi h + \partial_t h \quad .$$

Since $\dot{r}(t) = v_r$, $r\dot{\theta}(t) = v_\theta$ and $r\sin\theta\dot{\phi}(t) = V_\phi + v_\phi = r\sin\theta + v_\phi$, we obtain for the boundary condition at the upper edge of the atmosphere

$$v_r\Big|_{r=1+H_1+h} = \left(\frac{v_\theta}{r} d_\theta H_1(\theta) + (\underline{V} + \underline{v})\cdot\nabla h + \partial_t h\right)\Big|_{r=1+H_1+h} \quad . \tag{3.3}$$

Linearization yields the simpler condition

$$v_r\Big|_{r=1+H_1} = \left(\frac{v_\theta}{r} d_\theta H_1(\theta) + \partial_\phi h + \partial_t h\right)\Big|_{r=1+H_1} \quad . \tag{3.4}$$

In addition to these velocity boundary conditions we must require that the total pressure P + p vanishes at the displaced upper boundary (3.2). Using Eqs.(2.2), (2.9) and (3.2) we thus obtain the nonlinear pressure boundary condition

$$- \left[\frac{1}{2}(1+H)^2 + \frac{1}{\varepsilon(1+H)}\right] + \frac{1}{2}(1+H_1+h)^2\sin^2\theta +$$

$$+ \frac{1}{\varepsilon(1+H_1+h)} + p\Big|_{r=1+H_1+h} = 0 \quad . \tag{3.5}$$

For its linearization we make use of the equilibrium condition (2.10) with (2.9) and obtain

$$\left[\varepsilon(1 + H_1)\sin^2\theta - \frac{1}{(1 + H_1)^2}\right]h + \varepsilon p\Big|_{r=1+H_1} = 0 \quad . \tag{3.6}$$

4. EXPANSION WITH RESPECT TO r-1 AND ε

In trying to solve the linear stability Eqs.(1.6) - (1.9) it appears natural at first glance to try a solution with $v_r \equiv 0$. The most general solution of Eq.(1.6) would be $\underline{v} = \nabla r \times \nabla\psi$ in this case, $\psi(r,\theta,\phi,t)$ being a stream function. Inserting this and Eq.(2.1) in the r-component of the vorticity Eq.(1.8) leads to a second order differential equation for ψ which has essentially the dispersion relation for Rossby waves as a solvability condition. However, if one inserts $\underline{v}$ solutions thus obtained into the θ- and ϕ-components of Eq. (1.8) one finds that these are violated. This means that the assumption of strictly horizontal flow is in contradiction to momentum conservation, and we must admit $v_r \not\equiv 0$. The complications which are involved by this fact can be partially circumvented by employing an approximation scheme which will now be introduced.

In our problem, the independent variable r varies essentially in the region between the surfaces $r = r_0(\theta)$ and $r = r_1(\theta)$ given in Eqs.(2.5) with (2.7) and (2.8) with (2.12). Inserting both for H_0 and H_1 the largest possible value we obtain the range of variability $-\varepsilon/2 \leqslant r-1 \leqslant H$. Since for the planets considered in section 1 both limits are relatively small, we do not make much of a mistake if we expand all r-dependent quantities around their values at r = 1 up to the lowest significant order in r - 1. In doing so we set

$$\begin{aligned} p &= p^0(\theta,\phi,t) + p^1(\theta,\phi,t)(r - 1) \quad , \\ v_r &= v_r{}^0(\theta,\phi,t) + v_r{}^1(\theta,\phi,t)(r - 1) \quad , \end{aligned} \tag{4.1}$$

and similarly for the other velocity components. Furthermore, we expand all quantities which depend on H or ε with respect to these up to the lowest significant order, e.g.

$$h = h^0(\theta,\phi,t) + \varepsilon h^1(\theta,\phi,t) \quad . \tag{4.2}$$

This kind of expansion has already been used in deriving Eqs.(2.7) and (2.12), and we have seen that even under less favorable circumstances (data for Saturn), there is only a relatively small error involved. One should, however, be aware of the fact that the applicability of the theory presented will to a certain extent be limited through this.

Inserting the expansions (4.1) - (4.2) in the equations of our problem will greatly simplify these. Let us first deal with the boundary conditions. In expanding Eq.(3.1) we make use of Eq.(2.7) and obtain

$$v_r{}^0 = \frac{\varepsilon}{2}(1 - \sin^2\theta)v_r{}^1 + \varepsilon\sin\theta\cos\theta v_\theta{}^0 \tag{4.3}$$

to the lowest significant order. As was to be expected $v_r{}^0$ is a first order quantity. Expanding Eq.(3.3) and making use of Eq.(2.12) we obtain

$$v_r{}^0 + \left[H - \frac{\varepsilon}{2}(1 - \sin^2\theta)\right]v_r{}^1 = \varepsilon\sin\theta\cos\theta v_\theta{}^0 + \partial_\phi h^0 + \partial_t h^0 \ . \tag{4.4}$$

Inserting Eq.(4.3) in this we obtain

$$v_r{}^1 = \frac{1}{H}(\partial_t + \partial_\phi)h^0 \ . \tag{4.5}$$

Finally, expanding Eq.(3.6) we obtain

$$-h^0 + \varepsilon p^0 = 0 \ . \tag{4.6}$$

The pressure p^0 entering this equation can be obtained by returning e.g. to the ϕ-component of Eq.(1.9) and determining the corresponding lowest order approximation

$$\partial_\phi p^0 = -\sin\theta[(\partial_t + \partial_\phi)v_\phi{}^0 + \Omega\cos\theta v_\theta{}^0] \ . \tag{4.7}$$

Applying the same approximation procedure to Eqs.(1.6) - (1.8), we obtain

$$\sin\theta v_r{}^1 + \cos\theta v_\theta{}^0 + \sin\theta\partial_\theta v_\theta{}^0 + \partial_\phi v_\phi{}^0 = 0 \tag{4.8}$$

for the continuity equation. As a small quantity of first order (Eq.(4.3)) $v_r{}^0$ did not contribute to this equation. Note that in spite of the smallness of $v_r{}^0 + v_r{}^1(r-1)$, the derivative $v_r{}^1$ of v_r is equally as important as $v_\theta{}^0$ and $v_\phi{}^0$ in Eq.(4.8) and may by no means be neglected.

To lowest order in our expansion, the components of Eq.(1.7) in spherical coordinates are

$$\begin{aligned} \omega_r{}^0 &= \partial_\theta v_\phi{}^0 + \mathrm{ctg}\theta\, v_\phi{}^0 - \frac{1}{\sin\theta}\partial_\phi v_\theta{}^0 \\ \omega_\theta{}^0 &= -(v_\phi{}^1 + v_\phi{}^0) \quad , \quad \omega_\phi{}^0 = v_\theta{}^1 + v_\theta{}^0 \quad . \end{aligned} \tag{4.9}$$

Inserting these and Eqs.(2.1) and (2.3) in Eq.(1.8), in spherical coordinates we obtain to lowest order

$$\begin{aligned} (\partial_t + \partial_\phi)(\sin\theta\partial_\theta v_\phi{}^0 + \cos\theta v_\phi{}^0 - \partial_\phi v_\theta{}^0) \\ - 2\sin^2\theta v_\theta{}^0 = 2\sin\theta\cos\theta v_r{}^1 \end{aligned} \tag{4.10}$$

$$\begin{aligned} -(\partial_t + \partial_\phi)(v_\phi{}^0 + v_\phi{}^1) - 2\cos\theta v_\theta{}^0 \\ = 2\cos\theta v_\theta{}^1 - 2\sin\theta\partial_\theta v_\theta{}^0 - 2\cos\theta v_\theta{}^0 \quad , \end{aligned} \tag{4.11}$$

$$(\partial_t + \partial_\phi)(v_\theta{}^0 + v_\theta{}^1) = 2\cos\theta v_\phi{}^1 - 2\sin\theta\partial_\theta v_\phi{}^0 \quad . \tag{4.12}$$

Putting together the three boundary conditions (4.3), (4.5), (4.6), the pressure equation (4.7), the continuity equation (4.8), and the three vorticity equations (4.10) - (4.12), we have eight linear equations for the eight unknowns $v_r{}^0$, $v_r{}^1$, $v_\theta{}^0$, $v_\theta{}^1$, $v_\phi{}^0$, $v_\phi{}^1$, p^0 and h^0. It is obvious that all quantities except for one can be eliminated. Since Eqs.(4.11) - (4.12) can be considered as determining the quantities $v_\theta{}^1$ and $v_\phi{}^1$ and since these do not appear in the other equations, we may use Eqs.(4.5) - (4.8) and (4.10) in order to derive an equation for $v_\theta{}^0$ only. This will be done in the next section.

5. STABILITY EQUATION FOR v_θ^0

We shall discuss linear stability and wave propagation in terms of normal modes assuming that the ϕ and t dependence of all dependent variables is $\sim e^{i(m\phi+\omega t)}$. It is very simple at this point to perform the transition from our inertial frame r,θ,ϕ to rotating coordinates r',θ',ϕ'. Since r and θ do not change in this transition, we shall keep the old notation and that for z, respectively. For the time normalization chosen in section 1, ϕ transforms according to $\phi = \phi' + t$. From the identity $\exp(m\phi + \omega t) = \exp[m\phi' + (m + \omega)t]$ we get

$$\omega' = m + \omega \tag{5.1}$$

for the frequency in the rotating frame. It turns out to be convenient to introduce the abbreviation

$$\omega_m = \frac{\omega'}{m} = 1 + \omega/m \quad . \tag{5.2}$$

With this and our normal mode ansatz we get, e.g.,

$$(\partial_t + \partial_\phi)v_\phi^0 = im\omega_m v_\phi^0$$

and similarly for all other quantities. Equation (4.7) becomes

$$p^0 = \sin\theta\left(\frac{2i}{m}\cos\theta v_\theta^0 - \omega_m v_\phi^0\right) \quad . \tag{5.3}$$

Inserting this in Eq.(4.6) and Eq.(4.6) in Eq.(4.5) yields

$$v_r^1 = -\frac{2\omega_m \sin\theta}{R}(2\cos_\theta v_\theta^0 + im\omega_m v_\phi^0) \quad , \tag{5.4}$$

where we have introduced the abbreviation $R = 2H/\varepsilon$. In the original dimensioned variables, this quantity is given by the ratio

$$R = 2\,\frac{\tilde{H}}{\tilde{a}}\,\frac{\tilde{g}}{(\tilde{V}^2/\tilde{a})}\ , \tag{5.5}$$

i.e., R = (2 × gravity force × height of the atmosphere)/(centrifugal force × Earth radius). Formally, R looks like a Richardson number. Inserting the numbers of section 1, we have $R \approx 0.73$ for the Earth, $R \approx 0.2$ for Jupiter and $R \approx 0.05$ for Saturn. R is the only parameter which is left in the problem at this stage. Inserting Eq.(5.4) in Eq.(4.8) yields

$$v_\phi{}^0 = \frac{i}{m}\,\frac{(R-4\omega_m \sin^2\theta)\cos\theta\, v_\theta{}^0 + R\sin\theta\,\partial_\theta v_\theta{}^0}{R-2\omega_m{}^2\sin^2\theta}\ . \tag{5.6}$$

Inserting this back into Eq.(5.4), we get $v_r{}^1$ as a function of $v_\theta{}^0$ only;

$$v_r{}^1 = -\,2\omega_m \sin\theta\,\frac{(2-\omega_m)\cos\theta\, v_\theta{}^0 - \omega_m \sin\theta\,\partial_\theta v_\theta{}^0}{R - 2\omega_m{}^2\sin^2\theta}\ . \tag{5.7}$$

This result shows that in general $v_r{}^1$ cannot be neglected. Even if $\omega_m = 1$ and $v_\theta{}^0 = \sin\theta$, for which $v_r{}^1$ vanishes, it can easily be seen that Eq.(4.10) can never be satisfied. This demonstrates again that a complete neglect of the vertical velocity field is in contradiction with momentum conservation. Only for $R \to \infty$ or $|\omega_m| \to 0$ can $v_r{}^1$ be neglected. According to Eq.(5.5), $R \to \infty$ means that one is considering a very slowly rotating planet, which is not very interesting. The meaning of $|\omega_m| \to 0$ will be discussed in section 7b.

At this stage it turns out to be convenient to introduce the new independent variable

$$z = 1 - \cos 2\theta\ . \tag{5.8}$$

(Note that this z is different from the one in Fig. 1.) At the north pole we have $z = 0$ (corresponding to $\theta = 0$), and z increases monotonotically to $z = 2$ at the equator. From there, it returns monotonotically to $z = 0$ at the south pole. We shall see later that due to regularity conditions, it suffices to treat only one hemisphere.

The amplitude of $v_\theta{}^0$ will be denoted by y(z), setting

$$v_\theta{}^0 = y(z)e^{i(m\phi+\omega t)} \quad . \tag{5.9}$$

Inserting our results (5.6) - (5.7) into Eq.(4.10) and making use of the definitions (5.2) and (5.8), after some lengthy calculations one obtains

$$z(2-z)y'' + \left(4 - \frac{5}{2} z + \frac{z(2-z)\omega_m{}^2}{R - \omega_m{}^2 z}\right)y' - \frac{1}{2z}\Big\{m^2 - 1 + \tag{5.10}$$

$$+ \left(1 - \frac{1}{\omega_m}\right)z + \frac{z}{R}\Big[(4+4\omega_m - m^2\omega_m{}^2) - 2(1+\omega_m)z - \frac{\omega_m{}^2(2-z)(R-2\omega_m z)}{R-\omega_m{}^2 z}\Big]\Big\}y = 0 \quad .$$

This is the basic stability equation of our problem. It must be completed by some conditions arising from uniqueness and regularity requirements which technically play the role of boundary conditions. Note, however, that all physical boundary conditions have already been incorporated into Eq.(5.10).

6. UNIQUENESS AND REGULARITY CONDITIONS

Since in spherical coordinates the cones $\theta = \pi + \alpha$ and $\theta = \pi - \alpha$ are identical, our solutions must be the same for $\theta = \pi + \alpha$ and $\theta = \pi - \alpha$. This is automatically satisfied due to their dependence on $z = 1 - \cos 2\theta$. The singularity of spherical coordinates at the poles implies an additional condition: for any velocity field with nonvanishing velocity $\underline{v}_0$ at one of the poles we have

$$v_\theta{}^0 = |\underline{v}^0 - \underline{e}_r \cdot \underline{v}^0 \underline{e}_r| \cos(\phi - \phi_0) \quad , \tag{6.1}$$

whence we get the condition

$$y(0) = 0 \qquad \text{for } m \neq 1 \;, \tag{6.2}$$

i.e., for all modes which don't have the ϕ-dependence of Eq.(6.1).

Furthermore, the poles ($z = 0$) and the equator ($z = 2$) are singular points of our stability equation (5.10). For $z \to 0$, the asymptotic behavior of the solutions can be determined from

$$2zy'' + 4y' - \frac{1}{2z}(m^2 - 1) = 0 \;. \tag{6.3}$$

This equation has a regular solution $y \sim z^{(|m|-1)/2}$ and a singular solution $y \sim z^{-(|m|+1)/2}$. For physical reasons we must exclude the singular solution. The regular one satisfies exactly the uniqueness requirement (6.2), and it is useful to extract the asymptotic behavior at $z = 0$ by setting

$$y(z) = z^{\frac{|m|-1}{2}} g(z) \;. \tag{6.4}$$

Inserting this transformation into Eq.(5.10), we obtain the equation

$$(2-z)zg'' + \Big[2(|m|+1) - \Big(\frac{1}{2} + |m| + \frac{R-2\omega_m{}^2}{R-\omega_m{}^2 z}\Big)z\Big]g' \tag{6.5}$$
$$- \frac{1}{2}\Big\{\frac{|m|+m^2}{2} - \frac{1}{\omega_m} - \frac{(|m|-1)\omega_m{}^2(2-z)}{R-\omega_m{}^2 z} + \frac{1}{R}\Big[\Big(2 + \frac{R\omega_m(2-\omega_m)}{R-\omega_m{}^2}\Big)(2-z) - m^2\omega_m{}^2\Big]\Big\}g = 0$$

for $g(z)$. Again, of course, it has a singular solution at $z = 0$. The physical solution is regular and tends towards a finite value $\neq 0$ for $z \to 0$.

The asymptotic behavior of the solutions at the second singular point $z = 2$ of Eq.(6.5) can be obtained from

$$2(2-z)g'' - g' - \frac{1}{2}\left(\frac{|m|+m^2}{2} - \frac{1}{\omega_m} - \frac{m^2\omega_m{}^2}{R}\right)g = 0 \quad . \tag{6.6}$$

This equation has two regular solutions: $g \sim (2-z)^0$ and $g \sim (2-z)^{1/2}$. The last one has a derivative $g'(z) \sim (2-z)^{-1/2}$ which diverges and we shall therefore call it the singular solution. We shall now show that we have to choose the regular one again.

For physical reasons, we must require continuity of $v_\theta{}^0$ and $\partial v_\theta{}^0/\partial\theta$ across the equator. For $g \sim (2-z)^{1/2}$ we get $v_\theta{}^0 \sim (1 + \cos 2\theta)^{1/2}$, and putting $\theta = \pi/2 + \alpha$, we have $v_\theta{}^0 \sim |\alpha|$ for $\alpha \to 0$. Obviously $\partial v_\theta{}^0/\partial\theta$ becomes discontinuous and we must therefore exclude this solution. According to Eq.(6.6), the solution $\sim (2-z)^0$ has a finite derivative $g'(z)$ at the equator. For this reason, $\partial v_\theta{}^0/\partial\theta \sim 2g'(z)\sin 2\theta$ is continuous across the equator and becomes zero there.

7. SOLVABLE SPECIAL CASES

Equation (6.5) possesses polynomial solutions in all special cases for which it can be reduced to the simple form

$$(2-z)zg'' + (A + Bz)g' + Cg = 0 \quad , \tag{7.1}$$

A, B and C being constants. Inserting the ansatz

$$g = \sum_0^{\ell} c_\nu z^\nu \tag{7.2}$$

into Eq.(7.1) one obtains the recursion scheme

$$\begin{aligned} c_1 &= (C/A)c_0 \quad , \quad c_2 = -\frac{B+C}{2(A+2)}c_1 \quad , \\ c_{\nu+1} &= \frac{\nu(\nu-1) - \nu B - C}{(\nu+1)(A+2\nu)}c_\nu \quad . \end{aligned} \tag{7.3}$$

The closure condition $c_{\ell+1} = 0$ yields the dispersion relation

$$\ell(\ell-1) - \ell B - C = 0 \quad . \tag{7.4}$$

Note that our regularity conditions for g(z) are automatically satisfied for any $\ell < \infty$.

a) Special Case $R \to \infty$

For $R \to \infty$ (very slowly rotating planet), and on the assumption that ω_m stays finite, Eq.(6.5) assumes exactly the form of Eq.(7.1) with

$$A = 2(|m| + 1) \ , \quad B = -(|m| + \frac{3}{2}) \ ,$$

$$C = -\frac{|m| + m^2}{4} + \frac{1}{2\omega_m} \quad . \tag{7.5}$$

Using Eq.(5.2), the dispersion relation (7.4) becomes

$$\omega' = \frac{2m}{m^2 + (2\ell)^2 + 4\ell|m| + 2\ell + |m|} \quad . \tag{7.6}$$

This is essentially the well known dispersion relation for Rossby waves (see, e.g. [6]), the main difference being that we have a coupling term between the lateral and zonal wave numbers. This result is not very surprising since we mentioned earlier that for $R \to \infty$ we have $v_r^1 \to 0$, thus losing one of the main features of our treatment.

b) Special Case $|\omega_m| \to 0$ for Finite R

Assuming $|\omega_m| \to 0$, in Eq.(6.5) we may neglect all terms $\sim \omega^\nu$ with $\nu \geqslant 0$ and $2\omega_m^2$ or $\omega_m^2 z$ as compared to R, provided R stays finite. Since we shall find $\omega_m \sim 1/(|m|+2\ell)^2$, our assumption requires large values of m and ℓ in order to be valid, and we shall therefore keep the terms $\sim |m|$ or m^2. Again, Eq.(6.5) reduces to the structure of Eq.(7.1) with

$$A = 2|m| \ , \quad B = -|m| \ , \quad C = -\frac{|m|+m^2}{4} + \frac{1}{2\omega_m} \quad . \tag{7.7}$$

The dispersion relation becomes

$$\omega_m = \frac{2}{m^2 + (2\ell)^2 + 4\ell|m|} , \tag{7.8}$$

where the terms which are linear in ℓ and $|m|$ have now been neglected. Equation (7.8) is the limit of Eq.(7.6) for large ℓ and $|m|$ and represents again Rossby waves. Note that according to Eq.(5.5) we have also $v_r^1 \approx 0$ in this case. Again, this result is not very surprising, since for very large ℓ and m or very short wavelengths the usual approximations (neglection of v_r and β-plane approximation) should become valid.

c) Special Case $R \to 0$

We set

$$\mu = \sqrt{R} , \tag{7.9}$$

and shall restrict our consideration to modes for which ω_m depends linearly on μ, i.e.

$$\omega_m = \mu v . \tag{7.10}$$

Expanding Eq.(6.6) with respect to μ and keeping only two orders of μ in each coefficient, we get

$$\mu(2-z)zg'' + \mu[2(|m| + 1) - (\frac{1}{2} + |m| + \frac{1-2v^2}{1-v^2z})z]g'$$

$$+ \frac{1}{2}(\frac{1}{v} - \frac{2v(2-z)}{1-v^2z} - \frac{2(2-z)}{\mu})g = 0 . \tag{7.11}$$

We can obtain an approximate solution to this equation by employing a WKB approach. With the ansatz

$$g(z) = g_0 e^{\phi(z)} \quad , \tag{7.12}$$

where g_0 is a constant, we get the equation

$$\mu(2-z)z\phi'' + \mu(2-z)z\phi'^2 + \mu\left[2(|m| + 1) - \left(\frac{1}{2} + |m| + \frac{1-2v^2}{1-v^2 z}\right)z\right]\phi' +$$

$$+ \frac{1}{2}\left(\frac{1}{v} - \frac{2v(2-z)}{1-v^2 z}\right) - \frac{2-z}{\mu} = 0 \tag{7.13}$$

for ϕ'. If ϕ is expanded according to

$$\phi = \frac{\phi_0}{\mu} + \phi_1 + \cdots \quad , \tag{7.14}$$

ϕ_0 and ϕ_1 must obey the equations

$$z\phi_0'^2 - 1 = 0 \quad , \tag{7.15}$$

and

$$2(2-z)z\phi_0'\phi_1' + (2-z)z\phi_0'' + \left[2(|m| + 1) - \left(\frac{1}{2} + |m| + \frac{1-2v^2}{1-v^2 z}\right)z\right]\phi_0' +$$

$$+ \frac{1}{2}\left(\frac{1}{v} - \frac{2v(2-z)}{1-v^2 z}\right) = 0 \quad . \tag{7.16}$$

One of the solutions of Eq.(7.15) is

$$\phi_0 = 2\sqrt{z} \quad , \tag{7.17}$$

which will turn out to be the proper one for satisfying the regularity condition at $z = 0$. (The other solution, $\phi_0 = -2\sqrt{z}$, corresponds to the solution g which is singular at $z = 0$). After ϕ_0, from Eq.(7.17), has been inserted in Eq.(7.16), this can be brought into the form

$$\phi_1'(z) = -\frac{1/(4v)}{(2-z)\sqrt{z}} + \frac{v/(2\delta)}{\sqrt{z}} - \frac{v^2/2}{1+v\sqrt{z}} - \frac{(1+2|m|)/4}{z} + \frac{1/4}{2-z} \quad . \tag{7.18}$$

$$\delta = \frac{1}{\sqrt{2v}} - 1$$

Integrating this and combining the solutions for ϕ_0 and ϕ_1, we obtain the approximate solution

$$g = g_0 \frac{(1+v\sqrt{z})(2-z)^{\delta} e^{2\sqrt{z}/\varepsilon}}{(\sqrt{2}+\sqrt{z})^{1/(2\sqrt{2v})} z^{(1+2|m|)/4}} \tag{7.19}$$

It must now be ascertained that our approximate solution for g can be fitted to the regular solutions at $z = 0$ and $z = 2$.

Let us first consider $z = 0$. Since $\phi_1''(z)$ has been neglected in Eq.(7.16) and diverges for $z \to 0$, our solution can only be good for, e.g., $|\mu(2-z)z\phi_1''| \ll |2(2-z)z\phi_0'\phi_1'|$ which yields the condition $z/\mu^2 \gg 1/4$ for small values of z. It must therefore not disturb us that our WKB solution diverges for small values of z/μ^2.

An asymptotic representation of the regular solution at $z = 0$ can be obtained from Eq.(7.11) by series expansion. Collecting terms of the same order in μ, one obtains easily

$$g_{as}(z) = g_0^* \sum_{\nu=0}^{\ell} \frac{|m|!x^{\nu}}{\nu!(|m|+\nu)!} + O(\mu) \quad , \quad x = z/\mu^2 \quad , \tag{7.20}$$

where ℓ can go up to infinity and where the $O(\mu)$ term vanishes as $\mu \to 0$ so it can be neglected. The asymptotic behavior of the approximate solution (7.19) for $z \to 0$ is given by

$$g_{ap}(z) = \frac{e^{2\sqrt{z}}}{z^{(1+2|m|)/4}} , \qquad (7.21)$$

where the free factor g_0 has been chosen appropriately. It was found numerically, and there is some analytical evidence, that $g_{ap}(z)$ is well fitted by $g_{as}(z)$ up to arbitrary large values $x >> 1$ if only ℓ is chosen large enough. This is demonstrated in Fig. 2 for the special case $m = 1$. (Note that this behavior is independent of the eigenvalue v.) These findings support strongly that Eq.(7.19) represents the solution which is regular at $z = 0$ without further requirements on v.

Considering the singular point $z = 2$, it turns out to be favorable to rewrite Eq.(7.11) in terms of the original variable θ to obtain

$$\mu\ddot{g}(\theta) + \mu \frac{\cos\theta}{\sin\theta}\left(1 + 2|m| + \frac{4v^2\sin^2\theta}{1-2v^2\sin^2\theta}\right)\dot{g}(\theta)$$
$$+ 2\left(\frac{1}{v} - \frac{4v\cos^2\theta}{1-2v^2\sin^2\theta} - \frac{4\cos^2\theta}{\mu}\right)g(\theta) = 0 . \qquad (7.22)$$

Note that the singular point $z = 2$ is converted into a regular point $(\theta = \pi/2)$ by this transition. The problem is, of course, not fully removed in this way. Our approximate solution turns into

$$g = g_0 \frac{(1+\sqrt{2}v\sin\theta)\cos^{2\delta}\theta e^{(2\sqrt{2}\sin\theta)/\varepsilon}}{(1+\sin\theta)^{1/(2\sqrt{2}v)}\sin^{(|m|+1)/2}\theta} , \qquad (7.23)$$

and may become infinite at $\theta = \pi/2$ for Re $\delta < 0$. Furthermore, the terms neglected in the derivation of Eq.(7.23) may blow up even for finite g if $\ddot{g}(\theta)$ diverges. Since

$$g \sim (\theta - \pi/2)^{2\delta} = (\theta - \pi/2)^{2\delta_r} e^{i2\delta_i \ln(\theta-\pi/2)}$$

for $\theta \to \pi/2$, both possibilities are avoided by requiring Re $\delta > 1$. Inserting the definition for δ from Eq.(7.18), we get finally the condition

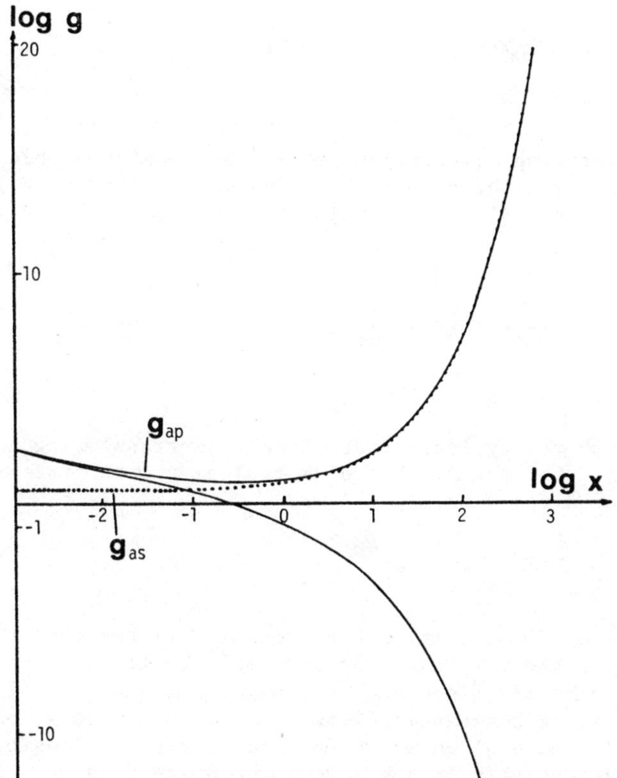

Fig. 2. Numerical evaluation of $g_{as}(x)$ (dotted) and $g_{ap}(x)$ (drawn) for $m = 1$ and $\ell = 32$. The factor g^* in g_{as} was chosen such that the curves do not intersect but only touch. This happens at $x = 456$. The descending drawn curve is the other WKB solution corresponding to $\phi_0 = -2\sqrt{z}$.

$$(v_r - \frac{1}{10\sqrt{2}})^2 + v_i^2 < (\frac{1}{10\sqrt{2}})^2 \quad . \tag{7.24}$$

v_r and v_t being the real and imaginary parts of v resp. Any value of v which is chosen inside the circle in the complex v plane defined through inequality (7.24) leads to a regular solution of our stability equation. This means that we have a continuous spectrum. Since v_i can assume positive and negative values there, we have thus obtained also a continuous spectrum of unstable solutions.

The largest growth rate compatible with inequality (7.24) and Eq.(7.10) with (5.2) is given by

$$\omega_i' = \frac{m/R}{10\sqrt{2}} \quad . \tag{7.25}$$

(Since in deriving Eq.(7.11) terms ~ m and ~ m^2 have been neglected, Eq.(7.25) is restricted to reasonably small values of m.) Returning to dimensional quantities, we obtain

$$\tilde{t} > \frac{10\sqrt{2}}{2\pi m/R} \text{ days} \approx \frac{2.25}{m/R} \text{ days} \tag{7.26}$$

from this for the typical growth times. Extrapolating this result to the value R ≈ 0.7 for the Earth yields typical growth times in the order of magnitude of a few days.

CONCLUSION

An essential outcome of our treatment is the fact that the spatial derivative of the vertical velocity field is tightly coupled to the horizontal velocity field and can in general not be neglected. There exist special circumstances (very slow planetary rotation or very short wavelength phenomena) under which a neglection is possible. For these one finds essentially Rossby waves, the dispersion relation being slightly modified as compared to that obtaind under the usual approximations. For long wavelength phenomena and rotation velocities such as those of the Earth, Saturn or Jupiter, the influences of the vertical velocity field become of major importance. Extrapolating slightly from the analytically tractable case R → 0 we expect the appearance of unstable wave motions with typical growth times of a few days in the case of the Earth. The rigorous application of our analysis to the Earth will require a numerical treatment of the stability equation which will be taken up as the next step.

The problem of stability depends on the parameter R = (2 × gravity force × height of the atmosphere)/(centrifugal force × Earth radius). In the transition from large R to R → 0 we expect a critical value R_{cr} for which instability occurs for the first time. It can be expected that this becomes a bifurcation point for either steady state equilibria or periodically oscillating states. If this expectation turns out to be correct, a future step to be taken will be a study of this nonlinear bifurcation problem.

REFERENCES

[1] J. T. Houghton, "The Physics of Atmospheres," Cambridge University Press, Cambridge, 1977, p.94.
[2] ibid., p.708.
[3] ibid., p.36
[4] NASA Technical Note SP-8069, 1971.
[5] NASA Technical Note SP-8091, 1972.
[6] J. T. Houghton "The Physics of Atmospheres," Cambridge University Press Cambridge, 1977, p.95.

Acknowledgement. The author's interest in the study of atmospheric waves was originally raised by G. Nicolis. The author expresses deep gratitude towards G. Nicolis and C. Nicolis for many stimulating discussions. An interesting discussion with each of R. Benzi, J. Egger, and L. Koschmieder respectively is also gratefully acknowledged.

PART IV

CLIMATE DYNAMICS

NONLINEAR PHENOMENA IN CLIMATE DYNAMICS

M. Ghil
Department of Atmospheric Sciences and
Institute of Geophysics and Planetary Physics
University of California, Los Angeles, CA 90024,
and Courant Institute of Mathematical Sciences
New York University, New York, NY 10012, U.S.A.

ABSTRACT. Climate dynamics is a new and rapidly developing geophysical discipline. Selected problems from this discipline are presented, starting with their physical motivation. Both equilibrium models and oscillatory models are introduced. Mathematical aspects emphasize multiple equilibria and their stability on the one hand, and nonlinear oscillations, entrainment and detrainment by periodic and quasi-periodic forcing, on the other.

Based on this nonlinear analysis, a specific prediction is made about the existence of combination tones in the power spectrum of time series from the Quaternary. Such combination tones appear in recent Fourier analyses of marine cores and ice cores with sufficiently high sedimentation rates. We also discuss the consequences of a complex power spectrum, with multiple lines superimposed on a continuous background, for the detailed prediction of future climates, or simulation of past climates.

1. INTRODUCTION

Climate dynamics is a relatively new member of the family of geophysical sciences. Descriptive climatology goes back, of course, at least to the ancient Greeks, who realized the importance of the Sun's mean zenith angle in determining the climate of a given latitude belt, as well as that of land-sea distribution in determining the regional, zonally asymmetric characteristics of climate. The general human perception of climate change is also preserved in numerous written records throughout history, starting with the floods described in the epic of Gilgamesh and in the Bible.

Only recently did the possibility of global *climate monitoring* present itself to the geophysical community, through ground-based observational networks and space-borne instrumentation. This increase in quantitative, detailed knowledge of the Earth's current climate was accompanied by the development of elaborate geochemical and micropaleontological methods for sounding the planet's climatic past.

Observational information about present, spatial detail, and about past, temporal detail were accompanied in the 1960s by an increase

C. Nicolis and G. Nicolis (eds.), Irreversible Phenomena and Dynamical Systems Analysis in Geosciences, 313–320.

of *computing power* used in the processing of climatic data, as well as in the modeling and simulation of the seasonally varying general circulation of the atmosphere and ocean. The knowledge thus accumulated led to an increase of insight which was distilled in simple models, in an attempt to analyze the basic ingredients of climatic mechanisms and processes.

In the following lecture, I shall describe a few simple models, and try to convey the flavor of the new, *theoretical* climate dynamics. As in every area of the exact sciences, the fundamental ideas suggested by simple models have to be tested by further observations and detailed simulations of the phenomena under study. It is hoped that this description of preliminary, theoretical results will stimulate the comparisons and verifications required to further develop the theory.

Theoretical climate dynamics as presented in this lecture is covered in Part IV, Chapters 10-12, of Ghil and Childress (1986; hereafter GC) and in Part V of Ghil *et al.* (1985). For this reason, only the main ideas are explicitly mentioned here, and extensive references to additional research and review papers are given.

2. RADIATION BALANCE AND EQUILIBRIUM MODELS

The major characteristics of a physico-chemical system, such as the climatic system, are given by its energy budget. The climatic system's *energy budget* is dominated by the short-wave radiation, R_i, coming in from the Sun, and the long-wave radiation, R_o, escaping back into space. The approximate balance betweeen R_i and R_o determines the mean temperature of the system. The distribution of radiative energy within the system, in height, latitude and longitude, determines to a large extent the distribution of climatic variables, such as temperature, throughout the system.

We consider in this section a spatially zero-dimensional (0-D) model of radiation balance, with mean global temperature as the only variable (Ghil, 1985). The dependence of the solar radiation's reflection on temperature, the so-called ice-albedo feedback, and the dependence of infrared absorption on temperature, the greenhouse effect, are discussed. Stationary solutions of this model and their linear and nonlinear stability are investigated, via a variational principle.

The same analysis is then carried out for a spatially one-dimensional (1-D) *energy-balance model*, in which surface air temperature is averaged with respect to longitude (zonally), but allowed to depend on latitude (Budyko, 1969; Ghil, 1976; Held and Suarez, 1974; North *et al.*, 1981; Schneider and Dickinson, 1974; Sellers, 1969). The main results are that the present climate is exponentially stable and that its mean temperature changes by roughly $1^{o}K$ for one percent change of insolation. Additional stable and unstable model solutions exist, but their connections to observed past climates appear to be rather tenuous.

3. GLACIATION CYCLES: PHENOMENOLOGY AND SLOW PROCESSES

3.1. Climatic Variability and Climate Modeling

The previous section dealt with the climatic system's radiation balance, which led to the formulation of equilibrium models. Slow changes of these equilibria due to small external forcing, internal fluctuations about an equilibrium, and transitions from one possible equilibrium to another one have also been studied (Benzi *et al.*, 1982; GC, Chapter 10; Nicolis, 1982).

Climatic records exist on various time scales, from instrumental records on the time scale of months to hundreds of years, through historical documents and archaeologic evidence, to geological proxy records on the time scale of thousands to millions of years. These records indicate that climate varies on all time scales in an irregular fashion. It is difficult to imagine that a model's stable equilibrium, whether slowly shifting or randomly perturbed, can explain all this variability.

A summary of *climatic variability* on all time scales appears in Mitchell (1976). The most striking feature is the presence of sharp peaks superimposed on a continuous background. The relative power in the peaks is poorly known; it depends of course on the climatic variable whose power spectrum is plotted. Furthermore, phenomena of small spatial extent will contribute mostly to the high-frequency end of the spectrum, while large spatial scales play an increasing role towards the low-frequency end.

Many phenomena are believed to contribute to changes in climate. *Persistent anomalies* in atmospheric flow patterns affect climate on the time scale of months and seasons (Ghil, 1986; GC, Chapter 6). On the time scale of tens of millions of years, plate tectonics and continental drift play an important role. Variations in the chemical composition of the atmosphere and oceans are essential on the time scale of billions of years, and significant on time scales as short as decades.

The appropriate definition of the *climatic system* itself depends on the phenomena one is interested in, which determine the components of the system active on the corresponding time scale. No single model could encompass all temporal and spatial scales, include all the components, mechanisms and processes, and thus explain all the climatic phenomena at once.

Our goal in this lecture is much more limited. We concentrate on the most striking phenomena to occur during the last two million years of the Earth's climatic history, the Quaternary period, namely on *glaciation cycles*. The time scale of these phenomena ranges from thousands to millions of years. We attempt to describe and model in the simplest way possible the components of the climatic system active on these time scales - atmosphere, ocean, continental ice sheets, the Earth's upper strata - and their nonlinear interactions.

We sketch the discovery of geological evidence for past glaciations, review *geochemical methods* for the study of deep-sea cores and describe the phenomenology of glaciation cycles as deduced from these cores (Duplessy, 1978; GC, Chapter 11; Imbrie and Imbrie, 1979). A near-periodicity of roughly 100,000 years dominates continuous records of isotope proxy data for ice volume, with smaller spectral peaks near 40,000 and 20,000 years. The records themselves are rather irregular and much of the spectral power resides in a continuous background (GC, Section 11.1; Hays *et al.*, 1976).

3.2. Cryodynamics and Geodynamics

We give a brief introduction to the *dynamics* of valley glaciers and *large ice masses*. The nonlinear visco-plastic rheology of ice is reviewed (Glen, 1955) and used in deriving the approximate geometry of ice sheets. The slow evolution due to small changes in mass balance of an ice sheet with constant profile is modeled next (Birchfield *et al.*, 1981; Oerlemans and van der Veen, 1984; Paterson, 1981). A simplified, but temperature-dependent formulation of the hydrologic cycle and of its effect on the ice mass balance is given. We study multiple equilibria of the ice-sheet model thus formulated and their stability, pointing out similarities with the study of energy-balance models in Section 2. A hysteresis phenomenon occurs in the transition from one equilibrium solution to another as temperature changes, due to a double *saddle-node bifurcation* (Ghil, 1984; GC, Section 11.2; see also lecture by G. Nicolis in this volume).

Next, we study the deformation of *the Earth's upper strata* under the changing load of ice sheets. The rheology of lithosphere and mantle is reviewed and approximated by a linear visco-elastic model. Post-glacial uplift data and their implications for this approximate rheology are outlined. A simple model of creep flow in the mantle is used to derive an equation for maximum bedrock deflection under an ice sheet, and for the way this deflection affects the mass balance of the sheet (GC, Section 11.3; Peltier, 1982; Turcotte and Schubert, 1982).

The equations derived and analyzed in this section for ice flow and bedrock response will lead, when combined with an equation for radiation balance from the previous section, to a system of differential equations which govern stable, self-sustained, periodic oscillations. Changes in the orbital parameters of the Earth on the Quaternary time scales provide small changes in insolation (Berger *et al.* 1984; GC, Section 12.3). These quasi-periodic changes in the system's forcing will produce forced oscillations of a quasi-periodic or aperiodic character, to be studied in the next section. The power spectra of these oscillations show the above mentioned peaks with periodicities near 100,000 years, 40,000 years and 20,000 years, as well as the continuous background apparent in the data.

4. CLIMATIC OSCILLATIONS

In Section 3 we reviewed some of the geological evidence for glaciation cycles during the Quaternary period. Large changes in global ice volume and changes of a few degrees in global mean temperature have occurred repeatedly over the last two million years. It is these changes we would like to investigate in the present section, with the help of very simple models.

These simple models do not represent the definitive formulation of a theory for climatic variability on the time scales of interest. They are used merely to illustrate some ideas we believe to be basic for an understanding of this variability, an understanding which is still in early stages of development. Other models and related ideas can be found in the references of Section 5, as well as in other contributions to this volume.

4.1. Self-Sustained Oscillations

We formulate and analyze a coupled model of two ordinary differential equations, for global temperature and global ice volume. The equations govern radiation balance (Section 2) and ice-sheet flow (Section 3), respectively. This model exhibits self-sustained oscillations with an amplitude comparable to that indicated by the records and a period of O(10 ka), where 1 ka = 1000 years. Phase relations between temperature and ice volume, and their role in the oscillation's physical mechanism, are investigated (Ghil, 1985; GC, Section 12.1). Stochastic perturbations of such self-sustained climatic oscillations have also been considered (Nicolis, 1984; Saltzman *et al.*, 1981).

Exchange of stability between equilibria (Section 2) and limit cycles (Section 3) in models with an arbitrary number of dependent variables and spatial dimensions is studied next. The distinction is made between a stable limit cycle which grows slowly in amplitude from zero as a parameter is changed (direct or *supercritical Hopf bifurcation*) and sudden jumps from zero to finite amplitude (reverse or *subcritical* Hopf bifurcation). Structural stability and the special role of limiting, structurally unstable, homoclinic and heteroclinic orbits is discussed (GC, Section 12.1; see also lectures by C. Nicolis and by G. Nicolis in this volume).

4.2. Forced Oscillations

We introduce the geometry and kinematics of *orbital changes* in the Earth's motion around the Sun, from the perspective of the small insolation changes they generate. Eccentricity, obliquity and precession of the Earth's orbit are defined. A few fundamental concepts of celestial mechanics and the associated mathematical methods are reviewed (Arnold, 1978; Brouwer and Clemence, 1961; Buys and Ghil, 1984;
Gallavotti, 1985; GC, Section 12.3; Ghil and Wolansky, 1986; Wolansky, 1985). We report currently accepted results on the periodicities of insolation changes during the Quaternary period: 19ka and 23 ka for precession, 41ka for obliquity and 100ka and 400ka for eccentricity. Their action on the climatic system's radiation balance and hydrologic cycle is modeled.

Finally, we take up the effects of this action upon the climatic oscillator above, augmented by a third equation, governing bedrock response to ice load (Section 3). The free oscillations of this model are found to differ but little from those of the previous one. Eccentricity forcing is shown to produce a very small or very large response according to whether the system operates in an equilibrium or in an oscillatory mode, i.e., the model exhibits *nonlinear resonance* (GC, Section 12.4).

We study in detail the internal mechanisms by which forcing at one or more frequencies can be transferred through the system to additional frequencies, as well as to the climatic spectra's continuous background. *Entrainment* results in the system's free frequency becoming locked onto an integer or rational multiple of a forcing frequency. Loss of entrainment leads to aperiodic changes in system response.

Combination tones are linear combinations with integer coefficients of the forcing frequencies. In addition to direct, nonlinear resonance at periodicities near 100ka, the model also produces a strong peak at 109ka as a difference tone between the two precessional frequencies of 1/19ka and 1/23ka. Most models of glaciation cycles do reproduce this peak well-known from many proxy records (e.g., contributions by G. E. Birchfield and B. Saltzman to this volume, and references therein, as well as in Section 5 here). The model under discussion here predicts, however, also a peak at the sum tone of the above frequencies, with a period of 10.4ka, and at the sum tone of each precessional frequency with the obliquity frequency, 1/41ka, with periods near 13ka and 15ka, respectively. Higher harmonics of the precessional frequencies, e.g., at 11.5ka and 9.5ka, are also predicted, along with higher harmonics of the sum tones above, e.g., at 5.2ka and 2.6ka, or at 6.5ka and 7.5ka.

These combination tones and harmonics result from the model's nonlinear, parametric response to orbital forcing. They are superimposed on a continuous background associated with aperiodic, irregular terminations of glaciated episodes (Ghil, 1985; GC, Section 12.5).

Four of the combination tones predicted by the model discussed here are found in deep-sea cores with sufficiently high sedimentation rates, and hence resolution, by Pestiaux and Duplessy (1985); among them are peaks at 13ka, 10.4ka and 9.5ka. A peak at 2.5ka, indistinguishable in practice from the 2.6ka predicted here, is also found in the well-dated upper part of ice cores from Greenland and the Antarctica (Benoist *et al.*, 1982; Dansgaard *et al.*, 1982).

4.3. Periodicity and Predictability

The search for periodicity in geophysical time series is motivated by the desire to understand, as well as to predict. Constant behavior is the most predictable; it is without surprises, but is seldom encountered in nature. The next most predictable type of behavior is purely periodic.

The consequences of multiple spectral lines and of the continuous background for predictability, or the lack thereof, are investigated. It is argued that irretrievable loss of predictive skill over a time interval O(100ka) is intrinsic in the aperiodic nature of Quaternary climate changes (Ghil, 1985; GC, Section 12.6).

The damped periodicity of lagged correlations shown in the references above is in fact model-independent. It can be obtained from the power spectrum of the data themselves by the Wiener-Khinchin formula, which connects spectral density and the correlation function. Hence, detailed simulation with arbitrarily small, prescribed errors of the Earth's paleoclimatic history is not possible, based on the limited amount and accuracy of proxy data available.

On the other hand, much can be learned about climatic mechanisms, and about orbital changes, by studying paleoclimatic records in the spectral domain. The distribution of power between the peaks and the continuous background, as well as the exact position of the peaks and their power relative to each other, give tell-tale indications about the climate's internal workings and the external changes which affect it.

Acknowledgments. It is a pleasure to acknowledge the continuing support of the NSF Climate Dynamics Program under grants ATM-8416351 and ATM-8514731. The organizers, lecturers and participants at the Advanced Study Institute provided emulation and stimulation for the writing of these lecture notes. They were typed with proficiency and good humor by Bianca Gola.

5. REFERENCES

V. I. Arnold 1978. *Mathematical Methods of Classical Mechanics.* Springer-Verlag, New York, 462 pp.

J. P. Benoist, F. Glangeaud, N. Martin, J. L. Lacoume, C. Lorius and A. Oulahman, 1982. 'Study of climatic series by time-frequency analysis'. In *Proc. ICASSP82,* IEEE Press, New York, pp. 1902-1905.

R. Benzi, G. Parisi, A. Sutera and A. Vulpiani, 1982. 'Stochastic resonance in climatic change'. *Tellus,* **34,** 10-16.

A. Berger, J. Imbrie, J. Hays, G. Kukla and B. Saltzman (eds.), 1984. *Milankovitch and Climate: Understanding the Response to Astronomical Forcing,* vols. I & II. D. Reidel, Dordrecht, 985 pp.

G. E. Birchfield, J. Weertman and A. T. Lunde, 1981. 'A paleoclimate model of Northern Hemisphere ice sheets'. *Quatern. Res.,* **15,** 126-142.

D. Brouwer and M. Clemence, 1961. *Methods of Celestial Mechanics.* Academic Press, New York, 598 pp.

M. I. Budyko, 1969. 'The effect of solar radiation variations on the climate of the earth'. *Tellus*, 21, 611-619.

M. Buys and M. Ghil, 1984. 'Mathematical methods of celestial mechanics illustrated by simple models of planetary motion'. In Berger *et al.* (1984), pp. 55-82.

W. Dansgaard, H. B. Clausen, N. Gundestrup, C. U. Hammer, S. F. Johnsen, P. M. Kristinsdottir and N. Reeh, 1982. 'A new Greenland deep ice core'. *Science,* **218,** 1273-1277.

J.-C. Duplessy, 1978. 'Isotope studies'. In Gribbin (1978), pp. 46-47.

G. Gallavotti, 1985. 'Stability near resonances in classical Hamiltonian systems'. In *Proc. Intl. Conf. Mathematical Problems from the Physics of Fluids,* Klim, Roma, pp. 55-64.

M. Ghil, 1976. Climate stability for a Sellers-type model. *J. Atmos, Sci.,* 33, 3-20.

M. Ghil, 1984. 'Climate sensitivity, energy balance models, and oscillatory climate models. *J. Geophys. Res.,* **89D,** 1280-1284.

M. Ghil, 1985. 'Theoretical climate dynamics: an introduction'. In Ghil *et al.* (1985), pp. 347-402.

M. Ghil, 1986. 'Dynamics, statistics and predictability of planetary flow regimes'. This volume.

M. Ghil and S. Childress, 1986. *Topics in Geophysical Fluid Dynamics: Atmospheric Dynamics, Dynamo Theory and Climate Dynamics.* Springer-Verlag, New York, 479 pp., in press.

M. Ghil and G. Wolansky, 1986. 'Non-Hamiltonian perturbations of integrable systems'. *Commun. Math. Phys.,* submitted.

M. Ghil, R. Benzi and G. Parisi (eds.), 1985. *Turbulence and Predictability in Geophysical Fluid Dynamics and Climate Dynamics.* North-Holland, Amsterdam/Oxford/New York/Tokyo, 449 pp.

J. W. Glen, 1955. 'The creep of polycrystalline ice'. *Proc. Roy. Soc. (London)*, **A228**, 519-538.

J. Gribbin (ed.), 1978. *Climatic Change*. Cambridge Univ. Press, Cambridge/New York/Melbourne, 280 pp.

J. D. Hays, J. Imbrie and N. J. Shackleton, 1976. 'Variations in the earth's orbit: pacemaker of the ice ages'. *Science*, **194**, 1121-1132.

I. M. Held and M. J. Suarez, 1974. 'Simple albedo feedback models of the ice caps'. *Tellus*, **36**, 613-628.

J. Imbrie and K. P. Imbrie, 1979. *Ice Ages: Solving the Mystery*, Enslow Publ., Short Hills, NJ, 224 pp.

J. M. Mitchell Jr., 1976. 'An overview of climatic variability and its causal mechanisms'. *Quatern. Res.*, 6, 481-493.

C. Nicolis, 1982. 'Stochastic aspects of climatic transitions - response to a periodic forcing'. *Tellus*, **34**, 1-9.

C. Nicolis, 1984. 'Self-oscillations, external forcings, and climate predictability'. In Berger *et al.* (1984), pp. 637-652.

G. R. North, R. F. Cahalan and J. A. Coakley, Jr., 1981. 'Energy balance climate models'. *Rev. Geophys. Space Phys.*, **19**, 91-121.

J. Oerlemans and C. J. van der Veen, 1984. *Ice Sheets and Climate*. D. Reidel, Dordrecht/Boston/Lancaster, 217 pp.

W. S. B. Paterson, 1981. *The Physics of Glaciers*, 2nd ed. Pergamon Press, Oxford, 380 pp.

R. Peltier, 1982. 'Dynamics of the ice age earth'. *Adv. Geophys.*, **24**, 1-146.

P. Pestiaux and J. C. Duplessy, 1985. 'Paleoclimatic variability at frequencies ranging from 10^{-4} cycle per year to 10^{-3} cpy: evidence for nonlinear behavior of the climate system'. *Quatern. Res.*, submitted.

B. Saltzman, A. Sutera and A. Evenson, 1981. 'Structural stochastic stability of a simple auto-oscillatory climatic feedback system'. *J. Atmos. Sci.*, **38**, 494-503.

S. H. Schneider and R. E. Dickinson, 1974. 'Climate modeling'. *Rev. Geophys. Space Phys.*, 12, 447-493.

W. D. Sellers, 1969. 'A climate model based on the energy balance of the earth-atmosphere system'. *J. Appl. Meteorol.*, **8**, 392-400.

D. L. Turcotte and G. Schubert, 1982. *Geodynamics: Application of Continuum Physics to Geological Problems*. J. Wiley, New York, 450 pp.

G. Wolansky, 1985. *Dissipative Perturbations of Intergrable Hamiltonian Systems, with Applications to Celestial Mechanics and to Geophysical Fluid Dynamics*. Ph.D. Theis, New York University, 212 pp.

CLIMATIC PREDICTABILITY AND DYNAMICAL SYSTEMS

C. NICOLIS
Institut d'Aéronomie Spatiale de Belgique
Avenue Circulaire 3
1180 Bruxelles
Belgique.

ABSTRACT. The applications of the methods of dynamical systems to the analysis of climatic variability are discussed. Climatic change is viewed, successively, as the transition between multiple steady states, as the manifestation of sustained oscillations of the limit cycle type and as a non-periodic behavior associated to a chaotic attractor. A method to reconstruct the climatic attractor from time series data independent of any modelling is presented. The results suggest that long term climatic change is described by a chaotic attractor of low fractal dimensionality.

1. INTRODUCTION

In this chapter we attempt to clarify the relative role of internally generated and external mechanisms of climatic change. Much of our discussion will be concerned with long term problems and, in particular, with the quaternary glaciations. However, similar ideas and techniques could most certainly be applied to short and intermediate scale problems as well.

The classical tool of the paleoclimatologist is the analysis of variance spectra of various climatic indices, such as the oxygen isotope record of ice or deep sea core sediments[1,2] (Figs. 1 and 2). Although the importance of this tool should not be underestimated, we believe that it is essential to develop alternative methods. Indeed, besides allowing one to identify the characteristic time scales involved in a problem, spectral analysis gives only limited information on the nature of the underlying system, since it refers to gross, averaged properties. As a matter of fact a striking theorem proved recently[3] asserts that a power spectrum does not even allow one to differentiate between noisy or deterministic evolutions!

The approach we shall adopt is motivated by the concepts and techniques of the theory of nonlinear dynamical systems. At the basis of our analysis is the idea that climatic change reflects the ability of a certain dynamical system to undergo instabilities and transitions between different regimes. We will examine some typical

C. Nicolis and G. Nicolis (eds.), Irreversible Phenomena and Dynamical Systems Analysis in Geosciences, 321–354.

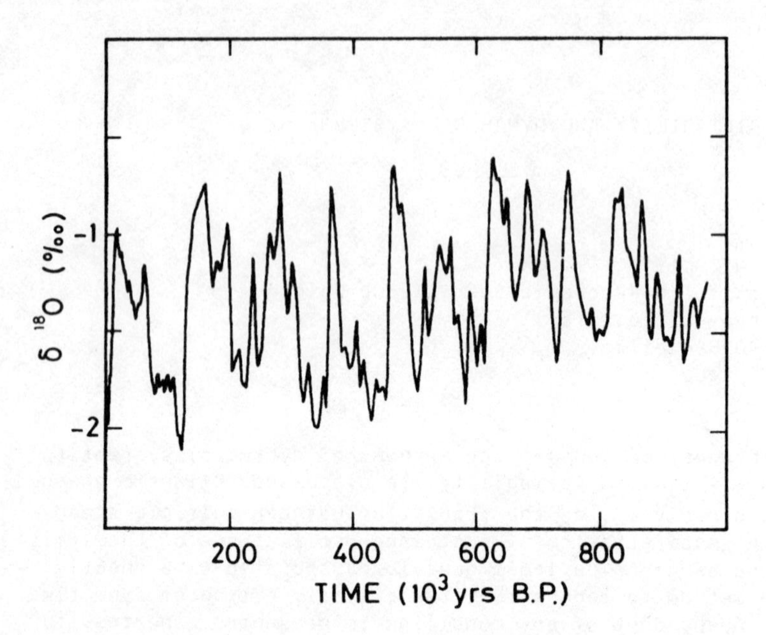

Fig. 1. Oxygen isotope record deduced from the V28238 deep sea core[1].

scenarios of such transitions. We will also assess the role of external periodic perturbations, arising from the earth's orbital variations, or of internally generated fluctuations, on these transitions. Next, we will propose a new method of analysis of the climatic record based on developments of dynamical systems theory, which will allow us to actually identify the dynamical regime represented by the data. This conclusion will finally be confronted with the predictions of mathematical models.

2. CLIMATIC VARIABILITY VIEWED AS TRANSITION BETWEEN MULTIPLE STEADY STATES

As shown in the chapter by G. Nicolis[4], the simplest attractor representing the long term behavior of a dynamical system is the point attractor. If only one such attractor is available the system will end up in a unique stationary state, and the problem of variability simply will not arise. We therefore inquire, in this section, on the possibilities arising from the coexistence of multiple point attractors in climate dynamics.

Energy balance models, a classs of climatic models treated more amply in the chapters by Ghil[5] and Saltzman[6], are known to generate quite naturally this behavior. To simplify matters as much

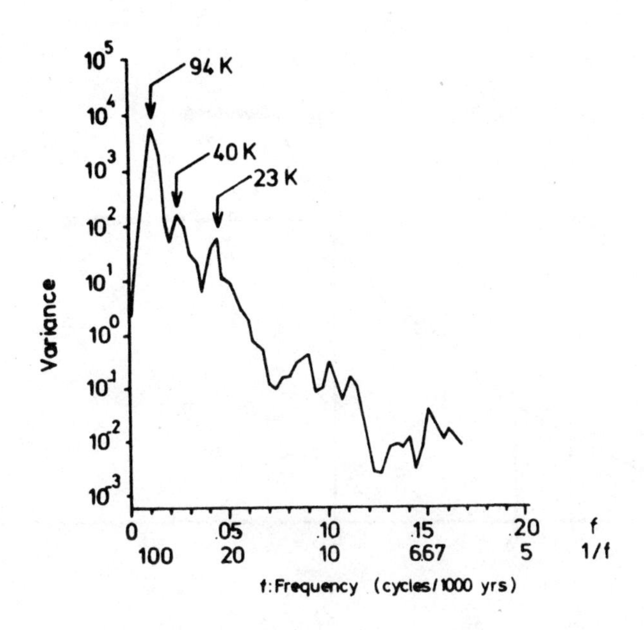

Fig. 2. Typical variance spectra deduced from the paleoclimatic record of the last million years.

as possible consider the case of globally averaged (also referred to as zero-dimensional 0-d) models :

$$\frac{dT}{dt} = \frac{1}{C}\left\{\begin{pmatrix}\text{incoming}\\ \text{solar energy}\end{pmatrix} - \begin{pmatrix}\text{outgoing}\\ \text{infrared energy}\end{pmatrix}\right\}$$

$$= \frac{1}{C}\,[Q\,(1 - \alpha(T)) - \varepsilon_B\,\sigma\,T^4] \qquad (1)$$

where T is the space averaged surface temperature, C the heat capacity, Q the solar constant, α the albedo, σ the Stefan constant and ε_B the emissivity factor representing the deviation from black body radiation. The surface-albedo feedback can be readily incorporated in this picture by modelling the albedo as a stepwise linear function[7]. The resulting energy balance is represented in Fig. 3.

For plausible parameter values it can give rise to two stable steady states, T_a and T_b, representing respectively a glacial and a more favorable climate, separated by an intermediate unstable one, T_o. If (as suggested by the record) the difference $T_a - T_b$ is small, the system could be further assumed to operate near a pitch-

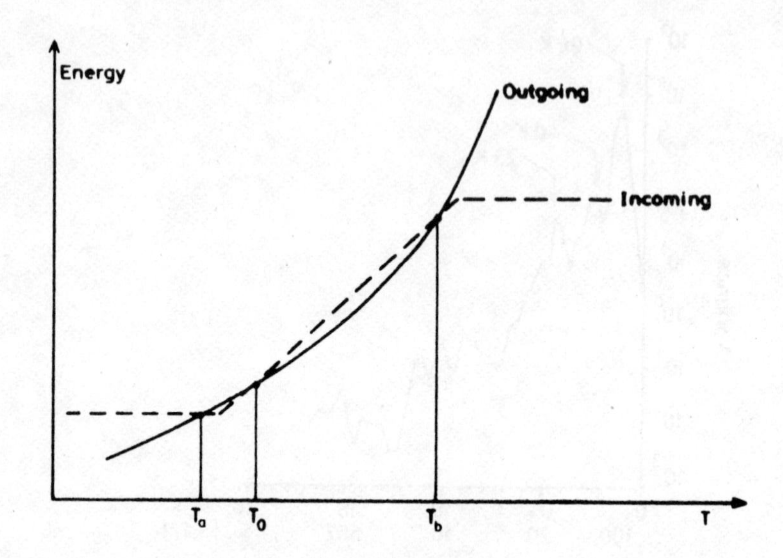

Fig. 3. Incoming and outgoing radiative energy curves as functions of T (global average temperature). Their intersections T_a, T_b and T_0 are the steady states predicted by eq. (1).

fork bifurcation point. As shown in the chapter by G. Nicolis such a system can be brought to the following universal, normal form[4] :

$$\frac{dx}{d\tau} = f(x, \lambda) = \lambda x - ux^3 \tag{2}$$

where x, τ ,u and λ are appropriately scaled combinations of initial variables, time and parameters.

Whether viewed in the form of (1) or (2), climatic change necessitates a transition between states a and b. Now, in the model elaborated so far no mechanism allowing for such a transition is present except for the trivial one, whereby the system initially at T_b (say) is perturbed and brought near T_a. Such massive perturbations are however hard to imagine. We therefore enlarge now our description by incorporating the effect of random fluctuations F(t) which, as explained in previous chapters, are always present in a complex physical system[4]. We model these fluctuations as a Gaussian white noise[8] :

$$\langle F(t) \rangle = 0$$

$$\langle F(t)\, F(t') \rangle = q\, \delta(t - t') \tag{3}$$

where q is the variance of fluctuations, and write the augmented energy balance equation as

$$\frac{dx}{d\tau} = f(x, \lambda) + g(x)\, F(\tau) \qquad (4)$$

where g(x) represents the coupling of the internal dynamics to $F(\tau)$.

This equation can be studied by the methods of the theory of stochastic processes. The main result is that the random forcing induces a diffusion-like motion between stable attractors which occurs on a time scale given by[9)]

$$\tau_{a,b} \sim \exp\left\{\frac{2}{q}\, \Delta U_{a,b}\right\} \qquad (5)$$

The quantity ΔU_a or ΔU_b is known as <u>potential barrier</u>. It is defined by

$$\Delta U_{a,b} = U_o - U_{a,b} \qquad (6a)$$

where U is the integral of the right hand side of eqs (1) or (2) :

$$U(x, \lambda) = - \int^{x} dx'\, f(x', \lambda) \qquad (6b)$$

By analogy to mechanics U can be referred to as <u>climatic potential</u>, since its derivative $\partial U/\partial x$ represents the "force", f responsible for the evolution. If as expected the variance of the fluctuations q is small and the barrier ΔU finite, τ_a or τ_b will be much larger than the local relaxation time and could well be in the range of glaciation time scales[9)]. In contrast, the local evolution in the vicinity of each attractor is given by the inverse of the first derivative of f, evaluated on a or b. For an energy balance model it should be, typically, of the order of the year. Still, eq. (4) cannot be considered as a satisfactory model of quaternary glaciations : the transitions between a and b occur at randomly distributed times, whereas the climatic record suggests that quaternary glaciations have a cyclic character bearing some correlation with the mean periodicities of the earth's orbital variations. Let us therefore study the response of our multistable model to both stochastic and periodic perturbations. Taking the simplest case of a sinusoidal forcing one is led to

$$\frac{dx}{d\tau} = f(x, \lambda) + g(x)\,[\varepsilon \sin\omega\tau + F(\tau)] \qquad (7)$$

where ε and ω are respectively the amplitude and frequency of the forcing. It should be pointed out that ε is very small, of the order of a fraction of percent.

The most striking result pertaining to eq. (7) is, undoubtedly, the possibility of <u>stochastic resonance</u>[10)] : when $\Delta U_a \simeq \Delta U_b$ and the forcing period, $2\pi/\omega$ is of the order of the characteristic passage times $\tau_{a,b}$ the response of the system is dramatically amplified. Specifically, the probability of crossing the barrier is substantially increased and the passage occurs with a periodicity equal to the periodicity of the forcing. This allows us to understand, at least qualitatively, how despite its weakness a periodic forcing may leave a lasting signature in the climatic record.

The interest of the description outlined in this section lies in its generality. Whatever the detailed features of a system might be, we know that near a simple bifurcation point (pitchfork or limit point), the dynamics will reduce to a universal form involving only a few parameters. It is therefore tempting to develop a modelling whereby these parameters are first tuned to reproduce some known properties (like for instance the past record), and subsequantly are used to make predictions about the future evolution.

3. CLIMATIC VARIABILITY AND SUSTAINED OSCILLATIONS

Our next step will be to account for the cyclic character of long term climatic changes like quaternary glaciations in a more straightforward manner. Indeed, according to the theory of dynamical systems, one-dimensional attractors in the form of limit cycles can account for periodic, sustained oscillations. Let us therefore explore the possibilities afforded by oscillatory climate models.

Large parts of the chapters by Ghil[5] and Saltzman[6] are devoted to the derivation of such models. Here, we shall first adopt a more general viewpoint, and use later on these models to illustrate the basic ideas.

Suppose that oscillatory behavior arises through a Hopf bifurcation, leading from a hitherto stable steady state to a self-oscillation of the limit cycle type [4]. We know that a dynamical sytem operating in the vicinity of such a bifurcation can be cast in a universal, normal form. Specifically, there exists a suitable linear (generally complex-valued) combination of the initial variables obeying to the equation[11]

$$\frac{dz}{dt} = (\lambda + i\omega_o) z - cz |z|^2 \tag{8}$$

Here t is a dimensionless time, λ the distance from the bifurcation point, ω_o the frequency of the linearized motion around the steady state, and $c = u + iv$ a combination of the other parameters occurring in the initial equations. In Appendix A a detailed illustration of the process of reduction to a normal form is provided, using Saltzman's model oscillator describing the interaction between sea-ice extent and mean ocean temperature[6].

Coming back to the normal form, eq. (8), we switch to radial and angular variables through

$$z = r e^{i\varphi} \tag{9}$$

Substituting into (8) we readily verify that the evolution can be separated into a radial part, which is independent of φ, and an angular part depending solely on r :

$$\frac{dr}{dt} = \lambda r - ur^3 \qquad (10a)$$

$$\frac{d\varphi}{dt} = \omega_o - vr^2 \qquad (10b)$$

The first of these equations allows us to evaluate analytically the radius of the limit cycle. Setting $dr/dt = 0$, which in the (r, φ) representation places us on the limit cycle, we obtain :

$$r_s = \left(\frac{\lambda}{u}\right)^{1/2} \qquad (11a)$$

This solution exists provided that λ/u is positive. To test its stability we study the evolution of a small perturbation $\delta\rho = r - r_s$. Linearizing eq. (10a) around r_s, by virtue of the theorem of linearized stability (see chapter by G. Nicolis[4]), one obtains :

$$\frac{d\delta\rho}{dt} = (\lambda - 3ur_s^2)\ \delta\rho$$
$$= -2\lambda\ \delta\rho \qquad (11b)$$

As long as $\lambda > 0$ (and thus automatically $u > 0$ also) this predicts an exponential relaxation to the limit cycle. In other words the variable r enjoys in this range asymptotic stability, in the sense that any perturbation that may act accidentally on r will be damped by the system.

The situation is entirely different for φ. Setting r equal to its value on the limit cycle, which is legitimate in the limit of long times (and provided of course that $\lambda > 0$, $u > 0$), one can integrate eq. (10b) straightforwardly to get

$$\varphi = \varphi_o + (\omega_o - \frac{\lambda v}{u})\ t \qquad (11c)$$

In other words φ increases continuously in the interval $(0, 2\pi)$ from the initial value φ_o. If φ_o is perturbed, this monotonic change will start all over again from the new value, and there will be no tendency to reestablish the initial phase φ_o[12].

In order to realize more fully the consequences of this property let us consider the following thought experiment (Fig. 4) Suppose that the system runs on its limit cycle $r = r_s$. At some moment, corresponding to a value $\varphi = \varphi_1$ of the phase, we displace the system to a new state characterized by the values r_o, φ_o of the variables r and φ. According to eq. (11b) the variable r will relax from r_0 back to the value r_s, as the representative point in phase space will spiral toward the limit cycle (cf. Fig. 4). On the other hand, according to eq. (11c) the phase variable φ will keep forever the memory of the initial value φ_0. In other words, when the limit cycle will be reached again, the phase will generally be different from the one that would characterize an unperturbed system following its limit

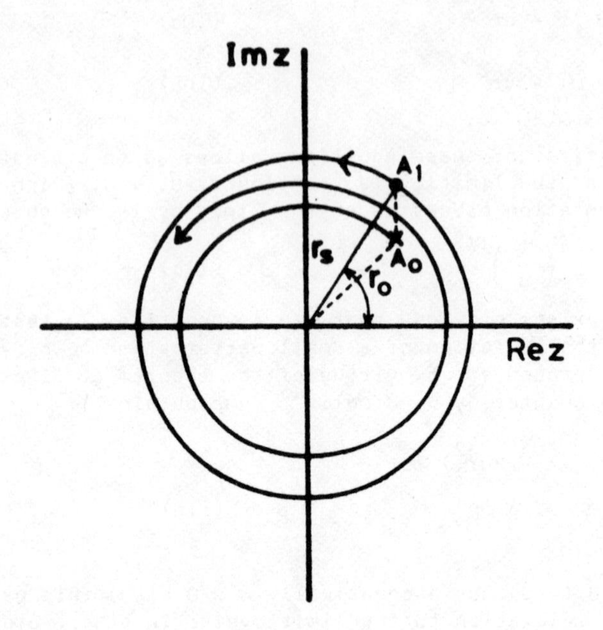

Fig. 4. Schematic representation of the evolution following the action of a perturbation leading from state A_1 on the limit cycle to state A_0, in the space of the variables of the normal form.

cycle during the same time interval. In as much as the state into which the system can be thrown by a perturbation is unpredictable, it therefore follows that the reset phase of the oscillator will also be unpredictable. In other words, our non-linear oscillator is bound to behave sooner or later in an erratic way under the action of perturbations. This is tantamount to poor predictability.

The above surprising property can be further substantiated by a stochastic analysis. As stressed repeatedly in this Institute complex physical systems possess a universal mechanism of perturbations generated spontaneously by the dynamics, namely the fluctuations[12]. Basically, fluctuations are random events. By modelling them as Gaussian white noises one is led to an augmented equation (8) in which the normal form variable z becomes itself a random process. The explicit form of this equation for Saltzman's model oscillator and its asymptotic solution are given in Appendix B. Here we summarize the most representative results.

i) one shows that the separation of the radial and phase variables is reflected, at the stochastic level, by the factorization of the probability distribution. In the limit of long times one obtains a stationary distribution,

$$P(r, \varphi) = P(\varphi/r)\, P(r) = \frac{1}{2\pi} P(r) \qquad (12)$$

ii) While the distribution of the phase variable is flat, the radial distribution $P(r)$ has the form

$$P(r) \sim \exp\left\{ - \frac{2}{Q}\; U(r) \right\} \qquad (13a)$$

where Q is an effective variance of fluctuations and $U(r)$ is the potential associated with the right hand side of eq. (10a) :

$$U(r) = -\left(\lambda \frac{r^2}{2} - u \frac{r^4}{4}\right) \qquad (13b)$$

It follows that φ is "chaotic", in the sense that the dispersion around its average will be of the same order as the average value itself. On the other hand, the dispersion of the radial variable around its most probable value is small, as long as the variance Q is small. Nevertheless, the mere fact that the probability of r is stationary rather than time-periodic implies that a remmant of the chaotic behavior of φ subsists in the statistics of r : if an average over a large number of samples (or over a sufficient time interval in a single realization of the stochastic process) is taken, the periodicity predicted by the deterministic analysis will be wiped out as a result of destructive phase interference[12].

In short, an autonomous oscillator cannot leave a marked signature on the long term climatic record : We have to look elsewhere to find an explanation of climatic changes believed to present a cyclic character, like the quaternary glaciations.

We shall now outline the analysis of a climatic oscillator forced by a weak periodic perturbation simulating, for instance, the earth's orbital variations. We shall show that under certain conditions, a stabilization of the phase of the oscillator can take place. This will remove unpredictability and guarantee the subsistence of periodic behavior over arbitrarily long times.

We shall illustrate this possibility by assuming the forcing to be additive. This is indeed the case for Saltzman's model as shown in Appendix C where further explicit results on the effect of forcing on this particular oscillator are outlined. The normal form, eq. (8) now reads[13] :

$$\frac{dz}{dt} = (\lambda - i\omega_o)\, z - cz|z|^2 + \varepsilon s \cos \omega_e t \qquad (14)$$

Here $2\pi/\omega_e$ is the external periodicity, ε an effective forcing amplitude assumed to be small, and s a complex-valued coefficient describing the coupling of the forcing with the normal form variable.

It will prove convenient to work with the equations for the real and imaginary parts of our variable z. Setting

$$z = x + iy \quad , \quad s = s_x + is_y$$

we obtain from eq. (14) :

$$\frac{dx}{dt} = \lambda x - \omega_o y - (ux - vy)(x^2 + y^2) + \varepsilon s_x \cos \omega_e t$$
$$\frac{dy}{dt} = \omega_o x + \lambda y - (vx + uy)(x^2 + y^2) + \varepsilon s_y \cos \omega_e t \quad (15)$$

A rather elaborate argument, which is not reproduced here, shows that phase stabilization - the phenomenon we are looking for- cannot be expected unless the system is in the immediate vicinity of bifurcation ($\lambda = 0$) and resonance ($\omega_o = k\omega_e$, k integer). Now, in this range standard perturbation techniques fail (see also reference[17]). The following singular perturbation scheme seems to satisfy all requirements. First, we express the vicinity of bifurcation and resonance through

$$\lambda = \bar{\lambda}\,\varepsilon^{2/3}, \quad \omega_e - \omega_o = \bar{\omega}\,\varepsilon^{2/3} \quad (16a)$$

Next, we introduce a fast time scale adjusted to the external forcing ($T = \omega_e t$) and a slow one ($\tau = \varepsilon^{2/3} t$) accounting for resonance :

$$\frac{d}{dt} = \omega_e \frac{d}{dT} + \varepsilon^{2/3} \frac{d}{d\tau} \quad (16b)$$

Finally, we expand x and y in perturbation series as follows :

$$x = \varepsilon^{1/3} x_1 + \varepsilon^{2/3} x_2 + \varepsilon x_3 + \dots$$
$$y = \varepsilon^{1/3} y_1 + \varepsilon^{2/3} y_2 + \varepsilon y_3 + \dots \quad (16c)$$

Substituting into eqs (15) we obtain, to order $\varepsilon^{1/3}$, a homogeneous system of equations :

$$\frac{dx_1}{dT} = -y_1, \quad \frac{dy_1}{dT} = x_1 \quad (17)$$

whose solution is a harmonic oscillation :

$$x_1 = A(\tau)\cos T + B(\tau)\sin T$$
$$y_1 = A(\tau)\sin T - B(\tau)\cos T \quad (18)$$

The integration coefficients A, B remain undetermined at this stage, They are expected to depend on the slow time scale τ which has not entered in eqs (17). To fix this dependence we consider the perturbation equations to the subsequent orders. As seen more explicitly in Appendix C these equations are now inhomogeneous. In order that they admit non-singular solutions we must make sure that certain solvability conditions are satisfied which, roughly speaking, guarantee that there is no risk of dividing a finite expression by zero[18]. For the problem under consideration there are two such solvability conditions which provide us with the following set of equations for the coefficients A and B :

$$\frac{dA}{d\tau} = \bar{\lambda}A - \bar{\omega}B - (uA + vB)\,(A^2 + B^2) + \frac{s_x}{2}$$

$$\frac{dB}{d\tau} = \bar{\omega}A + \bar{\lambda}B + (vA - uB)\,(A^2 + B^2) - \frac{s_y}{2} \qquad (19)$$

These equations can admit up to three real steady-state solutions which, by eqs (18), correspond to time-periodic solutions of the original system oscillating in resonance with the forcing. More importantly, since A and B are given once the parameters are specified, the phase of this oscillation relative to the forcing is well-defined[13)]. We have thus succeeded in reestablishing predictability thanks to this phase-locking phenomenon. Naturally, in order that phase-locking be physically relevant we must make sure that it corresponds to a stable solution of eqs (19). This is indeed so in a certain range of parameter values. On the other hand, beyond this range eqs (19) may give rise to a Hopf bifurcation for A and B. According to eqs (18) this corresponds to quasi-periodic solutions of the original system. The phase of the system relative to the forcing will now be a complicated function of time, and this will result in a rather loose predictability.

As in the preceding section, the interest of the description based on the normal form lies in its generality. For instance, ω_o, ω_e and s in eq. (14) can be tuned to the values suggested by the record and the astronomical theory, while parameters λ and c are related to the amplitude of the oscillation. If information on this latter quantity is available, eq. (14) can then be used to study different scenarios concerning, for instance, the effect of disturbances of various kinds on the evolution.

4. CLIMATIC VARIABILITY AND NON-PERIODIC BEHAVIOR

Let us contemplate once again the climatic record, Figs 1 and 2. We see a broad band structure, in which an appreciable amount of randomness is superimposed on a limited number of preferred peaks. So far the broad band aspect was discarded in our discussion. We now raise the question, whether it could not be accounted for by dynamical regimes more complex than the phase locked one analyzed in the previous section. As a byproduct such regimes could still lead to distinct peaks in some preferred frequencies, but this would by no means imply that the behavior is periodic in time. The question we have just raised will lead us to examine a class of oscillatory models in which the coupling with the external forcing introduces a completely aperiodic behavior[19,20)].

The normal form of a periodically forced oscillator in the vicinity of a Hopf bifurcation (eqs (14) - (15)) cannot account for this type of solution[12,13)]. On the other hand in the theory of dynamical systems one shows that such a behavior can arise near parameter values for which the system admits very special orbits known as homoclinic orbits, a flavor of which is given in the chapter by G.Nicolis[4]. Fig. 5 depicts a typical phase space portrait of a two-variable dynamical system involving a pair of homoclinic orbits. We observe two stable fixed points (x_a, 0) and (x_b, 0) around which the system

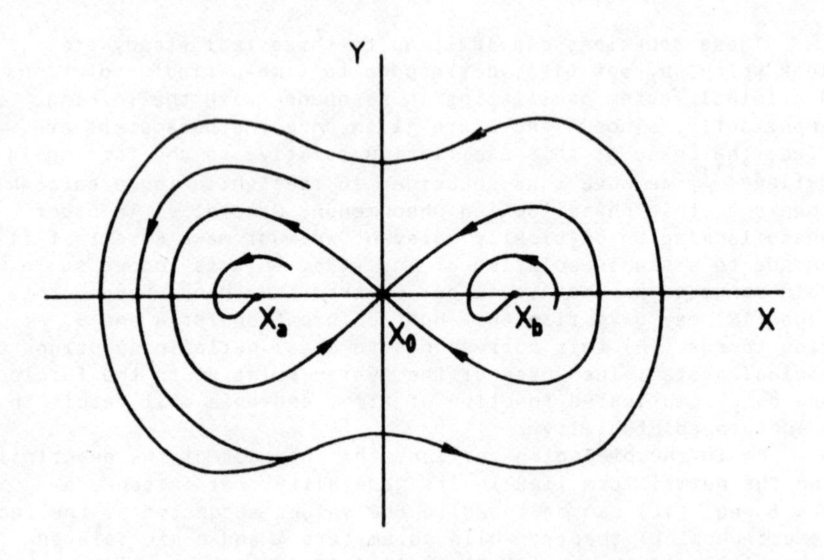

Fig. 5. Phase space portrait of a two variable system involving a pair of homoclinic orbits.

performs damped oscillations. An intermediate unstable point (x_0 = 0, 0) gives rise to a pair of unstable and a pair of stable trajectories known as separatrices, which eventually merge to form the double loop structure. These are precisely our homoclinic trajectories, which can also be viewed as infinite period orbits. Further away in phase space the system admits a large amplitude stable periodic solution of finite period.

As an illustration of the phase space portrait of Fig. 5 consider Saltzman's oscillator (see chapter by Saltzman[6]) and Appendices A to C). One has, in dimensionless variables,

$$\frac{d\eta}{dt} = \theta - \eta$$

$$\frac{d\theta}{dt} = b\theta - a\eta - \eta^2\theta \qquad (20)$$

where η and θ are, respectively, the (suitably scaled) deviations of the sine of latitude of sea ice extent and of the mean ocean temperature from a reference state. The system given by eqs. (20) may admit up to three steady states ($\theta_s = \eta_s = 0$ and $\theta'_s = \eta'_s = \pm(b-a)^{1/2}$).

The linearized equations (20) around the trivial state ($\eta = \eta_s + \delta n$, $\theta = \theta_s + \delta\theta$) are simply

$$\frac{d\delta n}{dt} = \delta\theta - \delta\eta$$

$$\frac{d\delta\theta}{dt} = b\delta\theta - a\delta\eta$$

They admit solutions which depend on time exponentially, with an exponent given by the characteristic equation (see chapter by G. Nicolis[4])

$$\omega^2 - \omega(b - 1) + a - b = 0 \tag{21}$$

When $a > b$ and $b = 1$ this equation admits a pair of purely imaginary solutions. For $b > 1$ and $a > b$ the real part becomes positive[12,13]. This is the range of Hopf bifurcation, which was discussed extensively in Section 3. On the other hand for $a < b$ the characteristic equation admits two real solutions of opposite sign. The reference state ($\theta_s = \eta_s = 0$) behaves then as a saddle, just like ($x_0 = 0$, 0) in Fig. 5. Both kinds of regimes merge for $a = b$, $b = 1$ for which values both roots of the characteristic equation vanish simultaneously.

We now set

$$\theta - \eta = \xi \tag{22}$$

Substituting into eqs. (20) we obtain

$$\frac{d\eta}{dt} = \xi$$

$$\frac{d\xi}{dt} = (b - 1)\,\xi + (b - a)\eta - \eta^3 - \eta^2\xi \tag{23}$$

Near the degenerate situation in which both characteristic roots vanish simultaneously one has

$$b - 1 = \varepsilon_1 \ll 1$$

$$b - a = \varepsilon_2 \ll 1 \tag{24}$$

It can be verified that for $\varepsilon_1 = 0.8\ \varepsilon_2$ eqs (23) generate the phase portrait of Fig. 5. As a matter of fact eqs (23) are no less than the normal form of any dynamical system operating near the degenerate situation in which both characteristic roots vanish simultaneously[19]. The validity of our conclusions extends therefore far beyond the specific model of eqs (20).

We now study the coupling between the above defined system and a weak periodic forcing. To simplify as much as possible we consider an additive periodic forcing acting on the second equation (23) alone :

$$\frac{d\eta}{dt} = \xi$$

$$\frac{d\xi}{dt} = \varepsilon_1\xi + \varepsilon_2\eta - \eta^3 - \eta^2\xi + s \sin \omega t \qquad (25)$$

where s and ω are, respectively, the amplitude and the frequency of the forcing. We perform the following scaling of variables and parameters :

$$\eta = x\mu \quad , \quad \xi = y\mu^2 \quad , \quad s = p\mu^4$$

$$\varepsilon_1 = \gamma_1\mu^2 \quad \varepsilon_2 = \gamma_2\mu^2 \quad , \quad \mu \ll 1$$

$$t = \tau\mu^{-1} \quad \omega = \Omega\mu \qquad (26)$$

Eqs (25) become

$$\frac{dx}{d\tau} = y$$

$$\frac{dy}{d\tau} = \gamma_2 x - x^3 + \mu(\gamma_1 y - x^2 y + p \sin \Omega\tau) \qquad (27)$$

These equations can be viewed as the perturbations (for $\mu \ll 1$) of a reference system described by

$$\frac{dx_o}{d\tau} = y_o$$

$$\frac{dy_o}{d\tau} = \gamma_2 x_o - x_o^3 \qquad (28)$$

Remarkably, this is a Hamiltonian system known as Duffing's oscillator[19]. For $\gamma_2 > 0$ its phase portrait is depicted in Fig. 6. We now have a continuum of periodic trajectories as well as a pair of homoclinic orbits existing for all positive values of the parameter γ_2. Going from eqs (28) to eqs (27) amounts therefore to inquiring how this phase space structure and, in particular, the infinite period homoclinic orbits are perturbed by both the "dissipative" terms $\gamma_1 y - x^2 y$ and by the periodic forcing.

Let us first formulate this problem analytically. Setting $x = x_o + \mu u$, $y = y_o + \mu v$, we obtain the following equations for the perturbations u, v :

$$\frac{du}{d\tau} = v$$

$$\frac{dv}{d\tau} = (\gamma_2 - 3x_0^2)u + \gamma_1 y_0 - x_0^2 y_0 + p \sin \Omega\tau \qquad (29)$$

We now are in a situation similar to that arising in the perturbative

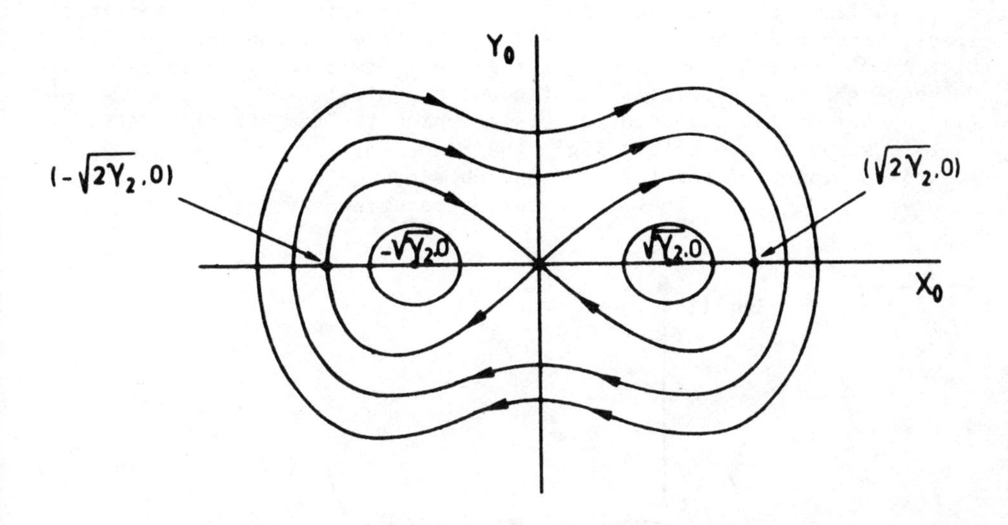

Fig. 6. Phase space portrait of Duffing's oscillator for $\gamma_2 > 0$.

analysis of eqs (15) : we have an inhomogeneous system of equations, which admits a non-singular solution provided that a solvability condition is satisfied. The argument is somewhat more elaborate than in Section 3 however, and will not be reproduced. We merely give the final result, known as Melnikov integral[21] :

$$\int_{-\infty}^{\infty} d\tau y_0 \, [\gamma_1 \, y_0 - x_0^2 \, y_0 + p \sin \Omega\tau] = 0 \qquad (30)$$

This relation allows us to identify a critical forcing amplitude, p_c beyond which the homoclinic orbit is destroyed by the periodic perturbation. One obtains, after a lengthy calculation[22] :

$$p_c = \frac{1}{3} \gamma_2^{3/2} \left(\gamma_1 - \frac{4}{5} \gamma_2\right) \frac{\mathrm{Ch} \dfrac{\Omega\pi}{(2\gamma_2)^{1/2}} + 1}{\mathrm{Ch} \dfrac{\Omega\pi}{2(2\gamma_2)^{1/2}}} \qquad (31)$$

In the theory of dynamical systems one shows that for $p > p_c$ a variety of complex non-periodic behaviors may arise[20]. In order to identify the nature of these regimes we turn to numerical simulations.

As stressed by other authors in this book, the most unambiguous way to characterize the type of regime displayed by a dynamical system is to perform a Poincaré surface of section. Remember that we are dealing with a forced system involving the two variables x and y. Effectively, such a dynamics takes place in a three-dimensional space, since one can always express the forcing through $b\sin\chi_e$, $d\chi_e/dt = \Omega$, thereby introducing its phase χ_e as a third variable. One can now map the original continuous dynamical system into a discrete time system by following the points at which the trajectories cross (with a slope of prescribed sign) the plane $\cos_e = C$, corresponding to a given value of the forcing. One obtains, in this way a Poincaré map (Fig. 7) that is to say, a recurrence relation

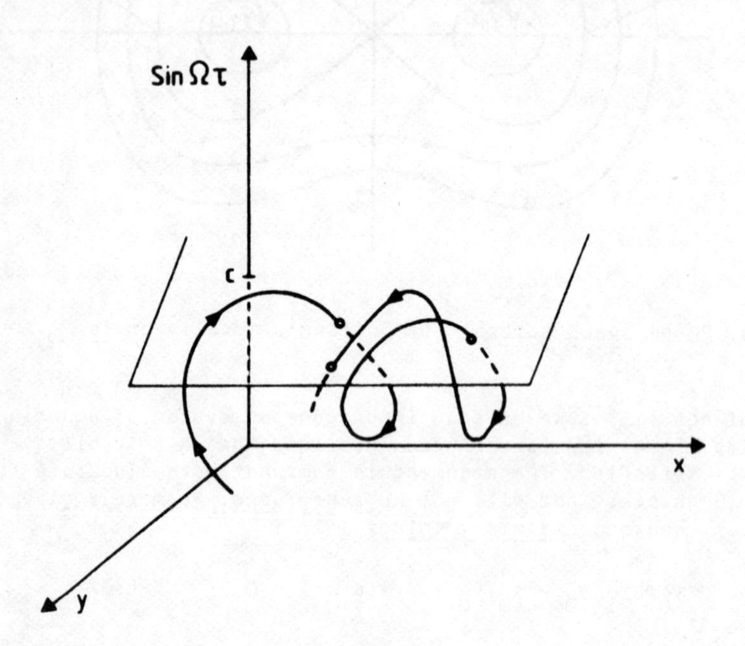

Fig. 7. Schematic representation of a Poincaré map for a forced system. As the system evolves, a representative trajectory cuts a plane of section C at discrete times τ_n. The study of the dynamics on the surface of the section gives valuable information of the qualitative behavior of the initial system.

$$x_{n+1} = f(x_n, y_n)$$

$$y_{n+1} = g(y_n, y_n) \qquad (32)$$

where n labels the successive intersections.

Suppose now that the trajectories of the continuous time flow tend, as $\tau \to \infty$, to an asymptotic regime. In the three-dimensional state space, this regime wil be characterized by an invariant object, the attractor. The signature of this object on the surface of the section will obviously be an attractor of the discrete dynamical system, eq. (32). Conversely, from the existence of an attractor on the surface of the section, we can infer the properties of the underlying continuous time flow.

Figure 8 depicts the Poincaré surface of section for values of p near the threshold p_c of eq. (31). We observe the coexitence of two attractors: A periodic one, with a period three times as large as the forcing period; and a quasi-periodic one, represented in the full phase space by a two-dimensional toroidal surface.

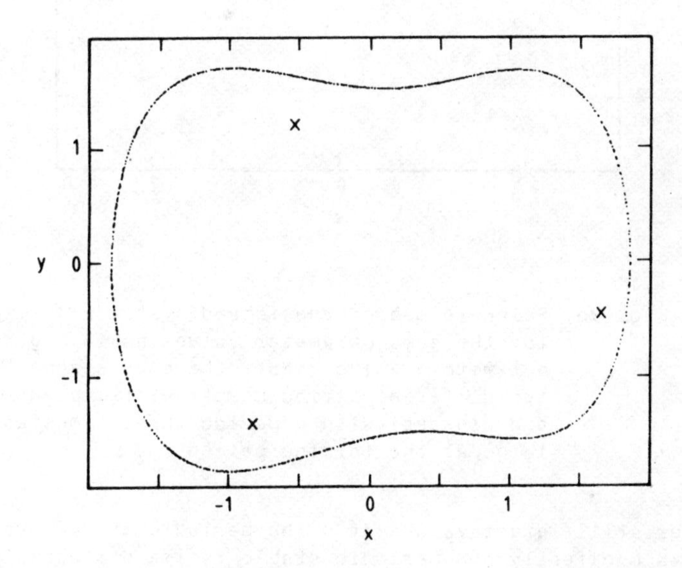

Fig. 8. Poincaré map of the forced system eqs. (27) for $\gamma_2 = 1$, $\gamma_1 = 1.01$, $\mu = 0.1$, $\Omega = 3$ and $q = q_c + 0.2$. Depending on the initial conditions the system evolves either to a quasi-periodic attractor (closed curve) or to a periodic one, with a period three times as large as the forcing period (points in crosses).

By increasing p further from p_c one observes the coexistence of two period three (Fig. 9a) and two period two attractors (Fig. 9b). As a matter of fact the attractors of a given period, differ by their phase relative to the phase of the forcing.

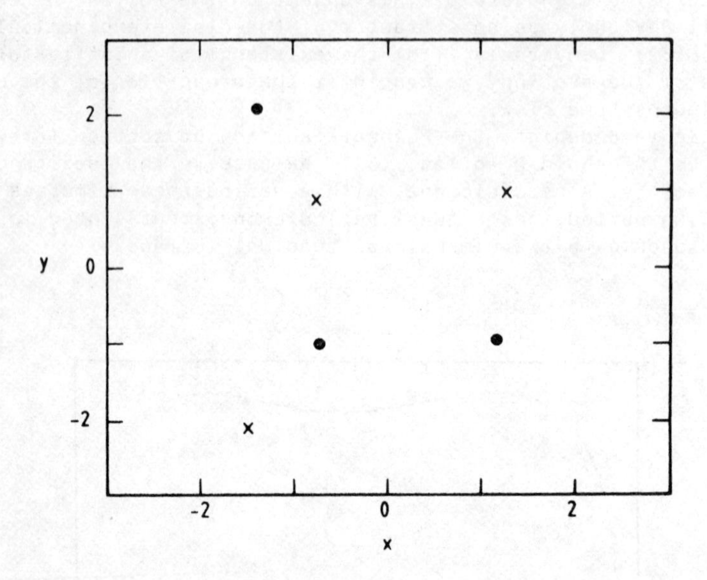

Fig. 9a. Poincaré map of the forced system eqs. (27) for the same parameter values as in Fig. 8 but with $q = 12q_c$. Note the coexistence of two different periodic attractors (points and crosses) with a period three times as large as the forcing period.

For still larger values of p the periodic attractors disappear, and a manifestly non-periodic stable regime dominates (Fig. 10a). We conjecture that we are in the presence of a chaotic attractor. This seems to be corroborated by the time dependence of the variables (Fig. 10b), as well as by the power spectrum, which exhibits an important broad band component along with a well-defined peak at the forcing frequency (Fig. 10c). More complex dynamical regimes which can be qualified as intermittent are also observed. For instance, after spending some time on a seemingly chaotic regime the system jumps on a period-three regime.

In short, we have identified an additional mechanism of climatic variability. The final question to which we now turn our attention is, which of the various regimes described in Sections 2 to 4 is best suited to interpret the climatic record?

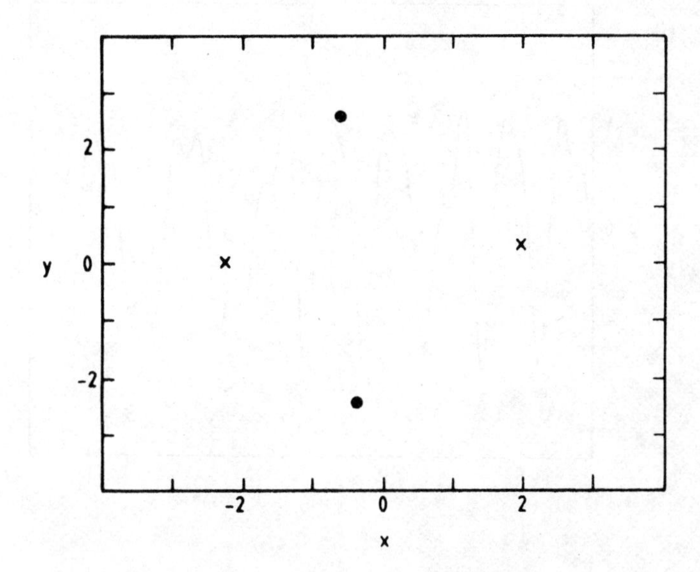

Fig. 9b. Same paremeter values with Fig. 9a. Note the coexistence of two additional periodic attractors (points and crosses) with a period two times as large as the forcing period.

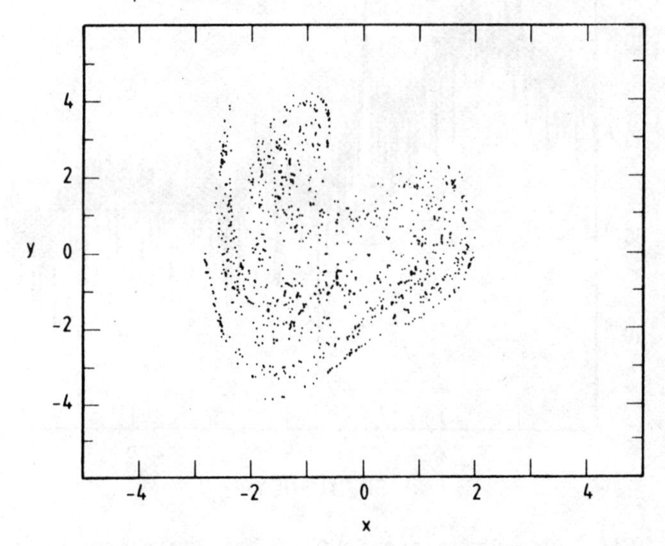

Fig. 10a. Non-periodic attractor of system (27) for the parameter values of Fig. 8 except $q = 20q_c$.

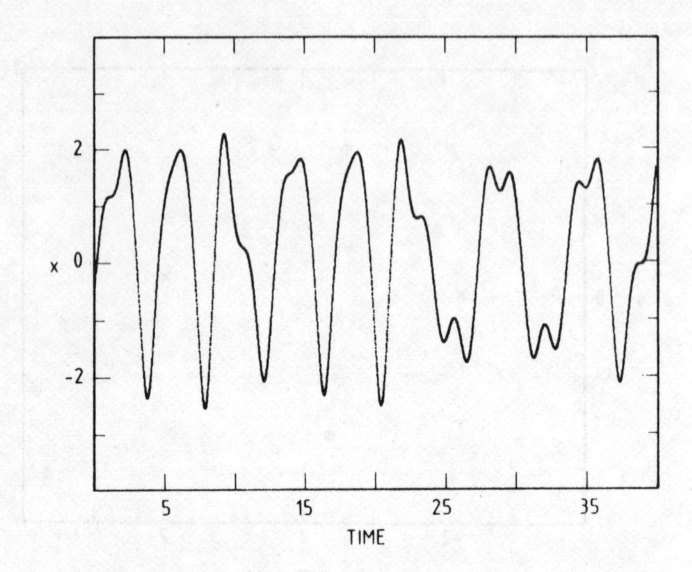

Fig. 10b. Time evolution of variable x for the parameter values used in Fig. 10a.

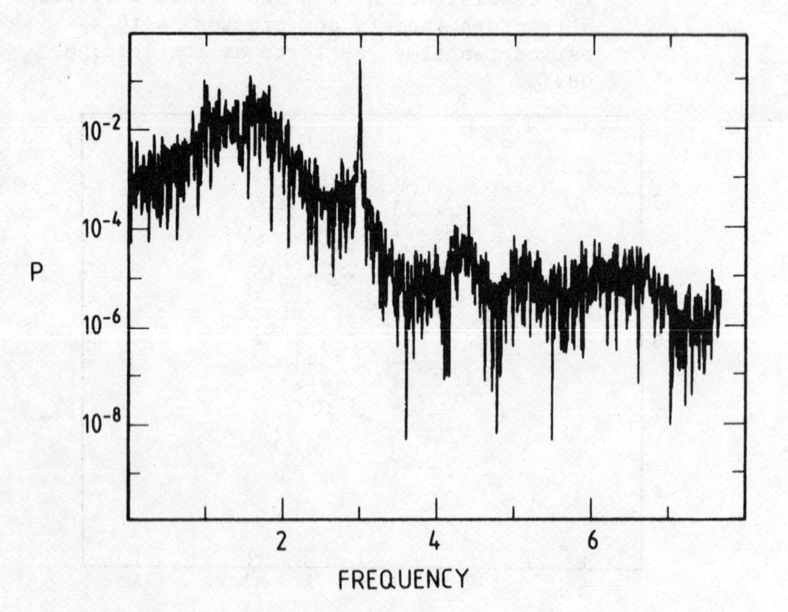

Fig. 10c. Power spectrum of the variable x whose time evolution is given in Fig. 10b.

5. RECONSTRUCTION OF THE CLIMATIC ATTRACTOR FROM TIME SERIES DATA

As pointed out in the Introduction much of our information on the climatic variability of the last million years is based on the oxygen isotope record[1] (see Fig. 1). We will now show that it is possible to extract from such a record information going far beyond the traditional spectral analysis. In particular, we will be able to characterize the nature of the dynamical regime involved as well as to determine the minimum number of variables needed for its description[23]. We will illustrate this new method of data analysis on a particular deep-sea core known as V28-238.

Let $X_o(t)$ be the time series available from the data, and $X_k(t)$, where $k = 0, 1, \ldots n-1$, the full set of variables actually taking part in the dynamics. X_k is expected to satisfy a set of first-order nonlinear equations, whose form is generally unknown but which, given a set of initial data $X_k(0)$, will produce the full details of the system's evolution.

By successive differentiations in time one can reduce this set to a single (generally highly nonlinear) differential equation of nth order in time for one of these variables. Thus instead of $X_k(t)$, $k = 0, 1, \ldots n-1$, we may consider $X_o(t)$, and its n-1 successive derivatives $X_o^{(k)}(t)$, $k = 1, \ldots, n-1$, to be the n variables of the problem spanning the phase space of the system. Now, both $X_o(t)$ and its derivatives can be deduced from a single time series (the one for $X_o(t)$ as provided by the data). We see therefore that, as anticipated at the beginning of this section, we have in principle sufficient information in our disposal to go beyond the "one-dimensional" space of the original time series and to take into account the multi-dimensional character of the system's dynamics.

Actually, instead of $X_o(t)$ and $X^{(k)}(t)$ it will be easier to work with $X_o(t)$ and the set of variables obtained from it by shifting its values by a fixed lag τ[24]. It suffices for this to choose τ in such a way that one keeps n linearly independent variables :

$$\begin{array}{llll} X_o(t_1) & , X_o(t_2), & \ldots\ldots\ldots\ldots\ldots\ldots & X_o(t_N) \\ X_o(t_1+\tau) & , X_o(t_2+\tau), & \ldots\ldots\ldots\ldots\ldots & X_o(t_N+\tau) \\ X_o(t_1+(n-1)\tau), & X_o(t_2+(n-1)\tau) & \ldots\ldots\ldots\ldots & X_o(t_N+(n-1)\tau) \end{array} \quad (33)$$

We will be interested in the structure of the trajectories of the dynamical system in the above defined <u>phase space</u>. The theory of dynamical systems shows that the structure of these trajectories is conditioned by two basic elements :

(i) The dimensionality of phase space, in other words, the number n of variables present.

(ii) The nature of the <u>attractors</u>, that is, of the asymptotically stable states attained in the course of time. The latter depends, in turn, on the dimensionality of the attractor.

The principal question to which we will try to answer is thus the following : Is it possible to obtain a lower bound on the number n of variables which capture the essential features of the long term evolution of the climatic system? And, concomitantly, is it possible to define a climatic attractor which represents this evolution, and to estimate its qualitative properties like its dimensionality?

The answer to these questions is in the affirmative[23]. It is based on a method recently developed by Grassberger and Procaccia[25]. Before describing the procedure that will enable us to obtain quantitative information on the nature of the climatic attractor and, in particular, on its dimensionality we present, in Fig. 11, a characteristic view of this object in a three-dimensional phase space. This view clearly exhibits the complexity of the underlying motion. However, the data are too coarse to draw any conclusion from this figure alone. It will be our task to characterize the complexity of the dynamics more sharply and, in particular, to assess its similarities and differences with random noise.

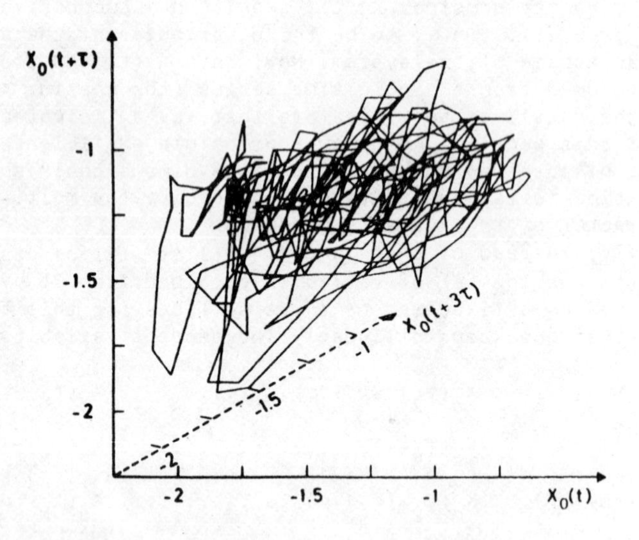

Fig. 11. Three-dimensional phase portrait generated by the time series of Fig. 1 ($\tau = \Delta t$, Δt = 2 Kyr).

Consider the set of N points obtained from our time series embedded in a phase space of n dimensions. For convenience we introduce a vector notation : $\vec{X}_i$ which stands for a point of phase space whose coordinates are $X_o(t_i), \ldots, X_o(t_i+(n-1)\tau)$.

A reference point $\vec{X}_i$ from these data is now chosen and all its distances $|\vec{X}_i - \vec{X}_j|$ from the N - 1 remaining points are computed. This allows us to count the data points that are within a prescribed distance, r from point X_i. Repeating the process for all values of i, one arrives at the quantity

$$C(r) = \frac{1}{N^2} \sum_{\substack{i,j=1 \\ i \neq j}}^{N} \theta(r - |\vec{X}_i - \vec{X}_j|) \tag{34}$$

where θ is the Heaviside function, $\theta(x) = 0$ if $x < 0$, $\theta(x) = 1$ if $x > 0$. The non-vanishing of this quantity measures the extent to which the presence of a data point X_i, affects the position of the other points. C(r) may thus be referred to as the (integral) correlation function of the attractor.

Suppose that we fix a given small parameter ε and we use it to define the site of a lattice which approximates the attractor. If the latter is a line, the number of data points within a distance r from a prescribed point should be proportional to r/ε. If it is a surface, this number should be proportional to $(r/\varepsilon)^2$ and, more generally, if it is a d-dimensional manifold it should be proportional to $(r/\varepsilon)^d$. We expect, therefore, that for r, relatively small C(r) should vary as

$$C(r) = r^d \tag{35}$$

In other words, the dimensionality d of the attractor is given by the slope of the log C(r) versus log r in a certain range of values of r :

$$\log C(r) = d\,|\log r| \tag{36}$$

This property remains valid for attractors of fractal dimensionality.

The above results suggest the following algorithm :

(1) Starting from a time series, we can construct the correlation function, equation (34) by considering successively higher values of the dimensionality n of phase space.

(2) Deduce the slope d near the origin according to equation (36) and see how the result changes as n is increased.

(3) If d reaches a saturation limit beyond some relatively small n, the system represented by the time series should possess an attractor. The saturation value d_s will be regarded as the dimensionality of the attractor. The value of n beyond which saturation is observed will provide the minimum number of variables necessary to model the behavior represented by the attractor.

This procedure has been applied to the analysis of the data pertaining to core V28-238. Figure 12 gives the dependence of log C(r) versus log r for n = 2 to n = 6. We see that there is indeed an extended region over which this dependence is linear, in accordance with equation (36). Figure 13 (points in circles) shows that the slope

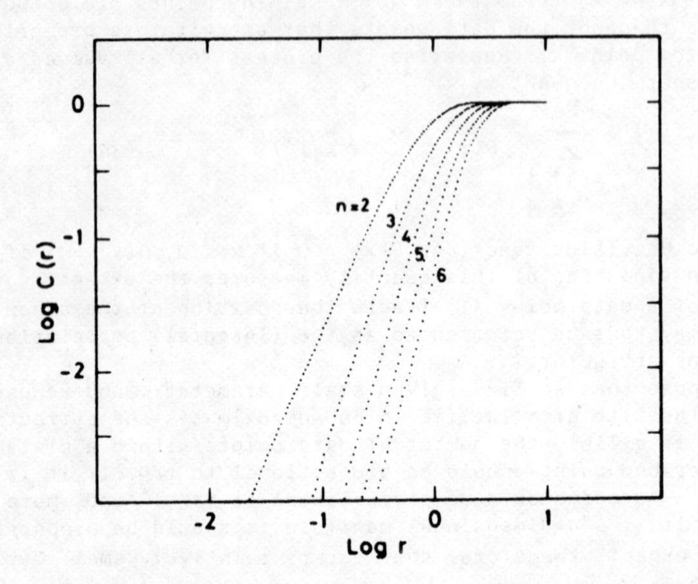

Fig. 12. Distance dependence of the correlation function on the climatic attractor. Parameter values as in Fig. 11, except that $\tau = 4\,\Delta t$.

reaches a saturation value at n = 4, which is about $d_s = 3.1$. The same plot also shows the way in which d varies with n if the signal considered is a Gaussian white noise : there is no tendency to saturate. In fact, in this case d turns out to be equal to n. It should be emphasized that the above results are independent of the choice of the time lag τ, provided that the latter is of the order of magnitude of the time scales pertaining to the long-term climatic evolution, and the linear independence of the variables is secured.

The existence of a climatic attractor of low dimensionality shows that the main features of long-term climatic evolution may be viewed as the manifestation of a deterministic dynamics, involving a limited number of key variables. The fact that the attractor has a fractal dimensionality provides a natural explanation of the intrinsic variability of the climatic system, despite its deterministic character[26]. Moreover, it suggests that despite the pronounced peaks of spectra in the frequencies of the orbital forcings, the actual behavior is highly non periodic[5].

This last aspect was further substantiated by the estimation of the largest positive Lyapounov exponent[4]. We know that there exist as many Lyapounov exponents as phase space dimensions. In this particular case this number is four. One of them is necessary equal to zero expressing the fact that the relative distance of initially close

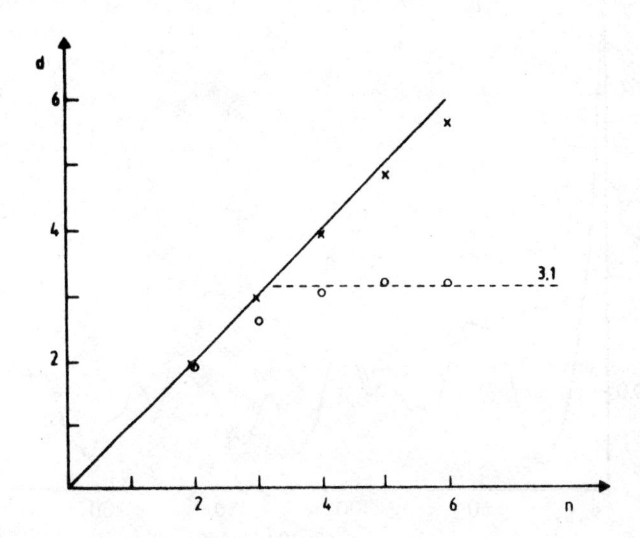

Fig. 13. Dependence of dimensionality, d on the number of phase space variables n, for the climatic attractor (O) and for a white noise signal (x) for the same parameter values as in Fig. 12.

states on a given trajectory varies slower than exponentially. Others are negative, expressing the exponential approach to the attractor. Finally, if the dynamics is chaotic, there will be at least one positive Lyapounov exponent witnessing the exponential divergence of nearby initial conditions on the stable attractor. Clearly, such a quantity is the ideal tool for describing the limits of predictability in climatic change. Note that in a well-behaved dissipative system the sum of all Lyapounov exponents must be strictly negative[4].

Recent algorithms[27] allow for the calculation of the largest positive Lyapounov exponent of a dynamical system from time series data. We have applied this procedure to the data set of Fig. 1 and found indeed a positive number σ, between 2.5×10^{-5} and 4×10^{-5} yr^{-1}. Its inverse, σ^{-1}, between 25 and 40 kyr gives the limits of predictability of the long term behavior of the system[28].

A further characterization of the nature of the underlying dynamical system is provided by the time correlation function of our data set. Figure 14 depicts this function. We observe a decay for small to intermediate times, a negative region and subsequently irregular oscillations around zero. This leads us to several important

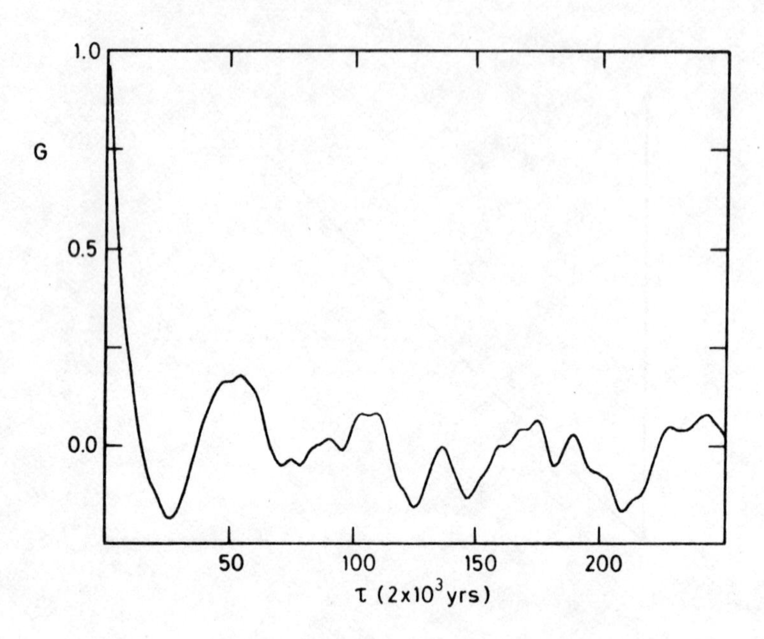

Fig. 14. Time correlation function of the variable X_o on the climatic attractor.

conclusions. First, the fact that the correlation function is not merely periodic or quasi-periodic, proves that climate dynamics cannot be viewed as a passive response to the orbital forcings. Second, the fact that the correlation function decays for short to intermediate times proves that we are dealing with a process displaying unpredictability. Barring white noise, the simplest example of such a process is a Markovian red noise generated, for instance, by an Ornstein-Uhlenbeck process. If however this was the case here, the correlation function would never go to zero and would decay strictly exponentially. Neither of these two properties holds for Fig. 14. We conclude that we deal with a process possessing <u>memory</u>[28]. Moreover, the limits of predictability provided by the inverse of the largest positive Lyapounov exponent correspond, roughly, to a time in the interval between the vanishing and the first minimum of the correlation function. We believe therefore that we have produced strong evidence that long term climatic variability is to be viewed as a chaotic dynamics possessing a low dimensional attractor characterized by an unstable dynamics of limited predictability.

6. CONCLUDING REMARKS

The principal idea which we tried to convey in the present chapter is that dynamical systems theory provides us with new ways to look at the long standing problem of climatic variability. For instance, modelling based on normal forms is a pragmatic tool for making rapid progress in problems involving a multitude of variables, which would remain otherwise intractable or would require extensive empirical parameterizations. Even more significant, perhaps, is the possibility to arrive at quantitative predictions of such quantities as attractor dimensions or predictability limits from time series data, independent of any modelling.

In the light of these possibilities problems related to intermediate and short range variability constitute a promising area of future investigations.

APPENDIX A

Consider Saltzman's model oscillator describing the interaction between sea-ice extent and mean ocean temperature[14])

$$\frac{d\bar{\eta}}{dt} = -\varphi_2\bar{\eta} + \varphi_1\bar{\theta}$$
$$\frac{d\bar{\theta}}{dt} = -\psi_1\bar{\eta} + \psi_2\bar{\theta} - \psi_3\bar{\eta}^2\bar{\theta} \qquad \text{(A1)}$$

Here φ_i, ψ_i are positive parameters, $\bar{\eta}$ is the deviation of the sine of the latitude of the sea-ice extent from the steady state and $\bar{\theta}$ is the excess mean ocean surface temperature. By performing the scaling transformation

$$\bar{\eta} = \left(\frac{\varphi_2}{\psi_3}\right)^{1/2}\eta \ , \ \bar{\theta} = \frac{\varphi_2^{3/2}}{\psi_3^{1/2}\ \varphi_1}\theta \ , \ \bar{t} = \frac{1}{\varphi_2}t \qquad \text{(A2)}$$

we can cast eqs. (A1) in a form displaying two dimensionless parameters :

$$\frac{d\eta}{dt} = -\eta + \theta$$

$$\frac{d\theta}{dt} = -a\eta + b\theta - \eta^2\theta \qquad \text{(A3)}$$

with

$$a = \frac{\psi_1\ \phi_1}{\varphi_2^2} \ , \quad b = \frac{\psi_2}{\phi_2}$$

The procedure transforming eqs. (A3) to the normal form can now be outlined. We first compute the characteristic roots of the linearized stability problem. Straightforward algebra gives :

$$\lambda \pm i\omega_o = \frac{1}{2}\left\{b - 1 \pm \left[(b-1)^2 - 4(a-b)\right]^{1/2}\right\} \qquad \text{(A4)}$$

At $\lambda = 0$ the system undergoes a bifurcation beyond which a limit cycle is expected to emerge. In general the full problem of exactly solving eqs (A3) in this range is intractable. We therefore limit ourselves to the case where $\lambda << 1$. To this end we cast the dynamics in a form corresponding to a full diagonalization of the linear part of eqs (A3). This is achieved by a linear tranformation T, which in the present case is a 2 x 2 matrix whose columns are the two right eigenvectors of the matrix coefficients of the linear part of (A3). This tranformation matrix T turns out to be

$$T = \begin{pmatrix} 1 & 1 \\ \lambda + i\omega_o + 1 & \lambda - i\omega_o + 1 \end{pmatrix} \quad \text{(A5)}$$

Operating on both sides of (A3) with the inverse matrix T^{-1} and introducing the new (complex) variables

$$\begin{pmatrix} z \\ z^* \end{pmatrix} = T^{-1} \begin{pmatrix} \eta \\ \theta \end{pmatrix} \quad \text{(A6)}$$

or more explicitly

$$\eta = 2\, R_e z$$

$$\theta = 2\,(R_e z - \omega_o\, \mathrm{Im} z) \quad \text{(A7)}$$

we obtain to the dominant order in λ, the following equation for the transformed variable z :

$$\frac{dz}{dt} = (\lambda + i\omega_o)\, z + \frac{i}{2\omega_o}\, (3 + i\omega_o)\, z|z|^2 \quad \text{(A8)}$$

giving rise to a radial and angular part of the form of eqs (10a) and (10b) respectively with $u = 1/2$ and $v = -3/(2\omega_o)$.

APPENDIX B

Let us incorporate the effect of fluctuations by adding random forces F_η, F_θ to the deterministic Saltzman oscillator[14), eqs (A3) :

$$\frac{d\eta}{dt} = -\eta + \theta + F\eta$$

$$\frac{d\theta}{dt} = -a\eta + b\theta - \eta^2\theta + F_\theta \quad \text{(B1)}$$

Repeating the procedure given in Appendix A, transforming eqs (B1) to the normal form we obtain :

$$\frac{dz}{dt} = (\lambda + i\omega_o)\, z + \frac{i}{2\omega_o} \left\{(3 + i\omega_o)\, z|z|^2 + (1 - i\omega_o)\, F_\eta - F_\theta\right\} \quad \text{(B2)}$$

Setting $z = x + iy$, substituting into eq. (B2) and separating real and imaginary parts one gets :

$$\frac{dx}{dt} = \lambda x - \omega_o y - \frac{1}{2}(x^2 + y^2)(x + \frac{3}{\omega_o} y) + F_x$$

$$\frac{dy}{dt} = \omega_o x + \lambda y - \frac{1}{2}(x^2 + y^2)(y - \frac{3}{\omega_o} x) + F_y \quad \text{(B3)}$$

where we have set $F_x = F_\eta/2$ and $F_y = (F_\eta - F_\theta)/2\omega_o$. We assume F_η, F_θ to define a multi-Gaussian white noise[8]:

$$< F_\eta(t)\, F_\eta(t')> = q_\eta \quad (t - t')$$

$$< F_\theta(t)\, F_\theta(t')> = q_\theta \quad (t - t')$$

$$< F_\eta(t)\, F_\theta(t')> = q_{\eta\theta} \quad (t - t') \quad \text{(B4)}$$

This allows us to write a Fokker-Planck equation for the probability distribution of the climatic variables

$$\frac{\partial P}{\partial t} = - \frac{\partial}{\partial x}\left\{\lambda x - \omega_o y - \frac{1}{2}(x^2 + y^2)(x + \frac{3}{\omega_o} y)\right\} - \frac{\partial}{\partial y}\left\{\omega_o x + \lambda y - \frac{1}{2}(x^2 + y^2)(y - \frac{3}{\omega_o} x)\right\} + \frac{q_x}{2}\frac{\partial^2 P}{\partial x^2} + \frac{q_y}{2}\frac{\partial^2 P}{\partial y^2} + q_{xy}\frac{\partial^2 P}{\partial x \partial y} \quad \text{(B5)}$$

where q_x, q_y, q_{xy} are the variances of F_x and F_y.

In general eq. (B5) is intractable. However, the situation is greatly simplified if one limits the analysis to the range in which the normal form (eqs (10)) is valid.

To see this it is convenient to introduce polar coordinates $x = r\cos\varphi$, $y = r\sin\varphi$. Note that the latter transformation being nonlinear, one should take into account the subtleties of Ito-Stratonivitch calculus[15]. One obtains after a lengthy calculation[16]:

$$\frac{\partial P(r, \varphi, t)}{\partial t} = - \frac{\partial}{\partial r}\left\{\lambda r - \frac{1}{2} r^3 + \frac{1}{2r} Q_{\varphi\varphi}\right\} P - \frac{\partial}{\partial \varphi}\left\{\omega_o + \frac{3}{2\omega_o} r^2 - \frac{1}{r^2} Q_{r\varphi}\right\} P + \frac{1}{2}\left| \frac{\partial^2}{\partial r^2} Q_{rr} + 2\frac{\partial^2}{\partial r \partial \varphi}\frac{Q_{r\varphi}}{r} + \frac{\partial^2}{\partial \varphi^2}\frac{Q_{\varphi\varphi}}{r^2}\right| P \quad \text{(B6)}$$

where

$$Q_{\varphi\varphi} = q_x \sin^2\varphi - 2q_{xy}\sin\varphi\cos\varphi + q_y\cos^2\varphi,$$

$$Q_{r\varphi} = - q_x \sin\varphi\cos\varphi + q_{xy}(\cos^2\varphi - \sin^2\varphi) + q_y \sin\varphi\cos\varphi,$$

$$Q_{rr} = q_x \cos^2\varphi + 2q_{xy} \sin\varphi \cos\varphi + q_y \sin^2\varphi, \qquad (B7)$$

and q_x, q_y and q_{xy} are suitable linear combinations of q_η, q_θ, $q_{\eta\theta}$.

To go further it is necessary to introduce a perturbation parameter in the problem. It is reasonable to choose it to be related to the weakness of the noise terms. Mathematically, we express this through the following scaling :

$$Q_{\varphi\varphi} = \varepsilon\bar{Q}_{\varphi\varphi},$$
$$Q_{r\varphi} = \varepsilon\bar{Q}_{r\varphi},$$
$$Q_{rr} = \varepsilon\bar{Q}_{rr}, \qquad (B8)$$

We next scale both the bifurcation parameter λ and the deviation of the radius r from its value on the limit cycle, $r = r_s$, by suitable powers of ε. We do this in order to be in accordance with the conditions of validity of the normal form (eqs. (10)). However, we should keep in mind that the qualitative predictions should still describe the general trend beyond the vicinity of the bifurcation point. Note that no scaling can be applied to the angular variable φ, as the latter increases in the interval $(0, 2\pi)$ and does not enjoy any stability property. Summarizing we write[16] :

$$\lambda = \bar{\lambda}\,\varepsilon^{2m}$$
$$r = r_s + \rho\varepsilon^k = (2\bar{\lambda})^{1/2}\,\varepsilon^m + \rho\varepsilon^k \qquad (B9)$$
$$\varphi = \varphi$$

The (non-negative) exponents k and m are chosen in such a way that both the drift and diffusion terms contribute to the evolution of P in eq. (B6). Indeed, should the diffusion terms be negligible, the probability P would be a delta function around the deterministic motion, and as a result the effect of fluctuations would be wiped out. If on the other hand the drift terms were negligible, P would exhibit a purely random motion similar to that of a Brownian particle in a fluid, and would tend to zero everywhere as $t \to \infty$. The best way to estimate the magnitude of these two terms is to introduce the conditional probability $P(\varphi/\rho, t)$ through

$$P(\rho, \varphi, t) = P(\varphi/\rho, t)\, P(\rho, t), \qquad (B10)$$

where

$$P(\rho, t) = \frac{1}{2\pi}\int_0^{2\pi} d\varphi\, P(\rho, \varphi, t). \qquad (B11)$$

Integrating eq. (B6) over φ and taking the scaling (B8) and (B9) into account, one can see that $\partial P(\rho,t)/\partial t$ is of order ε. $P(\rho,t)$ is therefore a slow variable : this is the probabilistic analogue of the fact that the variable r varies at the slowest of all time scales present in the problem. Using this property and dividing through eq. (B6) with $P(\rho,t)$ we obtain a closed equation for $P(\varphi/\rho,t)$ which to dominant order in ε reads :

$$\frac{\partial P(\varphi/\rho,t)}{\partial t} = -\frac{\partial}{\partial \varphi}\omega_o P(\varphi/\rho,t) \tag{B12}$$

This equation admits a properly normalized stationary solution,

$$P(\varphi/\rho,t) = 1/(2\pi). \tag{B13}$$

Introducing this into eq. (B10) and integrating eq. (B6) over φ, we can obtain a closed equation for the slow variable $P(\rho,t)$. From this equation one can see that the drift and diffusion terms are of the same order of magnitude if and only if

$$k = m = \frac{1}{4} \tag{B14}$$

The equation for $P(\rho,t)$ then reads :

$$\frac{\partial P(\rho,t)}{\partial t} = -\varepsilon^{1/2}\frac{\partial}{\partial \rho}\left[-2\bar{\lambda}\rho - 3\left(\frac{\bar{\lambda}}{2}\right)^{1/2}\rho^2 - \frac{1}{2}\rho^3 + \frac{\bar{Q}}{2(2\bar{\lambda}\varepsilon)^{1/2} + \rho}\right] P + \varepsilon^{1/2}\bar{Q}\frac{\partial^2}{\partial \rho^2} P \tag{B15}$$

where

$$\bar{Q} = \frac{1}{2}(q_x + q_y) . \tag{B16}$$

This equation admits a steady-state solution. Switching back to the original variables and parameters r, λ and Q through the inverse scaling (eqs. (B8) and (B9)), we can see that this steady-state is of the form

$$P_s(r) \sim r \exp\left\{\frac{2}{Q}\left(\frac{\lambda r^2}{2} - \frac{r^4}{8}\right)\right\} \tag{B17}$$

APPENDIX C

Consider Saltzman's oscillator forced by a weak periodic forcing[14) :

$$\frac{d\eta}{dt} = -\eta + \theta + \varepsilon s_\eta \cos \omega_e t$$

$$\frac{d\theta}{dt} = -a\eta + b\theta - \eta^2\theta + \varepsilon s_\theta \cos \omega_e t \tag{C1}$$

with $\varepsilon \ll 1$.
εs_i, ω_e are respectively the dimensionless amplitudes of the forcing and its frequency. By applying the same procedure as in Appendix A the normal form of eqs (C1) is:

$$\frac{dz}{dt} = (\lambda + i\omega_o) z + \frac{i}{2\omega_o} \left\{(3 + i\omega_o) z|z|^2 + \varepsilon \left[(1 - i\omega_o) s_\eta - s_\theta\right] \cos \omega_e t\right\} \tag{C2}$$

Setting $z = x + iy$ we obtain :

$$\frac{dx}{dt} = \lambda x - \omega_o y - \frac{1}{2} (x^2 + y^2) (x + \frac{3}{\omega_o} y) + \varepsilon s_x \cos \omega_e t$$

$$\frac{dy}{dt} = \omega_o x + \lambda y - \frac{1}{2} (x^2 + y^2) (y - \frac{3}{\omega_o} x) + \varepsilon s_y \cos \omega_e t \tag{C3}$$

where we have set

$$s_x = \frac{s_\eta}{2} \quad , \quad s_y = \frac{1}{2\omega_o} (s_\eta - s_\theta) \tag{C4}$$

We are interested in solving eqs (C4) in the case in which the system operates near the bifurcation point ($\lambda \ll 1$) and near resonance ($\omega_o \sim \omega_e$). Following the singular perturbation scheme, eqs (16), one gets the solutions for the first order in the expansion x_1 and y_1 given by eqs (18). The latter contain two unknown quantities A and B which may depend on the slow time scale τ.

The next order $\varepsilon^{2/3}$, leads to equations identical to (17) and therefore adds nothing new. To the order ε, on the other hand, we obtain :

$$\omega_e \left(\frac{\partial x_3}{\partial T} + y_3\right) = - \frac{\partial x_1}{\partial \tau} + \bar{\lambda} x_1 + \bar{\omega} y_1 - \frac{1}{2} (x_1^2 + y_1^2) (x_1 + \frac{3}{\omega_o} y_1) + s_x \cos T$$

$$\omega_e \left(\frac{\partial y_3}{\partial T} - x_3\right) = - \frac{\partial y_1}{\partial \tau} + \bar{\lambda} y_1 - \bar{\omega} x_1 - \frac{1}{2} (x_1^2 + y_1^2) \left(y_1 - \frac{3}{\omega_o} x_1\right) + s_y \cos T \tag{C5}$$

This is an inhomogeneous system of equations for x_3, y_3. It admits a solution only if a solvability condition expressing the absence of terms growing unboundedly in time, is satisfied. Such terms may arise by the following mechanism. To obtain (x_3, y_3), we have to "divide" the right-hand side of eqs. (C5) by the differential matrix operator

$$\omega_e \begin{pmatrix} \frac{\partial}{\partial T} & 1 \\ -1 & \frac{\partial}{\partial T} \end{pmatrix} \tag{C6}$$

but this is precisely the operator appearing in eqs. (17). According to this latter equation and eqs. (18), it possesses a non-trivial null space, that is, non-trivial eigenvectors corresponding to a zero eigenvalue. By dividing with such an operator, one may therefore introduce singularities, if the right-hand sides of eqs. (C5) contain contributions lying in this nul space. The solvability condition[18] allows one to rule out this possibility, by requiring that the right-hand sides of eqs. (C5) viewed as a vector, be orthogonal to the eigenvectors of the operator (C6). The latter are (cf. eq. (18)) :

$$(\cos T, \sin T) \quad \text{and} \quad (\sin T, -\cos T). \tag{C7}$$

The scalar product to be used is the conventional scalar product of vector analysis, supplemented by an averaging over T. The point is that we will dispose of two solvability conditions which will provide us with two equations for the coefficients $A(\tau)$, $B(\tau)$ appearing in the first order of the perturbative development (eqs. (18)). After a lengthy calculation one finds eqs. (19).

ACKNOWLEDGMENT

This work is supported, in part, by the European Economic Community under contract n° ST2J-0079-1-B (EDB).

REFERENCES

1. Shackleton, N.J. and Opdyke, N.D, Quat. Res. 3, 39-55 (1973).
2. Berger, A.L. and Pestiaux, P. Tech. Rep. N° 28, (Institute of Astronomy and Geophysics, Catholic University of Louvain, 1982).
3. Brock, W. and Chamberlain, G., SSRI W.P. n° 8419, Department of Economics, University of WI, Madison, WI (1984).
4. Nicolis, G. This volume.
5. Ghil, M. This volume.
6. Saltzman, B. This volume.
7. Crafoord, C. and Källén, E., J. Atmos. Sci. 35, 1123-1125 (1978).
8. Wax, N., Selected topics in noise and stochastic processes, New York, Dover (1954).
9. Nicolis, C. and Nicolis, G., Tellus, 33, 225-234 (1981).
10. Nicolis, C., Tellus, 34, 1-9 (1982).
    Benzi, R., Parisi, G., Sutera, A. and Vulpiani, A., Tellus, 34, 10-16 (1982).
11. Arnold, V., Chapitres supplémentaires de la théorie des équations différentielles ordinaires, Mir, Moscow (1980).
12. Nicolis, C., Tellus, 36A, 1-10 (1984).

13. Nicolis, C., Tellus, 36A, 217-227 (1984).
14. Saltzman, B., Sutera, A. and Hansen, A.R., J. Atmos. Sci., 39, 2634-2637 (1982).
15. Arnold, L., Stochastic differential equations : Theory and applications, Wiley, New York (1974).
16. See also Baras, F., Malek-Mansour, M. and Van Den Broeck, C., J. Stat. Phys. 28, 577-587 (1982).
17. Rosenblat, S. and Cohen, D.S., Stad. Appl. Math., 64, 143-175(1981).
18. Sattinger, D., Topics in stability and bifurcation theory, Springer-Verlag, Berlin (1973).
19. Guckenheimer, J. and Holmes, P., Nonlinear oscillations, dynamical systems and bifurcations of vector fields, Springer-Verlag, Berlin (1983).
20. Baesens, C. and Nicolis, G., Z. Phys. B., Condensed Matter, 52, 345-354 (1983).
21. Melnikov, V.K., Trans. Moscow Math. Soc. 12(1), 1-57 (1963).
22. Nicolis, C., Tellus, in press.
23. Nicolis, C. and Nicolis, G., Nature, 311, 529-532 (1984).
24. Ruelle, D. in Nonlinear phenomena in chemical dynamics, (eds Pacault A. and Vidal, C.), Springer, Berlin (1981).
25. See for instance, Grassberger, P. and Procaccia, I., Phys. Rev. Lett. 50, 346-349 (1983).
26. Lorenz, E.N., Tellus, 36A, 98-110 (1984).
Lorenz, E.N. This volume.
27. Wolf, A., Swift, J.B., Swinney, H.L. and Vastano, J.A., Physica 16D, 185 (1985).
28. Nicolis, C. and Nicolis, G., Proc. Nat. Acad. Sci., in press.

MODELING THE $\delta^{18}O$-DERIVED RECORD OF QUATERNARY CLIMATIC CHANGE WITH LOW ORDER DYNAMICAL SYSTEMS

Barry Saltzman

Department of Geology and Geophysics
Yale University
New Haven, CT 06511

ABSTRACT. The observed long-term changes in planetary ice mass appear to have occurred at rates generally too slow to be calculable deductively from first principles based on the fundamental hydrothermodynamical partial differential equations. A more inductive (or inverse) approach to a theory for these changes therefore seems necessary, perhaps best formulated in terms of low-order, multicomponent, nonlinear, stochastic-dynamical systems of equations governing the variables and feedbacks thought to be relevant from qualitative physical reasoning (e.g., "conceptual models"). A brief review of previous attempts to formulate such dynamical system models of the $\delta^{18}O$ record of the ice ages is given, and a recently developed prototype model based on three prognostic components is described in greater detail. These components are identified here with three variables influencing the $\delta^{18}O$ record of climatic change: snow-derived ice mass, mean surface temperature of the ocean (including pack ice fields), and deeper ocean (or thermocline) temperature. The effects of a representation of the external forcing resulting from variations in the earth orbital parameters on the near-100 k yr free behavior of the system are included in an attempt to account for the observed $\delta^{18}O$ record. It is shown that the major features of both the $\delta^{18}O$ and the associated surface water temperature variations including the 'spiky' behavior, can be accounted for, and side predictions are made concerning the concomitant evolution of the deep ocean temperature. The solution is shown to involve some rapid, high-energy transitions that may be amenable to a more quantitative treatment by a general circulation type model.

1. Introduction

Due to the generally small rates of change of ice mass and ocean temperature involved in the major Quaternary glacial variations, it is not feasible to account for these variations quantitatively from first principles based on deductions from the equations of geophysical fluid dynamics (Saltzman 1983). Instead, it seems promising to make a start

C. Nicolis and G. Nicolis (eds.), Irreversible Phenomena and Dynamical Systems Analysis in Geosciences, 355–380.

on the problem by forming simple dynamical systems that can account for the global aspects of the problem (both in the geographical and mathematical senses), particularly with regard to the sedimentary core $\delta^{18}O$ record that constitutes the primary continuous climate signal over the Quaternary. This $\delta^{18}O$ record is believed to be influenced mainly by global ice mass changes, and also by local changes in the ocean temperature where the foraminiferal material later deposited in the sediments is formed. In the above reference, and more recently in Saltzman (1985), we give general discussions of (1) the global constraints on energy and ice mass variations (emphasizing the non-equilibrium nature of the problem and the need for a qualitative, "inductive" approach) and (2) the formulation of climate as a "forced stochastic-dynamical system" based on a response-time distinction between the prognostic and diagnostic climatic variables. To a large degree, the problem of formulating a theory of the long term global ice-age variations as revealed by the $\delta^{18}O$ record reduces to one of discovering the relevant prognostic (i.e., dynamical) system governing the slow response-time variables - that is, to discover what we might call the slow climatic attractor.

In the next section we present a generalized model governing the variables determining the $\delta^{18}O$ record, followed by a brief review of some of the simpler one and two-variable models formulated in this search for the slow attractor. Next, a more detailed discussion is given of a three-variable model involving three slow variables that seems capable of accounting for a good deal of the variance of the observed $\delta^{18}O$ record over the last 500 k yrs.

2. A generalized, multi-variable, dynamical model for the variables determining the $\delta^{18}O$ record

In Fig. 1 we show a typical variation of $\delta^{18}O$ with time and the associated variations of ocean surface temperature T_s over the last 500 k yrs derived from two cores in the South Indian Ocean by Hays et al. (1976). The dominant 100 k yr oscillations over this period are

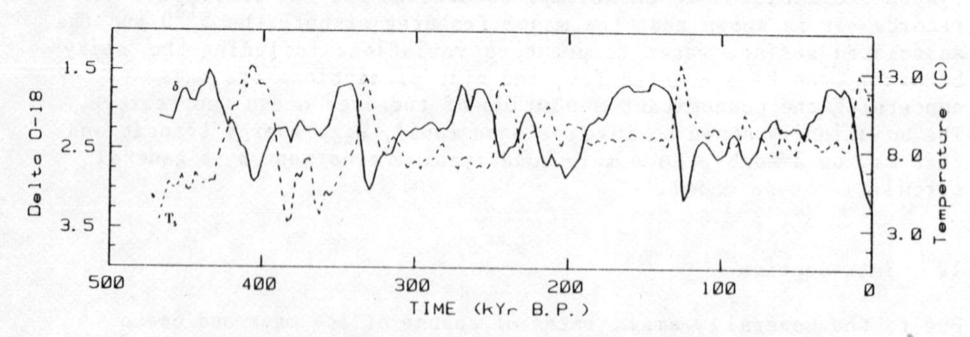

Fig. 1 Variation of $\delta^{18}O$ in units of ‰ (solid curve) and surface water temperature in °C (dashed curve) over the past 500 k yr, as determined from two South Indian Ocean cores RC11-120 and E49-18, from Hays, Imbrie and Shackleton (1976).

clearly exhibited, as well as the tendency for a relatively rapid collapse of $\delta^{18}O$ from its maximum value accompanied by a warm surface water "spike". As we have said, $\delta^{18}O$ is believed to be mainly a function of global ice mass I, and to some extent also a function of the temperature of the ocean T where the fossil planktonic fomanifera used to determine $\delta^{18}O$ variations were formed (Ruddiman, 1985). As in the above case, this ocean temperature is usually that of the upper mixed layer, which we denote by T_s, but can also be for a measure of deeper water temperature T_d if <u>benthic</u> foraminifera are used. Thus,

$$\delta(\lambda,\phi) = f\,[I, T_s(\lambda,\phi)] \qquad (1)$$

or

$$\delta(\lambda,\phi) = f\,[I, T_d(\lambda,\phi)] \qquad (2)$$

where for convenience, henceforth, we abbreviate $\delta^{18}O$ by δ in our mathematical developments ($\delta \equiv \delta^{18}O$), and λ and ϕ denote geographic longitude and latitude, respectively. We could further decompose the total ice mass I into a part ζ representing the mass of snow and ice sheets, and a part χ representing the snow-derived marine ice masses comprised mainly of ice shelves, icebergs, and sea-ice. That is, $I \approx \zeta + \chi$, where ζ and χ are the area integrals of the depth per unit area, $h_\zeta(\lambda,\phi)$ and $h_\chi(\lambda,\phi)$, of ice sheets and shelf ice, respectively, i.e., $(\zeta,\chi) \sim \int\int(h_\zeta, h_\chi)d\lambda\, d\phi$. ζ has a strong inverse effect on sea level, whereas χ has little effect. To a first order, we can write

$$\delta'(\lambda,\phi) = k_I I' + k_s T_s'(\lambda,\phi) + k_d T_d(\lambda,\phi) \qquad (3)$$

where k_I $(=\partial\delta/\partial I)$, k_s $(=\partial\delta/\partial T_s)$, and k_d $(=\partial\delta/\partial T_d)$ are constants, and the primes denote departures from standard values $\bar{I}$, $\bar{T}_s$, and $\bar{T}_d$ for which $\delta^{18}O$ equals a corresponding standard value in parts per thousand (‰). For shallow dwelling foraminifera $k_d = 0$, and for benthic foraminifera $k_s = 0$.

We now seek to formulate a simple set of dynamical equations for the variations of the two key variables I and $T_s(\lambda,\phi)$ that determine the $\delta^{18}O$ variations associated with the more commonly studied <u>shallow</u> planktonic species. These equations express qualitatively conservation of ice mass, and mixed layer energy at a point. In general these equations are of the following form, assuming the primes denote departures from a steady-state $(\hat{I}, \hat{T}_s, \bar{x}_i)$ corresponding to some fixed values of the external forcing parameters such as the earth orbital parameters:

$$\frac{dI'}{dt} = f_I^{(1)}(I, T_s, \ldots x_i \ldots, F(t)) \qquad (4a)$$

$$\frac{dT_s'}{dt} = f_s^{(1)}(I, T_s, \ldots x_i \ldots, F(t)) \qquad (4b)$$

$$\frac{dx_i'}{dt} = f_i^{(1)} (I \ , T_s \ , \ldots x_i \ldots , F(t)) \tag{4c}$$

where x_i denotes the many additional physical variables (indexed by i) that are relevant in determining the rates of change of I and T_s (e.g., all the atmospheric variables including temperature, wind, cloud, and CO_2 content, ocean currents, salinity and thermocline temperature, bedrock depression, and the geographic distributions h_ζ and h_χ), and F(t) denotes stochastic and deterministic forcing that can be additive or multiplicative.

Let us now postulate that I has a very much longer system response time than the mixed layer temperature $T_s(\lambda,\phi)$, and, moreover, that of all the many possible variables included in x_i the candidates most likely to have response times comparable to I are the separate components ζ and χ and their geographic distributions $h_\zeta(\lambda,\phi)$ and $h_\chi(\lambda,\phi)$, the mean deep ocean (or thermocline temperature) $T_d(\lambda,\phi)$ whose spatially average value we denote by θ, and the bedrock depression under the ice sheets $h_\epsilon(\lambda,\phi)$ whose global mean value we denote by ϵ . In addition, although at any point $T_s(\lambda,\phi)$ is a 'fast response' variable, we can define a new 'slow' variable which is the average of T_s over all of the world ocean not covered with shelf ice, including that part of the ocean which has undergone phase transition to sea or pack ice. This new variable, which we denote by τ , can have a very slow response time because of the positive feedbacks engendered by the 'ice albedo' effects and possible CO_2-surface temperature effects. Like $T_s(\lambda,\phi)$, all other variables, x_j are assumed to have relatively fast response times, say of periods less than several hundred years (a rough averaging period for the $\delta^{18}O$ data). It follows (see Saltzman 1983) that I , τ , θ, and ϵ , and the geographic distribution functions h_ζ, h_χ, h_ϵ, and T_d , are the prognostic, "carriers", for the long term dynamical evolution of the system, and T_s and x_j are the diagnostic variables assumed to be in quasi-static equilibrium with these prognostic variables.

As a first step, in order to form a simpler, global, model for the slow variables we shall assume further that we can consider the massive ice components ζ and χ as the single variable I and neglect the geographic variables (h_I , h_ϵ and T_d) in favor of their global mean values that are proportional to I, ϵ, and θ. This is tantamount to assuming that changes in the global values can be diagnostically related to the geographic changes and tend to be of the same sign. An aspect of this question is discussed in Saltzman (1984). Thus, our system (4a-c) becomes

$$\frac{dI'}{dt} = f_I^{(2)} (I', \tau', \theta', \epsilon', F(t)) \tag{5a}$$

$$\frac{d\tau'}{dt} = f_\tau^{(2)} (I', \tau', \theta', \epsilon', F(t)) \tag{5b}$$

$$\frac{d\theta'}{dt} = f_{\theta}^{(2)}\,(I', \tau', \theta', \epsilon', F(t)) \qquad (5c)$$

$$\frac{d\epsilon'}{dt} = f_{\epsilon}^{(2)}\,(I', \tau', \theta', \epsilon', F(t)) \qquad (5d)$$

and

$$T_s'(\lambda,\phi) = f_s^{(2)}\,(I', \tau', \theta', \epsilon', x'_j, F(t)) \qquad (5e)$$

$$x'_j = f_j^{(2)}\,(I', \tau', \theta', \epsilon', T_s', x'_k, F(t)) \qquad (5f)$$

Equations (5a-d) constitute the "dynamical" system governing the slow evolution of climate, and (5e, f) are fast-response "equilibrium" distributions obtainable from steady-state physical models (e.g., GCM s) providing we specify or solve simultaneously for h_{ζ}, h_{χ}, h_{ϵ}, and T_d (cf., Saltzman 1984). This system constitutes what we might call the "master" global climate model.

We seek the least complicated, physically plausible, representation we can make for f_I, f_{τ}, f_{θ}, ... and F(t) capable of yielding robust solutions for I , and T_s that are compatable with the observed δ^{18}0 and ocean surface temperature records derived from the sedimentary cores. This will constitute our theory for these records. The verification of this theory must depend on the side predictions made with regard to concomitant paleoclimatic variability determined from all sources and the "reasonableness" of the weightings deduced for the I and T_s variations needed to account for the δ^{18}0 record. Before discussing our particular representations we first discuss some of the simpler models that have been advanced recently.

3. The simplest low order models

3.1 Single-variable forced systems

The most severe reductions of the system (5) are based on the assumption that we can treat a single variable I to which all other variables are diagnostic, and on which δ^{18}0 solely depends. Thus, we have

$$\frac{dI'}{dt} = f_I\,(I', F(t)) \qquad (6)$$

By further assuming that f_I is of linear form we obtain an elemental model of the form

$$\frac{dI'}{dt} = -\,a\,I' + F(e,\mu,\pi\,;t) + R(t) \qquad (7)$$

where a^{-1} is a dissipative time constant, F represents additive external forcing due to the variations of the earth orbital parameters e (eccentricity), μ (obliquity), and π (precessional index), and R denotes additive stochastic forcing. In making the above assumptions one is essentially taking I as a measure of τ which in turn is closely related to ice extent or to global average surface temperature T_s. (In fact we can formally write

$\tilde{T}_s = (\frac{\sigma_c}{\sigma})\tilde{T}_c + (\frac{\sigma_w}{\sigma})\tau$, where σ is the area of the earth, c denotes

continental values including ice shelves and w denotes oceanic values.); hence (6) constitutes a zero-dimensional energy balance model in which we can substitute τ or T_s for I. Assuming a is positive, (7) has a single stable equilibrium corresponding to a standard (e.g., mean) set of values of the earth orbital parameters (e_o, μ_o, π_o).

Applications of (7) have been made by Imbrie and Imbrie (1980) for a = a large constant, and R = 0, so that

$$I = a^{-1} F(e, \mu, \pi ; t) \quad , \tag{8}$$

by Fong (1982) for $a \to 0$ (near-neutral stability) and R = 0, so that

$$\frac{dI}{dt} \approx F \quad , \tag{9}$$

and by Hasselmann (1976), Lemke (1977) and Fraedrich (1978), for F = 0 but $R \neq 0$.

In all these cases it was not possible to replicate the observed $\delta^{18}O$ record. Even if one were to assume the more general deterministic form of (6), where f_I can be highly nonlinear, it would not be possible to replicate the $\delta^{18}O$ curve by assuming that I is the quasi-static "equilibrium" response $I_E(e, \mu, \pi ; t)$ of a more general form than (8) (Saltzman, 1985). The problem of accounting for the $\delta^{18}O$ record is thus inescapably one of "dynamics".

A fundamental difficulty in accounting for the $\delta^{18}O$ record purely as a forced effect, even if one assumes the full nonlinear dynamical equation of the form (6), is that the 20 k yr precessional (π) and 40 k yr obliquity (μ) effects on the terrestrial radiative/thermal state and hydrologic cycle are so much greater than that of the 100 k yr (e) effect. Thus it would appear that only with very strong nonlinear rectification can a purely forced model yield the observed output. Although, in principle, this is not impossible, it has not yet been shown that such a plausible model can be constructed.

A notable attempt in this direction based on a modification of (7) was made in the paper by Imbrie and Imbrie (1980), where a^{-1} was assumed to be a two-valued function depending on whether dI/dt is positive or negative. A simple continuous approximation to their discontinuous nonlinear model can be written in the form

$$\frac{dI'}{dt} \approx -(a_o + a_1 I')I' + (a_2 + a_3 I')\, F\,(e, \mu, \pi\, ; t) \qquad (10)$$

where a_o, a_1, a_2, and a_3 are positive constants determined by two independent parameters, and F is expressed in terms of three additional parameters determining the phase and relative weighting of e, μ, and π. This model can produce solutions that fit the $\delta^{18}O$-curve with respectable accuracy for the past 400 k yrs, but is deficient in several respects: (1) the near 20 and 40 k yr variations induced by π and μ are indeed too strong compared to the dominant 100 k yr variability, (2) as a single component system there is no prediction made of concomitant variability of other features of the climatic system such as the temperature curve shown in Fig. 1, and, as noted above, (3) the tacit assumption is made that the $\delta^{18}O$ record is completely determined by ice mass (unless mixed layer temperature is exactly 180° out of phase with I, which is unlike the picture shown in Fig. 1).

Another one-component forced model worthy of special consideration is that of Sutera (1981) and Benzi et al (1982, 1983) (see also Nicolis 1982), based on an energy balance model (cf. Ghil 1976) in which the albedo dependence is formulated in such a way that reasonably-spaced steady states corresponding to glacial and nonglacial conditions are present, separated by an unstable "barrier" point. The relevant dynamical equation is of the general form

$$\frac{dI'}{dt} = a_o I' - a_1 I'^3 + F(t) + R(t) \qquad (11)$$

These authors showed that when stochastic forcing R is applied to this bistable system a new time scale is introduced, the "exit time" for transition between the glacial and non-glacial modes. Although the range of exit times can be broad, for reasonable parameter values the mean exit time can plausibly be near the dominant period shown in Fig. 1, 100 k yrs, and with the introduction of the eccentricity forcing the strength of the "barrier" can vary on this scale leading to a "resonant" response. It is only by this sort of stochastic forcing acting on a bistable system that a one-component model can be made to exhibit free variations with which external forcing can resonate. With the addition of the near-20 and 40 k yr forcing in addition to the 100 k yr eccentricity forcing one can expect that a fairly good approximation to the $\delta^{18}O$ curve can be obtained. Although this model is extremely simple, and still contains most of the deficiences of the Imbrie and Imbrie model, it possesses the important property of replicating the observed bimodality, with relatively rapid transitions, characteristic of the $\delta^{18}O$ record [see Fig. 2 for the histogram of the SPECMAP (Imbrie et al. 1984) representation of the $\delta^{18}O$ record over the last 500 k years]. The identification of this bimodality with a fundamental instability of the climatic system is stated in its most simple terms in this model.

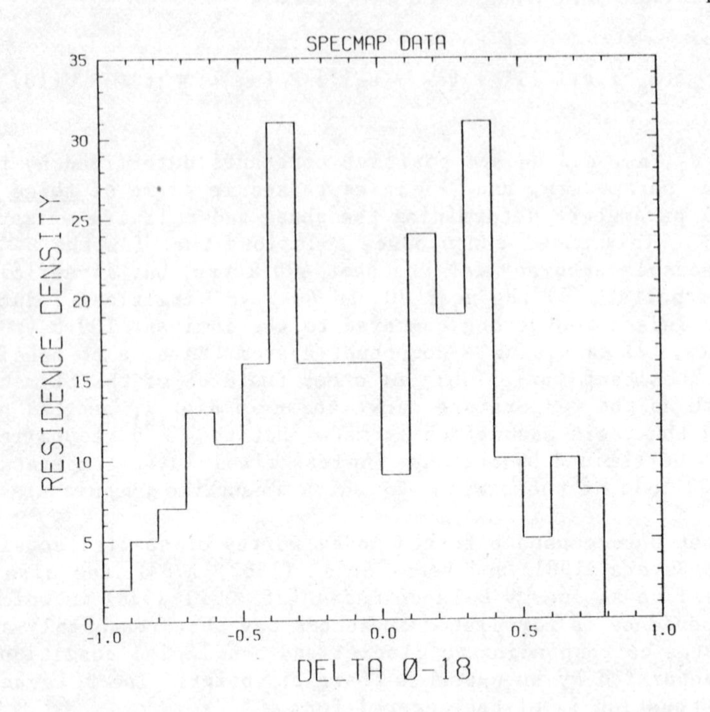

Fig. 2 Histogram of the SPECMAP $\delta^{18}O$ record (Imbrie et al. 1984) for 500 k yrs, normalized to amplitudes ± 1, showing on the ordinate (residence density) the number of times the values spaced every 2 k yrs lie within the limits shown on the abscissa.

Before proceeding to the two-component models, it is worth noting that an intermediate (one and a half-component) class of dynamical systems can be formed. In this scheme, a measure of the variation of the mean planetary temperature, say τ', is considered as an additional global variable, with I being identified with variations of the $\delta^{18}O$-record, and τ' with that of the global ice extent governed by an energy balance model in which the direct effect of earth-orbital or other forcing appears. However, while τ may affect I , there is no feedback of I on τ , i.e.,

$$\frac{dI}{dt} = f_I(I, \tau) \qquad (12a)$$

$$\frac{d\tau}{dt} = f_\tau^{(1)}(\tau; F(t)) \rightarrow \tau = f^{(2)}(F) \qquad (12b)$$

In essence, (12b) offers the possibility for nonlinearly transforming the observed external forcing $F(e, \mu, \pi; t)$ into one that more closely approaches a forcing with dominant period near the observed 100 k yrs. This type of model, which along with the diagnostic relation (5e), would appear to be of the kind discussed by North and Crowley (1985), is essentially equivalent to the one-component model (6). One unrealistic feature of this model, however, is the decoupling of ice mass and surface temperature; it seems plausible to expect that variations in planetary ice mass I (as well as deep water temperature θ) would have at least as much effect on sea ice extent or mean surface temperature τ as earth orbital forcing. In the following sub-section, we discuss some proposed models that contain such a coupled feedback.

3.2 Two-variable models

With a dynamical system comprised of at least two coupled variables we have the possibility for internally-generated, free, oscillatory behavior as well as for the forced oscillatory behavior we have just discussed. Since it is likely that more than one slow variable participates in long-term climatic change (e.g., the additional candidates we have enumerated in Section 2) it is highly probable that the spectrum of climatic change (e.g., Mitchell 1976) is created by some complex combination of both free and forced oscillatory variability. Although the first proposal for considering multi-component oscillatory dynamical systems to account for the ice ages appears to have been made by Eriksson (1968), the most detailed effort to apply this line of attack was made by Sergin (1979) during the mid 1970's.

A prototype of the two-variable systems treated most recently is based on considering the ice mass I to be dynamically coupled with a measure of mean ocean surface temperature or sea ice extent, τ. This prototypical system is of the general form,

$$\frac{dI'}{dt} = a_o I' + a_1 \tau' + N_I + F_I + R_I \qquad (13a)$$

$$\frac{d\tau'}{dt} = b_o I' + b_1 \tau' + N_\tau + F_\tau + R_\tau \qquad (13b)$$

where a and b are constants and N denotes nonlinear terms.

The model of Källén et al. (1979), as amplified by Ghil (1984), is of the form (13), characterized by the following sign choices of the coefficients: $a_o < 0$, $a_1 > 0$, $b_o < 0$ and $b_1 > 0$ (instability due to the ice-albedo feedback). These coefficient choices provide the conditions for self-sustained oscillations with τ maximum (or sea ice minimum) preceding an ice mass maximum. The assumption is made here that, on balance, large sea ice extent or cold surface temperature inhibits ice sheet growth through its negative effect on evaporation and the hydrologic cycle.

Another model of the form (13) was proposed and developed in a series of papers by Saltzman (1978, 1982) and Saltzman et al. (1981, 1982), and further elucidated by Nicolis (1984, 1985). As in the Källén et al. (1979) model $a_o < 0$ and $b_1 > 0$, but $a_1 < 0$ and $b_o > 0$. In the Saltzman et al. (1981, 1982) version and succeeding analyses by Nicolis (1984, 1985) the nonlinearity is of the form $N_I = 0$, $N_\tau = -b_2 I'^2 \tau'$, leading to a modified form of the van der Pol oscillator. The supposition in this model is that variations in τ are strongly related to variations in deep ocean (thermocline) temperature that depend critically on the ice-insulator effect of the large shelf and sea ice masses included in I (Newell 1974, Saltzman 1978). In this case the instability associated with $b_1 > 0$ is assumed to be related to a positive feedback associated with atmospheric CO_2 (Saltzman and Moritz, 1980) as well as to the ice-albedo effects associated with τ'. As noted, this model also leads to a self oscillating system but with τ (or θ) maxima preceding total ice mass (I) minima. To the degree that the mean surface temperature τ is related to the mixed layer temperature in the Indian Ocean portrayed in Fig. 1, it would appear that there is a tendency for ocean temperature maxima to precede ice mass minima as in this latter model. The free versions of both models, however, are incapable of showing the "spiky" character of the temperature curve shown in Fig. 1, a point we will return to later. Many of the forced properties of this model ($R \neq 0$, $F \neq 0$) are discussed in the above papers.

Before closing this brief review of two-variable systems, we note that another class of such models involving ice mass in the form of continental ice sheets, ζ , and the depression of the underlying bedrock, ϵ , have been suggested, notably by Oerlemans (1980), Peltier (1982) and Birchfield and Grumbine (1985). The "global" versions of these models are essentially of the form

$$\frac{d\zeta'}{dt} = a_o\zeta' + a_1\epsilon' + N_\zeta + F_\zeta \qquad (14a)$$

$$\frac{d\epsilon'}{dt} = b_o\zeta' + b_1\epsilon' + N_\epsilon \qquad (14b)$$

With the assumption that increased bedrock depression leads to reduced ice sheet elevation and hence to increased ice accumulation ($a_1 > 0$), and that $b_o > 0$, we have the possibility for unstable ice sheet growth under the restraint of linear dissipation, $b_1 < 0$, and of the nonlinear dissipative terms for large departures. Although we recognize the plausibility of some effects of this type, we shall not emphasize bedrock-ice sheet models here.

3.3 Difficulties with the one and two-variable models

With the inclusion of nonlinear terms, the simple models discussed in this section already have the potential to account for a good deal of the observed late Quaternary climatic variability recorded in the $\delta^{18}O$ records. As we have noted, the one-component forced models of Imbrie and Imbrie (1980) and the "stochastic resonance" models of Benzi et al. (1982) and Nicolis (1982) both seem capable of replicating key features of the $\delta^{18}O$ record. In addition, the two-component models illustrate a second major possibility, i.e., that free oscillations can account for the main 100 k yr oscillatory behavior observed in the $\delta^{18}O$ record (perhaps with the "phase locking" influence of earth-orbital forcing (Nicolis 1984)). It remains as a fundamental problem to determine whether forcing alone can generate this near-100 k yr variability or free oscillatory behavior is necessary. Stated in other related terms, it remains to be determined whether the long term oscillations are taking place near a stable or unstable equilibrium point.

In any event, there are several inherent deficiencies of one-variable purely periodically forced models, and two-variable, not periodically-forced (autonomous) models that make it unlikely that they can provide a fully adequate theoretical basis for an explanation of the $\delta^{18}O$ record of ice age variations:

(1) In these models, only by introducing stochastic forcing can one reproduce the red-like appearance of the $\delta^{18}O$-spectrum. In the absence of this stochastic forcing only a line-spectrum characteristic of a periodic or quasi-periodic response is obtainable. While this mechanism of stochastic forcing is acceptable, and in fact is a necessary feature of all models dealing with heavily time-averaged global climatic variables (cf. Saltzman 1983), the theory will be incomplete until the basic determinism underlying a good deal of the stochastic behavior so introduced is represented, leaving only an essential unrepresentable white noise component as a residual. Only with a minimum of three-components (possibly including a component due to non-autonomous periodic forcing) can such deterministic aperiodicity or "whiteness" be generated.

(2) Related to the above, only with at least a 3-component model can one internally, deterministically, generate very fast transitions between climatic states, e.g., the "sudden glaciations" that have been suggested, the signatures for which are only weakly apparent in the $\delta^{18}O$ record because of natural averaging processes (e.g., bioturbation). In fact, the T_s-curve accompanying the $\delta^{18}O$ record in Fig. 1 already gives ample testimony to the possibility for an extremely "spiky" behavior for this variable that may contribute significantly to the $\delta^{18}O$ record itself (equation (1)). This spikiness, which requires a broad continuous spectrum for its representation, is, in this sense, a realization of stochastic forcing. No one-variable, or unforced two-variable deterministic system is capable of yielding such a time record.

(3) In order to obtain a 100 k yr response from a two-variable

internal self-oscillatory model the "time constant" [e.g., $(-a_0)^{-1}$ of (13a), or $(-b_1)^{-1}$ of (13b)] must be of the order of about 40 k yr (e.g., Oerlemans 1980, Saltzman et al. 1982). Although it is conceivable that such a long time constant is appropriate in view of possible positive feedbacks, some shorter time constant (perhaps no larger than about 10 k yr) would seem to be more appropriate. As we show in the next section, with a three-variable system a free 100 k yr period can be obtained easily with a time constant of less than 10 k yr.

In the presence of a periodic forcing, however, a two-variable system essentially becomes at least a three-component autonomous dynamical system (e.g., C. Nicolis, this volume); with proper matching of the forcing to the free period of the two-variable system a chaotic response can be generated (Holmes, 1980) and additional longer periods (e.g., 100 k yr) in the forcing can be amplified through entrainment. Thus all of the features described in (1), (2), and (3) above are achievable, in principal, in a two variable system with periodic forcing. This possibility should be tested more fully (e.g., M. Ghil, this volume).

More generally, as noted in Section 1, it is reasonable to expect that more than two "slow" variables are participating dynamically in the major climatic changes. At a minimum we have the major snow-derived ice masses in the form of continental ice sheets and ice shelves, their underlying bedrock, the deep ocean or thermocline temperature and the mean ocean surface temperature and associated pack ice fields. For example, it seems unreasonable to exclude possible effects of slow variations of the deep water or thermocline temperature from (13b). The inclusion of some of these additional variables enables us at least to be able to offer some predictive consequences with regard to other slow response variables of the climatic system.

For these reasons we are led to consider a three-variable model. As is well known (e.g., Lorenz, 1963) such a three-variable system already contains so much possibility for dynamical richness that one need not hasten to go beyond to four or more, at least at the outset.

4. A three-variable system

Let us return now to our "master" system (5a-f) governing all the climatic variables including, specifically, the variations of I and T_s that largely determine the $\delta^{18}O$ record. We begin by specializing this system by assuming the diagnostic relations (5e, f) for the departure of $T_s(\lambda,\phi)$ from its equilibrium value can be approximated as a function of the global variables I, τ, and θ. We assume the following first order representation

$$T_s'(\lambda,\phi) = \alpha(\lambda,\phi)\, I' + \beta(\lambda,\phi)\tau' + \gamma(\lambda,\phi)\theta' \qquad (15)$$

where $\alpha(\lambda,\phi) = \partial T_s/\partial I$, $\beta(\lambda,\phi) = \partial T_s/\partial\tau$, and $\gamma(\lambda,\phi) = \partial T_s/\partial\theta$. Let us now define $\hat{T}_s$, $\hat{I}$, $\hat{\tau}$, and $\hat{\theta}$ as characteristic maximum departures from equilibrium values of T_s, I, τ, and θ, respectively (i.e.,

$T_s' = \hat{T}_s T_s^*$, $I' = \hat{I} I^*$, $\tau' = \hat{\tau}\tau^*$, and $\theta' = \hat{\theta}\theta^*$, where T_s^*, I^*, τ^*, and θ^* are nondimensional numbers between 0 and 1). Then we can write (15) in a nondimensional form

$$T_s^* = \alpha^*(\lambda,\phi)I^* + \beta^*(\lambda,\phi)\ ^* + \gamma^*(\lambda,\phi)\theta^* \qquad (16)$$

where $\alpha^* = \hat{I}\hat{T}_s^{-1}$, $\beta^* = \hat{\tau}\hat{T}_s^{-1}$ and $\gamma^* = \hat{\theta}\hat{T}_s^{-1}$. It seems reasonable to expect that $\alpha^*(\lambda,\phi)$, measuring the effect of polar ice sheet mass and extent, will tend to be a maximum in higher latitudes, and $\gamma^*(\lambda,\phi)$, measuring the effect of thermocline temperature, will tend to be a maximum in low latitudes and regions of upwelling. It is implied by (15) that direct earth orbital forcing F is less effective in determining the steady-state values of T_s than the prevailing state of τ and θ. This conjecture, as well as the general statement (16) might be testable with a coupled atmosphere-mixed layer GCM.

With (15) and (5a-c) we obtain a closed model for I and T_s based on a three-variable dynamical system governing I, τ and θ, providing we assume that the effect of bedrock depression ϵ is negligible compared to I, τ and θ in determining the slow variability of the climate system. This latter assumption is certainly contrary to that made in several other models where no reference is made to deep ocean (or thermocline) temperature, but great emphasis is placed on bedrock depression (e.g., Oerlemans 1980, Birchfield et al. 1981, Ghil and LeTreut 1981, LeTreut and Ghil 1983, Peltier 1982, Pollard 1982). Recognizing that, as shown by these authors, ϵ is in fact potentially important, and in any event that bedrock changes are a real feature of climatic evolution that must be accounted for, we view the ultimate inclusion of ϵ as a necessary further step toward a more complete global theory. At this point, however, we set down the basic three component system we have developed recently (Saltzman and Sutera 1984, Saltzman et al. 1984), the prototype for which was introduced in Saltzman et al. 1984b, and show with the aid of (15) or (16) the implications of this model for a theory of the $\delta^{18}O$ curve.

Although the three variables treated were previously identified with ζ, χ and θ, we assume here that the same dynamical system model can more plausibly be identified with I, τ and θ and we formally make the following transformations: $\zeta \rightarrow I$, $\chi \rightarrow -\tau$, and $\theta \rightarrow \theta$. This is for several reasons: (1) The shelf ice mass χ implied by the solution in Saltzman et al. (1984) seems to be unrealistic when comparing 20 k yr BP versus the present, whereas the values of area-averaged sea surface temperature beyond the ice shelves (the mass of which is now included in I) seem to be in line with those reported in CLIMAP (1982); (2) the Denton-Hughes (1981) scenario portrayed by the model through the emphasis on the χ-component is still highly controversial and remains to be substantiated satisfactorily; (3) the ice-albedo feedback assumed to provide a main drive for the instability of the system is readily identified with mean sea surface variations including the transformations to sea ice cover (e.g., Ghil, 1984); and more generally, (4) with τ as an explicit variable the linkage and continuity with the one and two component "energy balance" models discussed in Section 3 is more easily established.

Thus, as described in the above two references, the ordinary differential equations constituting the dynamical system are

$$\frac{dI'}{dt} = - a_1 (1+a_2 \tau')\tau' - a_3\theta' \tag{17a}$$

$$\frac{d\tau'}{dt} = - b_o I' + b_1 (1+b_2\tau'-b_3\tau'^2-b_4 I'^2)\tau' + b_5\theta' + F_\tau + R_\tau \tag{17b}$$

$$\frac{d\theta'}{dt} = c_o I' - c_1\tau' - c_2\theta' \tag{17c}$$

It can be verified that with the transformation $x' \rightarrow -\tau'$ embodied in (17a-c) the feedbacks implied all continue to be plausible from qualitative physical considerations [e.g., in (17b) high values of I favor decreasing sea surface temperature to first order].

Taking A , B , C as arbitrary scaling amplitudes of I', τ', and θ', $\hat{t}$ as a characteristic time scale (10 k yr), and $\hat{F}_\tau$ and $\hat{R}_\tau$ as characteristic amplitudes of deterministic and stochastic forcing, respectively, we can write (17a-c) in the nondimensional form,

$$\frac{dX}{dt^*} = - \alpha_1 Y - \alpha_2 Z - \alpha_3 Y^2 \tag{18a}$$

$$\frac{dY}{dt^*} = - \beta_o X + \beta_1 Y + \beta_2 Z + (\beta_3 Y - Y^2 - X^2)\, Y + F_\tau^* + R_\tau^* \tag{18b}$$

$$\frac{dZ}{dt^*} = X - \gamma_1 Y - \gamma_2 Z \tag{18c}$$

where $t^* = \hat{t}^{-1} t$, $X = A^{-1} I'$, $Y = B^{-1}\tau'$, $Z = C^{-1}\theta'$, $(F_\tau{}^*, R_\tau{}^*) = \hat{\tau}^{-1}\hat{t}\,(F_\tau, R_\tau)$, $\hat{I} = (b_1 b_4 \hat{t})^{-1/2}$, $\hat{\tau} = 2\times10^{-3/2} (b_1 b_3 \hat{t})^{-1/2}$, $\hat{\theta} = c_o(b_1 b_4 \hat{t})^{-1/2}$, and $\alpha_1 = (B \hat{t} A^{-1})a_1$, $\alpha_2 = (C \hat{t} A^{-1})a_3$, $\alpha_3 = (B^2 \hat{t} A^{-1}) a_1 a_2$, $\beta_o = (A \hat{t} B^{-1})b_o$, $\beta_1 = \hat{t}\, b_1$, $\beta_2 = (C \hat{t} A^{-1})b_5$, $\beta_3 = (B \hat{t}\,)b_1 b_2$, $\gamma_1 = (B \hat{t} C^{-1})c_1$, $\gamma_2 = \hat{t}\, c_2$. On the assumption that most of the variability exhibted by typical $\delta^{18}O$ curves is due to continental ice mass changes (Imbrie et al. 1984) the following values of these coefficients were chosen by Saltzman and Sutera (1984) to give a qualitatively plausible agreement between the free variations of ζ and the $\delta^{18}O$ record: $\alpha_1 = 0.40$, $\alpha_2 = 0.10$, $\alpha_3 = 0.0125$, $\beta_o = 10.00$, $\beta_1 = 3.77$, $\beta_2 = 20.00$, $\beta_3 = 0.01$, $\gamma_1 = 0.01$, and $\gamma_2 > 1.00$ ($c_2{}^{-1} \lesssim 10$ k yr if $\hat{t} = 10$ k yr).

<u>The equilibrium and its stability</u> - Neglecting variations in forcing, F and R , the equilibria of the system (17) (or 18) for the

coefficient values given above, consists of the obvious fixed point $(I', \tau', \theta') = (0,0,0)$ plus four imaginary (nonphysical) points.

The eigenvalues at this real equilibrium consist of 2 real positive values (+0.429 and +4.308) plus one real negative value (-1.967). (The two-component oscillatory models discussed in Section 3 are characterized by Hopf bifurcations in which two imaginary eigenvalues cross the imaginary axis at the bifurcation point). Thus we have here a "codimension two" system (Guckenheimer and Holmes 1983). It is this non-Hopf type bifurcation, with pure exponential type of instability, that gives rise to the long 100 k yr period free response in this model in spite of a first order dissipative time constant (c_2^{-1}) of less than 10 k yr. An analysis of this property of codimension 2 systems based on center manifold theory is now being purused in collaboration with Alfonso Sutera and Roberto Benzi.

Numerical solution of the $I-\tau-\theta$ system - the free solution for the above coefficient values is given in Saltzman and Sutera (1984), and the forced response to a representation of earth-orbital forcing F_τ together with an analysis of the sensitivity of the solution to changes in these coefficent values and to stochastic forcing R_τ is given in Saltzman, Hansen, and Maasch (1984). This forced solution is shown in Fig. 3 in terms of normalized values defined by $I^* = X\hat{X}^{-1}$, $\tau^* = Y\hat{Y}^{-1}$, and $\theta^* = Z\hat{Z}^{-1}$ where $\hat{X}$, $\hat{Y}$ and $\hat{Z}$ are the maximum absolute values of X, Y, and Z, respectively, so that I^*, τ^* and θ^* do not exceed $|1|$. Thus, $A = \hat{I}\hat{X}^{-1}$, $B = \hat{\tau}\hat{Y}^{-1}$, $C = \hat{\theta}\hat{Z}^{-1}$. The coefficients for the earth-orbital forcing variations were chosen to be physically plausible and to yield qualitative agreement between the observed SPECMAP $\delta^{18}O$ curve and the ice mass I. Note in Fig. 3 that along with this I-variation we also predict a quasi-catastrophic, spiky, collapse and regrowth of sea ice coinciding with the rapid deglaciation of the ice sheets and shelves as well as variations of deep ocean or thermocline temperature that lag I by about 5 to 10 k yr.

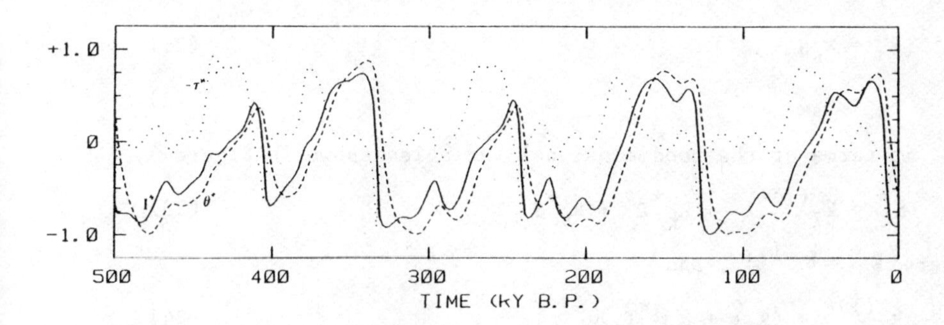

Fig. 3 Orbitally-forced solution of the three-component dynamical system (18a-c) in nondimensional units, I^* (solid curve), $-\tau^*$ (dotted curve), and θ^* (dashed curve).

5. Theory of the $\delta^{18}O$ variations

We assume now that the $\delta^{18}O$ variations determined from surface water foraminifera are not only a function of ice mass (I) with which we tried to achieve a qualitative fit, but also of T_s, as represented by (3). By finding the best least-squares fit between the observed $\delta^{18}O$ values (let us say the SPECMAP values) and our solution not only for I but also for τ and θ, we shall deduce the implications of our model concerning (i) the magnitudes of the coefficients α^*, β^* and γ^* of (16), (and hence of the T_s variations necessary to improve the explanation of the $\delta^{18}O$ variations), and (ii) the relative contributions of snow-derived ice mass and ocean surface temperature changes to the observed $\delta^{18}O$ departures at any time during the past 500 k yrs. From another viewpoint, we seek to find whether, given a relationship of the form (15) and accepted values of $(\partial\delta/\partial T)$ and $(\partial\delta/\partial I)$, our predicted variations of τ and θ, can, with physical plausibility, account for a greater amount of the observed variability of the SPECMAP $\delta^{18}O$ curve than was possible with I alone. By "physical plausibility" we mean, for example, that an improved fit to $\delta^{18}O$ will require that k_s in (3) be <u>negative</u>. If such an improved explanation of variance is possible, it would give some indication of the internal consistency of the model.

5.1 The best-fit solution for $\delta^{18}O$

Recalling our simplifying abbreviation, $\delta^{18}O \equiv \delta$, we have from (3) and (15)

$$\delta' = k_I^{(1)} I' + k_\tau \tau' + k_\theta \theta' \quad (19)$$

where

$$k_I^{(1)} = k_I + k_s \alpha \quad (20)$$

$$k_\tau = k_s \beta \quad (21)$$

$$k_\theta = k_s \gamma \,, \quad (22)$$

or, in terms of the nondimensional variables shown in Figure 3,

$$\delta^* = k_I^{(1)*} I^* + k_\tau^* \tau^* + k_\theta^* \theta^* \quad (23)$$

where $\delta^* = \hat{\delta}^{-1}\delta'$, and

$$k_I^{(1)*} = (k_I \hat{I} + k_s \alpha^* \hat{T}_s)\delta^{-1} \quad (24)$$

$$k_\tau^* = k_s \beta^* \hat{T}_s \delta^{-1} \quad (25)$$

$$k_\theta^* = k_s \gamma^* \hat{T}_s \delta^{-1} \quad (26)$$

The best least-squares fit of the SPECMAP estimate of $\delta^{18}O$

variations over the past 500 k yrs to the solution for I^*, τ^* and θ^* shown in Figure 3 yields the following characteristic values of the coefficients (24) - (26): $\bar{k}_I^{(1)*} = 0.75$, $\bar{k}_\tau^* = -0.22$, and $\bar{k}_\theta^* = -0.26$. In Figure 4 we show this best fit curve, δ^*, in comparison with the SPECMAP curve, δ_s^*.

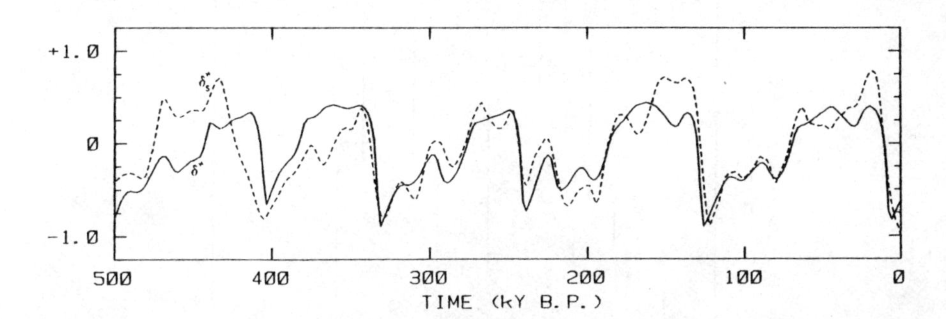

Fig. 4 Best-fit solution for $\delta^{18}O$, $\delta^* = 0.75\, I^* - 0.22\tau^* - 0.26\theta^*$ (solid curve), compared with SPECMAP $\delta^{18}O$ record of Imbrie et al., 1984 (dashed curve), both in nondimensional units.

5.2 Properties of the solution

Although the model coefficients in (18a-e) were chosen to give the relatively large positive value obtained for $k_I^{(1)*}$, we consider it to be a significant measure of the internal physical consistency of the model that k_τ^* and k_θ^* both turn out to be negative as would be expected. The correlation coefficient between the two curves, δ_s^* and δ^*, shown in Figure 4 is 0.74 (55 percent 'explained variance'), as compared with 0.68 (47 percent 'explained variance') between δ^* and I^* alone. We find generally good agreement between the two curves, the most obvious defect being the failure to replicate the high amplitudes of the last two δ^* maxima.

It is of particular interest that the fit to the $\delta^{18}O$ curve is markedly improved from that which is possible by considering I alone (Saltzman et al 1984) over the period between 125 and 60 k yr BP where the Dodge et al. (1983) sea-level results indicate that temperature and/or floating ice shelf effects are required. Another interesting feature of this best-fit curve is the minimum of $\delta^{18}O$ at 6 k yr BP; although the composite SPECMAP curve does not show this minimum there are many individual cores (e.g., V28-238, Shackleton and Opdyke 1973; V34-88 Prell, 1984) that do reveal this minimum associated with the well-known "climatic optimum".

As is to be expected from the similarity of the two curves shown in Figure 4, the best-fit solution for the 500 k yr period has a similar histogram to that shown in Fig. 2 for the SPECMAP data (see Fig. 5). The bimodality is again present with a pronounced minimum near $\delta^{18}O=0.00$ representing the fundamental instability of the system.

The spectrum of the best-fit solution δ^* is shown in Fig. 6,

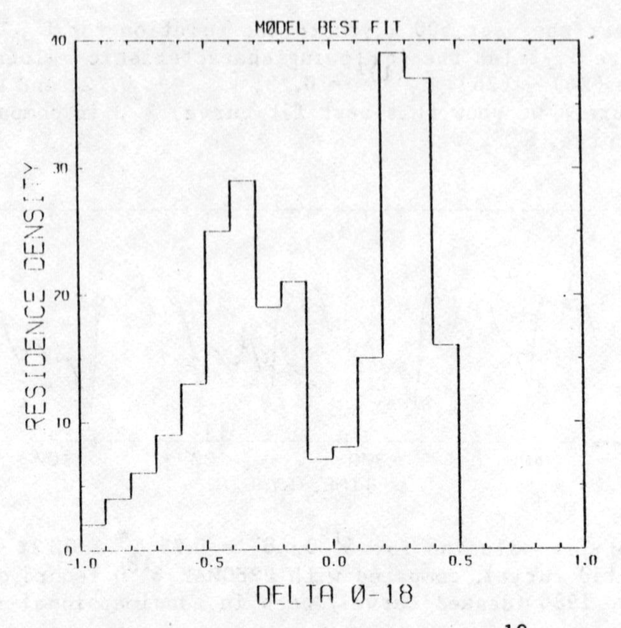

Fig. 5 Histogram of the best-fit solution for $\delta^{18}O$ shown in Fig. 4. As in Fig. 2, the ordinate (residence density) gives the number of times the δ^* values spaced every 2 k yrs lie within the limits shown on the abscissa.

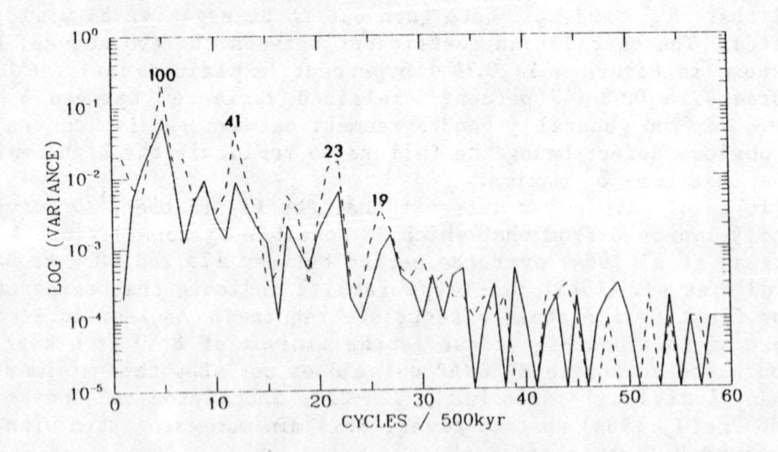

Fig. 6 Variance spectrum of the best-fit solution for $\delta^{18}O$ shown in Fig. 4 (solid curve) and of the SPECMAP $\delta^{18}O$ record (dashed curve), both for 500 k yr. Periods of the principal peaks are shown in k yrs.

along with the spectrum of the SPECMAP-$\delta^{18}O$ evolution over the past 500 k yr, δ_s^*. Although the model contains no stochastic forcing, due largely to the presence of periodic forcing a continuous red-like background spectrum is generated in addition to the observed peaks near 100, 41, 29, 23, and 19 k yr (Saltzman, Hansen and Maasch 1984). This best fit $\delta^{18}O$ spectrum, which includes contributions from the solutions for τ and θ as well as I, shows some improvement over that for I alone given in the above reference, but also shows systematically lower magnitude peaks than the δ^*-spectrum. This is related to the inability of the best fit curve to replicate the pronounced maxima of δ_s^* during the last two major glacial episodes. In this latter study it was already shown that a non-periodically forced "chaotic" or "strange" attractor is not necessary to generate deterministically a continuous spectrum close to the observed; a forced, three-variable, co-dimension two system of the kind we have formulated will generate enough "whiteness" through spikes and fast transitions to replicate the observations with good fidelity.

In calculations performed by Kirk A. Maasch using the algorithm of Grassberger and Procaccia (1983) it was found that the dimension of the attractor giving rise to the SPECMAP $\delta^{18}O$ record over the last 700 k yr is about 2.95. For the 1 million year record obtained from the individual core V28-238, Nicolis and Nicolis (1984) obtain a dimension of 3.1, from which they conclude that "the main long-term climatic evolution may be viewed as the manifestation of a deterministic dynamics involving a limited number of key variables" and, moreover, that the "fractal" dimensionality of the record implies that this determinism is of the "chaotic" variety producing the aperiodic signal. We have here illustrated the first of these conclusions, showing that a three-variable model with external periodic forcing can replicate the $\delta^{18}O$ curve.

5.3 The surface temperature, and its role in the $\delta^{18}O$ variations

We now use the best-fit coefficient values and the relations (24) - (25) to estimate (or "predict") as consequences of our model characteristic area average values of α^*, β^*, and γ^* (which we denote by $\bar{\alpha}^*$, $\bar{\beta}^*$, $\bar{\gamma}^*$) appearing in (16) and hence the typical surface ocean temperature variations such as shown in Fig. 1. For this purpose we adopt the following parameter values (Fairbanks and Matthews 1978, Dodge et al. 1983):

$$k_I = \left(\frac{\partial \delta}{\partial I}\right) = 3.06 \times 10^{-20} \quad kg^{-1}$$

which corresponds to a 0.11‰ change of $\delta^{18}O$ with a 10 m change of sea level,

$$k_s = \left(\frac{\partial \delta}{\partial T}\right) = -0.22‰ K^{-1} ,$$

and (e.g., Imbrie et al. 1984)

$$\hat{\delta} = 1.3‰$$

corresponding to the fact that δ^* in Fig. 4 typically is of magnitude $\pm$ 0.75. We also estimate that $\hat{T}_s \approx 4$ K (e.g., Fig. 1), and that $\hat{I} \approx 2.5 \times 10^{19}$ kg (Flint, 1971).

With the above values we find from (24) - (26) the following characteristic values of the coefficients in (16): $\bar{\alpha}^* = -0.267$, $\bar{\beta}^* = 0.332$, and $\bar{\gamma}^* = 0.378$. In Fig. 7 we show the curve obtained for the variations of T_s^* (dashed line) using these derived coefficients compared with the SPECMAP $\delta^{18}O$ curve. The similarity of this figure with Fig. 1 is striking.

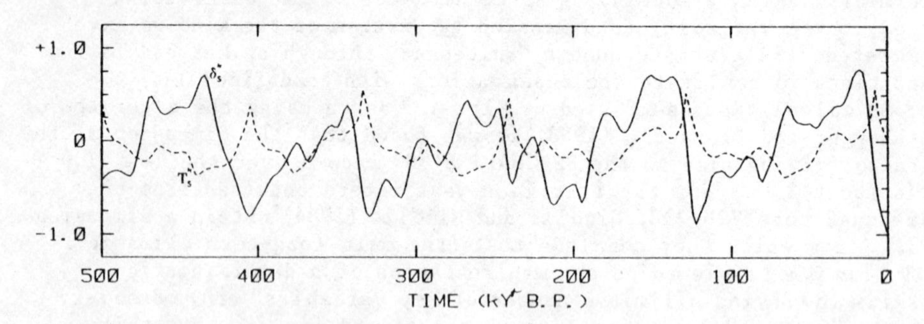

Fig. 7 Characteristic variation of surface temperature T_s^* (dashed curve), in nondimensional units, consistent with the best-fit $\delta^{18}O$ solution shown in Fig. 4. The SPECMAP $\delta^{18}O$ record is the solid curve. Note the similarity of this variation of T_s^*, relative to $\delta^{18}O$, with that shown in Fig. 1 for the South Indian Ocean cores.

Now, given this evolution of T_s^* we are in a position to use (3), or its nondimensional form,

$$\delta^*(t) = k_I^* I^*(t) + k_s^* T_s^*(t) , \qquad (27)$$

where $k_I^* = (k_I \hat{I} \hat{\delta}^{-1}) = 0.575$, and $k_s^* = (k_s \hat{T}_s \hat{\delta}^{-1}) = 0.662$, to calculate the relative contributions of snow-derived ice mass (I) and surface water temperature (T_s) to the computed value of $\delta^{18}O$. For example, for the present climatic state (t = 0 k yr BP) we have from Figs. 3 and 7, $I^*(0) = -0.90$, and $T_s^*(0) = 0.05$, respectively. From (27) we find that $\delta^*(0) = -0.55$, which, as expected, agrees well with the value shown in Fig. 4. It is implied therefore that 94 percent of δ' is due to the negative ice mass departure and the remaining 6 per cent is due to the positive surface temperature departure.

As another example, if we consider $t = t_1$ = 100 k yr BP we find $I^*(t_1) = -0.75$ and $T_s^*(t_1) = -0.15$, implying from (27) that $\delta^*(t_1) = -0.332$ (compare Fig. 4). Thus at t_1, the ice mass effect overestimates the $\delta^{18}O$ deficit by 30 percent, which amount is augmented by the negative anomaly in surface temperature. This is in the same direction as deduced by Dodge et al. (1983) for this time period, but in view of the sea level data reported, would seem to indicate that a significant component of I' is due to floating ice shelves measured by χ'. Of course, all these results could be altered substantially if our estimates of $\hat{I}$, $\hat{T}$, $\hat{\delta}$, k_I and k_s are markedly in error.

6. Summary and conclusions

After reviewing some of the simple one and two-component models of the ice-age climatic variations, we have presented a revised formulation of a three-component model discussed previously (Saltzman and Sutera 1984, Saltzman et al 1984). In essence we have re-identified our previous variable ζ with total snow-derived ice mass I (= $\zeta + \chi$), and χ with a global average temperature of the sea surface including its pack ice coverage. As before we take θ as a measure of the global deep ocean, or main thermocline, temperature. In introducing the variable τ we bring the model into closer alignment with simpler "energy balance" models that explicitly treat a variable like τ. This is particularly true if we choose the coefficients to permit multiple equilibria (e.g., Ghil 1976, Sutera 1981, Benzi et al. 1982, Nicolis 1982). From another viewpoint, with mean ocean surface temperature as an explicit variable it becomes more straightforward to treat the surface mixed layer temperature T_s as an explicit variable that can influence the $\delta^{18}O$ record derived from shallow foraminifera.

We have shown that the combination of I and T_s, as deduced from the model, can account for the major features of the observed $\delta^{18}O$ record and the associated surface temperature variations. The model itself is characterized by a free oscillatory response, with a dominant 100 k yr period, that has a 'saw-tooth' ice mass variation with rapid deglaciation that are phase-locked by earth-orbital forcing. The associated mean ocean temperature/pack ice variations, which represent the seat of instability of the system, have a characterisic spike of warmth centered near the deglaciation. Many other properties of the solution are shown to agree with the observations, notably the bimodal histogram, the red-like variance spectrum with significant peaks, and the dimensionality. The concomitant variations of the deep ocean or thermocline temperature are offered as side predictions to be verified by future paleo-oceanographic studies. If our model is correct, there should be a general warming of the bulk of the world ocean during glacial buildup with maximum temperature lagging the ice maximum and minimum temperature lagging the ice minimum (e.g., the bulk ocean should be cooling at present).

From a mathematical viewpoint, it is noteworthy that the eigenvalue structure of the model dynamical system is one of co-dimension two, permitting a long (100 k yr) period oscillation by exponential type instabilities, in spite of a relatively short first order response time of 6.5 k yr, and generating the spikes and rapid transitions that would not be possible in a one or two-component models. Thus, in partial answer to the question posed in the Introduction, we claim that the "slow climatic attractor" is that of a periodically-forced, three-component dynamical system of co-dimension two (Saltzman and Sutera 1984). The challenge is to find the proper identifications with the physical world for these three variables, and indeed, this has been the main subject of this paper.

Many further studies to improve and amplify our model are desirable. A few of these are:

1) Coupled atmosphere-ocean GCM studies of the diagnostics prescribed and solved for in Section 5, as well as of the rapid transitional periods deduced, which may uniquely involve high enough energy fluxes to be calculable,
2) Related to the above, more sharply focused studies of the physical properties and dynamics of special points of interest in the deduced evolution, such as the 6 k yr BP hypsothermal spike point,
3) Study of the mathematical properties of co-dimension two systems that lead to the properties exhibited by our system,
4) Study of the general effects of stochastic perturbations on this three variable, co-dimension two, system, in a similar manner to those studies made by Saltzman et al. (1981), Saltzman (1982), and Nicolis (1984) for the two-component oscillatory models,
5) Study of the roles of continental ice sheet mass ζ, shelf ice mass χ, and bedrock depression ϵ, as additional prognostic, slow-response, variables, and
6) Study of the role of atmosphere CO_2 as an <u>explicit</u>, probably diagnostic, variable, the implicit effects of which are now included primarily in the coefficients of equation (17b) for $d\tau/dt$.

In the near future we hope to be able to report on work in some of these areas that is now underway.

<u>Acknowledgements</u>

This review is based upon research supported by the Division of Atmospheric Sciences, National Science Foundation, under grant ATM-8411195 at Yale University. It is a pleasure to acknowledge the continuing fruitful discussions on the subject treated with Alfonso Sutera and also with Roberto Benzi. Many of the ideas expressed here owe their origin to these discussions. I am grateful also to Kirk Maasch for his aid in making the calculations and preparing the figures.

References

Benzi, R., G. Parisi, A. Sutera, and A. Vulpiani, 1982: Stochastic resonance in climatic change. <u>Tellus</u>, **34**, 10-16.

Benzi, R., G. Parisi, A. Sutera, and A. Vulpiani, 1983: The theory of stochastic resonance in climatic change. <u>SIAM J. Appl. Math.</u>, **43**, 565-578.

Birchfield, G.E. and R.W. Grumbine, 1985: On the 'slow' physics of large continental ice sheets, the underlying bedrock and the Pleistocene ice ages. J. Geophys. Res. (in press).

Birchfield, G.E., J. Weertman, and A.T. Lunde, 1981: A paleoclimate model of northern hemisphere ice sheets. Quatern. Res., **15**, 126-142.

CLIMAP Project Members, 1976: The surface of the ice-age earth. Science, **191**, 1131-1137.

CLIMAP Project Members, 1982: Seasonal reconstructions of the Earth's surface at the Last Glacial Maximum. Geol. Soc. Am. Map and Chart Series, **36**, (Text, Maps and Microfiche).

Denton, G.H. and T.J. Hughes, 1981: The last great ice sheets. J. Wiley & Sons, N.Y., pp. 484.

Dodge, R.E., R.G. Fairbanks, L.K. Benninger, and F. Maurrasse, 1983: Pleistocene sea levels from raised coral reefs of Haiti. Science, **219**, 1423-1425.

Eriksson, E., 1968: Air-ocean-icecap interactions in relation to climatic fluctuations and glaciation cycles. Meteor. Monogr., **8**, No. 30, 68-92.

Fairbanks, R.G. and R.K. Matthews, 1978: The marine oxygen isotope record in Pleistocene coral, Barbados, West Indies. Quatern. Res., **10**, 181-196.

Flint, R.F., 1971: Glacial and quaternary geology. Wiley, New York, 892 pp.

Fong, P., 1982: Latent heat of melting and its importance for glaciation cycles. Climatic Change, **4**, 199-206.

Fraedrich, K., 1978: Structural and stochastic analysis of a zero-dimensional climate system. Q. J. R. Meteorol. Soc., **104**, 461-474.

Ghil, M., 1976: Climate stability for a Sellers-type model. J. Atmos. Sci., **33**, 3-20.

Ghil, M., 1984: Climate sensitivity, energy balance models and oscillatory climate models. J. Geophys. Res., **89**, 1280-1284.

Ghil, M. and H. LeTreut, 1981: A climate model with cryodynamics and geodynamics. J. Geophys. Res., **86**, 5262-5270.

Grassberger, P. and I. Procaccia, 1983: Measuring the strangeness of strange attractors. Physica, **9D**, 189-208.

Guckenheimer, J. and P.J. Holmes, 1983: Nonlinear oscillations, dynamical systems and bifurcations of vector fields. Springer, New York.

Hasselmann, K., 1976: Stochastic climate models, part I. theory. Tellus, **28**, 473-485.

Hays, J.D., J. Imbrie, and N.J. Shackleton, 1976: Variations in the earth's orbit: Pacemaker of the ice ages. Science, **194**, 1121-1132.

Holmes, P.J., 1980: Averaging and chaotic motions in forced oscillations. S.I.A.M.J. Appl. Math. **38**, 65-80 [Errata and addenda, SIAM J. Appl. Math., 40, 167-168].

Imbrie, J. and J.Z. Imbrie, 1980: Modeling the climatic response to orbital variations. Science, **207**, 943-953.

Imbrie, J., N.J. Shackleton, N.G. Pisias, J.J. Morley, W.L. Prell, D.G. Martinson, J.D. Hays, A. McIntyre, and A.C. Mix, 1984: The orbital theory of pleistocene climate: support from a revised chronology of the marine ^{18}O record. Milankovitch and Climate Part I (A. Berger, J. Imbrie, J. Hays, G. Kukla, and B. Saltzman, eds.) Reidel, Dordrecht, 269-305.

Källén, E., C. Crafoord, and M. Ghil, 1979: Free oscillations in a climate model with ice-sheet dynamics. J. Atmos. Sci., **36**, 2292-2303.

Lemke, P., 1977: Stochastic climate models, part 3. application to zonally averaged energy models. Tellus, **29**, 385-392.

LeTreut, H. and M. Ghil, 1983: Orbital forcing, climatic interactions, and glaciation cycles. J. Geophys. Res., **88**, 5167-5190.

Lorenz, E.N., 1963: Deterministic non-periodic flow. J. Atmos. Sci., **20**, 130-141.

Mitchell, J.M., 1976: An overview of climatic variability and its causal mechanisms. Quaternary Res., **6**, 481-493.

Newell, R.E., 1974: Changes in the poleward energy flux by the atmosphere and ocean as a possible cause for ice ages. Quatern. Res., **4**, 117-127.

Nicolis, C., 1982: Stochastic aspects of climatic transitions - response to a periodic forcing. Tellus, **34**, 1-9.

Nicolis, C., 1984a: Self-oscillations and predictability in climate dynamics. Tellus, **36A**, 1-10.

Nicolis, C., 1984b: Self-oscillations and predictability in climate dynamics - periodic forcing and phase locking. Tellus, **36A**, 217-227.

Nicolis, C., 1984c: A plausible model for the synchroneity or the phase shift between climatic transitions. Geophys. Res. Let., **11**, 587-590.

Nicolis, C., 1985a: Correlation functions and variability in an oscillatory climate model. Geophys. Astrophys. Fluid Dynamics, **32**, 91-102.

Nicolis, C., 1985b: Correlation functions and variability in a periodically forced oscillatory climate model. Geophys. Astrophys. Fluid Dynamics, (in press).

Nicolis, C. and G. Nicolis, 1984: Is there a climatic attractor? Nature, **311**, 529-532.

Oerlemans, J., 1980: Model experiments on the 100,000-yr glacial cycle. Nature, **287**, 430-432.

Peltier, W.R., 1982: Dynamics of the ice age earth. Advances in Geophysics, 24, 1-146.

Pollard, D., 1982: A simple ice sheet model yields realistic 100 k yr glacial cycles. Nature, **296**, 334-338.

Prell, W.L., 1984: Monsoonal climate of the Arabian Sea during the Late Quaternary: A response to changing solar radiation. Milankovitch and Climate Part I (A. Berger, J. Imbrie, J. Hays, G. Kukla, and B. Saltzman, eds.). Reidel, Dordrecht, 349-366.

Ruddiman, W.F., 1985: Climate studies in ocean cores. Paleoclimate analysis and modeling (A.D. Hecht, ed.), J. Wiley & Sons, New York, Chapter 5, 197-257.

Saltzman, B., 1978: A survey of statistical-dynamical models of the terrestrial climate. Advances in Geophysics, **20**, 183-304.

Saltzman, B., 1982: Stochastically-driven climatic fluctuations in the sea-ice, ocean temperature, CO_2 feedback system. Tellus, **34**, 97-112.

Saltzman, B., 1983: Climatic systems analysis. Advances in Geophysics, **25**, 173-233.

Saltzman, B., 1984: On the role of equilibrium atmospheric climate models in the theory of long period glacial variations. J. Atmos. Sci., **41**, 2263-2266.

Saltzman, B., 1985: Paleoclimatic modeling. Paleoclimate analysis and modeling (A.D. Hecht, ed.), Chapter 8, J. Wiley & Sons, New York, 341-396.

Saltzman, B., 1985: Comments on a climatic 'equilibrium for the Quaternary. J. Atmos. Sci, **42**, .

Saltzman, B., A.R. Hansen, and K.A. Maasch, 1984: The late Quaternary glaciations as the response of a three-component feedback system to earth-orbital forcing. J. Atmos. Sci., **41**, 3380-3389.

Saltzman, B. and R.E. Moritz, 1980: A time-dependent climatic feedback system involving sea-ice extent, ocean temperature, and CO_2. Tellus, **32**, 93-118.

Saltzman, B. and A. Sutera, 1984: A model of the internal feedback system involved in late Quaternary Climatic Variations. J. Atmos. Sci., **41**, 736-745.

Saltzman, B., A. Sutera, and A. Evenson, 1981: Structural stochastic stability of a simple auto-oscillatory climatic feedback system. J. Atmos. Sci., **38**, 494-503.

Saltzman, B., A. Sutera, and A.R. Hansen, 1982: A possible marine mechanism for internally generated long-period climate cycles. J. Atmos. Sci., **39**, 2634-2637.

Saltzman, B., A. Sutera, and A.R. Hansen, 1984a: Earth-orbital eccentricity variations and climatic change. Milankovitch and Climate (A. Berger, J. Hays, J. Imbrie, G. Kukla, and B. Saltzman, eds.) D. Reidel Publ. Co., Dordrecht, Holland, 615-636.

Saltzman, B., A. Sutera, and A.R. Hansen, 1984b: Long period free oscillations in a three-component climate model. New Perspectives in Climate Modelling (A. Berger and C. Nicolis, Eds.), Elsevier Sci. Publ. Co., Amsterdam, 289-298.

Sergin, V. Ya., 1979: Numerical modeling of the glaciers-ocean-atmosphere global system. J. Geophys. Res., **84**, 3191-3204.

Shackleton, N.J. and N.D. Opdyke, 1973: Oxygen isotope and paleomagnetic stratigraphy of equatorial Pacific core V28-238: Oxygen isotope temperature and ice volumes on a 10^5 and 10^6 year scale. Quaternary Res., **3**, 39-55.

Sutera, A., 1981: On stochastic perturbation and long-term climate behaviour. Quart. J. Royal Met. Soc., **107**, 137-153.

ICE SHEET DYNAMICS AND THE PLEISTOCENE ICE AGES

G. Edward Birchfield
Department of Geological Sciences
Northwestern University
Evanston, Illinois
U. S. A.

ABSTRACT. A brief review of ice sheet physics is given in the very specific context of modelling of continental ice sheets of the late Pleistocene. The focus is on possible roles of the 'slow physics' of the ice sheets and bedrock response in explaining the prominent feature of the 100 ky response in the spectral representation of oxygen stable isotope record from deep sea sediments, which is taken as a first approximation of ice volume changes.

1. INTRODUCTION

The first part of the following lectures will review the features of one of the most important paleoclimate records, that of the stable oxygen isotopic analysis applied to deep sea sediments. A very brief review of the astronomical theory of the Ice Ages will be given including a presentation of the paradox of the spectral response of the oxygen isotopic record at 100 ky. As components of the climate system with the longest response times, and which offer an explanation of the '100 ky' problem, I will discuss briefly the dynamics of very large continental ice sheets and the underlying bedrock. I then present some modelling results directed toward the solution of this problem. In conclusion I will return briefly to the isotope record.

2. THE OXYGEN STABLE ISOTOPE RECORD IN DEEP SEA SEDIMENTS

One of the most important sources of paleoclimate information comes from the analysis of deep sea sediment cores collected from the basins of the world ocean. These sediments are frequently made up of a high percentage of calcium carbonate shells of microscopic animals, planktonic and benthic. In certain circumstances, these fossil shells have preserved the relative amount of the stable isotopes 18 of oxygen and 13 of carbon that existed when the shells were formed during the lifetime of the animals. When these animals precipitate their shells the relative amount of stable isotope of oxygen going into their shells depends on two properties of the water: the temperature and the relative amount present in the water.

The stable isotope anomaly is defined

$$\delta \equiv \frac{R - R_S}{R_S} \times 1000$$

C. Nicolis and G. Nicolis (eds.), Irreversible Phenomena and Dynamical Systems Analysis in Geosciences, 381–398.

where R is the relative amount of heavy to light oxygen in the sample, and R_S is the ratio for a known standard substance. An empirical equation has been developed relating δ_F, the isotope anomaly in the fossil shell and the in situ water temperature and isotope anomaly at the time the animal lived:

$$T = 16.9\text{–}4.4(\delta_F - \delta_W) + 0.10(\delta_F - \delta_W)^2 \tag{1}$$

or

$$\delta_F \equiv \delta_F(T, \delta_W) \approx \delta_W - \delta_T \qquad \delta_T \equiv (T - 16.9)/4.4$$

where T is the water temperature in °C, δ_W the anomaly of the water, and we neglect the last term in (1). We can define anomalies relative to the present ocean values T_O, δ_{FO}, δ_{WO} so that

$$\hat{\delta}_F = \hat{\delta}_W - \hat{\delta}_T \qquad \hat{\delta}_T \approx 0.23\,\hat{T} \tag{2}$$

where the ^ denotes relative values of the variables. The great usefulness of the stable isotope data as a proxy paleoclimatic indicator depends on the fact that $\hat{\delta}_T$ tends to be small relative to $\hat{\delta}_W$. This means that $\hat{\delta}_F$ is a measure of changes in ocean water isotope value.

On the time scale of the Pleistocene Ocean, the total amount of stable isotope in the hydrosphere is constant; surface evaporation from the ocean surface involves significant fractionation, with the vapor having preferentially isotopically lighter water. An increase in the mean ocean water value as indicated by an increase in $\hat{\delta}_F$ points to a net loss of water from the ocean surface through evaporation, with the isotopically light water mass stored on the continents as ice. Thus, insofar as $\hat{\delta}_T$ is small, the stable oxygen isotope record is a measure of changes in mean sea level or continental ice volume. Shackleton and Opdyke (1973) have estimated that a change of .1 per mil in $\hat{\delta}_F$ approximates 10 m sea level change. If the change in $\hat{\delta}_F$ is due to temperature only, a .1 per mil change corresponds to approximately .4 °C temperature change.

Two isotope records are illustrated in Figure 1; the remarkable similarity of these two records substantiates their representation of global climate changes; further, the fact that the similarity extends to a number of relatively small features in the two records is indication of the relative low level of noise and the fact that some global changes have occurred on relatively fast time scales.

It has been well established that the isotopic record for the last million years of the Pleistocene has been made up of similar fluctuations to the one seen in Figure 1. In the late Pleistocene this pattern appears as relatively long glacials of approximately 100 ky length, separated by brief interglacials; there is a distinctive saw tooth pattern to the record of the major glacials. Data of this kind and some other paleoclimate proxy records have been used to substantiate the astronomical theory of the Pleistocene Ice Ages.

3. THE ASTRONOMICAL THEORY OF THE ICE AGES

The theory is based on the existence of perturbations of the earth's orbit, that is, a) variations in the eccentricity of the orbit, b) variations in the tilt of the earth's axis to the plane of the ecliptic, c) precession of the longitude of perihelion relative to the moving equinox produced from the time varying gravitational attraction of the other planets of the solar system. These nearly periodic perturbations produce small fluctuations in the insolation received by the earth. The magnitudes are at most 10%

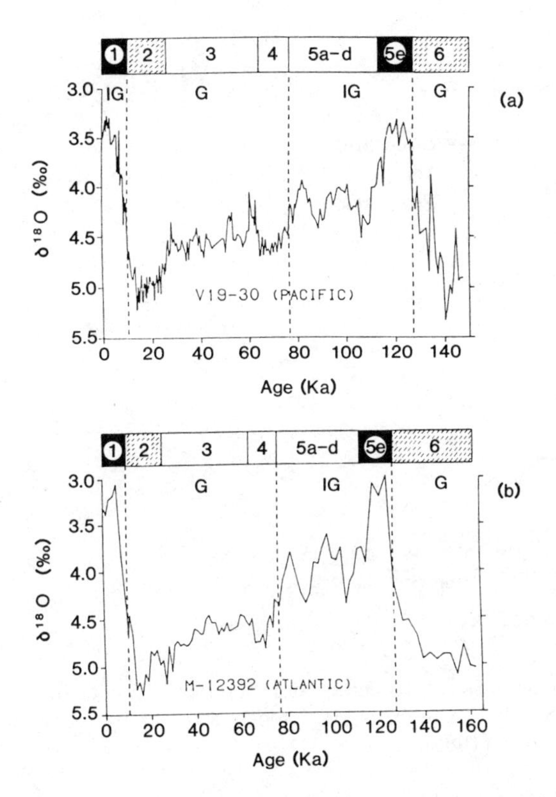

Figure 1. Two benthic oxygen isotope records versus time, from Shackleton et al. (1983). Isotope stages 1-6 are labelled at top (fully interglacial intervals marked with a solid pattern, full glacial intervals by a random dash pattern.)

and mostly smaller. They involve changes in the distribution of insolation with latitude, with the seasons, and to a lesser extent in mean annual radiation.

The spectral properties of each of the three obliquity perturbations are shown in the upper part of Figure 2. The eccentricity varies primarily near 413 and 100 ky. The tilt perturbations occur near 41 ky; the precessional variations occur near 19 and 23 ky. The periods of the perturbations change slightly with time and more importantly the relative amplitudes vary significantly with time.

The major achievement in support of the astronomical theory has been the verification of the high degree of similarity of the spectrum of the orbital perturbation forcing with the spectral response of the climate system as seen in the oxygen isotope record, for example in the lower part of Figure 2. The agreement not only includes the peaks associated with the precessional and orbital perturbations at 19, 23 and 41 ky respectively, but also in the 100 ky peak associated with the eccentricity. It is in this latter feature that the theory has been in a sense too successful. Although the insolation at any latitude is indeed perturbed by the three orbital features, the relative amplitude of the eccentricity changes are so small that they are scarcely detectable in the insolation, as seen in the typical insolation spectrum shown in Figure 3. Not only is this spectral peak the largest in the isotope record, but as shown by Imbrie et al. (1984), the phase correlation is also high with the eccentricity

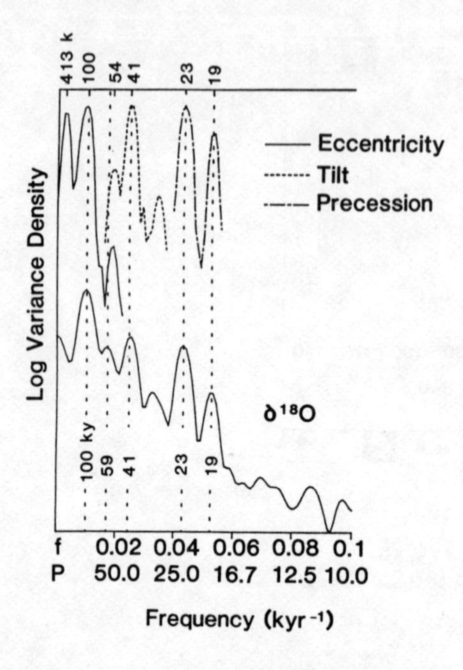

Figure 2. Variance spectra of orbital variations a_l and 'stacked' oxygen isotope record for the last 800000 years. The log of the variance is plotted versus frequency in cycles per thousand years. From Imbrie (1985).

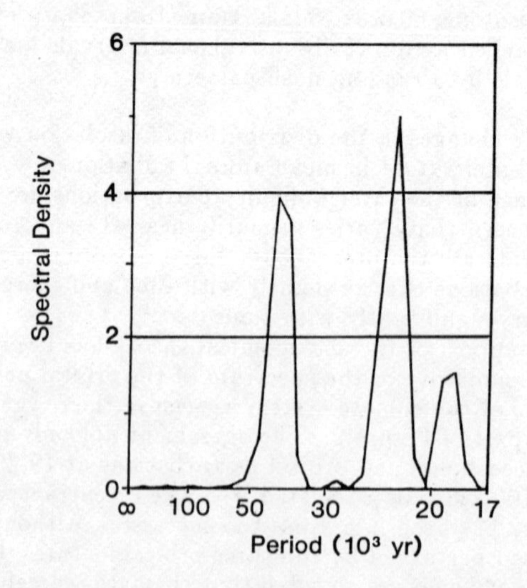

Figure 3. Spectral density on a linear scale as a function of period in thousands of years, for insolation at 60 ° N over the last 600000 years, from Birchfield et al. (1981).

perturbations.

It has been concluded that while the tilt and precessional spectral peaks in the isotope record represent a linear response of the climate system to the astronomical forcing, the spectral response at 100 ky, if related to the astronomical forcing at all, must represent a nonlinear response. Because of its large amplitude it presents a paradox of some importance.

Because the time scale of the continental ice sheets and of the underlying bedrock are the longest of all the components in the climate system and are the closest to those of the astronomical forcing, these components are promising places in the system to look for nonlinear processes which might resolve the paradox.

4. SIMPLE ICE SHEET DYNAMICS: A REVIEW

The dominance of viscous forces in large ice sheets assures that they are in near equilibrium. I consider here only two dimensional ice sheet models. Neglecting deviatoric stresses, I have the following dynamical equation:

$$-\rho_i\, gH\frac{\partial h}{\partial x} = \tau_{zx} \qquad H \equiv h + h' \tag{3}$$

where H is the total thickness of the ice sheet, h the elevation of the free surface, h' the elevation of the bedrock surface below sea level, ρ_i is the density of ice and τ_{zx} is the basal shear stress. This states that the downstream net hydrostatic pressure gradient force resulting from the slope of the free surface of the ice sheet, is balanced by the shear stress exerted on the bottom surface of the ice sheet.

I will discuss briefly two simple ice sheet models that have been used in various paleoclimate models. They differ primarily in the rheology assumed for the ice. In addition I will present very simple models of the underlying bedrock response to the overlying ice load.

4.1. Perfect Plastically Flowing Ice Sheet Models

The simplest ice sheet model assumes that the ice flows continuously and plastically, that is at all times

$$\tau_{zx} = \tau_0 \qquad -\rho_i\, gH\frac{\partial h}{\partial x} = \tau_0$$

where τ_0 is the yield stress, approximately 1 bar.

4.1.1. *Case A: No Bedrock Deformation Under Ice Load.* If the bedrock is perfectly rigid, then the total thickness of the ice H is simply h. The profile of the plastically flowing ice sheet is then

$$h = H = \sqrt{\lambda(L - x)} \qquad \lambda = \frac{2\tau_0}{\rho_i\, g}\,, \quad V = \frac{2}{3}\sqrt{\lambda}\, L^{3/2}$$

where L is the half width of the ice sheet assumed symmetric about the midpoint, $x = 0$ and V is the volume of the southern half of the ice sheet.

A schematic picture is shown in Figure 4. In this simple model mass balance on the north half is assumed to be maintained automatically by flux of ice into the polar ocean. In the southern half, there is net accumulation for $x < x_f$ and net ablation for $x > x_f$, where x_f is the firn line defined by the intersection of the

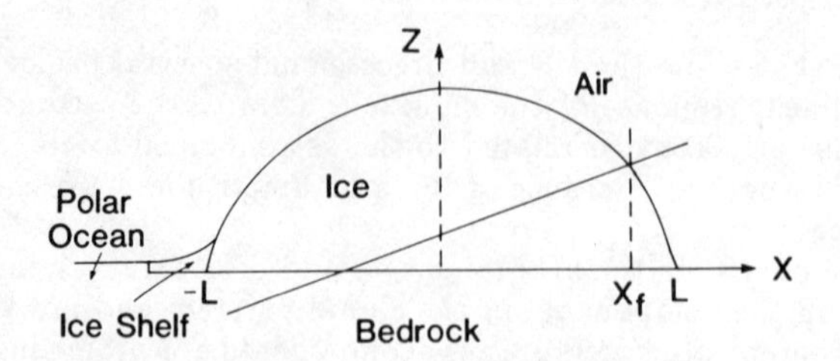

Figure 4. Schematic cross-section of a model continental ice sheet adjacent to a polar ocean. The sloping snow line intersects the ice sheet at x_f. Here bedrock is assumed perfecly rigid to the varying ice load.

sloping snow line with the free surface of the ice sheet. If the accumulation and ablation rates are taken as constants, $a, a' > 0$, respectively, then mass conservation for the southern half of the ice sheet is given by:

$$\frac{dV}{dt} = \sqrt{\lambda} L^{1/2} \frac{dL}{dt} = a x_f - a' (L - x_f)$$

This equation is closed by expressing x_f in terms of L making use of the equation for the snow line and its intersection with the profile of the ice sheet. Specifying the accumulation and ablation rates together with the time varying snowline position and an initial profile establishes an initial value prediction problem for the half width L. Weertman (1964) has shown that the approximate time to reach equilibrium is given by

$$t_e \approx \frac{3}{2} \frac{\lambda}{(aa')^{1/2} s}$$

where s is the slope of the snow line. Taking $\lambda \sim 20\ m$, a $\sim 1\ my^{-1}$, $a' \sim 2.4\ my^{-1}$, s $\sim 10^{-3}$, adjustment time is seen to be of the order of 10^4 years, which makes it indeed the slowest of the components of the climate system.

4.1.2. *Case B: Instantaneous Bedrock Adjustment.* A second version of the perfect plasticity ice sheet model assumes instantaneous bedrock response to the varying ice load and is illustrated in Figure 5. For bedrock adjustment, buoyancy forces must be in equilibrium

$$(\rho_R - \rho_i) h' \approx \rho_i h \tag{4}$$

Now $\rho_i / \rho_R \approx 1/3$, then $H \approx 3/2\ h$ and from (3)

$$h = \sqrt{\hat{\lambda}(L - x)} \qquad \hat{\lambda} = 2/3 \lambda$$

where ρ_R is the density of the bedrock. The profile of the free surface remains parabolic with a redefined parameter λ; the bedrock profile is also parabolic. The effect of the bedrock adjustment is to reduce slightly the ice volume and the time required for the ice sheet to reach equilibrium.

4.1.3 *Case C: Bedrock Adjustment With Time Delay, Simplest Case.* In this simplest of delayed bedrock response, it is assumed that the bedrock depression at a given latitude is a function only of the ice load at that point. Bedrock changes are then caused

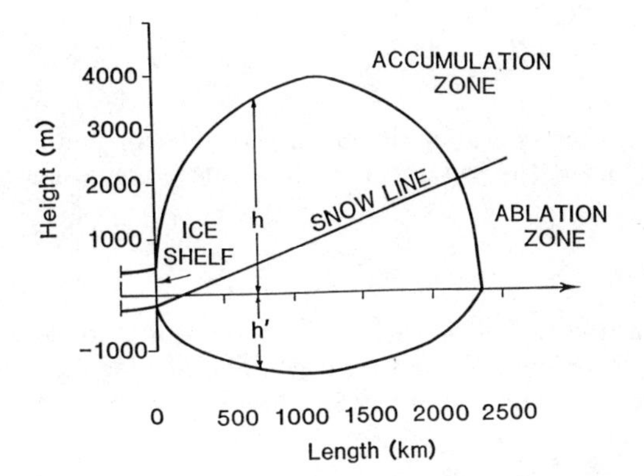

Figure 5. Schematic cross-section of a model continental ice sheet with bedrock response to the varying ice load.

by: a) the ice load, b) a restoring force proportional to bedrock displacement. That is,

$$\frac{\partial h'}{\partial t} = \nu H - \tau_l^{-1} h' \quad , \; \nu = \frac{\rho_i \tau_l^{-1}}{\rho_R} \tag{5}$$

where τ_l is the bedrock relaxation time. It is a consequence of the local loading that there is a *single* time scale associated with the bedrock response. In order to preserve the features of the perfect plasticity model, it is necessary to assume a spatially constant ratio of bedrock displacement to total ice sheet thickness; this can be written as

$$h'(x,t) \equiv \beta(t) h(x,t) \qquad 0 \leq \beta \leq 1/2$$

$$H = (1+\beta) h$$

$$0 < \tau_l < \infty$$

where the right limit case in the second inequality corresponds to the rigid bedrock case and the left limit to the instantaneous adjustment case. The profile is found as usual from (3)

$$h = \sqrt{\tilde{\lambda}(L-x)} \qquad V = 2/3 \tilde{\lambda}^{1/2} L^{3/2} \quad , \qquad \tilde{\lambda} \equiv \frac{\lambda}{1+\beta}$$

Mass balance is found as above

$$\tilde{\lambda}^{1/2} L^{1/2} \frac{dL}{dt} - \frac{1}{3} \tilde{\lambda}^{1/2} (1+\beta)^{-1} L^{3/2} \frac{d\beta}{dt} = a x_f - a'(L - x_f)$$

Utilizing the definition of h' the bedrock equation (5) becomes

$$h \frac{d\beta}{dt} + \beta \frac{dh}{dt} = \frac{\rho_i}{\rho_R} \left[1 - (1 - \frac{\rho_i}{\rho_R}) \beta \right] h \; \tau_l^{-1}$$

The model consists then of two ordinary nonlinear differential equations. This model is essentially that used in a series of zero-dimensional climate models of Ghil and others. See for example LeTreut and Ghil (1983).

4.2 Ice Sheet Model 2:

A straightforward integration in the vertical of the incompressible continuity equation together with boundary conditions at the top and bottom yields a prediction equation for ice sheet thickness, H:

$$\frac{\partial H}{\partial t} + \frac{\partial M}{\partial x} = A(x,t) \tag{6}$$

where M is the horizontal mass flux and A is the sum of the mean annual net mass gain per unit time at the top and bottom of the ice sheet. The rheology is based on 'Glenn's Law' which states that the shear strain rate is proportional to a power of the stress.

$$\frac{\partial u}{\partial z} = \hat{c}\,\tau_{zx}^{m} \ , \ m \approx 3$$

where u is the horizontal ice velocity, $\hat{c}$ a constant. Balance of forces can be applied at level z to give, as in (3)

$$\tau_{zx}(x,z,t) \approx -\rho_i\, g\,(h-z)\frac{\partial h}{\partial x}$$

Integration twice with respect to z expresses the horizontal mass flux in terms of the total thickness of the ice sheet and the slope of the free surface.

$$M = -cH^5\left(\frac{\partial h}{\partial x}\right)^3 \ , \ c = 1/5\hat{c}\,(\rho g)^3 \approx 1\times10^{-12} m^{-3} s^{-1}. \tag{7}$$

The constant of proportionality, $\hat{c}$, is adjusted to crudely allow for possible sliding at the base of the ice sheet.

4.2.1. *Case A: Coupling with 'Simple' Bedrock Model.* When (6) and (7) are combined with the simplest bedrock model (5), a coupled set of equations for the ice sheet thickness and bedrock displacement is obtained. This is the model used by Birchfield et al. (1981), for example.

4.2.2. *Case B: Coupled Ice Sheet Model and Two Layer Bedrock Model.* The upper layer of the bedrock model is an elastic 'lithosphere'; the lower layer of very great depth is a viscous 'asthenosphere'. There are then two modes of response: an external mode associated with the surface bedrock displacement and an internal mode associated with the interface between the lithosphere and asthenosphere. To solve numerically the model equations it is necessary to expand the displacements and the load in appropriate orthogonal functions. Although spherical harmonics are appropriate for the earth (Peltier (1982) has used Legendre functions), if the width of the ice sheet is sufficiently small compared to the radius of the earth, ordinary fourier transforms may be used. For a finite grid used in a numerical model, a possible representation of the bedrock displacement is

$$h'(x,t) = \sum_{l=0}^{N} h_l(t)\, e^{ik_l(x-\Delta x)} \ , \ k \equiv 2\pi l/L$$

where L is M times the grid spacing δx, M is the total number of points on the grid, and $N = M-1$.

The derivation of the equation for the bedrock displacement expansion coefficients is lengthy; the solution consists of the sum of an elastic response and a viscous response; there is a contribution to h' from the internal modes in addition to the external modes. If the elastic contribution is neglected (not always the best approximation) and the contribution from the internal modes is neglected (a good approximation) and for not too small ice sheets, the bedrock equations reduce to

$$\frac{\partial h_l'}{\partial t} = \eta_l H_l - \tau_l^{-1} h_l' \quad , \qquad \eta_l = \frac{\rho_i}{\rho_R} \tau^{-1}$$

where the H_l are the coefficients for the ice load and τ_l is the relaxation time for the l th wavelength in the bedrock displacement, h_l' . Thus there is a spectrum of response times, one for each wavelength. Figure 6 shows a typical spectrum for the external and internal modes for the two layer bedrock model. For wavelengths longer than 1000 km the viscous contribution dominates; for shorter wavelengths the elastic response dominates the relaxation times for the external modes. Note that for a model with no lithosphere, i.e., a purely viscous single layer model in which the ice sheet floats on the 'asthenosphere', the relaxation time continues to increase as the wavelength decreases. It is a result of some importance that in either model, there are scales for which there are relatively long relaxation times, that is of the order of 10,000 years.

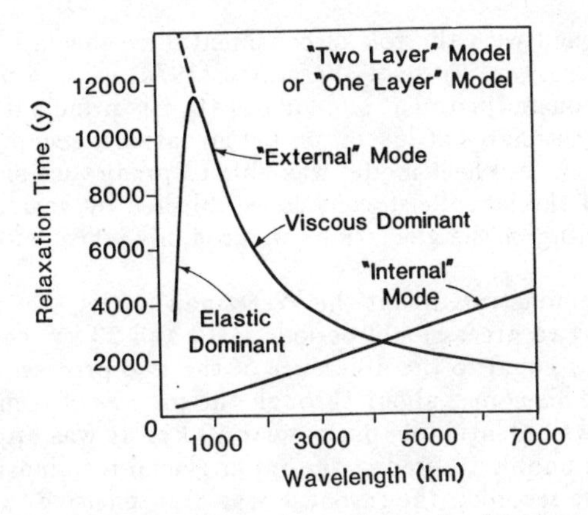

Figure 6. Typical relaxation time of the modes for a two-layer earth model versus horizontal wavelength. The relaxation time of modes for a single viscous layer model is shown schematically as the right side of the bell shaped curve (extended to short wavelengths by the dashed line).

4.3. Parameterization of the Accumulation Function

The complete hydrological cycle in the atmosphere is involved in determining the net mass balance at the surface of the ice sheet; complicated thermodynamics at the base of the ice sheet also enter into basal melting. The following shows examples of two simple schemes for incorporating some of the most important features of the surface mass balance on the ice sheet.

4.3.1. *Scheme A.*

$$A[h(x),t] = \begin{bmatrix} a[1-bh(x,t)] & x < x_f(t) \\ -a'[1-bh(x,t)] & x \geq x_f(t) \end{bmatrix} \tag{9}$$

where a, a' and b are positive constants. This is the scheme used by Weertman (1976), Birchfield et al. (1981), Peltier (1982) and others. The term involving b allows in a very crude sense for the effect of temperature decrease with elevation in the accumulation and ablation zones.

4.3.2. *Scheme B.*

$$A(x,h,t) = \begin{bmatrix} a[1+bT(x,h,t)] & T \leq 0 \\ -a'[1+b_1 T(x,h,t)] & T > 0 \end{bmatrix} \tag{10}$$

$$T(x,z,t) = \gamma[S(x-x_0(t)-z]$$

where T is the 'temperature', γ the 'lapse rate', x_0 the sea level position of the snow line $T=0$, and S, b, b_1 are constants. In addition to the elevation effect the temperature and hence accumulation and ablation decrease with latitude.

5. RESULTS OF SOME PRELIMINARY EXPERIMENTS

A. The first experiments concerned with the role of continental ice sheets in the astronomical theory of the ice ages were those of Weertman (1976) using a perfect plasticity model with instantaneous adjustment similar to (4). His principal conclusions were that with reasonable estimates of insolation perturbations, accumulation rates, ablation rates, such a simple ice sheet model was able to predict ice sheets of the lateral and vertical extent of the late Pleistocene as established by various geological records; further that the timing of the glaciations was roughly consistent with the geological record.

B. In Birchfield (1977) I demonstrated that the Weertman model, when forced by a perturbed snow line at the two precessional periods of 19 and 23 ky, can result in a large response at a frequency equal to the difference of the two precessional frequencies, that is, near 100 ky. This comes about through the process of combination tones; the model also produces well identifiable peaks near 10 ky. It was apparent, however, that such a model was unable to predict the rapid glacial terminations apparent in the proxy ice volume records; the response was also sensitive to the relative amplitudes of the two forcing periods.

C. In Oerlemans (1980) a model similar to the power law rheology ice sheet and the simplest bedrock model described above (i.e., with a single response time scale) was used to demonstrate that the bedrock response was important to the long period model response. Oerlemans showed that if the snow line is perturbed with a period at approximately the mean precessional period, the response depends sensitively on the bedrock response time, specifically that if this response time was greater than about 10 ky, then what appears to be subharmonic response occurred at a period of approximately 100 ky, as seen in Figure 7. From the above discussion this time scale is unrealistically large except for a narrow range of wavelengths of bedrock response.

D. At about the same time Birchfield et al. (1981) utilized the Glenn power law rheology together with the simplest bedrock model with a relaxation time scale of ca 3000 years, which is consistent with the rebound data from the sites of the Laurentide and Fennoscandian ice sheets, to carry out experiments with astronomical forcing for

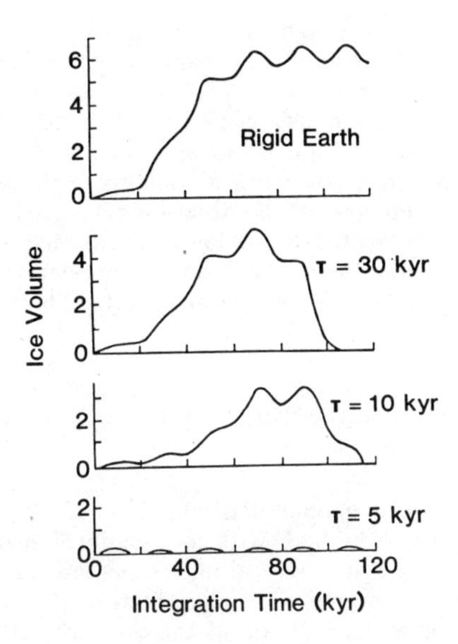

Figure 7. Model ice volume versus time of a continental ice sheet forced by a periodically moving snow line of 20000 years period, for different bedrock relaxation times, from Oerlemans (1980).

insolation at various northern latitudes. Their conclusions were that a reasonable reproduction of the proxy ice volume record was possible with the model, although the model was sensitive to the latitude of insolation , and the accumulation and ablation rates used; it was not excessively sensitive to, for example, the flow law constant in (7). In the spectral response of the model, although there was a significant response near 100 ky, the 40 ky year peak was much larger and clearly inconsistent with the isotope record. The argument was put forward as a partial explanation for this discrepancy in terms of the presence in the ice volume record and absence in the model of a 'red noise' spectral response which if added to the model response would produce a record more similar to that of the oxygen isotope records. Such a red noise spectrum does indeed appear in the 'low order models' of LeTreut and Ghil (1983), Saltzman et al. (1984) which include the relatively fast components of the climate system, the atmosphere and ocean. One additional conclusion from these experiments was that with bedrock delay, rapid terminations appeared in the ice volume record.

Pollard (1984) coupled the power law rheology model to an atmospheric energy balance model; his bedrock model differed from the above in that he assumed that mantle flow response to ice sheet loading was confined to a relatively thin layer. For realistic ice sheet dimensions, however, this model also has very long relaxation times relative to those indicated by sea level changes. With these long relaxation times for the bedrock, his model did show a strong spectral response near 100 ky. It again demonstrated the importance of bedrock adjustment in the rapid glacial terminations.

Conclusions fairly well established from these model experiments are the following.

1. Precessional forcing at 19 and 23 ky through combination tones produces spectral response at 100 ky, but the amplitude appears considerably smaller than that indicated in the isotope records.

2. Bedrock displacement with time delay is necessary to give rapid glacial terminations which is part of the 'sawtooth' phenomenon responsible for the response at 100 ky. This bedrock delay mechanism appears to work as follows: with large enough bedrock depression and sufficiently slow rebound in the ablation zone of the ice sheet, ablation is accelerated by the warm temperatures at the lower elevations; with rapid melting near the leading edge, mass flux into the ablation zone is accelerated due to the increased surface slope. This mechanism appears operative with relaxation times as low as 3300 years.

6. RECENT MODEL EXPERIMENTS CONCERNED WITH NONLINEAR RESPONSES

Peltier (1982) coupled the power law ice sheet model of Birchfield et al. (1981) to a sophisticated visco-elastic earth model which he had, with others, developed for estimating mantle viscosity. The earth model has several layers; associated with each interface there is, as in the two layer model above, a set of normal modes with a corresponding set of relaxation times, some of which are of the order of 100 ky. The bedrock displacement at the surface of the earth is then made up of a weighted sum of contributions from each of these modal solutions.

To simplify his model equations, both written in the spectral domain, he neglects the elastic response and the contribution from all the internal modes to the surface bedrock displacement and in addition, assumes a not too large ice sheet. Most importantly, by making a drastic simplification in the spectral representation of the accumulation rate function A in (9), he reduces each of his spectral model equations to the form of the simple damped harmonic oscillator equation, in which nonlinear terms are written as forcing terms.

He asserts that for sufficiently long bedrock relaxation times, free oscillations are underdamped, but the relaxation time required for oscillations near 100 ky is so unrealistically high, he concludes that the contribution from those internal modes which have time scales near 100 ky also must be included in the surface response. He thus proposes that the spectral response in the isotope record is caused by a resonant response to the astronomical forcing.

Birchfield (unpublished manuscript), however, demonstrated that as Peltier's oscillator equation stands, linear oscillatory response must be overdamped for any bedrock relaxation time. Further Weertman and Birchfield (1984) demonstrated that the contribution to the bedrock sinking from those internal modes with long relaxation times, was so small as to be negligible.

Although it is clear that Peltier's original argument for resonant oscillations of the ice sheet-bedrock system is incorrect, the presence of nonlinear terms in the harmonic oscillator equation does indeed leave open the question of nonlinear oscillations and their possible role in the spectral response near 100 ky. Hyde and Peltier (1985) present the results of a number of experiments with a similar model to that described above. They demonstrate that for certain parameter ranges the model, when forced at a single harmonic period, displays prominent subharmonic oscillations at an integral multiple of the forcing period. Although most of their experiments are concerned with showing subharmonic response at 100 ky, for different parameter values other periods appear. Because the response shows the slow growth and rapid decay of the ice volume similar to the isotope records, they have called their model a relaxation oscillator. See Figure 8 for an example of 100 ky response to 20 ky periodic forcing.

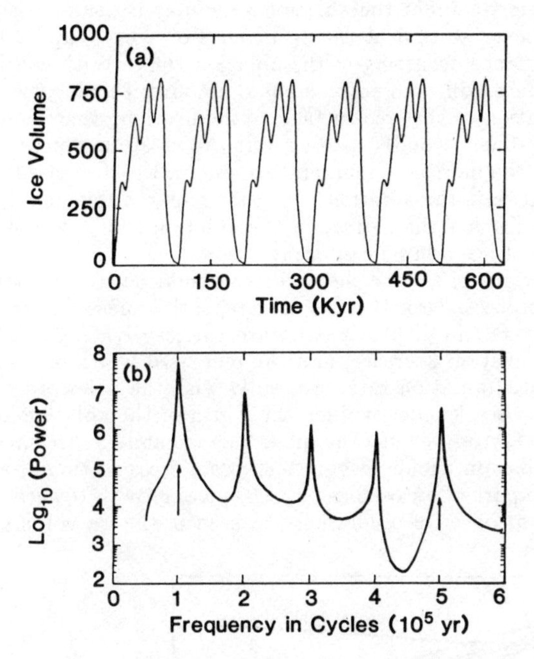

Figure 8. Subharmonic response of the Peltier-Hyde (1984) ice sheet, bedrock model, forced by a periodically moving snow line at 20000 years. Upper picture shows ice volume versus time; lower shows spectral response variance versus frequency; right arrow is the forcing period, left arrow the lowest subharmonic (100 ky).

The mechanism of rapid deglaciation described above appears to play a crucial role in the subharmonic response. As the ice sheet grows from small to large size, at each cycle of advancing snow line, the bedrock displacement is deeper than in the previous cycle, particularly in the ablation zone. At the beginning of the advance before collapse, the bedrock is so deep, that the ablation near the leading edge is so great due to the elevation effect in (9) that the total mass loss more than balances the gain due to the advancing snow line; it appears that the rapid melting in the forefront of the ablation zone, increases the slope sufficiently to enhance mass flow into the depressed region, further increasing the mass loss. By the time the snow line starts to retreat the ice volume has already dropped significantly and the ice sheet collapses with this last retreat.

In Birchfield and Grumbine (1985), an analagous study of free nonlinear oscillations of a coupled ice sheet - bedrock model has been made. We have focussed on the response of the system to a forcing constant in time. The ice sheet model is an improved numerical model of the Glenn rheology and the bedrock model is a modified form of the two layer model described above. We find, for certain parameter ranges, that the coupled model with constant snow line position displays self-sustained oscillations of period around 50 ky. These oscillations are, as is the subharmonic response above, closely associated with the interaction of the bedrock and the upper surface of the ice sheet in the ablation zone.

Considering an ice sheet rapidly advancing southward and approaching mass balance, the bedrock displacement under the ablation zone may lag sufficiently the approach to overall mass balance, so that at the occurrence of mass balance, there is still significant bedrock subsidence occurring in the ablation zone. With continued bedrock sinking, the free surface will then sink an equal amount; this has two effects: first, with a drop of the elevation of the free surface at the firn line, there is a northward displacement of the firn line; secondly, throughout the ablation zone, the ablation rate will increase due to the increased temperature due to the decreased elevation. Both factors tend to increase the ablation over that required for mass balance. The increased melting lowers further the surface in the ablation zone; with this positive feedback, rapid decrease of ice volume can occur.

When the volume and width of the ice sheet decrease sufficiently, two restoring 'forces' enter to reverse the process. First there is compensating mass flux from the accumulation zone, which lowers and displaces poleward the ice divide; the increased surface temperature due to elevation decrease, and the increased width of the accumulation zone, increase the accumulation rate. Secondly with the poleward retreat of the ice sheet towards the fixed sea level snow line, the width of the ablation zone will eventually decrease. The two factors reverse the mass balance and the ice sheet again grows toward its initial equilibrium profile. Figure 9 shows a cross-section of an ice sheet undergoing the collapse part of an oscillation. A typical growth towards mass equilibrium and the subsequent onset of oscillations, as seen in the ice volume, is shown in Figure 10.

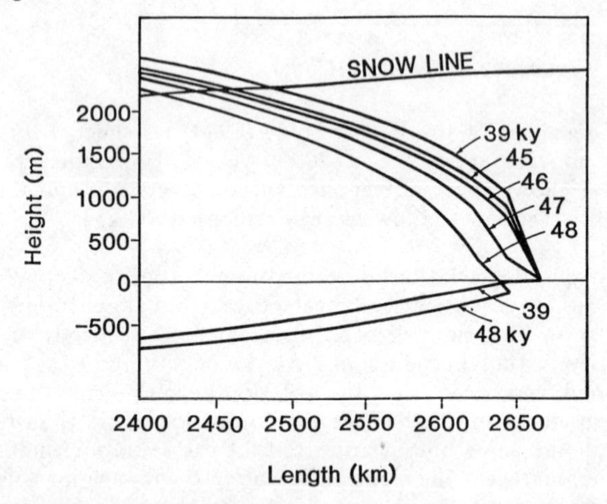

Figure 9. Example of collapse part of a nonlinear 'free' oscillation cycle in Birchfield, Grumbine (1985) model of a continental ice sheet and bedrock response. Cross-section sequence starts at time = 39 ky and ends at 48 ky.

Many parameter variations were made; the period of the oscillations was found to be remarkably insensitive to these changes and in no case approaches 100 ky. The dependence of the 'equilibrium' ice volume on the position of the snowline, holding all other parameters fixed, is shown in Figure 11; this includes delineation of the range over which self-sustained oscillations appear.

Preliminary conclusions can be made from these two studies. With preliminary testing of the Birchfield-Grumbine model with periodic forcing, we have yet to observe the striking subharmonic oscillations described by Hyde and Peltier (1985). Although the two physical models are similar, there are significant differences in detail; the methods of numerical solution are quite different. In spite of these, I am confident that our model will, for the proper choice of parameters, display subharmonic oscillations similar to those found in the Peltier-Hyde model.

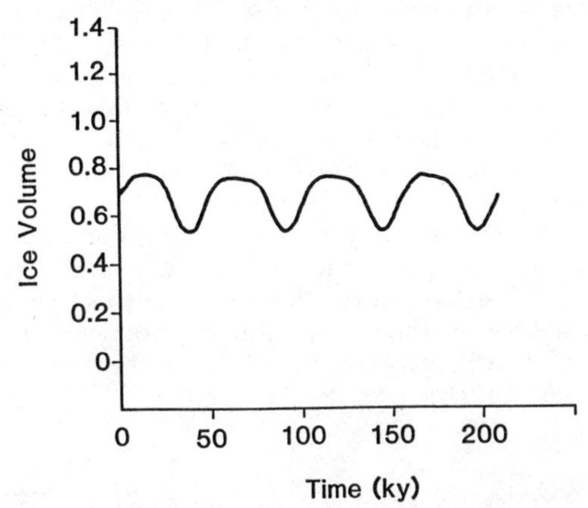

Figure 10. An example of free nonlinear oscillations of a continental ice sheet volume versus time for the Birchfield-Grumbine (1985) ice sheet-bed rock model.

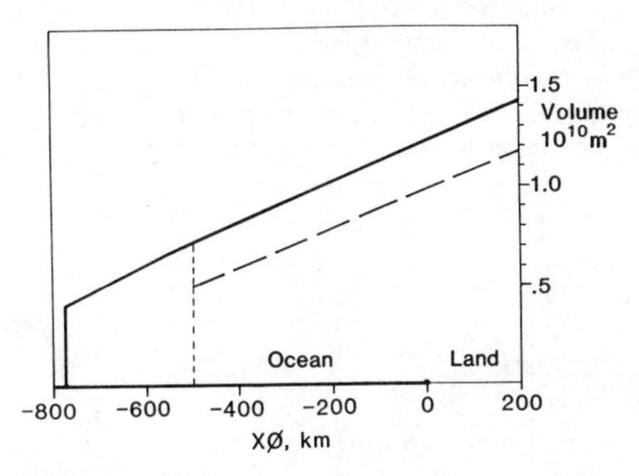

Figure 11. Equilibrium response of the Birchfield-Grumbine (1985) model ice volume versus sea-level snowline position, x_0. For $x_0 > -550$ km, one solution oscillates between the upper solid curve and the lower dashed curve; for $-775 < x_0 < -550$ km two steady states exist, one with zero volume (extending to $x_0 = 0$) and the sloping line to the left of the vertical dashed line.

Such physical processes may play important roles in explaining the spectral response near 100 ky observed in the isotope record. What is not resolved, however, is: a) how are the 'free' oscillations related to the 'subharmonic' oscillations? Is it possible for both kinds of oscillations to appear for the same parameter ranges, that is, in one case with a fixed snow line and in the other with a periodically varying snow line, or are they mutually exclusive responses of the model? b) more importantly, whether the range of parameters for either of these processes falls near the range appropriate for the climate system of the earth.

In the Hyde-Peltier experiments, for example, an unusually low mean bedrock density of ca. 2400 kg/m^3 has been used; this results in unrealistically large bedrock displacements. Another unrealistic feature is that, for almost all times, nearly all of the icesheet north of the ice divide lies in an artificially created ablation zone, created for mimicking the loss of mass to the polar ocean. This almost certainly means that continuously the mass loss on the north side of the ice sheet exceeds that on the south side (certainly the width of the ablation zone on the north side far exceeds that on the south side). While there are indeed possible important interactions with a marine icesheet component in the long time scale response of the system, this part of their model response is not presented in their discussion in the context of marine ice physics, or as having any role in the physics of the subharmonic response associated with the ablation zone on the south side of the ice sheet.

In the Birchfield-Grumbine model, on the other hand, the presence of self-sustained oscillations usually requires an unrealistically large ice sheet to be present. It is also inherent in the oscillatory mechanism for the ice sheets to be near equilibrium. It is possible, although not yet certain, that with astronomical forcing, the ice sheets growth and decay over the late Pleistocene were far from being in equilibrium, as is suggested by the work of Held (1983).

7. THE ISOTOPE RECORD AGAIN

There is increasing evidence that glacial to interglacial temperature changes in the deep ocean in the late Pleistocene may be as large as 2° C. See for example Duplessy, et al. (1985), Birchfield (1985). Thus there is evidence that as much as .4 per mil of the isotope record may be due to temperature changes. Further there is the work of Chappell (1983) on estimating sea level changes from raised beach levels on the coast of New Guinea. His first approximation to sea level changes has been constructed as seen in Figure 12.

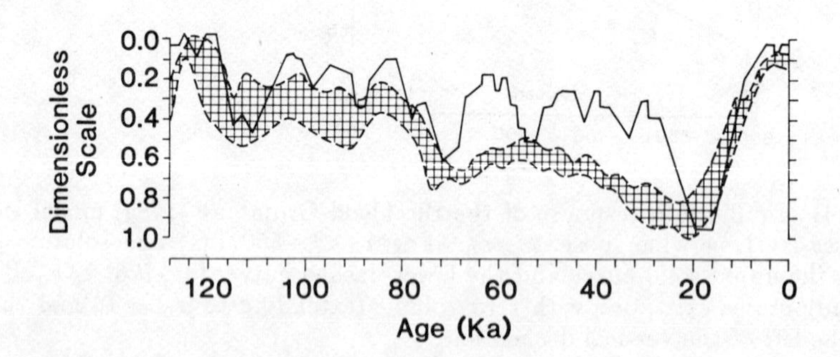

Figure 12. New Huon Peninsula (New Guinea) detailed sea-level record (solid line) for the last 120000 years; hatched area is the envelope of planktonic, benthonic oxygen isotope records for V28-238, from Chappell (1983).

A striking feature of the record, for example, is that for stages 5a and 5c, sea level stands remarkably close to the level of the last interglacial and that of the present.

This is in contrast to the isotopic record where these stages, though defined as high sea level stands, fall well below the values indicated in Chappell's record.

If there is as much as a 2 °C change from glacial to interglacial, this can have an effect not only on the interpretation of the isotope record as a sea level - ice volume record as a time series, but it can also affect the spectral properties of that time series. If indeed stage 5 was characterized by significantly higher sea level stands, as seen in Figure 12, the record loses to a considerable extent the sawtooth character seen in the isotope record. It is this sawtooth character of the isotope record for the last few 100 ky, which is responsible for the very strong spectral response near 100 ky. Depending on the phasing of the temperature changes over the extent of a major glacial, it is possible that the corrected ice volume record may display somewhat less response at 100 ky than does the isotope record.

I summarize:

a. The possible importance of a red noise background must be considered in interpretation of the relative spectral response at the astronomical periods; an overestimation of the volume response at 100 ka may be the consequence of ignoring this.

b. There is some indication that the ice volume record may differ sufficiently from the isotope record for the late Pleistocene due to deep ocean temperature changes, that its spectral properties may be significantly altered from those of the isotope record. Specifically, depending on the time sequencing of the deep ocean temperature changes, there may be less of a prominent response near 100 ky.

c. There are at least two nonlinear mechanisms in the 'slow physics' of the continental ice sheets and underlying bedrock which in model studies produce significant spectral response near 100 ky. First, there is a possible combination tone produced response to the dual precessional period forcing, which appears to be an underestimate of the proxy ice volume record. Secondly there is under proper circumstances a very strong subharmonic response of the system to simple periodic forcing at the mean precessional period.

d. Further studies are needed in order to more clearly ascertain: 1) Deep ocean temperature changes effect on the proxy ice volume record; 2) What the role of 'slow physics' of the ice sheets and bedrock is in the late Pleistocene climate changes.

8. ACKNOWLEDGEMENTS

The research of Birchfield and Grumbine was supported inpart by the Climate Dynamics Section of the National Science Foundation, grant #ATM 8306251. Appreciation is extended to Isabelle Muszynski for critically reading the manuscript.

9. REFERENCES

Birchfield, G. E., 1977, 'A study of the stability of a model continental ice sheet subject to periodic variations in heat input,' *J. Geophys. Res.,* **82,** 4909-4913.

Birchfield, G. E., J. Weertman, A. Lunde, 1981, 'A paleoclimate model of Northern Hemisphere ice sheets,' *Quat. Res.,* **15,** 126-142.

Birchfield, G. E., R. W. Grumbine, 1985, '"Slow physics" of large continental ice sheets and underlying bedrock, and its relation to the Pleistocene ice ages,' to appear *J. Geophys.*

Chappell, J., 1983, 'A revised sea-level record for the last 300,000 years from Papua New Guinea,' *Search,* **14,** 101.

Held, I., 1982, 'Climate models and the astronomical theory of the ice ages,' *Icarus,* **50,** 449-461.

Hyde, W. T., W. R. Peltier, 1985, 'Sensitivity experiments with a model of the ice age cycle: the response to harmonic forcing,' submitted to *J. Atm. Sci.*

Imbrie, J., J. D. Hays, D. G. Martinson, A. McIntyre, A. C. Mix, J. J. Morley, N. G. Pisias, W. L. Prell, N. J. Shackleton, 1984, 'The orbital theory of Pleistocene climate: support from a revised chronology of the marine δO^{18} record,' *Milankovitch and Climate,* part 1, Berger et al., eds., pp. 269-305, D. Reidel, Boston.

Imbrie, J., 1985, 'A theoretical framework for the Pleistocene ice ages,' *J. Geol. Soc. London,* **142,** 417-432.

Le Treut, H., M. Ghil, 1983, 'Orbital forcing, climatic interactions, and glaciation cycles,' *J. Geophys. Res.,* **88,** 5167-5190.

Oerlemans, J., 1980, 'Model experiments on the 100,000-yr glacial cycle,' *Nature,* **287,** 430-432.

Peltier, W. R., 1982, 'Dynamics of the ice age earth,' *Advances in Geophysics,* **24,** 2,146.

Peltier, W. R., W. T. Hyde, 1984, 'A model of the ice age cycle,' *Milankovitch and Climate,* part 2, Berger, et al., eds., pp. 565-580, D. Reidel, Boston.

Pollard, D., 1984, 'Some ice-age aspects of a calving ice-sheet model,' *Milankovitch and Climate,* part 2, Berger et al., eds., pp. 541-564, D. Reidel, Boston.

Saltzman, B., A. R. Hansen, K. A. Maasch, 1984, 'The late quaternary glaciations as the response of a three-component feedback system to earth-orbital forcing,' *J. Atm. Sci.,* **41,** 3380-3389.

Shackleton, N. J., N. D. Opdyke, 1973, 'Oxygen isotope and palaeomagnetic stratigraphy of equatorial Pacific core V28-238: oxygen isotope temperature and ice volumes on a 10^5 and 10^6 year scale,' *Quat. Res.,* **3,** 39-55.

Shackleton, N. J., J. Imbrie, M. A. Hall, 1983, 'Oxygen and carbon isotope record of East Pacific core V19-30: implications for the formation of deep water in the late Pleistocene North Atlantic,' *Earth Planet. Sci. Lett.,* **65,** 233-244.

Weertman, J., 1976, 'Milankovitch solar radiation variations and ice age ice sheet sizes,' *Nature,* **261,** 17-20.

Weertman, J., G. E. Birchfield, 1984, 'Ice-sheet modelling,' *Ann. Glaciol.,* **5,** 180-184.

A RELAXATION OSCILLATOR MODEL OF THE ICE AGE CYCLE

W. R. Peltier
Department of Physics
University of Toronto
Toronto, Ontario
Canada M5S 1A7

ABSTRACT. Oxygen isotope dilution histories of the oceans, which have been inferred through mass spectrometric analysis of foraminiferal tests from deep sea sedimentary cores, demonstrate that throughout the Pleistocene period the volume of continental ice has been a highly oscillatory function of time. The 10^5 year cycle which dominates this variability has proven rather difficult to reconcile with the conventional astronomical theory of ice ages which is otherwise strongly supported by the data. This paper describes a new nonlinear model of ice age climate which incorporates an explicit description of ice sheet accumulation and flow and of the physics of the isostatic sinking of the earth under the weight of the ice. The model is shown to explain the dominant 10^5 year cycle as a subharmonic resonant relaxation oscillation which characterizes its response to realistic astronomical forcing.

1. INTRODUCTION

Of the many problems which remain to be solved in the general area of planetary climatology, one of the more interesting is surely that associated with identifying the causative agency responsible for the large scale glaciation and deglaciation events to which the northern hemisphere continents in high latitudes have been subjected throughout the Pleistocene period. During these past two million years of geological history, all of the Canadian landmass, and much of northwestern Europe, has been covered episodically by vast continental ice sheets having thicknesses of 3-4 km. Given that the last such event ended only 6-7 thousand years ago, it should not be too surprising that there remains a considerable body of evidence of it which is easily extracted from the geological record (terminal moraines of known age, etc.). Indeed, the Earth's dynamic response to this last deglaciation event still continues today, a consequence of the fact that there was a large and essentially viscous component of the deformation of the "solid" earth induced by the ice sheet loads, and of the fact that the effective viscosity of the Earth's mantle - which determines the rate at which

C. Nicolis and G. Nicolis (eds.), Irreversible Phenomena and Dynamical Systems Analysis in Geosciences, 399–416.

gravitational equilibrium in the system is restored by viscous flow following unloading - is extremely high.

As one example of the available evidence of this so called isostatic adjustment process, Figure 1 shows a flight of raised beaches which now exist in the Richmond Gulf of Hudson Bay, near what was the centroid of the Laurentide ice sheet which covered all of northern North America 18000 years ago. Each of the individual horizons visible in the photograph represents a past location of sea level (a "strandline"), the age of which may be determined by application of ^{14}C dating to materials which were open systems (e.g. mollusc shells, etc.) when the horizon was at sea level. The relative sea level curve for this site, which is obtained by plotting the height of each horizon above present day sea level as a function of its ^{14}C age, is shown in Figure 2. These data demonstrates that the land has been rising out of the sea at this location ever since deglaciation, and that the rate of emergence has been an exponentially decreasing function of time, with a characteristic time constant near 2×10^3 years. In the past 8000 years the amount of emergence at this site has been about 250 metres. The purpose of this paper is to demonstrate that this postglacial rebound process is not only crucial to understanding the observed response of the Earth to deglaciation, but probably also to understanding the mechanism responsible for the appearance and disappearance of the ice sheets themselves.

Figure 1. A flight of raised beaches from the Richmond Gulf of Hudson Bay.

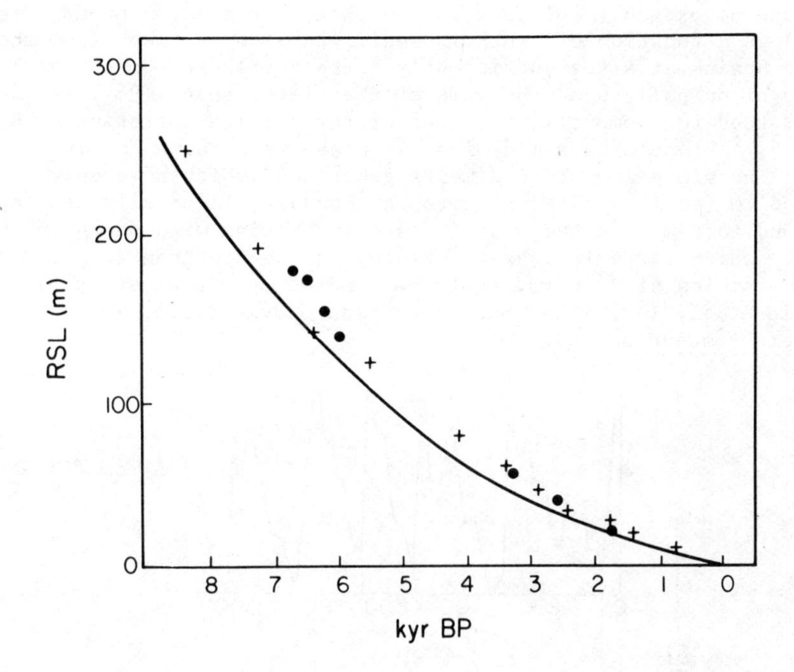

Figure 2. The relative sea level curve obtained from the Richmond Gulf beaches. The age of each sample has been determined by ^{14}C dating.

I have previously discussed in some detail (e.g. Peltier 1982) other signatures of the Earth's response to deglaciation which can also be attributed to this cause. These include, besides the relative sea level data illustrated above, the free air gravity anomalies observed over once ice covered regions (Wu and Peltier 1983), the so called non-tidal component of the acceleration of planetary rotation which has recently been confirmed by laser ranging to the LAGEOS satellite (Yoder et al. 1983, Rubincam 1984, Peltier 1982, Peltier 1983, Peltier and Wu 1983, Wu and Peltier 1984), and the secular drift of the north rotation pole with respect to the surface geography which has been extracted (e.g. Dickman 1977) through analysis of the polar motion time series for the time period 1900-1982 produced by the International Latitude Service.

Since our purpose here is to establish the crucial role played by this physics in controlling the appearance and disappearance of the ice sheets themselves, we will require reasonably precise knowledge of the history of Pleistocene ice volume fluctuations in order to test the model of this dynamical system which we shall produce. The best available proxy information controlling the variations of continental ice volume which have occurred over timescales of order 10^6 years or more consist of measurements of the concentration of the comparatively rare isotope of oxygen (^{18}O) relative to that of the more abundant isotope (^{16}O) as a function of depth in sedimentary cores taken from the deep ocean basins at sites sufficiently distant from the continental margins. Based principally upon the work of Shackleton (e.g. 1967) it has been understood for some time that the preferential evaporation of $H_2\,^{16}O$ over $H_2\,^{18}O$ would have made ice age ocean water anomalously rich in ^{18}O. Therefore in-core sedimentary horizons which have anomalously high $\delta^{18}O$ (as the relative isotopic abundance is usually denoted) correspond to times in the past (of age increasing with depth in the core) during which large ice sheets existed on the continents. Figure 3 shows four examples of $\delta^{18}O$ vs. depth data based upon analyses published by Imbrie et al. (1973) and Shackleton and Opdyke (1973, 1977) for three different ocean basins.

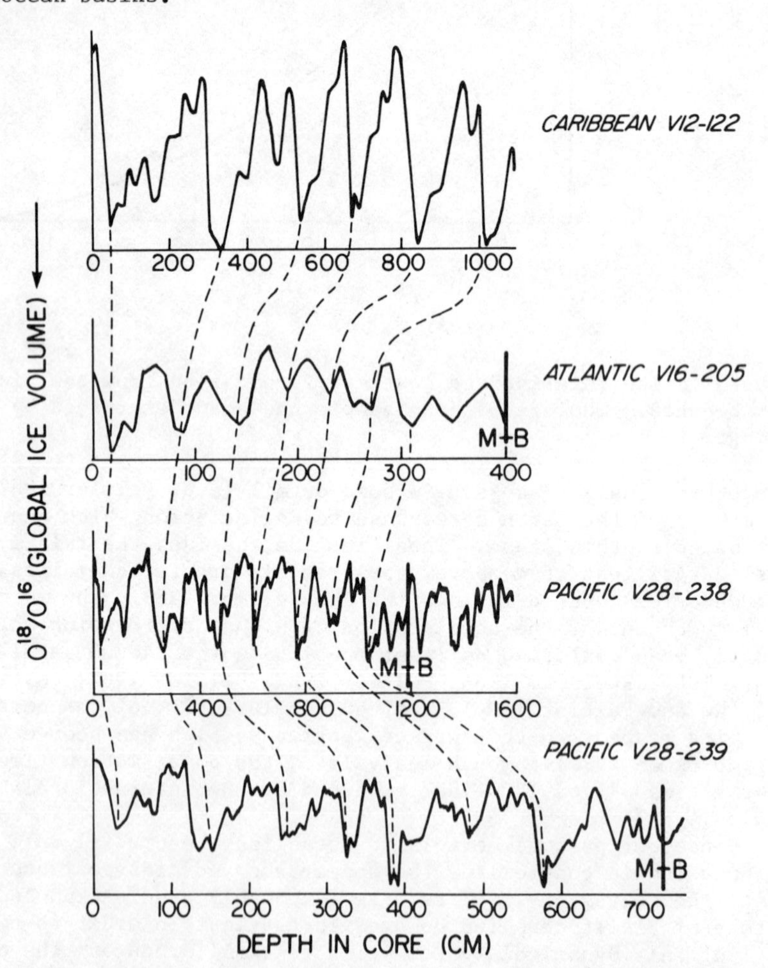

Figure 3. Oxygen isotope data from four different sedimentary cores. The down core variations of the $^{18}O/^{16}O$ ratio is proxy for the time variation of continental ice volume.

Also marked on the last three of the four data sets shown is the depth in the core corresponding to the time of the last reversal of the earth's magnetic field, the so called Brunhes-Matuyama boundary of age 730,000 ± 11,000 years (e.g. Cox and Dalrymple 1967). Assuming that the sedimentation rate at each site has remained constant over the time spanned by the core allows one to convert each of these "depth-series" into time series to which conventional techniques of analysis may be applied.

Rather than illustrating the results obtained through such analysis by reference to a number of individual cores, Figure 4(a) shows the

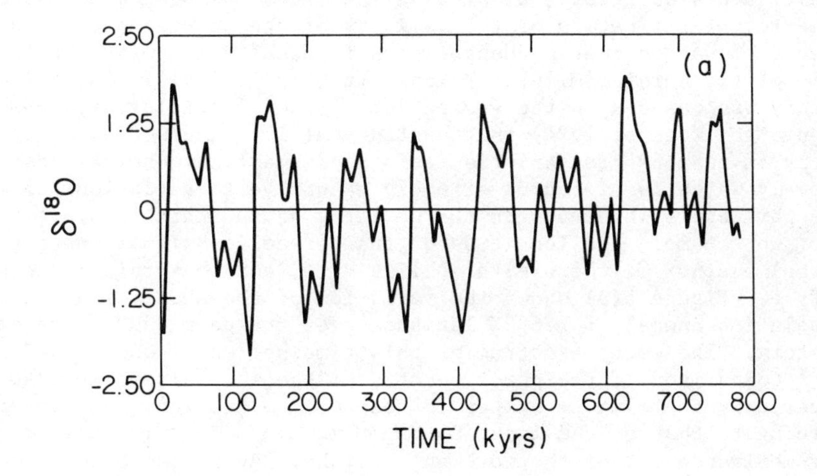

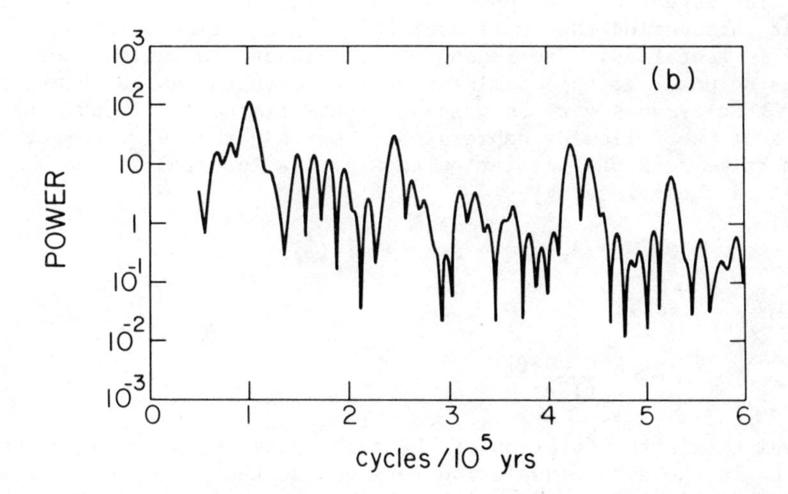

Figure 4. (a) The SPECMAP composite $\delta^{18}O$ record of Imbrie et al. (1984). (b) The power spectrum of the SPECMAP record.

composite history produced by Imbrie et al. (1984) by stacking and smoothing four individual long records. This record has been called the SPECMAP record and I will use it in all that follows as the best available proxy record of past fluctuations of continental ice volume. Figure 4(b) shows the power spectrum of this ice volume time series, inspection of which demonstrates that the variability is dominated by five strong spectral lines corresponding to periods of 100 k years, 41 k years, 24.1 kyrs, 23.4 kyr, and 19.2 kyrs. Given that each of these periods is of astronomical significance it is not surprising that the paper of Hays et al. (1976), which first revealed them as dominant constituents of records of $\delta^{18}O$ variability, has been widely construed as direct verification of the validity of the so-called astronomical theory of paleoclimatic change which is usually associated with the name of the Serbian Milutin Milankovitch (e.g. 1941). The Milankovitch theory was based upon the suggestion of the climatologist Köppen (e.g. Köppen and Wegener 1924) that continental ice sheets should grow and decay in response to variations of the summertime seasonal insolation anomaly which is of course strongly sensitive to variations of all of the parameters which govern the geometry of the Earth's orbit around the sun: namely the longitude of the perehelion with respect to the vernal equinox $\hat{\omega}$, the orbital obliquity ε, and the orbital eccentricity e. Figure 5(a) shows the variation of the summertime seasonal insolation anomaly for 65°N latitude over the past 800 kyr of earth history. The power spectrum of this time series is shown in Figure 5(b).

Comparison of the power spectra of the ice volume fluctuations delivered by the climate system in response to the input astronomical forcing to that of the input itself, immediately reveals the enigma which has motivated all of the work which I shall summarize here. Inspection of Figure 5(b) demonstrates that the spectrum of the astronomical input contains essentially no power at the period of 10^5 years even though it is at this period that well over 60% of the variance in the $\delta^{18}O$ record is contained. The reason why the input insolation anomaly contains no power at this dominant period, even though earth orbital eccentricity does vary on precisely this timescale, is obvious on the basis of the following expression which gives (e.g. Berger 1978) the summertime seasonal insolation anomaly δQ_s in terms of the orbital parameters themselves.

$$\delta Q_s = \Delta\, R_s\, \Delta\, \varepsilon + m\, \Delta\, (e \sin(\hat{\omega})) \tag{1}$$

where

$$m = \frac{2TS\,\cos(\phi)}{\pi^2(1-e^2)^{\frac{1}{2}}}$$

in which T is the duration of the tropical year, S the solar constant, and R_s is the total insolation received at the top of the atmosphere over the caloric summer. Although the orbital eccentricity e has dominant characteristic periods of 100 kyr and 400 kyr it appears in (1) as a modulation of the $\sin\hat{\omega}$ term which describes the precession of the equinoxes. It is because of the modulation of the $\sin\hat{\omega}$ term by e that

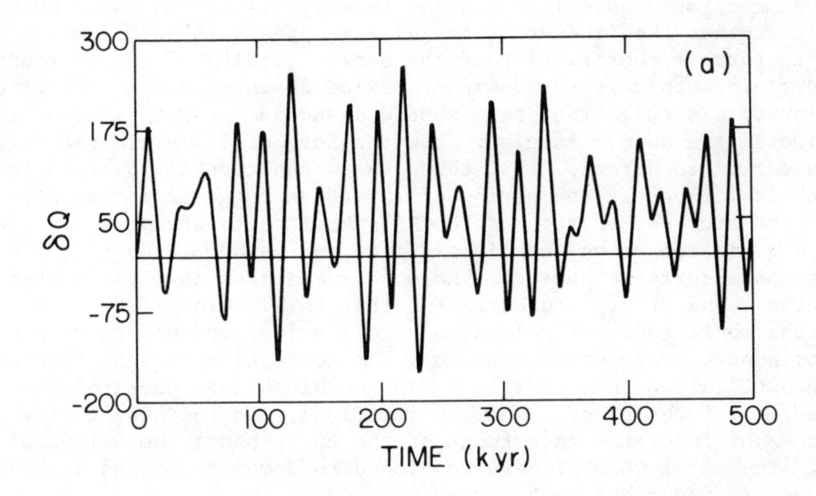

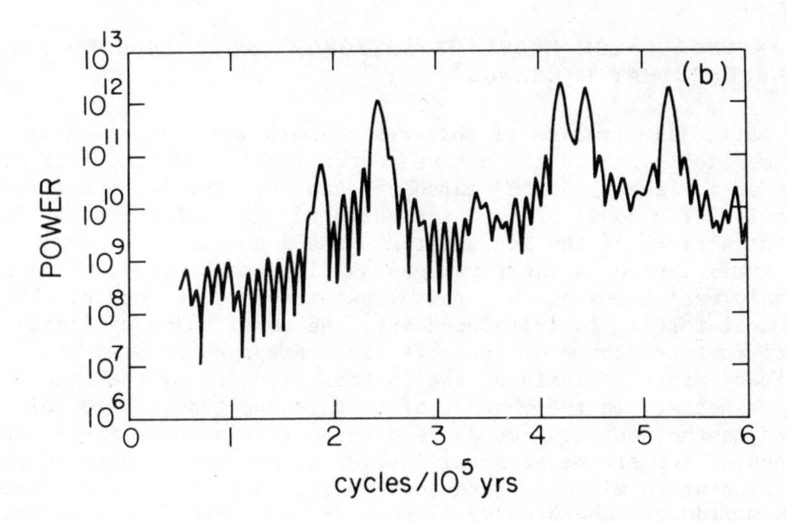

Figure 5. (a) Summertime seasonal insolation anomaly for 65° North Latitude computed from the formulae in Berger (1978). (b) Power spectrum of the summertime seasonal insolation anomaly.

the triplet of lines with the closely spaced periods of 24.1 kyr, 23.4 kyr, and 19.1 kyr appears in the spectrum of the insolation anomaly time series. The single precessional line with period 22 kyr is split by the eccentricity modulation. The fact that the power spectrum of the ice volume fluctuations shown in Figure 4(b) contains this same triplet of lines must be interpreted as rather strong circumstantial evidence to the effect that the climate system is in fact responding to the summertime seasonal insolation anomaly (1) as suggested by Köppen. The fact that the line at period 41 kyr appears in both input and output is further confirmation of the basic validity of the astronomical theory since this is the dominant period on which variations of orbital obliquity ε occur. That this should induce a separate line of the same period in the output is clear from the form of (1) since the forcing at this period is direct. That there is no substantial power in the astronomical input at the period of 10^5 years suggests rather strongly that the simplest version of the Milankovitch theory does not provide any explanation of why continental ice volume should oscillate with this characteristic period. The work summarized in this paper has been devoted to an attempt to establish that this dominant 10^5 year cycle is a subharmonic resonant relaxation oscillation produced in response to a dominantly precessional forcing. My contention is that this relaxation oscillation is supported by the combined influences of the nonlinearity of the process whereby ice flows, and the essentially linear process of isostatic adjustment of the earth under the weight of the ice. The model which I have developed to demonstrate and analyse this process is described in the next section.

2. AN OSCILLATOR MODEL OF ASTRONOMICALLY INDUCED PLEISTOCENE CLIMATIC CHANGE

A schematic illustration of this new climate model is shown in Figure 6. The model contains two active and explicity modelled ingredients, namely an ice sheet and the earth beneath it. The ice sheet is assumed to consist of a zonal ring of ice surrounding a polar ocean. Expansions and contractions of the ice sheet are assumed to occur in response to the small variations of summertime seasonal insolation produced by long timescale variations of the orbital parameters $\hat{\omega}$, ε, and e. This astronomical forcing is introduced into the model using a climate point - climate surface scheme of the sort first proposed by Weertman (1976) in his steady state analysis of the factors determining the size of ice age ice sheets. In the absence of surface ice the climate point is defined as the southern boundary of the perennial snow field. Fluctuations of summertime seasonal insolation δQ_S are assumed to cause this climate point to migrate in the north south direction by a distance $\delta x = \alpha\delta Q_S/(dQ/dx)$ where dQ/dx is the current north-south insolation gradient and α is a meteorological factor associated principally with ice-albedo feedback. Energy balance analyses by North et al. (1983) suggest $1 < \alpha < 3$ with a preferred value near 2. In the presence of an ice sheet, initially produced by the buildup of snow north of the climate surface, after it has made an excursion outland, the boundary separating the accumulation area from the ablation area (see Figure 6) is located considerably to the south of the climate point. This is essentially because temperature decreases with increasing atmospheric height so that an increase of elevation of the ice sheet pushes the critical isotherm separating accumulation from ablation further south. In the model it is assumed that the slope of the climate surface is identical with the slope of the 0^oC isotherm in a standard atmosphere,

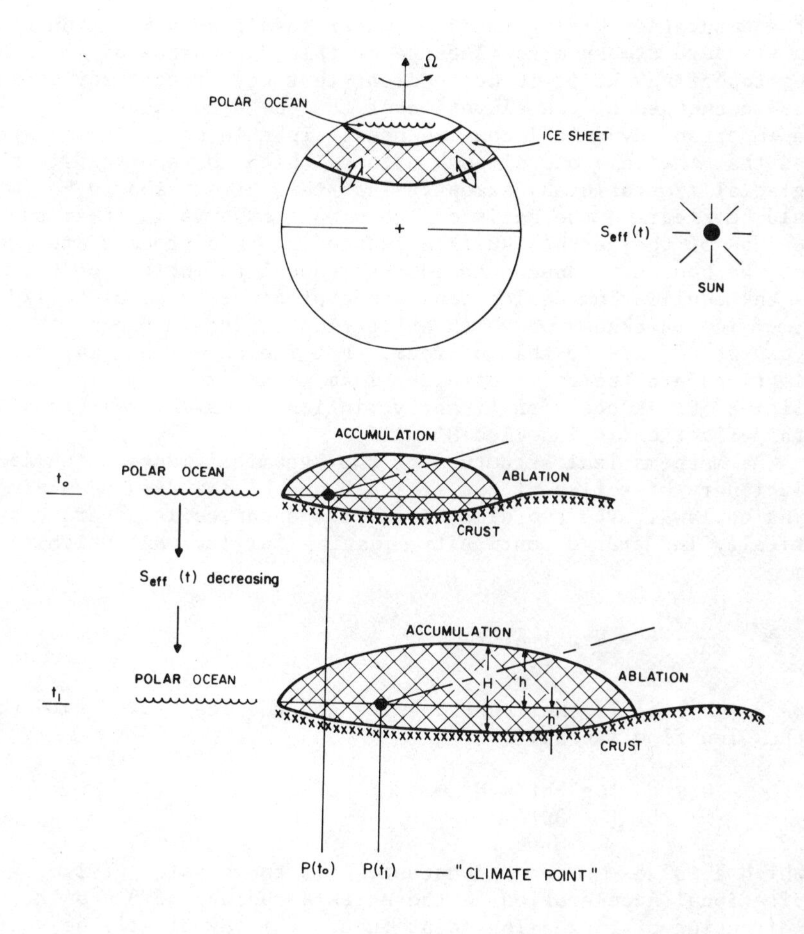

Figure 6. Schematic diagram of the new paleoclimate model.

that is s=.0067 km/rad. All of the meteorological and hydrological processes which determine the precipitation and melting fields are assumed to be parameterizable in the form

$$A(\theta,t) = \max\,\{a[1-bh(\theta)],\,0\},\ \text{above the climate surface} \qquad (2)$$

$$= \min\,\{a^1[1-bh(\theta)],\,0\},\ \text{below the climate surface}$$

The parameter $h(\theta)$ in (2) is the instantaneous elevation of the ice sheet above sea level (see Figure 6), a and a' are respectively the sea level accumulation and ablation rates, and b is an extremely important parameter which describes the rate of decrease of accumulation and ablation rates with increasing topographic height. This describes the so called "elevation desert effect". The parameter b which controls its strength is estimated on the basis of the observed rate of decrease of the saturation mixing ratio of water in air as a function of height in a standard atmosphere. Because of this dependence of A upon the ice sheet topography h, it is quite clear that any process which contributes to changes of the elevation of the ice sheet above sea level will have important dynamical consequences. In this model it is hypothesized that the only one of such effects which is important is the effect of glacial isostatic adjustment. That this effect should be important should be clear on the basis of the simple observation that the deflections of the earth's surface induced by this process are such as to allow the buoyancy induced thereby to cancel the weight of the load. Thus the equilibrium deflections are approximately equal to 1/3 the maximum ice thickness where this fraction is determined as the ratio of the density of ice to that of rock. For ice sheets 3-4 km thick the deflections are therefore of order 1 km which from (2) given b is sufficient to induce significant variations of A. In Figure 6 the earth deflection is labelled $h'(\theta)$.

The mathematical structure of the dynamical model which links the evolutionary histories of the fields h and h' is based upon simple conservation laws. The two dimensional ice dynamics is governed by the vertically integrated continuity equation for ice mass which has the form

$$\frac{\partial H}{\partial t} = -\underline{\nabla}\cdot U + A(\theta t) \tag{3}$$

where $H = h + h'$ is the ice thickness and the ice flux $\underline{U}$ is determined by the Glen flow law as:

$$\underline{U} = \frac{\hat{c}}{5}\left(\frac{\rho g}{a}\frac{\partial h}{\partial \theta}\right)^3 H^5\,\hat{\theta} \tag{4}$$

in which $\hat{c}$ is an empirical constant, ρ is the density of ice, g the gravitational acceleration, a the earth's radius and $\hat{\theta}$ a unit vector in the direction of increasing colatitude. (3) may clearly be rewritten as a nonlinear diffusion equation as:

$$\frac{\partial h}{\partial t} = \frac{1}{a\sin\theta}\frac{\partial}{\partial\theta}\left[\sin\theta\,K(\theta)\,\frac{\partial h}{\partial\theta}\right] - \frac{\partial h'}{\partial t} + A(\theta t), \tag{5}$$

in which the diffusion coefficient K is given by

$$K = \frac{\hat{c}}{5}\left(\frac{\rho g}{a}\right)^3\left(\frac{\partial h}{\partial\theta}\right)^2 H^5(\theta). \tag{6}$$

As described in Peltier (1982) a separate evolution equation for the field $h'(\theta,t)$ may be obtained from the theory of glacial isostasy which delivers a prediction of $h'(\theta t)$ in the form of a convolution of the ice thickness field $H(\theta,t)$ with an appropriate viscoelastic Green's function u_r as:

$$h'(\theta,t) = \iiint \delta\Omega' dt' \rho H(\theta',\lambda',t')\ u_r\ (\theta/\theta',\lambda/\lambda',t/t') \qquad (7)$$

Theory originally developed in Peltier (1974) shows that the Green's function u_r may be represented in the form of a normal mode expansion as:

$$u_r\ (\theta/\theta',\lambda/\lambda';t/t') = \frac{a}{m_e} \sum_{\ell=0}^{\infty} q_\ell\ (t-t')P_\ell\ (\cos\ ) \qquad (8)$$

where the surface load "Love numbers" q_ℓ have Dirichlet expansions:

$$q_\ell(t) = \sum_{j=1}^{M} r_j^\ell\ e^{-s_j^\ell t} + q_\ell^E\ \delta(t) \qquad (9)$$

where the r_j^ℓ and s_j^ℓ are the amplitudes and inverse decay times of the M normal modes of viscous gravitational relaxation which are required to synthesize the response of the degree ℓ component of the deformation spectrum.

All of the results which have been obtained to date with this model have been obtained on the basis of the assumption that an adequate approximation to the isostatic adjustment process could be obtained with M=1 and the complete neglect of the elastic component of the response determined by the q_ℓ^E. Under these assumptions Peltier (1982) showed that (5) and (7) reduce to the following infinity of coupled nonlinear evolution equations for the fields $h(\theta,t)$ and $h'(\theta,t)$:

$$\frac{\partial h_\ell}{\partial t} = B_{\ell mn} K_m h_n - C_\ell r^\ell h_\ell + (s^\ell - C_\ell r^\ell)\ h_\ell' + A_\ell \qquad (10a)$$

$$\frac{\partial h_\ell'}{\partial t} = C_\ell r^\ell h_\ell - (s^\ell - C_\ell r^\ell)\ h_\ell' \qquad (10b)$$

In (10) the numbers h_ℓ and h_ℓ' are the Legendre coefficients in the spherical harmonic expansions of the h and h' fields, $C_\ell = 3\rho/2\ell+1)\rho_e$ where ρ_e is the effective density of the earth, and the $B_{\ell mn}$ are spectral interaction coefficients defined as:

$$B_{\ell mn} = -\frac{(2\ell+1)}{2a} \int_{-1}^{+1} (1-x^2)\ \frac{\partial P_\ell}{\partial x}\ P_m \frac{\partial P_n}{\partial x}\ dx \qquad (11)$$

where P_ℓ is the Legendre polynomial of degree ℓ and x is its argument. When the expansions of h and h' are truncated at some finite number of terms, 48 for the solutions to be described below, the resulting finite system can be integrated forward in time from prescribed initial conditions and a synthetic history of ice volume variations constructed from the field H = h+h'. The synthetic may then be compared to the SPECMAP record. The numerical methods employed to do this have recently been described in Hyde and Peltier (1985). Typical results are described in the following section.

3. RESULTS FROM TYPICAL INTEGRATIONS OF THE MODEL

Figure 7 shows the results obtained from an integration of equations (10) with the astronomical forcing approximated by a simple sinusoidal variation with period 20 kyr.

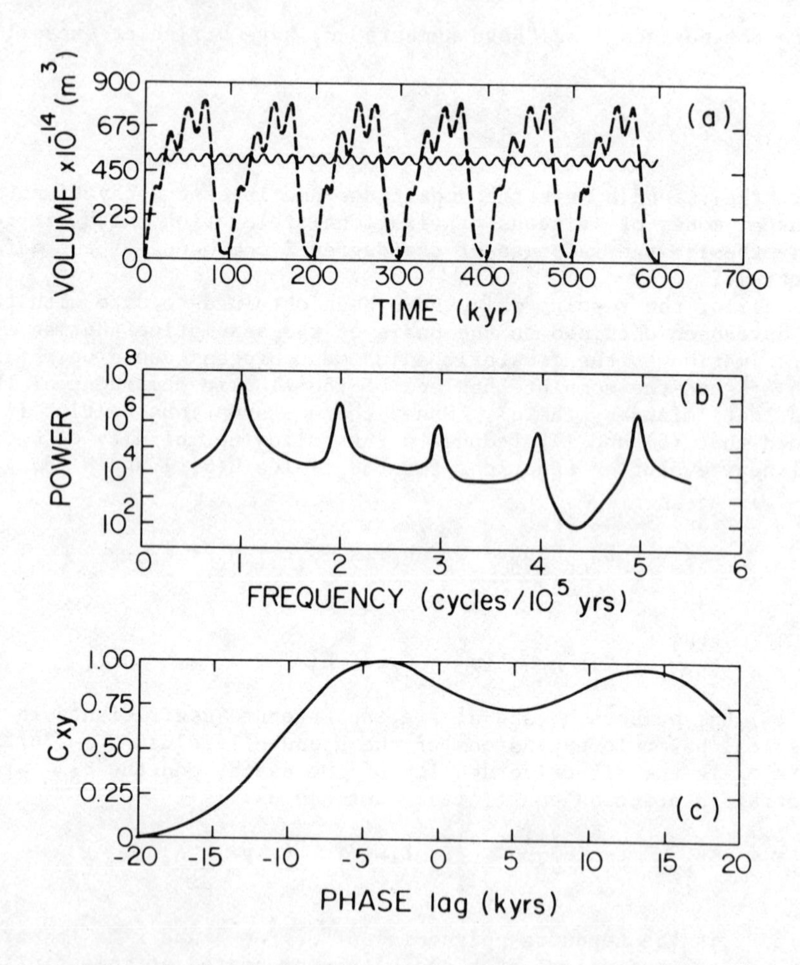

Figure 7. (a) Synthetic ice volume time series produced by the climate model in response to 20 kyr harmonic forcing. (b) Power spectrum of the synthetic in (2). (c) Cross correlation vector between the 20 kyr harmonic input and the synthetic ice volume output in (2).

The dashed line in plate (a) of this Figure is the synthetic ice volume time series while the thin solid line shows the 20 kyr periodic forcing. Inspection of this Figure shows that the model response is dominated by an ice volume oscillation whose period is precisely 10^5 yr. Each pulse in the output time series is characterized by a slow oscillatory growth to maximum amplitude lasting about 80 kyr followed by a fact collapse which is strongly reminiscent of the terminations which were shown to be characteristic of the SPECMAP record of Figure 4(a). Plate (b) of Figure 7 shows the power spectrum of the synthetic $\delta^{18}O$ record. Because the response is exactly periodic the power spectrum is seen to consist of strong lines at the fundamental period of 10^5 yrs and all of its harmonics. One of these harmonics, the fifth with period 20 kyr, coincides with the period of the forcing which explains its anomalously high amplitude in the line spectrum. Plate (c) of Figure (7) shows the normalized cross correlation vector of the 20 kyr input with the ice volume output. Inspection shows that the ice volume response lags the astronomical forcing by 4-5 kyr at the period of 20 kyr. This agrees very closely with the phase delay at this period which was extracted from the SPECMAP stack by Imbrie et al. (1984). It is important to realise that this rather encouraging result was obtained with the parameters in (2) fixed to the values $a = 1.2$ m yr^{-1}, $a^1 = -3.6$ m yr^{-1}, and $b = 2.3 \times 10^{-4}$ m^{-1}, while the motion of the climate point produced by the 20 kyr input was assumed to be of amplitude 490 km about a point located 140 km into the Arctic ocean. The slope of the climate surface was assumed to be $s = .0067$ in accord with the slope of the zero degree isotherm in a standard atmosphere. The amplitude of oscillation assumed for the climate point implies a meteorological feedback factor of $\alpha = 1.6$ which is with the range suggested by the energy balance modelling results of North et al. (1983).

The pulse shape of the 10^5 year oscillation which this climate model supports is highly suggestive of a relaxation oscillation of the sort which occurs in a simple dc circuit consisting of a resistor and a capacitor in series when a discharge tube is connected across the capacitor and set to fire when the voltage approaches its saturation value. The nonlinear physics in the climate model which is analogous to the discharge tube and which is responsible for the rapid terminations which characterize the deglaciation phase of each 10 year glacial cycle is illustrated in Figure 8. This shows ice sheet and earth deflection profiles from 71 kyr to 79 kyr through one 10^5 yr cycle of the harmonically forced experiment. Note that as the ice sheet begins to melt back from its maximum amplitude state, its southern flank begins to extend into the low warm earth depression which has been produced by the isostatic adjustment process. In this depression, according to (2), the ablation rate increased rapidly and this in turn causes the southern face of the ice sheet to become extremely steep. From (4) this steep slope strongly amplifies the flow of ice from north to south, so strongly that the ice sheet completely disintegrates within a half precessional cycle (10 kyr). This highly nonlinear amplification of the retreat by ice physics is analogous to the action of the discharge tube in the simple dc electrical circuit which supports a similar

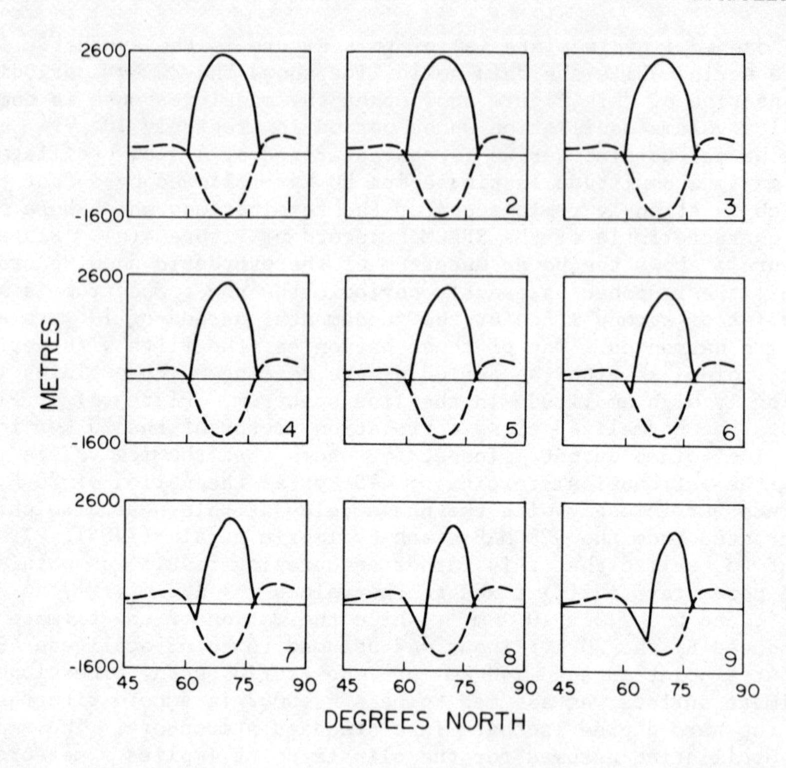

Figure 8. Profiles of ice sheet height and earth deflection from 71 kyr to 79 kyr (a-f inclusive) at 1 kyr intervals in one 10^5 periodic glaciation cycle of the harmonically forced experiment of Figure 7. Note the melt back into the isostatically produced depression which marks the onset of the termination.

relaxation oscillation. In the climate model subject to periodic forcing the nonlinear relaxation oscillation inevitably has a period which is some multiple (subharmonic) of the period of the forcing.

Although limitations of space do not allow here a detailed discussion of the sensitivity of model response to plausible variations of its parameters (Hyde and Peltier 1985) it is important for completeness sake to demonstrate that the basic 10^5 yr subharmonic resonant relaxation oscillation which the model has been shown to support continues to exist when the simple 20 kyr harmonic forcing is replaced by realistic Milankovitch forcing. That this is in fact the case is demonstrated in Figure 9 which shows in plate (a) the result obtained from a 10^6 yr Milankovitch experiment in which the astronomical input δQ_s has been taken equal to that at 65°N latitude, the meteorological

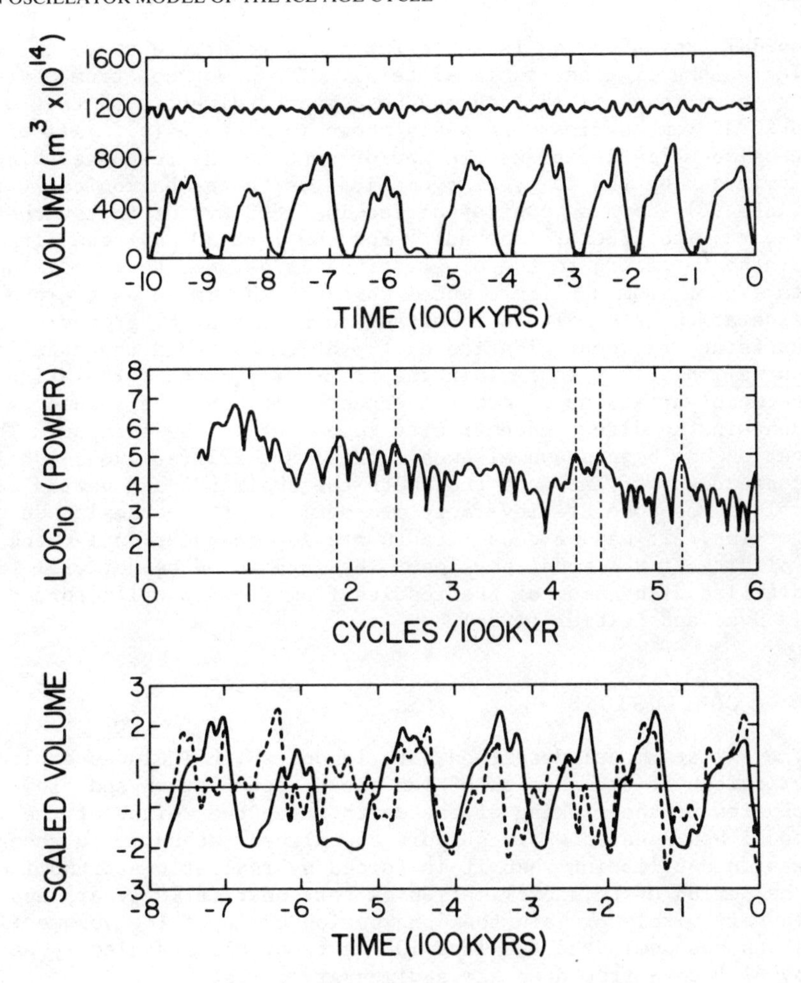

Figure 9. (a) Synthetic ice volume time series produced by the climate model in response to realistic Milankovitch forcing consisting of the summertime seasonal insolation anomaly for 65°N latitude. (b) Power spectrum of the synthetic which should be compared to the power spectrum of the SPECMAP record shown in Figure 4(a). (c) Overlay of the synthetic ice volume record (solid) with SPECMAP (dashed).

feedback parameter set at $\alpha = 2.0$, and the mean location of the climate point taken to be 170 km north of the model Arctic coast. Inspection of plate (a) shows that the model again delivers roughly 10^5 yr cyclic oscillations in spite of the fact that the forcing is highly nonsinusoidal (this forcing is shown for phase reference above the synthetic ice volume time series in plate a). The power spectrum of the response is shown in plate (b) and may be compared directly to that of the SPECMAP time series previously shown in Figure 4(b). All of the precession-obliquity peaks are now present in the response which is still dominated by the 10^5 year cycle just as in the harmonically forced case. Plate (c) shows an overlay of the last 800 kyr of the synthetic (solid) record with SPECMAP (dashed). For the last 500 kyr the fit of the synthetic record to the observations is as good as we have any reason to expect from a climate model which is as simple as the one under consideration here. It is important to note that by providing such a satisfactory reconciliation of the SPECMAP record the model has been shown implicitly to explain the previously noted correlation in SPECMAP between variations of orbital eccentricity and ice volume even though it contains no direct eccentricity component of the forcing. This explanation has been shown elsewhere (Hyde and Peltier 1985, 1986) to be a consequence of the fact that both the amplitude and period of the ice volume response are inversely proportional to the amplitude of the forcing. It may be construed to provide an additional check on the validity of the model physics. The interested reader will find a more detailed discussion of the results from realistically forced experiments in Hyde and Peltier (1986).

4. CONCLUSIONS

The climate model described here is one which includes explicit descriptions of the physics of ice sheet accumulation and flow and the physics of the sinking of the earth under the weight of the ice. The model has been shown to support a nonlinear subharmonic resonant relaxation oscillation when it is forced by realistic astronomical input. The period of this oscillation is very near to 10^5 years and appears to very nicely explain the long period cycle of ice volume fluctuations which has dominated the last 10^6 years of climate history as revealed by $\delta^{18}O$ data from deep sea sedimentary cores.

REFERENCES

Berger, A., 'Long term variations of daily insolation and Quaternary climate changes,' J. Atmos. Sci., 35, 2362-2367 (1978).

Cox, A., and Dalrymple, G.B., 'Statistical analysis of geomagnetic reversal data and precision of potassium-argon dating,' J. Geophys. Res., 72, 2603-2614 (1967).

Dickman, S.R., 'Secular trend of the earth's rotation pole: consideration of motion of the latitude observatories,' Geophys. J.R. astr. Soc., 57, 41-40 (1977).

Hays, J.D., Imbrie, J., and Shackleton, N.J., 'Variations in the earth's orbit: Pacemaker of the ice ages,' Science, 194, 1121-1132 (1976).

Hyde, W.T., and Peltier, W.R., 'Sensitivity experiments with a model of the ice age cycle: the response to harmonic forcing,' J. Atmos. Sci., 42, 2170-2188, (1985).

Hyde, W.T., and Peltier, W.R., 'Sensitivity experiments with a model of the ice age cycle: the response to realistic astronomical forcing,' J. Atmos. Sci., submitted.

Imbrie, J., Van Donk, J., and Kipp, N.G., 'Paleoclimatic investigation of a late Pleistocene Caribbean deep-sea core: Comparison of isotopic and faunal methods,' Quat. Res., 3, 10-38 (1973).

Imbrie, J., Shackleton, N.J., Pisias, N.G., Morley, J.J., Prell, W.L., Martinson, D.G., Hays, J.D., McIntyre, Andrew, Mix, A.C., 'The orbital theory of Pleistocene climate: support from a revised chronology of the marine $\delta^{18}O$ record, 'In Milankovitch and Climate (A. Berger, J. Imbrie, J. Hays, G. Kuhla, and B. Saltzman eds.), pp. 269-305 (1984).

Milankovitch, M., Canon of Insolation and the Ice-Age Problem, K. Serb. Akad. Geogr. Spec. Publ. No. 132, 484 pp. [Translated by the Israel Program for Scientific Translations, Jerusalem, 1969, U.S. Department of Commerce], (1941).

North, G.R., Mengel, J.G., and Short, D.A., 'Simple energy balance model resolving the seasons and the continents: applications to the astronomical theory of the ice ages,' J. Geophys. Res., 88, 6576-6586 (1983).

Peltier, W.R., 'The impulse response of a Maxwell earth,' Rev. Geophys. Space Phys., 12, 649-669 (1974).

Peltier, W.R., 'Dynamics of the ice age earth,' Advances in Geophys., 24, 1-146 (1982).

Peltier, W.R., 'Constraint on deep mantle viscosity from LAGEOS acceleration data,' Nature, 304, 434-436 (1983).

Peltier, W.R., and Wu, P., 'Continental lithospheric thickness and deglaciation induced true polar wander,' Geophys. Res. Lett., 10, 181-184 (1983).

Rubincam, D.P., 'Postglacial rebound observed by LAGEOS and the effective viscosity of the lower mantle', J. Geophys. Res., 89, 1077-1087 (1984).

Shackleton, N.J., 'Oxygen isotope analyses and Pleistocene temperatures re-addressed,' Nature, 215, 15-17 (1967).

Shackleton, N.J., and Opdyke, N.D., 'Oxygen isotope and paleomagnetic stratigraphy of equatorial Pacific core V28-238: Oxygen isotope temperatures and ice volumes on a 10^4 and 10^5 year timescale,' Quat. Res., 3, 39-54 (1973).

Shackleton, N.J., and Opdyke, N.D., 'Oxygen isotope and paleomagnetic evidence for early northern hemisphere glaciation,' Nature, 270, 216-219 (1977).

Weertman, J., 'Milankovitch solar radiation variations and ice age ice sheet sized,' Nature, 261, 17-20 (1976).

Wu, P., and Peltier, W.R., 'Pleistocene deglaciation and the earth's rotation: a new analysis,' Geophys. J. R. astr. Soc., 76, 753-791 (1984).

Yoder, C.F., Williams, J.G., Dickey, J.O., Schultz, B.E., Eanes, R.J., and Tapley, B.D., 'Secular variation of earth's gravitational harmonic J_2 coefficient from LAGEOS and nontidal acceleration of earth rotation,' Nature, 303, 757-762 (1983).

CLOUD-ICE-VAPOR FEEDBACKS IN A GLOBAL CLIMATE MODEL

Volker Jentsch
Max-Planck-Institut für Meteorologie
2000 Hamburg 13, Bundesstraße 55
Federal Republic of Germany

ABSTRACT. We are concerned with a globally averaged, time dependent climate model based on a simple hydrological cycle and the heat balance of an equivalent atmosphere and ocean. The domains are coupled by exchange of heat and moisture at their interface. Clouds and precipitation are related to humidity and the temperature of the atmosphere, while ice formation is related to the temperature of the ocean. The model is formulated as an initial value problem and integrated until an asymptotic equilibrium state is reached. The stability of the system against perturbations decreases and its sensitivity increases if variable vapor/cloud cover/ice cover is allowed to feed back into the radiation budget. The model also shows that clouds tend to cool rather than warm the surface.

1. INTRODUCTION

For a climate in long-term equilibrium, the globally averaged incoming solar energy and outgoing terrestial radiation must balance each other. This balance strongly depends on variable cloud cover, cloud type, cloud height, as well as water vapor and ice extent. Therefore, clouds, vapor and ice are important ingredients in climate theory and have to be considered in any valid treatment of the long-term climatic variability.

The problem can be tackled in a number of ways, ranging from the sophisticated three-dimensional general circulation models to the one-dimensional energy balance and radiative-convective models. One-dimensional models combined with a hydrological cycle have been persued by several authors, notably Weare and Snell (1974), Paltridge (1975), Sellers (1976) and Roads and Vallis (1984). Yet we believe that a number of problems relevant to climate variability have not been treated in sufficient detail so far. These include (a) the relative importance of cloud-ice-vapor feedbacks (b) the stability and sensitivity to perturbations of internal and external parameters (c) the time evolution towards asymptotic equilibrium (d) the importance of atmosphere-ocean coupling.

In this paper, we present a globally averaged, time dependent model which in some respects resembles that given by Petukhov (1974). Our model retains some of the detailed radiative computations characteristic of ra-

C. Nicolis and G. Nicolis (eds.), Irreversible Phenomena and Dynamical Systems Analysis in Geosciences, 417–437.

diative convective models but at the same time it incorporates the ice albedo-temperature feedback characteristic of energy balance models. We also include cloud-temperature and water vapor-greenhouse feedbacks in a simple, but highly nonlinear fashion. More specifically, we deal with separate heat budgets for an equivalent atmosphere (including clouds) and ocean (including sea ice), supplemented by the continuity equation for water vapor. The temperatures are calculated from the balance between radiative and convective heat fluxes, while water vapor is due to evaporation and precipitation. The ocean surface is made up of water and ice, the extent of which is governed by the temperature of open water. The surface temperature is an area weighted average of predicted water temperature and parameterized ice temperature. Cloud cover and precipitation are coupled through predicted atmosphere temperature and humidity. Cloud height and lapse rate are held constant; if the temperature of the clear sky changes, so does the effective cloud temperature. Ice cover, cloud cover and humidity all feed back into the radiation budget. Atmosphere and ocean are nonlinearly coupled by exchange of heat and moisture at their interface.

Although this model does not allow for horizontal transports of heat and vapor, it is useful for gaining insight into the interactions of positive and negative feedback mechanisms. In addition, the model identifies certain critical parameters upon which the temperature distribution reacts sensitively . Furthermore, it may be used to interpret the results of multi-dimensional models. In a forthcoming paper, we shall concentrate on the latitudinal structure of climatic feedbacks, by using temperature dependent transport coefficients and an idealized circulation structure for the atmosphere. In addition, we shall employ an extra equation for ice temperature and ice formation.

The paper is organized as follows. Section 2 evaluates the basic formulas. The free parameters of the model are determined so that the system closely simulates the present day climatology. In section 3, we investigate the stability of the model by following the trajectories in 3-dimensional phase space. It will be shown that the model becomes intransitive if the solar constant falls just below its present day value. The intransitivity is brought about by large amplitude perturbations in the temperatures of the ocean and atmosphere. In contrast, the specific choice of initial humidity does not affect the stability properties of the system. There exist two stable attractors (i.e., fixed points); because of the time scale separation between ocean and atmosphere, no limit cycle exists. Special attention is paid to the time stationary response of the system due to changes in solar constant. However, cloud height and air-sea coupling may be subject to systematic and/or stochastic variations, so the sensitivity of the climatic variables to changes in these parameters is also investigated. Several features of our model, such as the sensitivity of surface temperature to changes in solar constant and CO_2 concentration, as well as the specific form of cloud-temperature coupling, resemble those found in 3-dimensional models. This apparent agreement does not, of course, validate our model; it merely says that it can simulate certain basic aspects of the climatic system. The results are summarized and discussed in section 4.

2. THE MODEL

The globally integrated balance equations for heat and humidity in the atmosphere-ocean system are written as

$$c_a\dot{T}_a = H + R_a + L_v C \qquad (1)$$

$$c_o\dot{T}_o = -H + R_o - L_v E + L_f F \qquad (2)$$

$$c_q\dot{q} = E - P \qquad (3)$$

Here, $c_a\dot{T}_a$ and $c_o\dot{T}_o$ are the time rate of change of heat in the atmosphere and ocean, respectively, and $c_q\dot{q}$ is the rate of change of water vapor. The climatic variables are air temperature (T_a), water temperature (T_o) and specific humidity (q), all of which are functions of time t. The corresponding inertia coefficients are $c_a = 10^7 Wsm^{-2}K^{-1}$, $c_o = 20\,c_a$ and $c_q = 3x10^3\ kgm^{-2}$. The sensible and latent heat fluxes between surface and overlying air are denoted by H and L_vE, respectively, where

$$H = \lambda(T_s - T_a) \qquad (4)$$

and

$$E = (\lambda/c_p)(q_{sat}(T_s) - q) \qquad (5)$$

Here, λ is the transfer coefficient to be determined below, c_p is the specific heat at constant pressure, q_{sat} (T_s) is the saturation humidity at surface temperature T_s,

$$T_s = xT_o + (1-x)T_i \qquad (6)$$

where x and 1-x are fractional areas of open water and sea-ice, respectively; T_i is the ice temperature and T_f = 273K is the freezing temperature of water:

$$T_i = \begin{cases} T_a \text{ if } T_a \leq T_f \\ T_f \text{ if } T_a > T_f \end{cases} \qquad (7)$$

Furthermore, R_a and R_o are the (net) radiative fluxes (short wave absorption plus long wave emission) of the two domains; $R = R(T_a, T_o, q)$. Because of the energy balance, condensation rate C must equal precipitation rate P. In (2), we have neglected the heat stored in sea - ice which is small in comparison to that stored in the ocean. Freezing rate F is related to ice mass m_i via $F = \dot{m}_i$, where $m_i = \rho_i h_i(1-x)$, and ρ_i, h_i are density and height (fixed at 1 m) of sea-ice, respectively. Thus, $F = -\rho_i h_i \dot{x}$, or equivalently,

$$L_f F = -c_i\dot{x} \qquad (8)$$

where $c_i = 3x10^8 Wsm^{-2}K^{-1}$, and $L_{f,v}$ is the latent heat of fusion (f) and vaporization (v), respectively. Equations (2) and (3) must be handled with some care. When $T_o(t)$ in (2) approaches T_f, then T_o is set equal to zero, and H is replaced in (1) with $H = R_o+L_vE$, evaluated at $T_o= T_f$. Similarily, if relative humidity $rh(t) = q(t)/q_{sat}(T_a(t))$ approaches unity, then $\dot{q}$ in (3) is set equal to zero, and E is evaluated at $q = q_{sat}(T_a)$. Adding (1)-(3) yields the total enthalpy of the atmosphere-ocean-ice system, which reduces to the statement of global radiative balance, $R_a+R_o = 0$, in equilibrium state.

The shortwave radiation scheme used here is similar to that of Lacis and Hansen (1974). The amount of solar energy absorbed by the atmosphere is

$$A_a/Q = 1-r_a-\tau(\tau r_s+(1-r_s))/(1-r_s r_a) \quad (9)$$

where Q is the annually and globally averaged insolation at the top of the atmosphere (the "solar constant"), r_a is the reflectivity, $\tau = 1-r_a-\kappa$ is the transmissivity and κ is the absorptivity of the atmosphere. The denominator $1-r_s r_a$ accounts for successive reflections between atmosphere and ground. The optical properties of the atmosphere are weighted by fractional cloud cover c, i.e., $r_a = cr_{cloudy}+ (1-c)r_{clear}$, $\kappa = c\kappa_{cloudy}+ (1-c)\kappa_{clear}$. The reflectivity is dominated by clouds and is of the order of 0.5; however, there is also some reflection and scattering due to aerosols and air molecules approximated by $r_{clear} = 0.08$. The absorptivity of the clear atmosphere is taken as

$$\kappa_{clear} = \kappa_o+\kappa_1(q) \quad (10)$$

where κ_o is the average absorption due to CO_2, O_3 and aerosols, and κ_1 represents the absorption effect of water vapor, as given by formula (20) in Lacis and Hansen (1974). Extra absorption of solar energy in clouds is taken into account by $\kappa_{cloudy} = \kappa_{clear}+0.04$ (Paltridge, 1974). The actual values for r_{cloudy} and κ_o are determined below.

By the same token, the solar energy absorbed by the surface is

$$A_s/Q = \tau(1-r_s)/(1-r_s r_a) \quad (11)$$

where $r_s = xr_o+ (1-x)r_i$ is the area weighted surface albedo, composed of water albedo r_o ($r_o = 0.1$) and ice albedo r_i which may vary between 0.4 and 0.8. For definiteness, we set $r_i = 0.6$. By means of (9) and (11), we obtain for the planetary albedo

$$\alpha_p = r_a+\tau^2 r_s/(1-r_s r_a) \quad (12)$$

which reduces to $\alpha_p \approx r_a+r_s(1-2r_a)$, if $\kappa, r_s r_a \ll 1$.

Our longwave scheme resembles that of Adems (1961) and Paltridge (1974). We assume that surface and clouds radiate as black bodies at temperatures T_s and T_c, respectively, where $T_c = T_a-\Gamma h_c$ is the average cloud temperature, Γ is the lapse rate and h_c is the average cloud height. The surface receives radiation from the atmosphere and (through the atmospheric window) from the clouds, and emits radiation according to $\sigma T_s{}^4$. The net longwave flux leaving the surface is then

$$E_s = -\sigma T_s^4 + \varepsilon_1 \sigma T_a^4 + c(1-\varepsilon_1)\sigma T_c^4 \quad (13)$$

where ε_1 represents the greenhouse effect,

$$\varepsilon_1 = \varepsilon_{10} + \varepsilon_{11} \sqrt{e} \quad (14)$$

The first term in (14) is the emissivity of the atmosphere below the clouds resulting from, e.g., CO_2; the second term is due to water vapor, where $e = 1.6 \times 10^3 q$ is the vapor pressure in mbar. Numerical values are $\varepsilon_{10} \approx 0.6$ and $\varepsilon_{11} \approx 0.05$ (Sellers, 1965).

The radiation emitted to space under clear sky is composed of upward radiation of the sky itself, i.e., $\varepsilon_2 \sigma T_a^4$ plus that portion of surface emission which is directly transmitted through the window, i.e., $(1-\varepsilon_1)\sigma T_s^4$. The effective upward emissivity ε_2 of the atmosphere is little affected by water vapor (because of its rapid decrease with height) and is therefore taken to be constant. The net energy loss under cloudy conditions is the upward radiation of the clouds themselves, i.e., $(1-\varepsilon_3)\sigma T_c^4$ plus the contribution of the residual atmospheric gases above cloud level, i.e., $\varepsilon_4 \sigma T_a^4$. Both ε_3 and ε_4 are modeled as functions of cloud height; if clouds are placed near the ground (fog), then $\varepsilon_3 = \varepsilon_1$, and $\varepsilon_4 = \varepsilon_2$. On the other hand, if we deal with high clouds, absorbing gases above cloud level are virtually absent, implying that $\varepsilon_3 \approx \varepsilon_4 \approx 0$. These constraints are met by simply setting $\varepsilon_3 = \varepsilon_1 e^{-\alpha h_c}$ and $\varepsilon_4 = \varepsilon_2 e^{-\beta h_c}$. The empirical parameters ε_{10}, ε_2, α and β are adjusted such that they fit the present day global climate (see below).

The net longwave flux leaving the planet is thus

$$E_p = (1-c)E_{clear} + cE_{cloudy} \quad (15)$$

where

$$E_{clear} = -\varepsilon_2 \sigma T_a^4 - (1-\varepsilon_1)\sigma T_s^4 \quad (16)$$

and

$$E_{cloudy} = -(1-\varepsilon_3)\sigma T_c^4 - \varepsilon_4 \sigma T_a^4 \quad (17)$$

Combination of (13), (15)-(17) yields the net longwave loss E_a of the atmosphere, namely, $E_a = E_p - E_s$.

Finally, the radiation balance of the two domains is $R_a = E_a + A_a$ and $R_s = E_s + A_s$, respectively. To close the set of equations, we must establish relations for precipitation, ice and clouds. Since, on the global scale, condensation occurs even for undersaturation, we set (see Fig. 1)

$$c(rh) = \bar{c}((rh-rc)/(\overline{rh}-rc))^{c_o} \quad (18)$$

If, however, rh becomes too small, say $rh < rc$, then c is set equal to zero. In (18), $\bar{c}$ and $\overline{rh}$ are current values of cloudiness and relative humidity, and exponent

$$c_o = \ln(1/\bar{c})/\ln((1-rc)/\overline{rh}-rc)) \quad (19)$$

ensures that $c \leq 1$ for $rh \leq 1$. In order that the derivative of c with respect to rh be continuous at rh = rc, rc < o.5. In what follows, rc is fixed at o.4.

Precipitation depends, among other things, on humidity and cloudiness. This may be simply expressed by

$$P = P_o qc \tag{20}$$

where P_o is the precipitation strength (in units of kg $m^{-2}s^{-1}$) to be fixed later. Therefore, P = o, if c = o or q = o.

We further assume that ice cover solely depends on sea temperature T_o. It is modeled in terms of

$$x = \tanh[x_o((T_o-T_f)/(\overline{T}_o-T_f))^{x_1}] \tag{21}$$

Here $x_o = \operatorname{arctanh}(\overline{x})$ ensures that $x(\overline{T}_o) = \overline{x}$, where $\overline{x}$, $\overline{T}_o$ are current values of x, T_o. Also, (2o) states that x = o, if $T_o = T_f$, and $x \rightarrow 1$, if $T_o \gg \overline{T}_o$. In order that dx/dT_o be finite at $T_o = T_f$, the slope determining factor x_1 must be ≥ 1. A reasonable value for x_1 appears to be 1.5 (see Fig. 2).

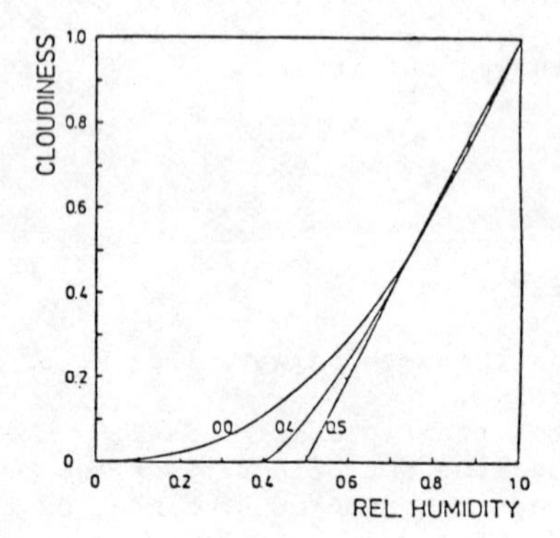

Figure 1. Cloudiness as function of rh, for cut-off humidity rc = o.o, o.4, o.5.

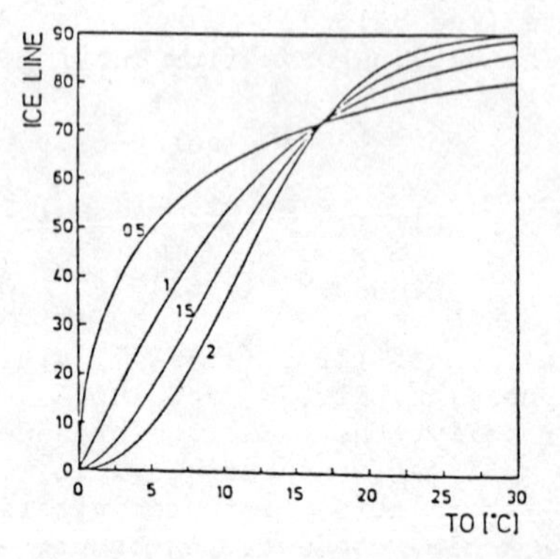

Figure 2. Ice line $\sin^{-1}(x)$ in degrees of latitude as function of T_o, for x_1= o.5, 1, 1.5, 2.

Model calibration

In the preceeding section, we have fixed those parameters which are either fairly well known or otherwise of minor importance. We are thus left with parameters which are either poorly known or effect the results significantly. They are determined by minimizing the weighted sum of residuals between prescribed and calculated global quantities. The latter are obtained from the time independent, algebraic set of non-linear equations (1)-(3). The results of the least square fit are:

(a) parameter values

$\kappa_o = 0.0702$, $r_{cloudy} = 0.4715$, $\varepsilon_{10} = 0.6151$, $\alpha = 0.2921$, $\beta = 3\alpha$,

$\varepsilon_2 = 0.3866$, $\lambda = 8.9833\ Wm^{-2}K^{-1}$, $P_o = 7.86x10^{-3} kg\ m^{-2}s^{-1}$.

(b) global climatology

The globally averaged climate is listed in Table I. Numbers are adopted from Sellers (1965), Paltridge and Platt (1976), Oort (1983) and references given therein. Temperatures and Bowen ratio H/L_vE are representative for the southern hemisphere. Since more clouds reduce the outgoing infrared radiation, $\zeta = (1-c)E_{clear} / cE_{cloudy}$ should be greater than unity. Rewriting E_p in the usual notation, i.e., $E_p = -A+Bc$, then $A/Bc \approx 0.08$ (van der Dool, 1980), corresponding to $\zeta = 1.2$. Also included in Table I is the calculated globally integrated climate. It differs by 3% at most from the prescribed one (For comparison: the uncertainty in determining the solar constant is of the order of 1%).

TABLE I

Prescribed and computed globally averaged climate. The solar constant is $Q_o = 340\ Wm^{-2}$.

	T_a[°C]	T_o[°C]	rh	c	P[m/a]	x	ζ	H/L_vE	α_p
prescribed	14.8	16.8	0.75	0.5	1.0	0.95	1.2	0.13	0.31
computed	14.8	16.79	0.75	0.5	0.97	0.95	1.2	0.134	0.309

	A_a/Q_o	A_s/Q_o	E_s/Q_o	E_a/Q_o	E_p/Q_o
prescribed	0.23	0.46	-0.69	-0.48	-0.21
computed	0.226	0.465	-0.691	-0.484	-0.207

(c) radiative balance

The global decomposition of short and longwave radiation is depicted in Fig. 3. It fits well into corresponding diagrams of various textbooks (see e.g, Paltridge and Platt, 1976 and Gill, 1982).

In summary, we conclude that our model is capable of reproducing the principal properties of the globally integrated climate.

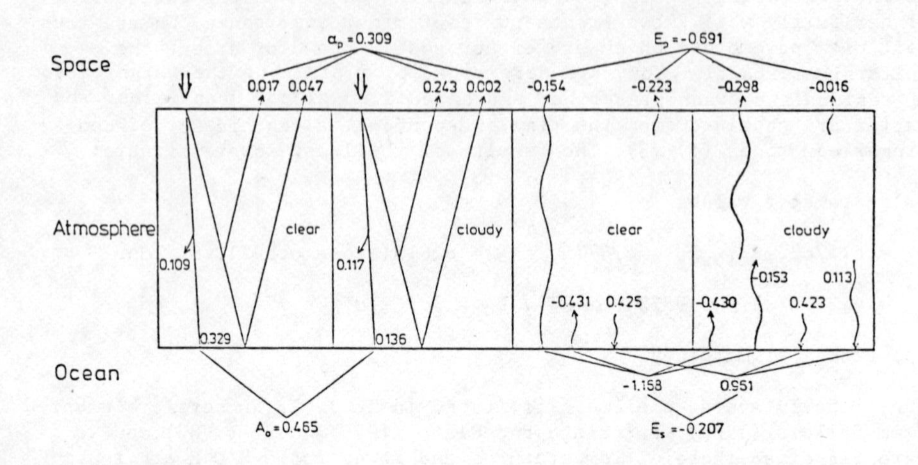

Figure3. Global decomposition of model short wave (left hand panel) and long wave radiation. Incoming solar radiation is normalized to Q_o = 34o Wm^{-2}. Energy balance of the atmosphere (surface) is achieved by adding (substracting) sensible and latent heat (o.o31 Q_o and o.227 Q_o , respectively).

3. RESULTS

The coupled, nonlinear, first order differential equations (1)-(3) were integrated until total energy gain balances total energy loss. In other words: if $|R_a+R_s| < \varepsilon Q$, where ε is a small machine dependent number (approximately equal to 10^{-14} for simple precision of the CDC 83o), we may be certain that an asymptotic equilibrium state has been achieved.

Equations (1)-(3) are integrated by an explicite Runge-Kutta method of variable order and variable step size. A variable step size is useful for the following reasons. A crude analysis of the system yields time constants $c_q c_p/\lambda = O(1$ day), $c_a/\lambda = O(1o$ days) and $c_o/\lambda = O(1oo$ days), associated with q, T_a and T_o, respectively. The ratio of the time scales is thus of $O(1o^2)$. Problems of this sort are called stiff; they are most efficiently treated by using a variable step size such that (see, e.g., Gear, 1971) (a) the eigensolutions with small time constants which decay rapidly are approximated stably (b) the eigensolutions with large time constants which decay slowly and may contribute significantly to the solution are approximated accurately.

In the following, the variable order, variable time-step technique is used to study the time dependent behaviour of the system.

Stability analysis

To start with, let us investigate the (internal) stability of the system to perturbations in initial values. The model is found to be asymptotically stable for present day conditions, no matter how large the perturbations may be. However, if the solar constant is lowered by 1%, we discover yet another stable solution, which represents the well-known "deep-freeze" branch of solution present in most global models with ice-albedo feedback. A survey of phase trajectories in two-dimensional subspace is given in Fig. 4. Fixpoints (marked by squares) are situated at $T_a = -3.8^oC$, $T_o = 4.8^oC$ and $T_a = 13.1^oC$, $T_o = 15.6^oC$, respectively. They are separated by an unstable saddle point (marked by a circle) at $T_a \approx 0.5^oC$, $T_o = 6.5^oC$. The nature of the bifurcation is indicated in Fig. 5 and 6. Initial values are set at q = o and $T_o = 6.5^oC$. Both values represent large negative perturbations of the current climate state. Two experiments are performed: the first, labeled "10", starts at $T_a = 10^oC$; the second, labeled at "-10", starts at $T_a = -10^oC$. In the former, the model returns to the current equilibrium, in contrast to the latter, which approaches a cold state of climate. Because stability properties of the system are of great importance, we shall study the time development of the bifurcation in some detail.

After one day (see Fig. 5), humidity has increased from zero to $\approx 10^{-3}$, whereas cloudiness still has not overcome its cut-off value. In the next ten days or so, there is a significant change in air temperature associated with a rapid increase in cloudiness. After twenty days have elapsed the system has settled at a quasi-equilibrium, where the 20° temperature difference between the two experiments has shrunk to 1°C. Note that the amount of humidity created by "10" is slightly larger than that created by "-10"; in contrast, there appears to be no difference in cloudiness. So far, the time evolution has been governed by exchange of sensible and latent heat between surface and air. The ocean is still little affected, because of its large heat capacity; however, the coupling between the two heat reservoirs implies that $\dot{T}_o \approx (-c_a/c_o)\, \dot{T}_a$, as can be seen in Fig. 5.

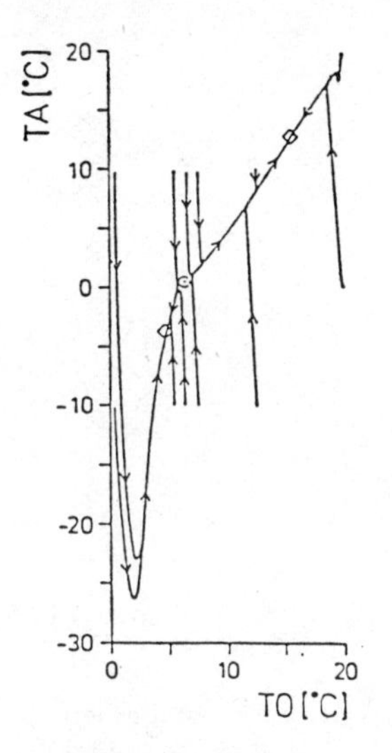

Figure 4. Phase portrait in (T_a, T_o)-plane. Time direction is marked by arrows.

Before proceeding, let us look at Fig. 6. It shows the net radiation loss versus net radiation gain as time elapses. Initially, the outgoing IR flux decreases because humidity increases; for the same reason, solar energy absorption increases.

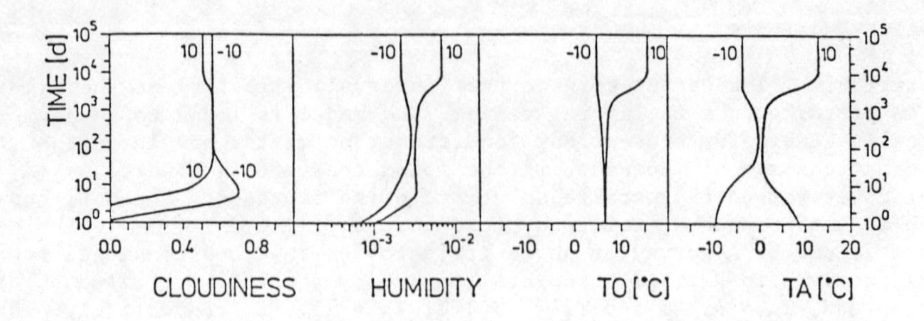

Figure 5. Time evolution of climatic variables towards equilibrium, for initial temperatures $T_a = \pm 10°C$.

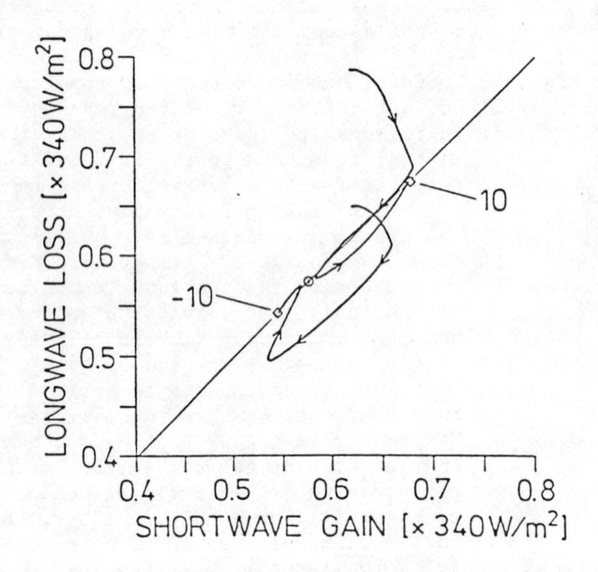

Figure 6. Phase portrait of net energy loss $|E_p|$ versus net energy gain (A_p) of the entire system. The two stable fixed points are marked by the squares; the unstable fixed point is marked by the circle.

The first turning point marks the onset of cloud formation: "10" moves down along the bisector, whereas "-10" turns once again (because of a decrease in cloudiness) until it reaches a point close to that occupied by "10". This state corresponds to the plateau-like situation revealed by Fig. 5 and is maintained over a period of several hundred days. There remains a small, but significant imbalance between loss and gain; "10" lies slightly below the bisector, whereas "-10" lies above. In other words: "10" and "-10" persist in a state of "positive" and "negative" energy balance, respectively. The reason for this is the difference in humidity (see Fig. 5); the higher amount of humidity for "10" more than compensates the higher radiation temperature, when compared to "-10". The question arises as to whether or not the radiation budget has influenced the temperature of the system.

For the time period considered so far, temperature variations are of course dominated by sensible and latent heat exchange. Hence, if there is a back-reaction of radiation on temperature, humidity and so forth, it must be weak. Consider once again Fig. 5. Due to its larger surface temperature, there is more humidity available for "10" than for "-10". However weak, the positive vapor-temperature coupling will thus be more relevant for "10" than for "-10". In effect, "-10" has just a bit too little humidity as to make $T_o < o$ switch into $T_o > o$. That there exists something like a "critical" humidity (which is, incidentally, quite close to that belonging to "-10") is confirmed by another experiment, which starts at $T_a = -10$, but with a one degree warmer surface and therefore generates more humidity. As in the case "-10", ocean temperature first decreases but increases, after humidity has passed the critical level. Eventually, we recover the same situation as in case "10".

However, why is the sign of $\dot{T}_o$ so important? Because $\dot{T}_o > 0$ implies $\dot{x} > o$; so increasing temperature gives rise to a retreat of ice, which in turn reinforces the temperature growth. Therefore "10" restores the original equilibrium, while "-10" does not. Coming back to Fig. 5, we observe that ice feedback sets in at $t \approx 10^3$ days. This timescale can be estimated from $\dot{x}/x = (dT_o/dx)\dot{T}_o$. Since the eigensolutions with large time constants have already decayed, associatiated with terms like H and L_vE, we may assume that $\dot{T}_o$ is governed by radiative balance, i.e., $c_o' \dot{T}_o = A_p + E_p$, where $c_o' = c_o + c_i dx/dT_o \approx 1.1 c_o$, and $A_p = Q(1-\alpha_p)$. Representative values of A_p and E_p are identical with the last turning point in Fig. 6. It follows that the characteristic timescale for growth or decay of ice is indeed of $O(10^3)$ days. The stepwise change in temperature is caused by ice feedback, in cooperation with vapor-temperature and cloud-temperature coupling (for a discussion of the cloud feedback see below). Asymptotic equilibrium is achieved after $\approx 10^4$ days (27 years). Of course, there is a corresponding statement about the radiation budget for each statement made about temperature. Considering once again Fig. 6, we observe that after having come close together, "-10" eventually moves over to the left, reaching its equilibrium from above (indicating that loss prevails over gain). On the other hand, "10" moves to the right, reaching its equilibrium from below.

In summary, it is seen that the bifurcation is initiated by the vapor-greenhouse effect, but is ultimately realized by ice feedback, with contributions from vapor and cloud feedback.

An alternative view of the two trajectories discussed in detail above is in three-dimensional (q, T_a, T_o)-space, as given in Fig. 7.

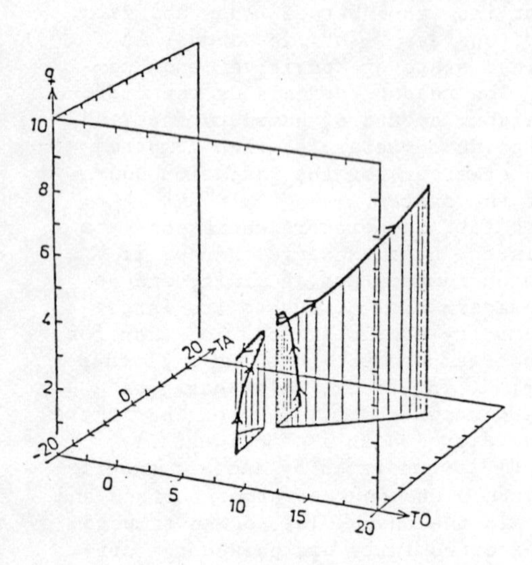

Figure 7. Phase space portrait showing the trajectories for initial temperatures $T_a = \pm 10°C$. Humidity q is given in g/kg.

Sensitivity experiments

In the following, we study (a) the transient response of the model to changes in the solar constant (b) steady state solutions as functions of solar constant and cloud height.

(a) Transient climate response

Fig. 8 shows the evolution of the model from present condition (see section 2) towards equilibrium, for stepwise change of the solar constant. An increase of the solar constant ($Q/Q_o > 1$) causes a gradual warming of air and water (or surface) such that the difference between the two becomes smaller (represented by sensible heat flux H). Also increased are evaporation (not shown), specific humidity and rainfall. The opposite holds true, if $Q/Q_o < 1$. The sharp decrease in temperature, humidity and rainfall at $Q/Q_o = 0.96$ is due to the ice feedback, as already encountered in the stability analysis. The time scales involved are of $O(10^2)$ days and $O(10^3)$ days. The former corresponds to c_o/λ, whereas the latter is governed by longwave damping and ice albedo feedback and increases for decreasing temperature.At first glance, the response of cloudiness to changes in Q may appear puzzling. On the global scale, the model predicts that cloud cover and humidity vary in opposite sense. According to (18), cloudiness depends on relative humidity $r = q/q_{sat}$ rather than specific humidity; both $\dot{q}$ and $\dot{q}_{sat}$ are positive but $\dot{q}_{sat}$ is greater than $\dot{q}$, due to its strong (exponential) dependence on air temperature. Therefore, the net effect is such than $\dot{c}$ (or $\dot{rh}$) is negative. These results can be easily checked by means of (3). Assume that $c_q\dot{q} << \lambda^*q+p$. If in addition, c is linearized about $rh = \overline{rh}$, then (3) becomes a quadratic equation in rh, yielding

$$rh = -\frac{\lambda^*-\lambda' rc}{2\lambda'} \pm \left\{\left(\frac{\lambda^*-\lambda' rc}{2\lambda'}\right)^2 + \frac{\lambda^* q_{sat}(T_o)}{\lambda' q_{sat}(T_a)}\right\}^{1/2} \tag{22}$$

where $\lambda' = P_o\bar{c}/(\overline{rh}-rc)$ and $\lambda^* = \lambda/c_p$. Since $\lambda^* > \lambda' r_c$, the positive sign must be taken. It follows

$$\dot{rh} = -\gamma\dot{T}_a/T_a^{\ 2} + \gamma'\dot{T}_o/T_s^{\ 2} \tag{23}$$

where $\gamma > o$, and $\gamma' = \gamma dT_o/dT_s$.

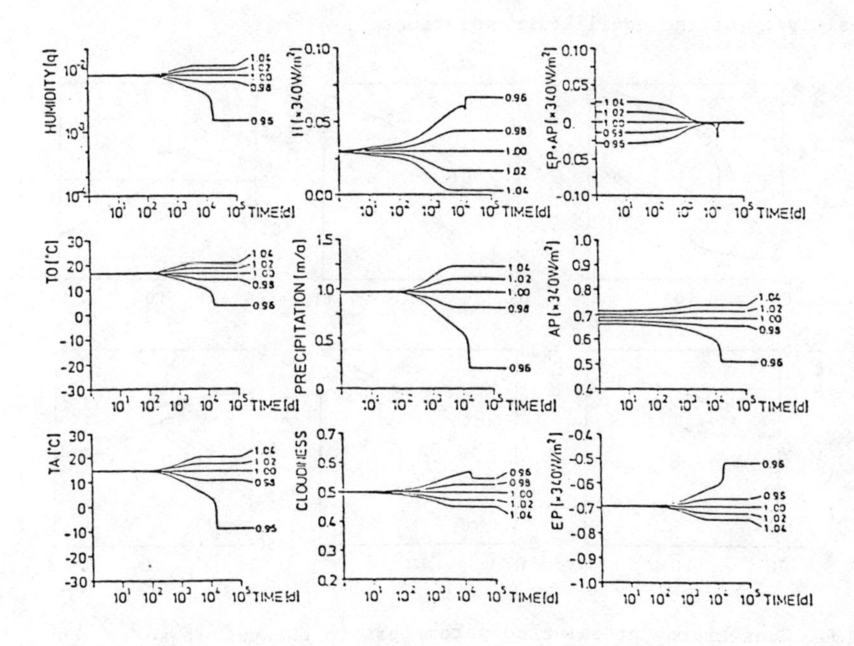

Figure 8. Time evolution of the model climate towards equilibrium, for selected solar constants (in fractions of 340 Wm^{-2}). The horizontal line represents the present day climate.

If the first term in (23) dominates over the second, relative humidity increases (decreases) if the temperature decreases (increases). However, Q/Q_o = o.96 also shows the opposite effect. Cloudiness first increases due to a decrease in temperature, but decreases, when ice feedback becomes important. In view of (23), this implies that the second term dominates the first, which turns out to be true, when T_o significantly differs from T_s (that is, when x becomes small).

For completeness, Fig. 8 also shows shortwave and longwave fluxes (A_p and E_p, respectively) as well as total radiation E_p+A_p as functions of time. The latter approaches zero from above for $Q/Q_o > 1$ and from below for $Q/Q_o < 1$. Because in equilibrium, the relations between temperature, humidity and cloudiness are dictated by the global radiation balance, it is worthwhile in this context to list the response of E_p and A_p to changes of q, T_a, T_o. The longwave flux varies according to $\partial E_p/\partial T_a < o$, $\partial E_p/\partial T_o < o$, but $\partial E_p/\partial q > o$. Note that $E_p < o$ in our notation. Similarily, $\partial A_p/\partial T_o > o$ and $\partial A/\partial q > o$ apply for the shortwave flux. In addition, $\partial A_p/\partial c < o$ but $\partial E_p/\partial c > o$. These relations follow from (12), (15)-(17) and are illustrated by Fig. 8. It is interesting to note that increasing atmospheric reflectivity, i.e., increasing cloud amount, tends to cushion the ice-albedo effect, because of the reduced transmissivity of the atmosphere.

(b) Sensitivity of the equilibrium solutions

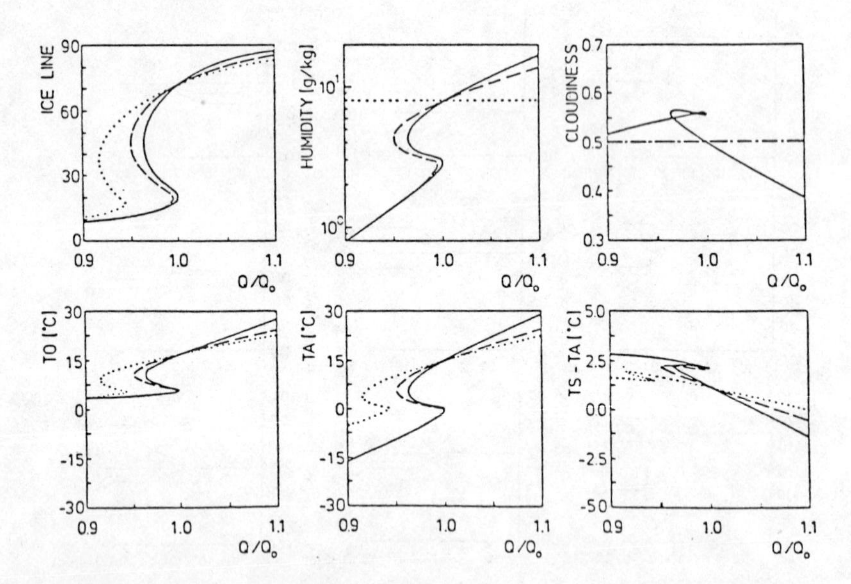

Figure 9. Sensitivity of selected parameters to changes of solar constant Q (normalized to Q_o = 34o Wm^{-2}). Steady-state solutions are obtained for variable humidity and cloud cover (solid lines), fixed cloud cover (dashed lines) and fixed humidity and cloud cover (dotted lines).

By means of the qualitative relations between global radiation, temperature, humidity and cloudiness established above, it is an easy matter to interpret the sensitivity of the time-stationary model to changes in external parameters. Fig. 9 shows selected climatic quantities as functions of the solar constant ranging from 0.9 Q_o to 1.1 Q_o. As in all other experiments carried out so far, cloud height is fixed at 3 km. Equations (1)-(3) are solved for variable cloud cover and humidity (solid curves), variable humidity and fixed clouds (dashed curves) and fixed clouds and humidity (dotted curves). All curves pass through the reference state $Q = Q_o$. Also, all solutions show the well-known multiple structure, with two stable roots representing small and large ice extent, respectively, and an unstable one in between (which has been classified as unstable saddle point in Fig. 4, 6 and 7). The stable branch is identified by North's (1981) "slope-stability" theorem; in our case, we conclude that $\partial T_{a,o}/\partial Q > 0$ and $\partial q/\partial Q > 0$ are necessary and sufficient conditions for stability. A 1% reduction in solar constant diminishes the temperature (in degrees Celsius) of air, surface and water by 1.8, 1.4 and 1.2, respectively; humidity and precipitation decrease by 11% and 7%, respectively and cloudiness is increased by 2%. The numbers depend on the parameterization of the ice-temperature relationship (see Fig. 2); the steeper the slope $\partial x/\partial T_o$, represented by parameter x_1, the more sensitive is the model to changes in Q-value. The bifurcation point seperating the "warm" climate from the "cold" climate is situated at 0.955 Q_o, 0.962 Q_o and 0.975 Q_o, corresponding to $x_1 = 1$, $x_1 = 1.5$ (current value) and $x_1 = 2$, respectively. Therefore, changing the slope of the ice-line by a factor ≈ 2 affects the bifurcation point by 2% at most.

Perhaps more interesting is the response of cloud cover to changes in Q. Cloudiness decreases with increasing temperature above the bifurcation and increases with increasing temperature below the bifurcation. Recalling that clouds cool rather than warm the system (because $r_{cloudy}/r_{clear} > E_{clear}/E_{cloudy}$ implies $|\partial A/\partial c| > |\partial E_p/\partial c|$), it follows that cloud-temperature coupling is positive in the former case, while it is negative in the latter. The positive feedback effect is clearly seen by comparison of the solid and dashed curves: the temperature is higher for variable clouds, if $Q > Q_o$ ($c_{var} < c_{fix} = 0.5$) and vice versa, if $Q < Q_o$ ($c_{var} > c_{fix}$). Nevertheless, it is just the ability of clouds to trap terrestrial radiation that keeps the sensitivity of the model to cloud cover fairly low. In contrast, water vapor acts in one direction only, for it reduces both the longwave loss (and to a lesser degree) the reflected part of solar energy. Thus, the sensitivity is greatly decreased, if humidity is fixed at its present value (dotted curve). An additional point deserves attention. Up to $Q/Q_o \approx 1.05$, the surface is warmer than the overlying air; however, if $Q/Q_o > 1.05$, sensible heat is carried from air to water rather than water temperature.This is related to the fact that air temperature responds more strongly to changes in Q than water temperature (see Fig.9). However, what makes T_a more sensitive than T_o? The only plausible explaination is the latent heat flux, because of its strong temperature dependence. The fact that the ocean looses heat by evaporation and the atmosphere gains heat by condensation implies that the response to changes of the solar constant is smaller for the ocean than for the atmosphere. This can be expressed in a more quantitative way. Differentiating (2) and the sum of (1) and (2) with respect to Q, we obtain

$$\frac{\partial T_a/\partial Q}{\partial T_o/\partial Q} \approx 1 + \frac{L_v \, \partial E/\partial T_s}{\lambda + 4\sigma\varepsilon_1 T_a^3 + 4c\sigma\,(1-\varepsilon_1)\,T_c^3} \tag{24}$$

showing that the temperature variation of the saturation value is indeed responsible for the difference in sensitivity of the reservoirs.

As outlined in the introduction, not only the magnitude of the solar input, but also the height of the clouds may significantly affect the results of the model. Cloud height enters the radiation balance via cloud temperature and emissivity (see section 2); according to (15) - (17), we have

$$\partial E_p/\partial h_c = c\left\{4\Gamma\sigma T_c^{\,3}\,(1-\varepsilon_3) + \beta\varepsilon_4\sigma T_a^{\,4} - \varepsilon_3\alpha\sigma T_c^{\,4}\right\} \tag{25}$$

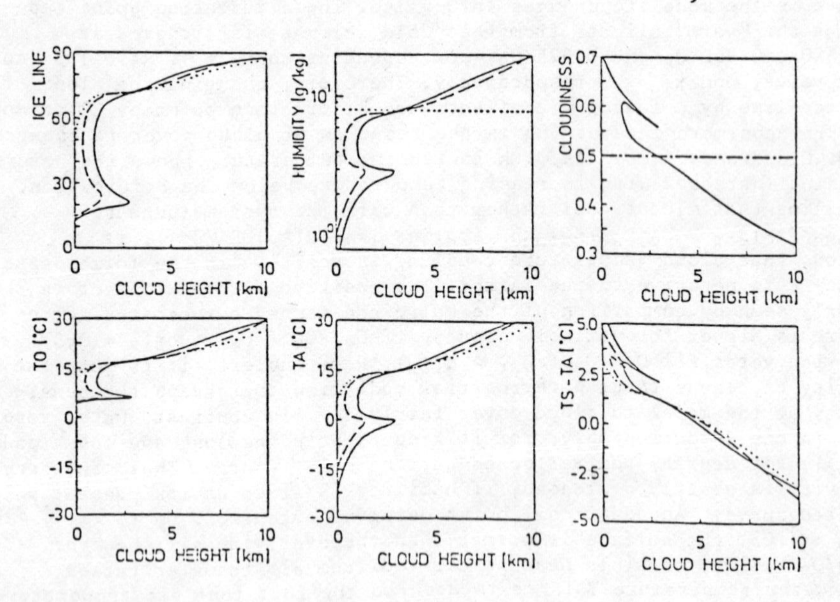

Figure 10. Sensitivity of selected quantities to changes of cloud height, for present solar constant. The different signatures of the curves are explained in Fig. 9.

Formula (25) tells that the longwave loss decreases if the cloud temperature decreases. Note, however, that the effect is reduced by the height dependent emissivity. Cloud height induced changes are illustrated in Fig. 10.

If the cloud height is increased above 3 km (assumed to be representative for present conditions) temperature and humidity increase while cloud cover decreases. The solution shows multiple structure, if the cloud layer is shifted below 3 km. As in Fig. 9, the climate may remain warm or become cold, depending on which initial temperature is taken.

Note that the unstable branch of the equilibrium solution is defined by $\partial T/\partial h_c < o$, (where T stands for T_a and T_o) in accordance with $\partial T/\partial Q < 0$ for solar constant variations. If cloud height approaches ≈ 1.1 km, the only possible stable state is that of an ice-covered earth. So ≈ 300 % reduction in cloud height produces the same effect as ≈ 4 % reduction in solar constant.

Variable clouds greatly enhance the cloud height-temperature coupling, as can bee seen by comparison of the solid and dashed curves. As in Fig. 9, the sensitivity of the model is further decreased in the absence of vapour feedback (dotted line).

Moreover, the model is remarkably sensitive to changes in the transfer coefficient λ. This is depicted in Fig. 11. As expected, the temperature difference between surface and air decreases as λ increases. Perhaps less obvious are the individual temperature changes. In the limiting case $\lambda = 0$, the hydrological cycle is cut off, and each reservoir behaves according to its radiation balance. If $\lambda > 0$, the surface looses heat due to evaporation and sensible heat transport which is absorbed by the atmosphere; hence air temperature increases while surface temperature decreases. The increase of T_a is stopped when cloudiness comes into play. More cloudiness causes more water to precipitate, which in turn reduces the humidity. By contrast, precipitation decreases, when the relative decrease of humidity outweighs the relative increase of cloudiness. At the same time, the temperatures are decreased due to the vapor and cloud feedback, until at $\lambda \approx 22\ Wm^{-2}K^{-1}$, ice feedback overtakes and forces the climate into an ice-covered state. It may be interesting to compare the sensitivity of the model to changes in solar constant, cloud height and sea-air coupling, respectively. As noted above, a 9 % decrease in surface temperature is equivalent with a 1 % decrease of the solar constant; the same effect is obtained by a 23 % increase in λ (reference value 9 $Wm^{-2}K^{-1}$) and a 38 % decrease in h_c (reference value 3 km).

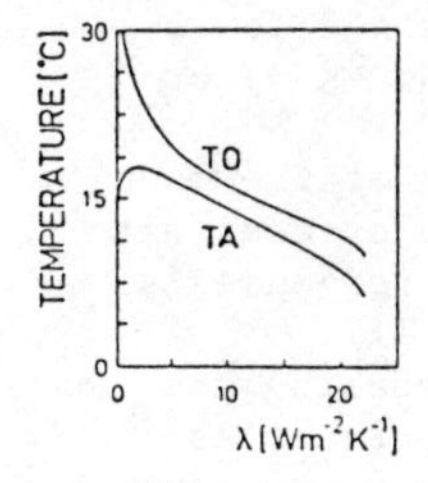

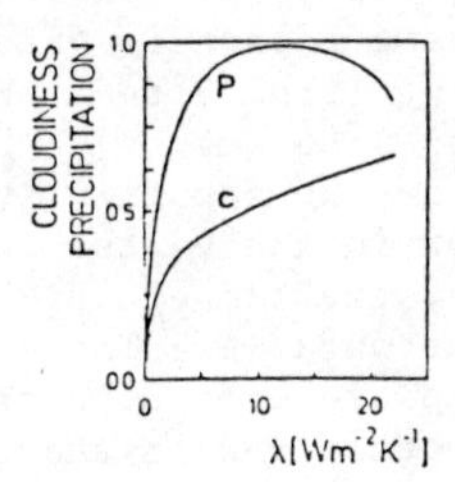

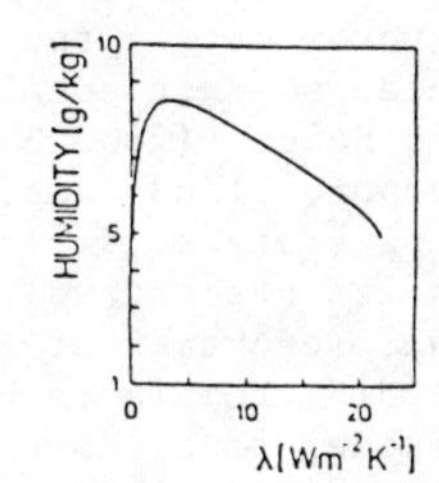

Figure 11. Response of selected quantities to changes of sea-air coupling coefficient λ. Only the stable branch of the solution is plotted.

Finally, we wish to study the surface temperature response to a doubling in the atmospheric CO_2 concentration. CO_2 enters our model via the downward emissivity and (to a much lesser extent) the absorptivity of the atmosphere (labeled ε_{10} and κ_o in section 2). Concentrating on the greenhouse feedback, we conclude that ε_{10} increases by ≈ 2.5% for doubled CO_2. This in turn, gives rise to a warming of the system, such that temperature differences are $\Delta T_s = 1^oC$ and $\Delta T_a = 1.2^oC$ in the no-feedback case, while $\Delta T_s = 2.1^oC$ and $\Delta T_a = 2.5^oC$ in the feedback case. Thus, a doubling in CO_2 is approximately equivalent to a 2% increase in solar constant. These values are fairly close to those predicted by radiative·convective models (see, e.g., Ramanathan, 1981 and Hansen et al, 1981).

4. Summary and conclusions

The theoretical results that have been discussed in this paper include the following.

o The stability of the model to perturbations of the equilibrium state, with fixed external parameters, is governed by the ice albedo-temperature feedback. However, it is also affected by the water vapor - greenhouse feedback, and the cloud cover-temperature feedback.

o The time scale on which the system approaches equilibrium is of the order of 3 years. It is governed by the damping terms (outgoing longwave flux) and temperature dependent amplification terms (incoming shortwave flux).

o Changes of the solar constant have a stronger impact on the equilibrium temperature of the atmosphere than on the temperature of the ocean. As a rule, a temperature increase (decrease) is coupled with a decrease (increase) of the sensible heat flux and an intensification (reduction) of the hydrological cycle.

o Clouds tend to cool the system (by increasing the planetary albedo), rather than warm the system (by increasing the cloud-greenhouse effect). Therefore, a temperature decrease is associated with an increase of cloud cover (positive feedback). However, at low temperaures ($T_a < 0^oC$), the cloud feedback is negative, implying that cloud cover increases when temperatures increase. This, in turn, leads to a screening of the ice-covered surface and thus a weakening of the ice feedback.

o Not only changes of the solar constant, but also changes of the cloud height and sea-air interactions may cause significant changes of the global temperature. A 1 % decrease of the solar constant leads to a 1.4^oC cooler surface; approximately the same effect is obtained for a 23 % increase in nonlinear sea-air coupling (from 9 to 11 $Wm^{-2}K^{-1}$) or a 38% decrease in cloud height (from 3 to 1.86 km).

o Feedbacks (ice, vapor, cloud) amplify the sensitivity of the model to changes in external parameters. If, for instance, the solar constant is changed by ± 1% from its present value, the temperature response is more than doubled, if all feedbacks are at work (see Table II).

The sensitivity studies remain incomplete without discussing the influence of those parameters that have been fixed at the outset. As already noted, the functional dependence of ice coverage on water tem-

TABLE II

The contributions from cloud (c), ice (i) and vapor (v) feedback and combinations of them to changes in surface temperature, caused by ± 1 % and ± 2 % changes in the current solar constant. "n" denotes the no-feedback case.

	n	i	c	v	i+c	i+v	c+v	i+c+v	
ΔT_s [°C]	0.61	0.74	0.74	0.77	0.89	0.97	1.03	1.34	1%
	-0.62	-0.77	-0.74	-0.78	-0.93	-1.04	-1.03	-1.47	-1%
	1.22	1.44	1.46	1.54	1.75	1.89	2.06	2.59	2%
	-1.24	-1.58	-1.49	-1.56	-1.93	-2.19	-2.05	-3.16	-2%

perature mainly affects the transition to an ice-covered climate. This transition is also affected by the numerical value of the ice albedo. However, the results prove to be insensitive to the particular cloud cut-off parameter. Not surprisingly, because cloudiness ranging from 0.4 to 0.6 is virtually independent of rc (see Fig. 1). Therefore, whether or not our results are affected by the particular type of cloud parameterization remains inconclusive. However, if there exists a relation between global cloudiness and global relative humidity, it should not be radically different from that adopted in this paper. In summary, we may conclude that there exists a large number of different parameter sets that fit the present day climate equally well. Each parameter set will yield a slightly different measure for the stability and sensitivity of the model. However different, none will change the global picture, provided the parameters are physically reasonable.

One of the most controversial points in our model is certainly the treatment of the hydrological cycle. To our knowledge, there is no globally average cloudiness change documented for which changes in absolute and relative humidity, as well as precipitation are also documented. Therefore, a parameterization for global cloudiness and precipitation remains highly speculative. Furthermore, a proper identification of the cloud feedback is complicated by the fact that both cloud height and cloud cover may change simultaneously. If a decrease in cloudiness also means a decrease in cloud height (as suggested by Wetherald and Manabe (1980) for equatorial regions), then the warming effect due to the increase of solar absorption would be offset by the cooling effect due to the increase of upward terrestrial emission. If, on the other hand, changes in cloud cover and changes in cloud height work in the same direction, then the positive cloud feedback would be enhanced rather than compensated.

Despite the apparent simplicity of our model, we are able to simulate the present climate in most of its aspects. Furthermore, the changes in temperature, cloud cover and precipitation to changes in solar constant and CO_2 concentration resemble those found in GCM type models (see e.g., Schneider et al, 1979; Wetherald and Manabe 1980). These authors found for upper and middle tropospheric clouds, that an increase in temperature is coupled with a decrease in cloudiness (in contrast, radiative convective models show just the opposite tendency; see Wang et al, 1981).

The next step towards a more realistic treatment of the climatic feedback problem is the use of a one dimensional, latitudinal dependent model. Such a model would allow for transport of heat and moisture and thus introduces some kind of cell structure in the model atmosphere. On the other hand, it would essentially keep the simplicity of radiative and convective heat fluxes of this model. Furthermore, the feedbacks acting at low and high latitudes may be opposite in sign (Wetherald and Manabe, 1981; Roads and Vallis, 1984), so that their net effect on the global climate might become smaller than suggested by this model.

ACKNOWLEDGEMENTS

I wish to thank E. Maier-Reimer, K. Herterich and K. Hasselmann for valuable discussions. I also thank V. Schlien for his help in programming.

REFERENCES

Adem, I., 'On the theory of the general circulation of the atmosphere' Tellus, 14, 102, 1962.
Gear, C.W., 'Numerical initial value problems in ordinary differential equations', Prentice Hall, New York, 1971.
Gill, A.E., 'Atmosphere-ocean dynamics', Academic Press, New York, 1982.
Hansen, I., Johnson, D., Lacis, A., Lebedeff, S., Lee, P., Rind, D., Russell, G., 'Climate impact of increasing atmospheric carbon dioxide', Science, 213, 957, 1981.
Lacis, A.A., and J.E. Hansen, 'A parameterization for the absorption of solar radiation in the earth's atmosphere', J. Atmos. Sci., 31 118, 1974.
North, G.R., 'Energy balance climate models', Reviews of Geophys. and Space Phys., 19, 91, 1981.
Oort, A.H., 'Global atmospheric circulation statistics', 1958-1973, NOAA, Rockville, 1983.
Paltridge, G.W., 'Global cloud cover and earth surface temperature', J. Atmos. Sci., 31, 1571, 1974.
Paltridge, G.W., 'Global dynamics and climate change - a system of minimum entrophy exchange', Quart. J.R. Met. Soc., 101, 475, 1975.
Paltridge, G.W., and C.M.R. Platt, 'Radiative processes in meteorology and climatology', Elsevier, New York, 1976.
Pethukov, V.K., 'The long-period process of heat and moisture exchange in the presence of broken clouds', Izv. Atmos. Oceanic Phys., 11, 133, 1974.

Ramanathan, V., 'The role of ocean-atmosphere interactions in the CO_2 climate problem', J. Atmos. Sci., 38, 918, 1981.
Roads, J.O. and G.K. Vallis, 'An energy balance climate model with cloud feedbacks', Tellus, 36A, 236, 1984.
Saltzman, B., and R.E. Moritz, 'A time dependent climatic feedback system involving sea-ice extent, ocean temperature, and CO_2', Tellus, 32, 93, 1980.
Schneider, S.H., Washington, W.H., and R.M. Chervin, 'Cloudiness as a climatic feedback mechanism', J. Atmos. Sci., 35, 2207, 1978.
Sellers, W.D., 'Physical climatology', The University of Chicago press, Chicago, 1965.
Sellers, W.D., 'A two-dimensional global climate model', Mon. Wea. Rev., 104, 233, 1976.
Van den Dool, H.M., 'On the role of cloud amount in an energy balance model of the earth's climate', J. Atmos. Sci., 37, 939, 1980.
Wang, W.-C., Rosow, W.B., Yao, M.-S., and M. Wolfson, 'Climate sensitivity of a one-dimensional radiative-convective model with cloud feedback', J. Atmos. Sci., 38, 1167, 1981.
Weare, B.C. and F.M. Snell, 'A diffusive thin atmosphere structure as a feedback mechanism in global climate modeling', J. Atmos. Sci., 31, 1725, 1974.
Wetherald, R.T. and S. Manabe, 'Cloud cover and climate sensitivity', J. Atmos. Sci., 37, 1485,1980.

SEA ICE-OCEAN INTERACTION

Peter Lemke
Max-Planck-Institut für Meteorologie
2000 Hamburg 13, Bundesstraße 55
Federal Republic of Germany

ABSTRACT. A Model for the seasonal variation of the oceanic mixed layer and the pycnocline is coupled to a thermodynamic sea ice model and applied to the Southern Ocean. Results are shown for a standard run, a polynya experiment and a stochastically forced integration.

1. INTRODUCTION

Long term climate fluctuations arise through the interaction of the atmosphere with the slow components of the climate system, i.e. ocean, ice and biosphere. Mixed layer and sea ice play an important role in climate dynamics since they represent the interface between atmosphere and the deep ocean, thereby linking time-scales of a few days (atmosphere) with those of a few decades to centuries (deep ocean). There is a variety of models describing the time evolution of the mixed layer properties in the open ocean. For a review see Niiler and Kraus (1977). Following the Arctic Ice Dynamics Joint Experiment (AIDJEX) 1975-1976, the structure of the boundary layer below sea ice was also investigated in greater detail (McPhee, 1975; McPhee and Smith, 1976; McPhee, 1978; Lemke, 1979; Morison and Smith, 1981; Lemke and Manley, 1984). The structure of the upper ocean near the ice edge and below the pack ice is of special importance in climate research since in these regions the atmosphere is directly coupled to the deep ocean by deep convection and bottom water formation.

There is a variety of sea ice models (Semtner, 1976; Parkinson and Washington, 1979; Hibler, 1979). In this paper we will discuss a simple thermodynamic sea ice model, since the main emphasis is on the interaction between sea ice and mixed layer, i.e. the prognostic determination of the vertical oceanic heat flux. Our model is strictly one-dimensional. Advective effects in the ocean are simply incorporated in a specified upwelling velocity and a net surface freshwater flux. Because of these two specified boundary conditions a stable seasonal response of the coupled model is maintained, although major distortions of the standard seasonal cycle may occur due to short-time disturbances, as shown in section 5.

C. Nicolis and G. Nicolis (eds.), Irreversible Phenomena and Dynamical Systems Analysis in Geosciences, 439–451.

2. THE ONE-DIMENSIONAL MIXED LAYER-PYCNOCLINE MODEL

In our model it is assumed that the vertical structure of temperature and salinity is described by a constant in the mixed layer and an exponential shape in the pycnocline (see Fig. 1)

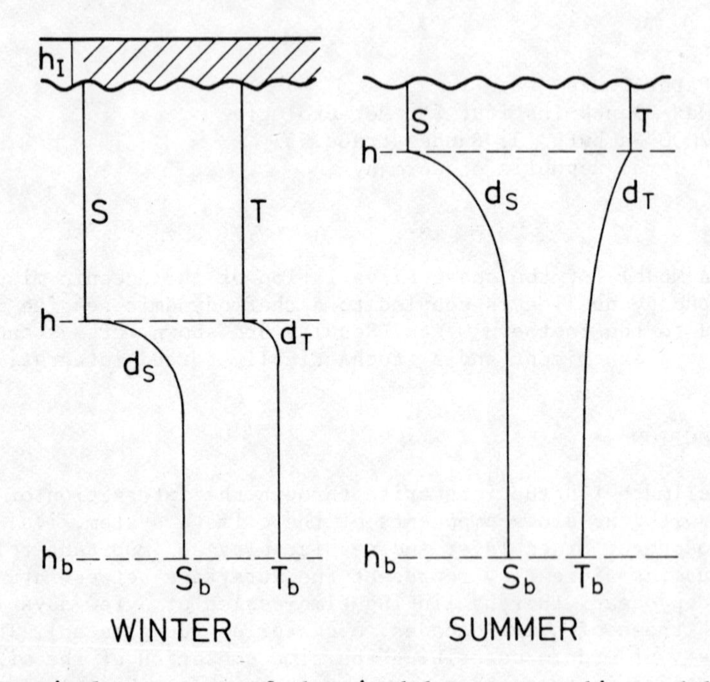

Fig. 1. Vertical structure of the mixed layer-pycnocline model.

$$\left.\begin{aligned} T(z) &= T \\ S(z) &= S \end{aligned}\right\} \quad o > z > -h \tag{1}$$

$$\left.\begin{aligned} T(z) &= T_\infty + (T - T_\infty)\ \exp[(z+h)/d_T] \\ S(z) &= S_\infty + (S - S_\infty)\ \exp[(z+h)/d_s] \end{aligned}\right\} \quad -h > z > -h_b$$

The lower level of our model h_b is set at 3000 m. Generally d_s, $d_T \ll h_b - h$, so that $T_b = T(-h_b) \approx T_\infty$ and $S_b = S(-h_b) \approx S_\infty$. The evolution of h, T, S, d_T and d_s is determined from prognostic equations, and T_b and S_b are given as boundary conditions. The prognostic equations of our one-dimensional model which describes only vertical mixing processes are derived from the conservation of heat and salt, potential energy considerations and a parameterization of the entrainment flux (see also Lemke and Manley, 1984 and Lemke, 1986).

With the assumption $h_b - h \gg d_s$ the salt content of the two layer system is given by

$$H_s = (S - S_b)(h + d_s) + S_b h_b \tag{2}$$

The salt balance in our one-dimensional system states that the change of salt content is balanced by the effective salt fluxes at the sea surface Q_S and by upwelling W.

$$\dot{H}_s = \dot{S}(h + d_S) + (S - S_b)(\dot{h} + \dot{d}_S) = Q_S + W(S_b - S) \tag{3}$$

The upwelling term is used to describe the net effect of the oceanic circulation in order to balance the access of precipitation over evaporation at high latitudes. A similar equation follows for temperature from heat conservation.

$$\dot{H}_T = \dot{T}(h + d_T) + (T - T_b)(\dot{h} + \dot{d}_T) = Q_T + W(T_b - T) \tag{4}$$

Like in other mixed layer models we will assume that the rate of change of the mixed layer salinity and temperature is dominated by the appropriate surface and entrainment fluxes.

$$\dot{S} = (Q_s + B_S)/h \tag{5}$$

$$\dot{T} = (Q_T + B_T)/h \tag{6}$$

B_s and B_T are the entrainment fluxes of salt and heat, respectively.

According to Kraus-Turner type models the closure for the mixed layer depth is taken from potential energy considerations, i.e. wind and ice keel stirring provide the energy $\tilde{K}$ needed to balance the increase of the potential energy due to the surface and entrainment fluxes, Q and B, respectively. In wintertime convection provides additional energy for the entrainment process. Other turbulent kinetic energy sources for the mixed layer evolution besides surface stress and convection are neglected. Additionally, we will assume that there is always enough turbulence in the pycnocline to provide energy required for maintenance of the exponential profile below the mixed layer.

The potential energy balance for the entrainment phase of the annual cycle is then given by (Lemke and Manley, 1984)

$$\tilde{K} - \varepsilon = \frac{h}{2} g (B - Q) \tag{7}$$

where g is the gravitational acceleration and ε is a dissipation term which is parameterized in terms of the active turbulence generating processes: wind stirring and convection.

The surface and entrainment buoyancy fluxes are given by

$$Q = \beta Q_s - \alpha Q_T \tag{8}$$

$$B = \beta B_S - \alpha B_T \tag{9}$$

where α and β denote the expansion coefficients of the density with respect to temperature and salinity. The entrainment salt, B_S, and heat fluxes, B_T, are parameterized in terms of a turbulent length scale δ and the entrainment velocity w_e such that

$$B_S = E_S w_e \tag{10}$$

and $$B_T = E_T w_e \tag{11}$$

where $$E_S = (S_b - S)(1 - \exp(-\delta/d_S))$$

and $$E_T = (T_b - T)(1 - \exp(-\delta/d_T)) \tag{12}$$

δ may be considered to be a measure for the thickness of the entrainment zone. Inserting (8) through (11) into (7) yields for the entrainment velocity

$$w_e = (2KD_1 + hQD_2)/(hE) \tag{13}$$

where $K = \tilde{K}/g$

and $E = \beta E_S - \alpha E_T$

D_1 and D_2 describe the depth-dependent dissipation of mechanical and convective energy input at the surface. This depth dependence is assumed to be exponential:

$$D_1 = \exp(-h/h_w) \tag{14}$$

$$D_2 = \begin{cases} 1 & Q < o \\ \exp(-h/h_c) & Q > o \end{cases}$$

where h_w and h_c are the scale depths of dissipation which together with the turbulent length scale δ are determined from a least squares fit of the model to observations. Rearranging (3), (4), (5), (6), (10) and (11) leads to

$$\dot{S} = Q_S/h + (E_S/h)w_e \tag{15}$$

$$\dot{T} = Q_T/h + (E_T/h)w_e \tag{16}$$

$$\dot{d}_S = d_S/(S_b - S)\dot{S} - \exp(-\delta/d_S)w_e \tag{17}$$

$$\dot{d}_T = d_T/(T_b - T)\dot{T} - \exp(-\delta/d_T)w_e \tag{18}$$

where we have used the fact that the actual change of the mixed layer depth $\dot{h}$ is given by the entrainment velocity and the upwelling w

$$\dot{h} = w_e - w \tag{19}$$

Equations (13) and (15) through (19) apply for the entrainment phase of the annual cycle ($w_e > o$). If the right hand side of (13) becomes negative, the stress induced energy at the surface is insufficient to overcome the stabilizing effect of the surface buoyancy flux. This phase of no entrainment occurs during the period of increased heating (melting). The mixed layer then retreats to an equilibrium depth $\hat{h}$ which is given by the Monin-Obukhov length determined from (13) with $w_e = o$ or

$$2KD_1 + \hat{h}Q = o \tag{20}$$

During the retreat phase the mixed layer and the pycnocline are treated to be decoupled such that

$$\Delta S = Q_S \Delta t/\hat{h} \qquad , \ \Delta T = Q_T \Delta t/\hat{h} \tag{21}$$

and

$$\begin{aligned} \Delta d_S &= d_S/(S_b - S)\Delta S - \Delta h \\ \Delta d_T &= d_T/(T_b - T)\Delta T - \Delta h \end{aligned} \tag{22}$$

where Δh denotes the diagnostic retreat of the mixed layer. (22) follows from heat and salt balance below the mixed layer.

For given fluxes at the sea surface and with given boundary conditions at $z = -h_b$, the evolution of the one-dimensional mixed layer-pycnocline model due to vertical mixing processes can now be described.

3. COMPARISON WITH AIDJEX DATA

The contribution of the heat flux to Q and E in (13) is negligible under the sea ice cover. The density and therefore the mixed layer dynamics is solely determined by the salinity. Therefore we have integrated the salt-part of the model (eq. (13), (15) and (17)) for 15 years starting with typical end of winter conditions, and have fitted the equilibrium seasonal cycle to salinity profiles obtained during the Arctic Ice Dynamics Joint Experiment (AIDJEX) 1975-1976 by tuning the three parameters h_w, h_c and δ. For details of the least squares fit see Lemke and Manley (1984) where similar models have been tested against AIDJEX data.

The optimal values of the model parameters have been obtained as $\delta = 5$ m, $h_c = 50$ m, and $h_w = 7$ m. The optimal fit of the model equilibrium seasonal cycles for S, h and d_S to the detrended AIDJEX data from Camp Caribou is shown in Fig. 2. Generally, the root mean square deviation between best fit model and data is about 10% to 15% of the observed amplitude of the annual cycles of the profile parameters.

4. SEA ICE-OCEAN COUPLING

In polar regions the pycnocline is generally warmer than the mixed layer. This fact represents the basis of a special interaction between the heat and the salt budgets of ice covered oceans. Once sea ice is formed due to heat loss to the atmosphere, brine is released. This leads to convection and entrainment of warm water into the mixed layer, which in turn is used to melt some ice again thereby modifying the surface salt flux. The heat storage of the deep ocean delays the winter freeze-up and reduces the maximum sea ice extent considerably as shown in Lem-

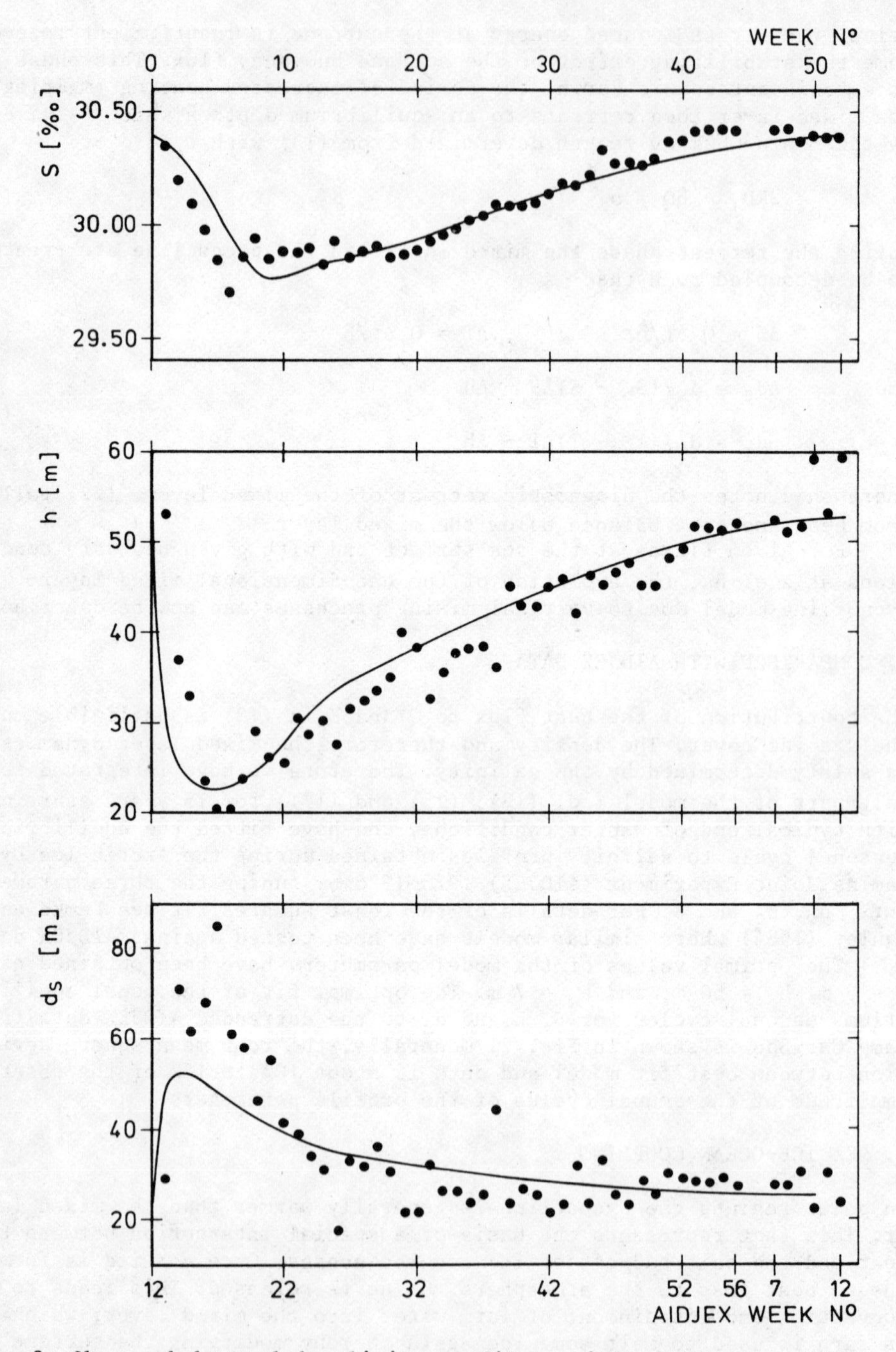

Fig. 2. Observed detrended salinity profile variables for Camp Caribou (dots) and optimally fitted model equilibrium annual cycles.

ke (1986a). The heat flux at the sea ice-ocean interface, Q_T, is given by the entrainment heat flux (11) which is lost to the melting of ice.

$$Q_T = -B_T = -E_T w_e \tag{23}$$

The total surface salt flux is accordingly determined by the freezing of ice due to heat loss to the atmosphere, Q_i, and by the melting of ice due to the oceanic heat flux (23).

$$Q_s = (S - S_I)\dot{h}_I = (S - S_I)[Q_i/(\rho L) - c/(\rho L)E_T w_e] \tag{24}$$

Where S_I is the salinity of sea ice taken to be 5 $^o/oo$, h_I is the sea ice thickness, ρ is the density of sea ice, L is the latent heat of fusion and $c = 4.26 \cdot 10^6 J/m^3 {}^oC$. Q_i is calculated from a thermodynamic sea ice model (Parkinson and Washington, 1979). Effects of sea ice dynamics (Hibler, 1979) will be included in a later publication. It is clear from (23) and (24) that the total surface buoyancy flux, Q, determined from (8) includes two terms which are proportional to the entrainment velocity. Together with (7), (10) and (11) this leads to a reformulation of the entrainment rate w_e.

$$w_e = (2KD_1 + hQ^*D_2)/(h(E + E^*)) \tag{25}$$

where $$E^* = \beta c/(\rho L)(S - S_I)E_T - \alpha E_T \tag{26}$$

Q^* contains only the atmospherically induced freezing term of (24)

$$Q^* = \beta(S - S_I)Q_i/(\rho L) \tag{27}$$

Since E^* is positive the entrained oceanic heat flux always leads to a reduction of the entrainment rate. Equations (21) to (25) apply only for the ice covered ocean. In the open ocean adjacent to the sea ice edge, Q_T is determined from the surface energy balance, and Q_s is specified as evaporation minus precipitation.

5. RESULTS

5.1. Standard experiment

The model equations derived in sections 2 and 4 represent a complex system of non-linear differential equations, which is forced by seasonally varying boundary conditions supplied by a surface energy balance model (sea ice model). A typical seasonal response of the model variables for the Southern Ocean (66 S) is shown in Fig. 3. During summer when there is no ice and the mixed layer temperature rises from freezing (-1.96°C) up to 0°C, the mixed layer salinity and depth increase only slightly. With the occurance of sea ice around day 140 there is a strong deepening of the mixed layer due to the brine rejection and subsequent pronounced increase of the surface salinity. In spring when the ice starts to melt the mixed layer suddenly retreats to its minimum depth due to the considerable freshwater flux at the

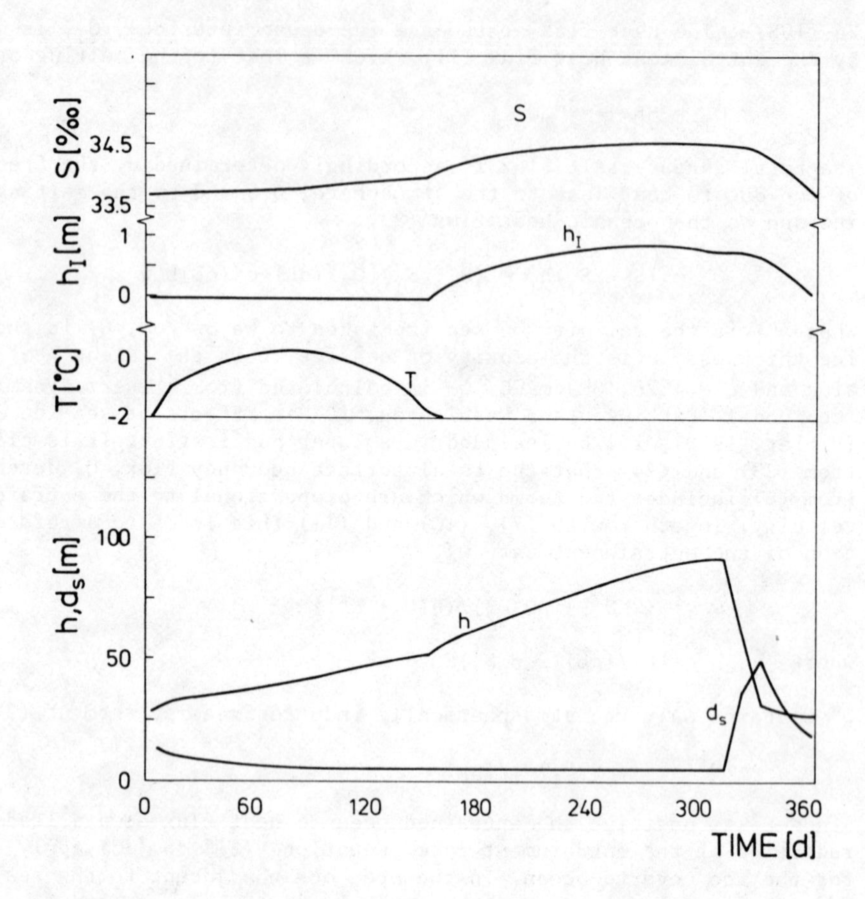

Fig. 3. Mean seasonal cycle of the model sea ice thickness h_I, the mixed layer depth h, salinity S and temperature T, and the pycnocline shape d_S for the Southern Ocean at 66 S.

surface. The pycnocline (d_S) is rather sharp during most of the year except for the retreat phase where d_S significantly increases. The modelled amplitudes of the mixed layer properties quantitatively agree with observations in the Antarctic seasonal sea ice zone. The results of a twenty-year integration of our standard model for the mixed layer depth, salinity, temperature and sea ice thickness are shown in Fig. 4. It is seen that the model reaches equilibrium after 10 years of integration. For further details and a comparison with the sea ice variations in a fixed mixed layer model see Lemke (1986a).

5.2. Polynya experiments

It has long been speculated which mechanism may lead to the occurance of the large polynya in the Weddell Sea. We will discuss two

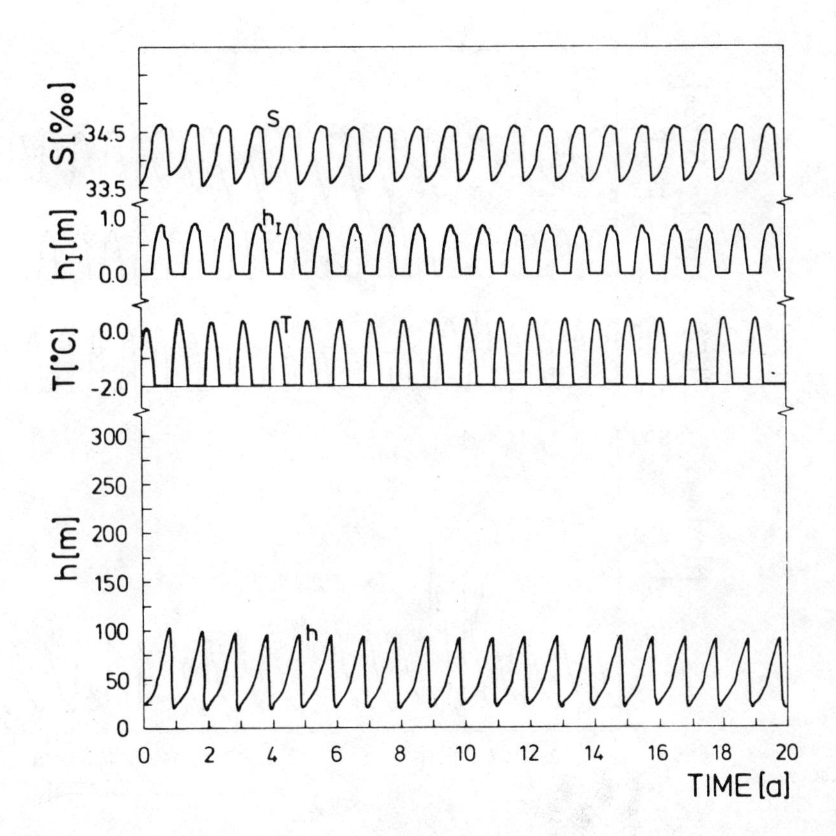

Fig. 4. Results for the model sea ice thickness h_I and the mixed layer depth h, salinity S and temperature T from a 20 year standard integration.

mechanisms which both destabilize the oceanic stratification allowing stronger entrainment of warm water and a subsequent reduction of the sea ice thickness. During a cruise in the Weddell Sea Gordon and Huber (1984) observed large warm subsurface eddies which travelled from the east into the Weddell Sea. These warm cells lifted the mixed layer base by about 40 m and increased the pycnocline temperature, allowing intenser entrainment of warm and salty water.

In our first poynya experiment we model the warm cells by fixing the mixed layer depth at 40 m for 50 days in the fourth year of integration (see arrow in Fig. 5) and by increasing T_b by 1°C. d_S and d_T were fixed at 10 m. After the 50 days T_b is again relaxed to the standard values of 0.6°C. Fig. 5 shows that although the disturbance by the warm eddy lasts only a short time (50 days) the model takes about 10 years of integration time to reach the standard equilibrium seasonal cycle. The major response of the sea ice thickness takes place in the

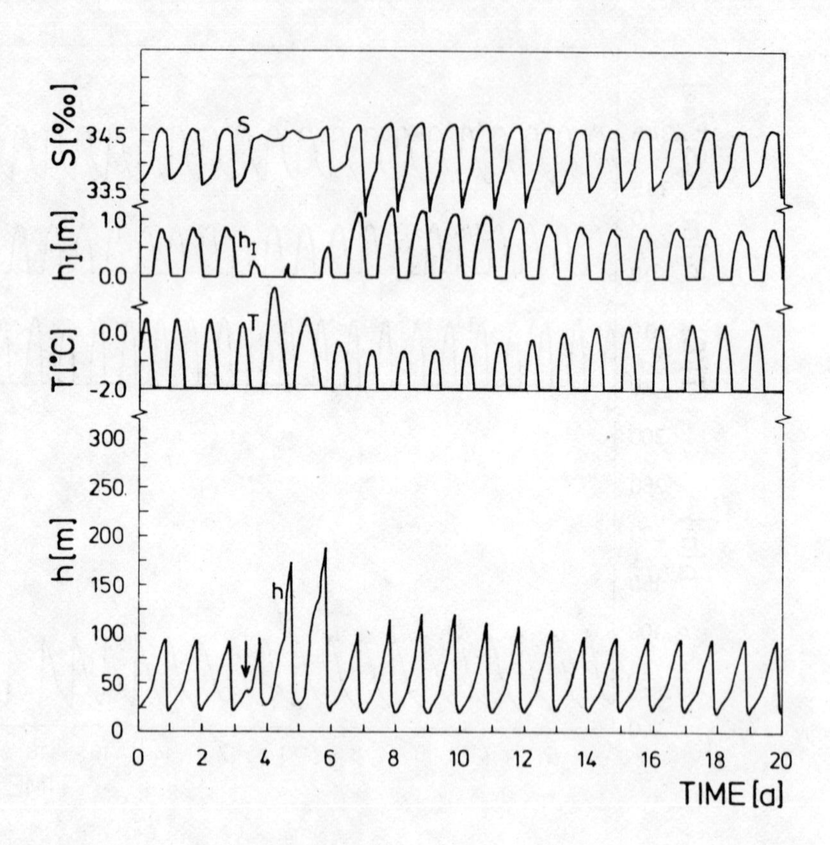

Fig. 5. Results for the model sea ice thickness h_I and the mixed layer depth h, salinity S and temperature T from a 20 year perturbation experiment. During a 50 day perturbation (see arrow) the mixed layer was fixed at 40 m, the profile parameters d_S and d_T at 10 m, and the deep ocean temperature T_b was increased by 1°C.

year after the disturbance. Due to the enhanced entrainment of salty water during the occurance of the eddy the mean annual salinity is significantly increased. This leads to stronger entrainment of warm water and a drastic reduction of the sea ice thickness in the following four years. The standard equilibrium seasonal cycle is finally reached due to the balancing of net freshwater flux (precipitation = 30 cm/year) and upwelling. The warm eddy is most effective in early to mid-winter when the buoyancy fluxes are strong enough to allow deep convection. The occurance in late winter leads to a moderate response since the buoyancy fluxes and accordingly the convection are weaker.

Another mechnism for creating a polynya is a divergent sea ice drift which reduces the mean sea ice thickness and allows access freezing and subsequent increase of the surface salinity. In our second polynya experiment we therefore reduce the sea ice thickness for 40 days in the fourth year to 10 cm as soon as it starts to grow thicker. Although the disturbance lasts only for a short time the response is similar to the previous polynya experiment.

5.3. STOCHASTIC FORCING EXPERIMENT

In order to investigate the response of the coupled sea ice-ocean model to the natural variability of the atmosphere a red noise forcing with zero mean, a standard deviation of 4.5°C, and a correlation time of several days is superimposed on the mean annual cycle of the surface air temperature which enters in the surface energy balance of the sea ice model. For more details see Lemke (1986b). A twenty-year integration is shown in Fig. 6. It is apparent that the shorttime fluctuations of the air temperature introduce long-term variations in the response of the coupled model.

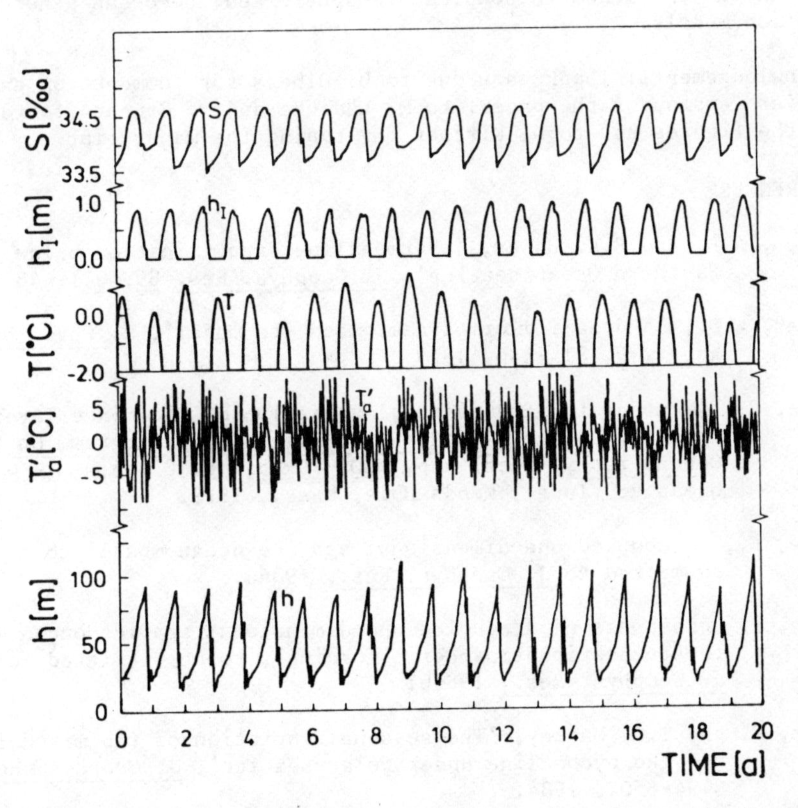

Fig. 6. Results for the model sea ice thickness h_I and the mixed layer depth h, salinity S and temperature T from a 20 year stochastically forced integration, during which the surface air temperature fluctuations T_a' were superimposed on the mean seasonal cycle in the surface energy balance.

6. CONCLUSIONS

A coupled one-dimensional sea ice-mixed layer-pycnocline model has been presented, which can be used for climate studies in conjunction with atmospheric and oceanic general circulation models. The model is able to describe the special interaction between the heat and salt budgets in ice-covered oceans, which arises through the fact that the ocean is warmer under the mixed layer than at the surface, where the temperature is at freezing. This delicate interaction is described by seasonally forced non-linear differential equations, which for a weakly stable density stratification in the ocean exhibit a pronounced sensitivity to short-time perturbations. This sensitivity is reflected in nature by the accurance of polynyas (ice free areas) within the pack ice. In order to understand this interaction better, the effects of advection in the ocean, the influence of sea ice dynamics (rheology) and the feedback via the atmosphere have to be investigated in greater detail. This in the end leads to coupled atmosphere-sea ice-ocean general circulation models.

Acknowledgements: Thanks are due to D. Olbers for comments on an earlier version of the paper, to M. Lüdicke and M. Grunert for drafting the figures and to U. Kircher for typing the manuscript.

REFERENCES

Gordon, A.L. and B.A. Huber, 'Thermohaline stratification below the Southern Ocean sea ice', J. Geophys. Res. 89, 641-648 1984.

Hibler, W.D., 'A dynamic thermodynamic sea ice model', J. Phys. Oceanogr. 9, 815-846, 1979.

Lemke, P., 'A model for the seasonal variation of the mixed layer in the Arctic Oean.' Woods Hole Summer Study Programm on Polar Oceanography, Tech. Rep. WHOI-79-85, pp. 82-96, Woods Hole Oceanogr. Inst., Woods Hole, Mass., 1979.

Lemke, P., 'A coupled one-dimensional sea ice-ocean model' to be submitted to J. Geophys. Res., 1986a.

Lemke, P., 'Response of a coupled one-dimensional sea ice-ocean model to stochastic atmospheric forcing', to be submitted to J. Geophys. Res., 1986b.

Lemke, P. and T.O. Manley, 'The seasonal variation of the mixed layer and the pycnocline under polar sea ice', J. Geophys. Res., 89 6494-6504, 1984.

McPhee, M.G., 'AIDJEX oceanographic data report', AIDJEX Bull., 39, 33-77, 1978.

McPhee, M.G. and J.D. Smith, 'Measurement of the turbulent boundary layer under pack ice', J. Phys. Oceanogr., 6, 696-711, 1976.

Morison, J. and J.D. Smith, 'Seasonal variations in the upper Arctic Ocean as observed at T-3', Geophys. Res. Lett., 8, 753-756, 1981.

Niiler, P.P. and E.B. Kraus, 'One-dimensional models of the upper ocean', in Modelling and Prediction of the Upper Layers of the Ocean', editedby E.B. Kraus, pp. 143-172, Pergamon, New York, 1977.

Parkinson, C.L. and W.M. Washington, 'A large-scale numerical model of sea ice', J. Geophys. Res., 84, 311-337, 1979.

Pollard, D., M.L. Batteen and Y.-J. Han, 'Development of a simple upper-ocean and sea ice model', J. Phys. Oceanogr., 13, 754-786, 1983.

Semtner, A.J., ' A model for the thermodynamic growth of sea ice in numerical investigations of climate', J. Phys. Oceanogr., 6, 379-389, 1976.

[illegible]

[illegible]

[illegible] and [illegible]

[illegible] the [illegible] of [illegible]

[illegible]

[illegible]

[illegible]

SOME ASPECTS OF OCEAN CIRCULATION MODELS

Peter Lemke
Max-Planck-Institut für Meteorologie
Bundesstraße 55
2000 Hamburg 13
Federal Republic of Germany

ABSTRACT. A brief review of the characteristics of oceanic general circulation models (OGCMs) is presented, followed by a discussion of some aspects of the response of particular OGCMs to climatological and to variable atmospheric forcing. Furthermore the possibility of multiple steady states for the oceanic circulation is addressed.

1. INTRODUCTION

The ocean represents an important component of the climate system. It affects climate conditions through transport of heat, and storage of heat and CO_2. The mean oceanic poleward heat transport is comparable with the atmospheric heat transport. Furthermore the ocean significantly redistributes heat zonally through circulation gyres. Due to its large heat capacity the ocean considerably attenuates the response to external forcing. The seasonal cycle in maritime regions is small compared with continental climate conditions. The long relaxation time of the ocean leads to an amplification of the response to internal "weather" forcing at low frequencies (red noise response). Besides heat the ocean also stores CO_2. Therefore the ocean significantly retards the increase of atmospheric CO_2 content through CO_2 uptake at high latitudes.

2. CHARACTERISTICS OF OCEAN CIRCULATION MODELS

The large-scale ocean circulation is driven by the exchange of heat, water and momentum with the atmosphere (Fig. 1). The windstress directly couples into the equations of motion. Heat fluxes and evaporation minus precipitation modify the heat and salt budgets, respectively. Through the equation of state these budgets affect the equations of motion, which in turn change the heat and salt budgets through the advection and diffusion of heat and salinity. Finally the bottom topography and the shape of the ocean basin determine the character of the oceanic circulation.

Although the basic physical laws governing a binary fluid system (Navier-Stokes equations (momentum balance), continuity, equation

C. Nicolis and G. Nicolis (eds.), Irreversible Phenomena and Dynamical Systems Analysis in Geosciences, 453–466.

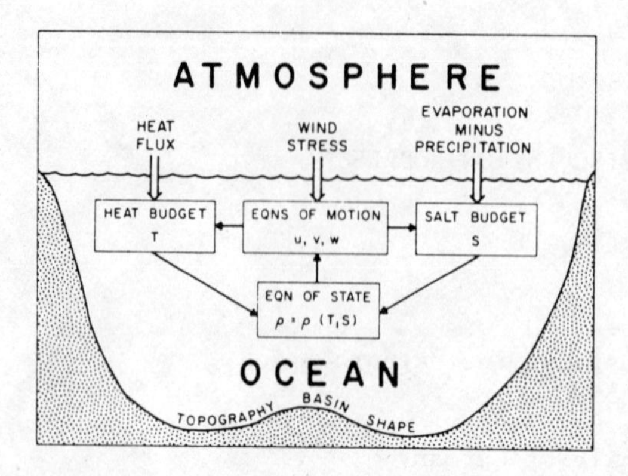

Fig. 1. A schematic diagram showing the various elements in large scale general circulation models of the ocean (from Holland, 1979).

of state and conservation of heat and salt) are well known, the description of the oceanic circulation is far from being complete. There are two basic reasons for this. First, the observational data set for model construction and verification is rather limited. Secondly, for oceanic purposes it is of no use to solve the Navier-Stokes equations which contain the full physics on all scales. Presently used approximations of these equations for climatic space and time scales are still in the state of discussion. Furthermore, the equations used in the present numerical models of the oceanic circulation are nonlinear and the solution is rather complicated.

Extensive reviews of oceanic general circulation models are given by Pond and Bryan (1976) and Holland (1977, 1979).

In most prognostic, three-dimensional large scale ocean models the momentum balance is given by

$$\underset{A}{\underset{\sim}{v}_t} + \underset{NA}{(\underset{\sim}{v} \cdot \nabla)\underset{\sim}{v}} + \underset{NA}{w\underset{\sim}{v}_z} + \underset{C}{\underset{\sim}{f} \times \underset{\sim}{v}} = \underset{P}{-\frac{1}{\rho_o}\underset{\sim}{\nabla} p} + \underset{HF}{A_m \nabla^2 \underset{\sim}{v}} + \underset{VF}{K_m \underset{\sim}{v}_{zz}} \qquad (1)$$

and

$$p_z = -g\rho \qquad (2)$$

where $\underset{\sim}{v}$ is the horizontal velocity vector, w is the vertical velocity, $\underset{\sim}{f}$ is the Coriolis parameter, p is the pressure, ρ is the density and A_m and K_m are the horizontal and vertical coefficients of eddy viscosity. Continuity states that

$$\underset{\sim}{\nabla} \cdot \underset{\sim}{v} + w_z = 0 \qquad (3)$$

The equation of state

$$\rho = \rho(T, S, p) \qquad (4)$$

relates the density to temperature T, salinity S and pressure. Eq. (4) is rather complicated but can be expressed in tables or various analytic approximations. Finally, the conservation of heat and salt is given by

$$(T,S)_t + \underset{\sim}{v}\cdot\underset{\sim}{\nabla}(T,S) + w(T,S)_z = \underset{HD}{A_{T/S}}\nabla^2(T,S) + \underset{VD}{K_{T/S}}(T,S)_{zz} + C \qquad (5)$$

where $A_{T/S}$ and $K_{T/S}$ are the horizontal and vertical eddy diffusion coefficients for temperature and salinity, respectively. C symbolically represents the convective adjustment process. Eqs. 1 through 5 are generally referred to as the primitive equations.

There is a large variety of models using various approximations of the above equations: low resolution primitive equation models (Bryan and Lewis, 1979), eddy resolving quasi-geostrophic models (Holland, 1978, Holland et al, 1984), large-scale geostrophic models (Hasselmann, 1982, Maier-Reimer, 1985), high resolution primitive equation models (Cox, 1985) and quasi-isopycnic models (Bleck and Boudra, 1981; Oberhuber, 1985). These models differ in the applied geometry (sector, regional, global, bottom topography), the physical processes resolved (wind driven gyre dynamics, equatorial dynamics, thermocline ventilation, water mass formation), the governing equations used (primitive equations, quasi-geostrophic, large scale geostrophic) and the numerical techniques applied (the vertical structure in levels, isopycnic layers or modes, the horizontal structure in gridpoints, spcectral components or finite elements and the time-stepping i.e. explicit, implicit, etc.).

Many features of the real ocean are apparent in the results of present numerical models. A schematic drawing of the thermohaline circulation found in large scale numerical ocean models is given in Fig. 2. The basic features are wind driven anticyclonic surface gyre with

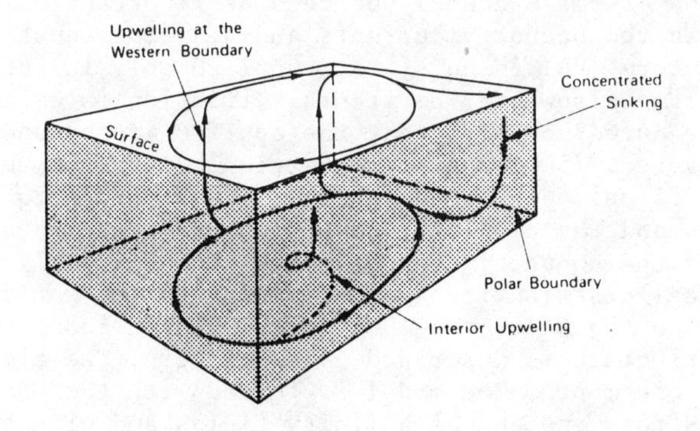

Fig. 2. A schematic drawing of the ocean circulation found in numerical models (from Bryan, 1975).

western boundary current, a cyclonic gyre at greater depth, sinking (deep convection) at high latitudes and upwelling in the interior, at the equator and at the western boundaries.

Numerical models of the oceanic circulation have been applied to determine the equilibrium response to climatological atmospheric forcing or to a step function change in the surface forcing (Bryan, et al., 1982). Recently variability studies have been undertaken by calculating the response of the ocean to time dependent atmospheric forcing (Willebrand, et al., 1980; Bryan et al., 1984). Three examples of these applications will be discussed in sections 3 and 4.

3. EQUILIBRIUM RESPONSE TO CLIMATOLOGICAL ATMOSPHERIC FORCING

In this section the equilibrium response of a large-scale geostrophic ocean model to climatological atmospheric forcing is discussed. This model was suggested by Hasselmann (1982) and implemented by Maier-Reimer (1985).

The large-scale geostrophic model is based on the fact that in the interior ocean the flow is essentially geostrophic, i.e. there is a balance between the Coriolis force and the horizontal pressure gradient (terms C and P in eq. (1)). In the final stage this model is planned to consist of the interior ocean model (purely geostrophic) which is connected to separate models for regions in which the geostrophic balance is no longer valid, i.e. the boundary currents, the equatorial regions and the surface mixed layer. These separate models are still in the process of development.

In the present version of the global large scale geostrophic ocean model the non-geostrophic regions are approximated by linear friction regimes. The model differs from the primitive equations only through certain approximations in equations (1) and (5). The acceleration term (A) and the non-linear advection terms (NA) in eqn. (1) are generally neglected. The vertical friction term (VF) is only applied in the surface layer (wind stress). The horizontal friction (HF) is included everywhere (like the terms C and P) but becomes important only in the large shear zones in the boundary currents and near the equator. The vertical diffusion term (VD) in eq. (5) is applied only in the surface layer and is neglected elsewhere. Horizontal diffusion is omitted everywhere. It should be noted however, that the applied vector upwind scheme used to solve eq. (5) leads to a numerical diffusion, which is comparable to the diffusion included in other ocean cirulation models. The time derivative and the nonlinear advection terms are retained in eq. (5), as well as the convective overturning (C).

The geographical distribution and bathymetry of the world oceans are fully included in the model. The horizontal resolution is 500 km and the vertical structure is described by ten layers. The time step used in eq. (5) is one month. The model is forced with the observed monthly mean wind stress and annual salinity fields and with the seasonal cycle of surface air temperature generated by an atmospheric GCM which uses a swamp ocean (Manabe, 1980). A simple parameterization for sea ice is included in the model.

The results of the circulation model after a 200 year integration are shown in the following figures. The sea surface velocities are

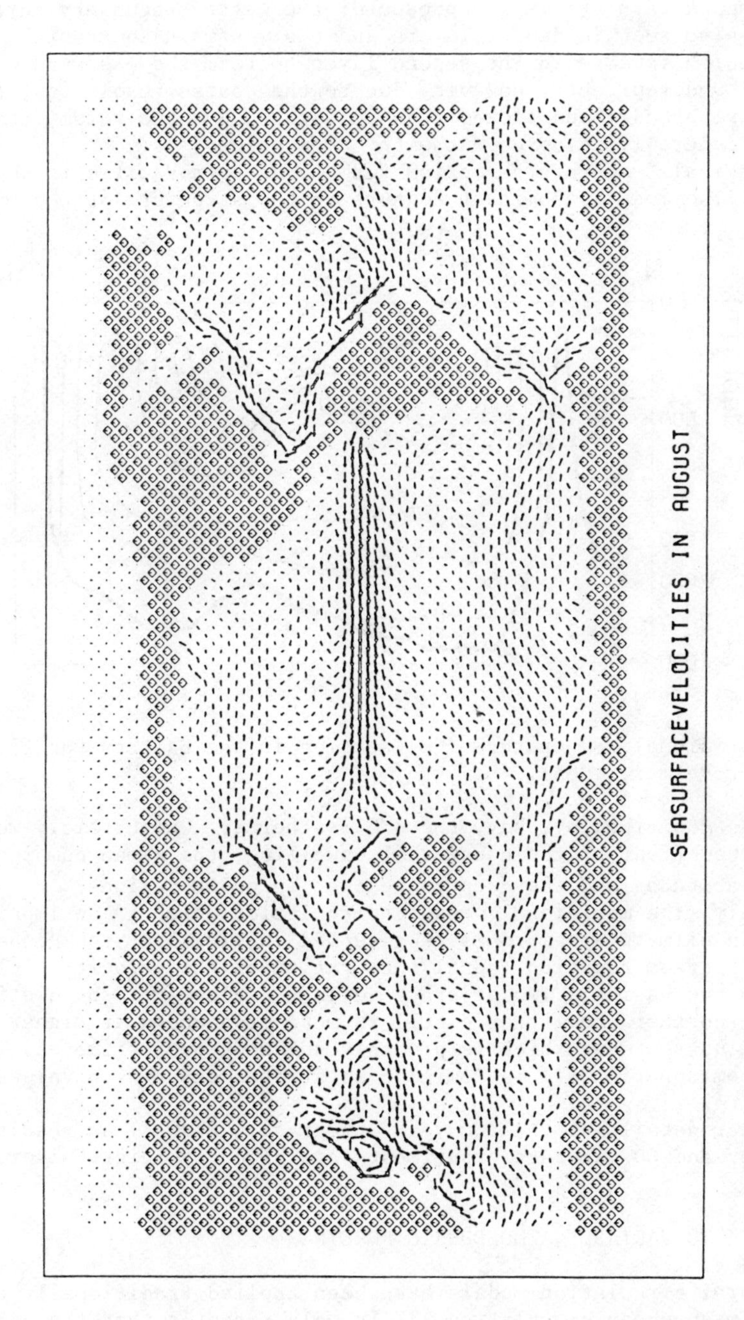

Fig. 3. Sea surface velocities in August. The length of the arrows is proportional to the square root of the velocity. The absolute maximum is 60 cm/s in the Kuroshio (From Maier-Reimer, 1985).

displayed in Fig. 3. The main features of the observed ocean surface circulation pattern are well reproduced: the western boundary currents which are also seen in deeper layers and the westward Equatorial Currents which reverse in the second layer to form the eastward Equatorial Undercurrents. However, due to the coarse resolution, the currents are broader and slower than in nature, but the volume transports are generally in agreement with observations.

A vertical temperature section for the eastern Pacific is shown in Fig. 4. The most predominant feature is the structure of the main

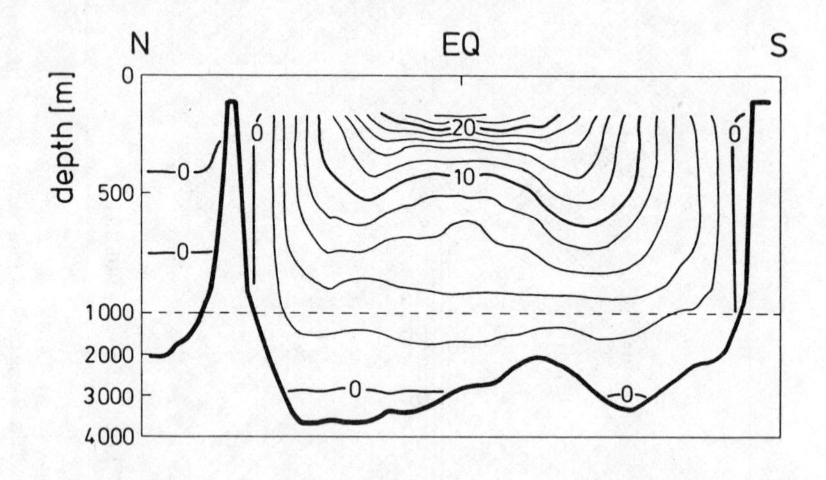

Fig. 4. Meridional temperature (°C) section in the Eastern Pacific. (From Maier-Reimer, 1985).

thermocline described by, say, the 8°C isotherme, which fairly well reflects the effect of the equatorial upwelling, the Ekman pumping at moderate latitudes and the deep convection in polar regions.

Finally, the meridional oceanic circulation which is an important quantity in climate dynamics is represented in Fig. 5 by the global meridional stream function (zonal integration of the velocity field over all ocean basins). Two separate cells are clearly to be distinguished. The northern hemispheric cell is more pronounced at higher levels and ranges up to 60Sv, which is probably somewhat overestimated. The southern hemispheric cell is dominant at greater depth with values up to 50Sv.

Further details of the 200 year integration as well as results from tracer and CO_2 storage experiments are given by Maier-Reimer (1985).

4. RESPONSE TO VARIABLE ATMOSPHERIC FORCING

Ocean general circulation models have been applied traditionally to study the mean ocean circulation. It is only recently that the variability of the oceanic circulation has been addressed in numerical

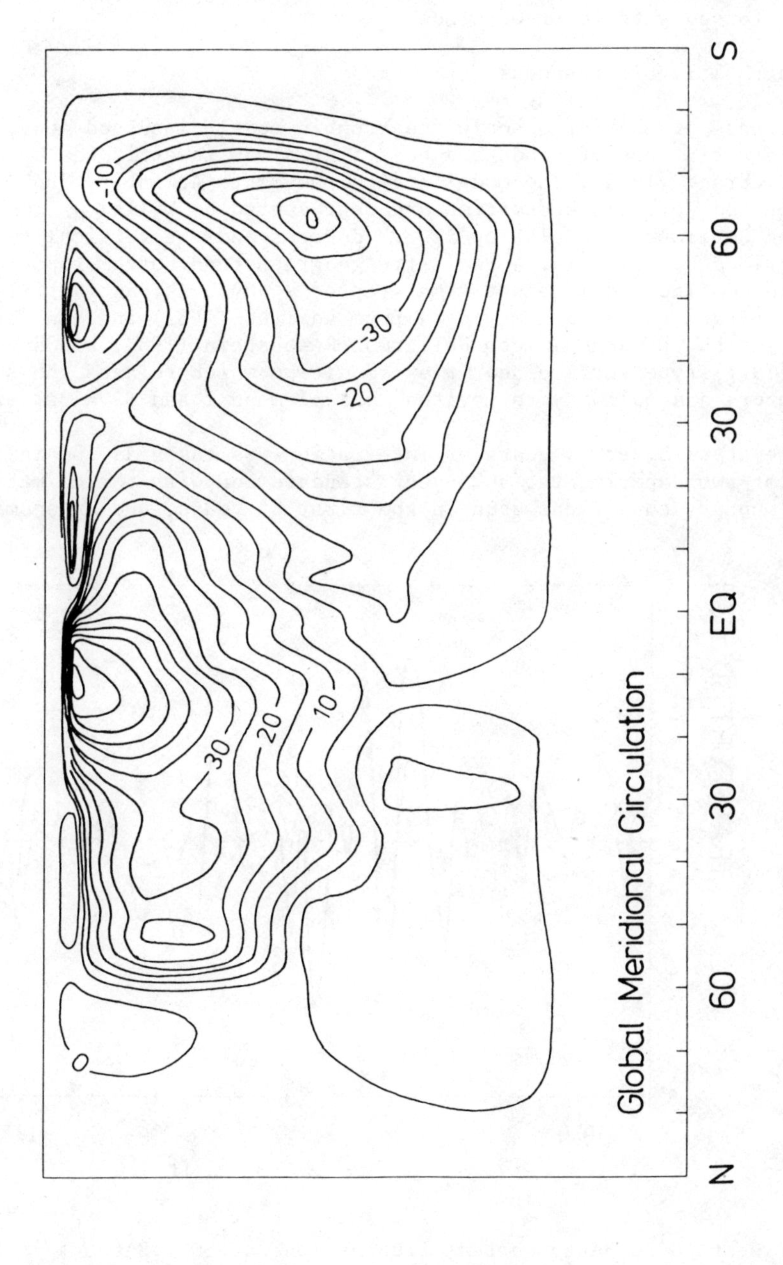

Fig.5, Global meridional stream function. Units are $10^6 m^3 s^{-1}$ [Sv] (From Maier-Reimer, 1985).

models. In this chapter we will discuss two examples, a large scale geostrophic model for the North Atlantic forced with variable winds and heat fluxes, and a primitive equation model for the equatorial Pacific forced with observed winds.

4.1 North Atlantic response

The response of the large scale geostrophic model discussed above to observed variations of winds and heat fluxes was investigated by Olbers and Willebrand (1985). The model covers the North Atlantic. The southern and northern boundaries are represented by Walls at 30N and 80N. The horizontal resolution is 2.2 degrees and the vertical structure is described by 6 layers. A realistic geography and bottom topography is included. The model uses a time step of 7 days. It was initialized with temperature and salinity data from Levitus (1982) and run for 25 years with the annual mean Hellerman-Rosenstein (1983) winds and a Haney (1971) type surface heat flux condition which relaxes the surface temperature and salinity to Levitus' annual mean T and S values within 5 days.

The state after 25 years of integration was taken as the initial state for two experiments, a 25-year standard run with the climatological boundary conditions used in the first 25 years, and an anomaly

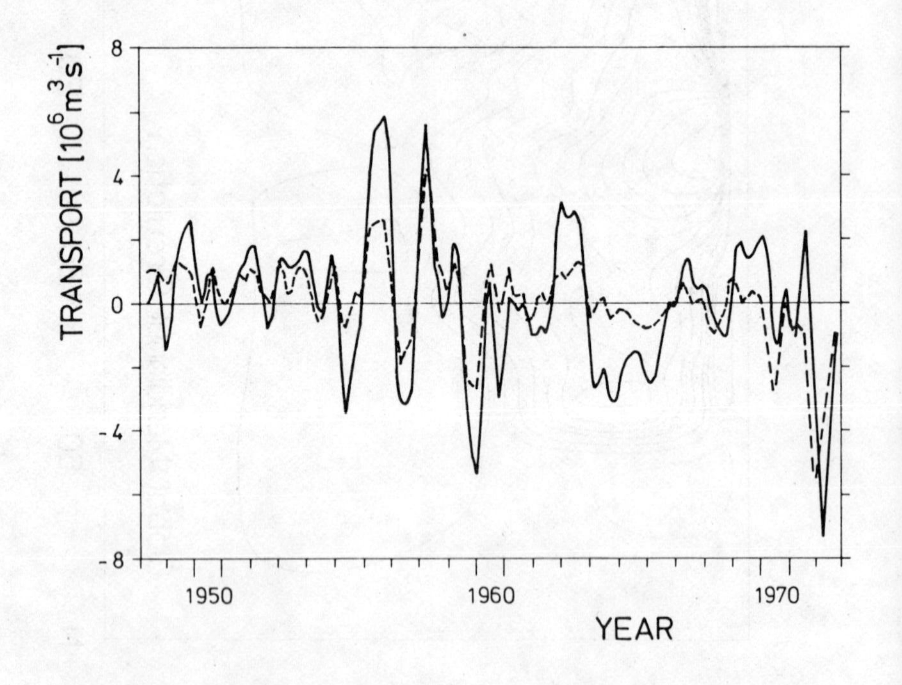

Fig. 6. Western boundary current fluctuations at 30N. Barotropic transport for variable wind and SST-forcing with (solid line) and without (dashed line) bottom torque term. (from Olbers and Willebrand, 1986).

experiment in which observed wind stress and sea surface temperature (SST) anomalies for the period 1948-1972 (Bunker and Goldsmith, 1979) were superimposed on the annual mean values. The seasonal variation was removed from the anomalies. Thus, the forcing and the model response as well do not include the seasonal cycle.

The difference between standard and anomaly experiment is displayed in Fig. 6, which shows the western boundary current fluctuations at 30N. The solid line represents the barotropic transport for variable wind and SST-forcing for the full baroclinic model including bottom topography effects. The amplitude of the fluctuations is about 2-5 Sv.

Elimination of the bottom torque term in the barotropic equations, i.e. a purely homogeneous ocean with bottom topography on the contrary reduces the response by a factor of 2 (dashed line in Fig. 6). This means that density variations through the vertically integrated baroclinic pressure gradient significantly enhance the variability of the barotropic transport.

The above values of the full model response are far smaller than the variations of the flat-bottom Sverdrup transport (determined from the curl of the wind stress anomalies) which amount to 10-20Sv.

4.2 Equatorial Pacific response

The equatorial Pacific exhibits the most important interannual climate variation known as El Niño. During El Niño events anomalous warm surface waters appear for several months over the entire equatorial zone, causing disastrous economic consequences for the fishing and guano industries along the South American coast. The occurance of El Niño has been empirically related to the equatorial wind anomaly fields (Wyrtki, 1975), to the Southern Oscillation (Rasmusson and Carpenter, 1982; Wright, 1977) and to North American weather patterns (Horel and Wallace, 1981).

The physical mechanism behind the El Niño phenomenon is not yet fully understood. Several simple models have been proposed. But it is generally believed that more sophisticated models like oceanic and atmospheric general circulation models are necessary to describe this coupled ocean-atmosphere phenomenon more realistically.

In order to investigate the oceanic part of the interaction loop Latif (1985) has forced a primitive equation ocean model for the equatorial Pacific with 32 years of observed winds. The model uses the full Eqs. (1) through (5) with the exception that the horizontal diffusion in Eq. (5) is omitted and salinity effects are neglected. A variable grid is applied with higher resolution (50 km) near the equator and the coasts. Vertically there are 13 levels, most of which are placed within the thermocline. Bottom topography is not included. The time step is two hours. The model is forced at the surface with 32 years (1947-1978) of observed wind stress (Barnett, 1983) and a heat flux which is parameterized according to Haney (1971) with a constant forcing temperature of 26°C and a relaxation time of about 30 days.

The period of forcing data used is characterized by pronounced interannual signals, associated with warm as well as with cold events. The heavy line in Fig. 7a represents the observed SST anomalies near

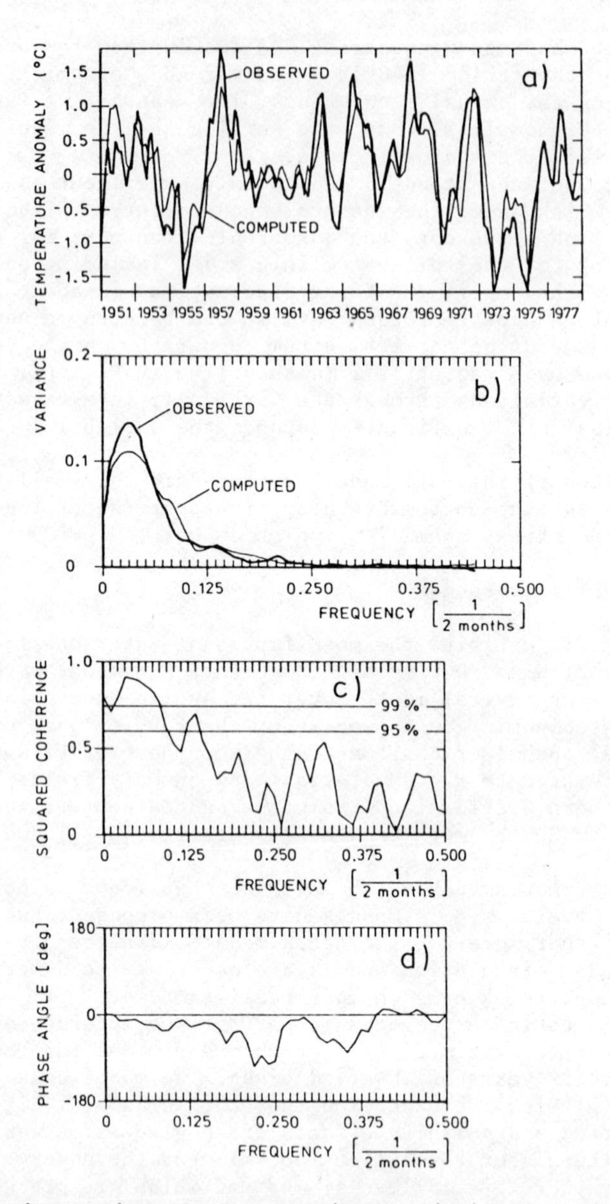

Fig. 7. Time series of observed (heavy line) and computed (thin line) SST anomaly near the date line on the equator (a), variance spectra (b), coherence spectrum (c) and phase spcectrum (d) of the two time series (From Latif, 1985).

the dateline on the equator. It is seen that the most prominent warm events occured during the years 1957, 1965, 1968 and 1972, whereas the years 1955, 1973 and 1975 must be classified as "cold years". It is also seen that the model response (thin line in Fig. 7a) is in remarkable agreement with the data. The model reproduces not only the observed warm events associated with the El Niños, but also the pronounced cold events. Fig 7b shows that most of the observed variance occurs within the frequency range of 2-8 cycles per 16 years with a maximum at 3 cycles per 16 years. The computed variance spectrum compares rather well. In the range of highest variance the coherence between observed and simulated time series is above the 99% confidence level (Fig. 7c), while the phase angles vanish (Fig. 7d). Generally the time and space structure of the equatorial model response to observed winds is in good agreement with observations.

5. MULTIPLE STEADY STATES

Shortly after the first theoretical concepts of the oceanic circulation have been developed, Stommel (1961) discovered the importance of the difference between the temperature and salinity boundary conditions at the ocean surface for the character of the thermohaline circulation. The model investigated was a simple two-box model, with hydraulic connections. Each box was forced by temperature and salt fluxes described by a Rayleigh law, i.e. the fluxes were proportional to the difference between the actual box values and externally prescribed reference values for temperature and salinity, respectively. The proportionality constants correspond to relaxation times of the surface temperature and salinity fields. If these constants were taken to be different for temperature and salinity, the model was capable of responding with several different steady states for the same steady forcing. For equal constants only one steady solution was possible.

Recently several other box-models have been explored with respect to their stability characteristics (Rooth, 1982; Walin, 1985; Welander, 1986). Rooth constructed a three-box model representing the ocean from pole to pole: one equatorial box and two smaller, equal polar boxes. Upper hydraulic connections exist between the polar boxes and the equatorial container, and a single deep connection joins the two polar boxes. The boundary conditions were given by a Rayleigh flux law for the temperature, with equal fluxes for both polar containers, and a fixed freshwater forcing, creating equal salinity fluxes from the two polar basins into the equatorial basin. A nonlinear equation of state was used. The model results show that a possible symmetric solution (equal conditions in both polar containers) could become unstable, leading to an asymmetric steady state, with a pole-to-pole flow in the deep connection.

Bryan (1985) has investigated these properties with a three-dimensional numerical ocean circulation model, which consists of two 60° wide sector ocean basins extending in the meridional direction from pole to pole. Both basins have a flat bottom at 5 km depth. The model retains the full dynamical and thermodynamical complexity of the world ocean circulation (the primitive equations (1) through (5)). The model resolution is 3.75^o in longitude and 4.5^o in latitude, and has 12 levels. It is forced with a zonal windstress and zonally uniform surface temperature and salinity. The boundary conditions were taken to be symmetric with respect to the equator.

Two experiments were conducted. A standard run with equatorial symmetric inital conditions shows an equatorially symmetric meridional circulation: two equator-to-pole circulation cells, which are stable to small perturbations. In a perturbation experiment a positive 2 o/oo salinity anomaly was introduced into the uppermost level of the model, north of 45N. Otherwise the initial conditions remained equatorially symmetric. As the model is integrated forward in time with fixed equatorially symmetric boundary fluxes from the standard run, the symmetric circulation decays and a single pole to pole meridional circulation gyre develops which remains stable. Associated with this change in the meridional circulation pattern is a dramatic change in the poleward heat transport. In the real world this would probably cause a significant climatic change. It should be noted here that there is evidence from paleoclimatic data (ice cores) for substantial changes in the basic oceanic circulation characteristics (Broecker, et al., 1985).

6. CONCLUSIONS

Existing ocean GCMs are able to reproduce most of the main features of the steady world ocean circulation, especially the wind-driven subtropical gyres, the equatorial circulation system and the overall temperature structure. The variability in numerical ocean models has been addressed only recently. The good agreement of variations on time-scales up to a few years produced by equatorial ocean models with observation seems promising. On longer time scales the situation is less satisfactory, mainly because observations for forcing and verification of ocean models are not available. In general the response of ocean models to variable boundary conditions and finite perturbations (multiple steady states) needs to be explored in greater detail. Comparison with nature will be a little easier in the near future with the collection of satellite data on surface winds and ocean surface topography.

Acknowledgements. Thanks are due to E. Maier-Reimer, D. Olbers and J. Willebrand for comments on the original manuscript and for providing material prior to publication. I am also grateful to M. Lüdicke for drafting the figures and to U. Kircher for typing the manuscript.

References

Barnett, T.P, 'Interaction of the Monsoon and Pacific trade wind system at interannual time scales. Part I: The equatorial zone', Mon. Weather Rev., 111, 756-773, 1983.

Bleck, R. and D.B. Boudra, 'Initial testing of a numerical ocean circulation model using a hybrid (quasi-isopycnic) vertical coordinate', J. Phys. Oceanogr., 11, 755-770, 1981.

Broecker, W.S., D.M. Peteet and D. Rind, 'Does the ocean-atmosphere

system have more than one stable mode of operation?', Nature, 315, 21-26, 1985.
Bryan, F.O., 'Maintenance and variability of the thermohaline circulation', Ph.D. Thesis, Princeton University, Princeton,N.J., 1986.
Bryan, K., 'Three-dimensional numerical models of the ocean circulation', In: Numerical Models of Ocean Circulation. National Academy of Sciences, Washington, D.C., 1975.
Bryan, K. and L.J. Lewis, 'A water mass model of the world ocean', J. Geophys. Res., 84, 2503-2517, 1979.
Bryan, K., F.G. Komro, S. Manabe and M.J. Spelman, 'Transient climate response to increasing atmospheric CO_2', Science, 215, 56-58,1982.
Bryan, K., F.G. Komro and C. Rooth, 'The ocean's transient response to global surface temperature anomalies.', Geophysical Monograph, 29, Maurice Ewing Volume 5, American Geophysical Union, 1984.
Cox, M.D, 'An eddy-resolving numerical model of the ventilated thermocline', J. Phys. Oceanogr., 15, 1312-1324, 1985.
Haney, R.L., 'Surface thermal boundary condition for ocean circulation models', J. Phys. Oceanogr., 1, 241-248, 1971.
Hasselmann, K, 'An ocean model for climate variability studies', Prog. Oceanogr., 11, 69-92, 1982.
Hellerman, S. and M. Rosenstein, 'Normal monthly windstress over the world ocean with error estimates', J. Phys. Oceanogr., 13, 1093-1104, 1983.
Holland, W.R., 'Oceanic general circulation models', In: The Sea, Vol 6, Wiley, New York, pp 3-45, 1977.
Holland, W.R. 'The role of mesoscale eddies in the general circulation of the ocean. Numerical experiments using a wind-driven quasi-geostrophic model', J. Phys. Oceanogr., 8, 363-392, 1978.
Holland, W.R., 'The general circulation of the ocean and its modelling', Dyn. Atmos. Ocean. 3, 111-142, 1979.
Holland, W.R., T. Keffer and P.B. Rhines, 'The dynamics of the oceanic general circulation: The potential vorticity field', Nature, 308, 698, 1984.
Horel, J.D. and Wallace, J.M., 'Planetary-scale atmospheric phenomena associated with the Southern Oscillation', Mon. Weather Rev., 1o9, 813-829, 1981.
Latif, M. 'Regional response differences in tropical ocean circulation experiments', Submitted to J. Phys. Oceanogr.,1985.
Levitus, S., 'Climatological Atlas of the World Ocean', NOAA Prof. Paper, 13, Rockville, Md., 1982.
Maier-Reimer, E., 'A large scale ocean circulation model'. Internal report, Max-Planck-Institut für Meteorologie, 1985.
Manabe, S. and B.J. Stouffer, 'Sensitivity of a global climate model to an increase of CO_2 concentration in the atmosphere', J. Geophys. Res. 85, 5529-5554, 1980.
Oberhuber, J.M., 'About some numerical methods used in an ocean general circulation model with isopycnic coordinates', Proceedings of the NATO ASI Advanced Physical Oceanographic Modelling, June 1985, Banyuls-Sur-Mer, France.
Olbers, D.J. and J. Willebrand, 'Response of the North Atlantic Circulation to interannual wind and heat flux variations', in prepara-

tion, 1986.

Pond, S. and Bryan, K., 'Numerical models of ocean circulation', Rev. of Geophys. Space Phys., 14, 243-263, 1976.

Rasmusson, E.N. and Carpenter, T.H., 'Variations in tropical sea surface temperature and surface wind fields associated with the Southern Oscillation/El Niño', Mon. Weather Rev., 110, 354-384, 1982.

Rooth, C.G.H., 'Hydrology and ocean circulation', Prog. Oceanogr., 11, 131-149, 1982.

Stommel, H., 'Thermohaline convection with two stable regimes of flow', Tellus, 13, 224-230, 1961.

Walin, G., 'The thermohaline circulation and the control of ice ages', Palaeoeogr., Palaeoclimatol., Palaeoecol., 50, 323-332, 1985.

Welander, P., 'Thermohaline effects in the general oceanic circulation', Proceedings of the NATO-Institute on "Large scale Transport Processes in the Ocean and Atmosphere" in Les Houches, France, February 1985. (D. Reidel, 1986)

Willebrand, J., S.G.H. Philander and R.C. Pacanowski, 'The oceanic response to large scale atmospheric disturbances', J. Phys. Oceanogr., 10, 411-429, 1980.

Wright, P.B., 'The Southern Oscillation patterns and methanisms of the teleconnections and the persistence', HIG-77-13, Hawaii Inst. of Geophys., University of Hawaii, Honolulu, Hawaii, 1977.

Wyrtki, K. 'Equatorial currents in the Pacific 1950 to 1970 and their relations to the trade wind field', J. Phys. Oceanogr., 4, 372-380, 1974.

CLIMATE VARIABILITY: SOME RESULTS FROM A SIMPLE LOW-ORDER GCM

H. N. Dalfes
National Center for Atmospheric Research
Boulder, 80307-3000
U.S.A.

ABSTRACT. The shortcomings of stochastic models of climate variability are briefly reviewed. A climate model consisting of a low-order semi-spectral atmospheric general circulation model coupled to a mixed layer "ocean" is described. Numerical experiments done with this model generate climate variability in all resolved time scales. Resulting surface temperature time series are analysed in space frequency domains. Observed features are compared with those produced by simple stochastic climate models.

1. INTRODUCTION AND MOTIVATION

The climate system can be considered as composed of several subsystems, each with different characteristic space and time scales and exchanging mass, energy and momenta at all space and time scales. This multiplicity of space and time scales of the climate system makes of any attempt of "synchronous" integration of all relevant equations an insurmontable task with the present day computer technology.

The climate spectra on the other hand, as estimated from instrumental and proxy records, display variability at all time scales, and can viewed as consisting of two components: accumulations of variance in narrow band regions (*i.e.* discrete spectrum) and a background continuum. This continuum has a marked "red" character, *i.e.* variance is preferentially concentrated toward the low end of the resolved frequency scales (see *e.g.* Kutzbach and Bryson, 1974).

The so-called *stochastic models* of climate variability have addressed both components of spectra. Here, we are only interested in results derived with these models for global and zonal surface temperature variances and shapes of the continuous part of their spectra. The approach (which was first suggested by Mitchell (1966), but later formalised by Hasselmann (1976)) is based on the hypothesis of characteristic time scale separation between interacting climate subsystems. This simplification permits a *stochastic* parameterization of forcings of the "fast" subsystem on the "slow" subsystem (here we are

C. Nicolis and G. Nicolis (eds.), Irreversible Phenomena and Dynamical Systems Analysis in Geosciences, 467–473.

only considering the case of a climate system consisting of two subsystems). The statistics of these random forcings can be made to depend on the state of the slow system only in fairly arbitrary ways; *i.e.*, lacking a statistical thermodynamical analogy, it is very difficult to develop parameterizations that will take into account realistic feedbacks from the slow system on the statistics of the fast system forcings.

Despite of above mentioned restrictions, applications of these ideas in the context of zonally averaged energy balance models (Lemke, 1977; Robock, 1978 and Dalfes *et al.*, 1983) lead to interesting results that, in very broad terms, confirm the basic idea of Hasselmann, but they also generated some fundamental questions. It has been observed that surface temperature variances generated as a result of integrations of the "fast" system noise with these simple climate systems depend on variance of his noise, on the global sensitivity of the "slow" component (as expressed, for example, by the sensitivity to solar irradiance changes), but also on the schemes used to introduce this noise in the equation(s) governing the dynamics of the "slow" system. But most importantly, it has been concluded that shapes of variance spectra of surface temperature time series depend strongly on the latitude considered *and* on the scheme used to introduce the stochastic noise.

These studies have shown the need to move up in the hierarchy of climate models, *i.e.* to use a more complexe, more realistic model to test the basic assumptions that went into stochastic models. We will present below some of the preliminary results with such a model consisting of a simple atmospheric general circulation model coupled to a mixed layer "ocean". A more detailed description of the model and its statistical properties will be deferred to a forthcoming paper (Thompson and Dalfes, 1986).

2. MODELS AND EXPERIMENTS

Experiments described in this paper are done with a climate model consisting of a low-order semi-spectral atmospheric general circulation model (AGCM) coupled through surface fluxes to the simplest possible "ocean" model, *i.e.* to a series of isolated (no oceanic heat or momemtum transport) zonal slabs of water of a depth of 50 m. The AGCM is a global model based on primitive equations. The vertical dimension is measured in σ coordinates and the horizontal fields are represented on a finite difference grid in latitude and with Fourier components in longitude. The version used in the experiments has 3 layers of equal $\Delta\sigma$ and we are considering only wavenumbers 0, 3 and 6 in longitudinal direction. The "physics" of the model is mostly zonally symmetric and moisture fields are prescribed.

As described above, this climate model represents an "all-ocean" planet; consequently there is a full hemispheric symmetry: any interhemispheric differences in the statistics are indirect indications of the level of confidence in the respective estimates.

Here we will discuss two experiments: first one consists of an integration of the AGCM alone (the "swamp" case). For the second experiment AGCM is coupled to the simple "ocean" described above (the "mixed layer" case). In both cases, the model is integrated 3 times for 56 years and the first 8 years are discarded to avoid transient effects. Hence each case is represented by an ensemble of three 48 year time series of zonal surface temperatures sampled at 3 day intervals.

Results are analyzed in time and frequency domains. Descriptive statistics are computed for 3-daily sample time series as well as for annually averaged series. Frequency spectra are estimated via discrete Fourier transform followed by ensemble and frequency averaging.

3. RESULTS

3.1. Global and zonal statistics of variability

Surface temperature variability generated in these experiments is displayed in Table 1. Relative proportions of variance in 3-daily samples and annually averaged time series already give coarse indications of the variance spectra for each experiment: in the case of the "swamp" experiment annual averaging, as expected, reduces the standard deviation almost by half. On the other hand such an averaging affects very little the variability in the case of the mixed layer experiment, most of the variance in this case being concentrated at time scales longer than a year.

TABLE 1. Comparison of global climatic variability levels generated in the experiments expressed as standard deviations of globally averaged surface temperatures.

	Standard deviations (K)	
Experiment	3-daily samples	Annual averages
"Swamp"	0.496	0.241
Mixed layer	0.237	0.195

The order of magnitude of the variability in the mixed layer case compares well with those obtained by Lemke (1977) or by Dalfes *et al.* (1983) (hereafter referred as DST) when an additive scheme is used to introduce the "atmospheric noise" into the climate equations. For the multiplicative (parametric) forcing case of DST, variability level is at least an order of magnitude lower. Though Robock (1978) reports a high level of variability with a scheme similar to the multiplicative case of DST, the disagrement between these results can be, to a certain extent, attributed to a difference in deterministic global sensitivity between these models.

The meridional profile of surface temperature variability as it can be seen on Fig.1 is somewhat different than those obtained by DST with a stochastic model. Both approaches agree in that they both generate low tropical variability: for the case DST, this result is not surprising since local stochastic forcing at the tropics was explicitly inhibited. In the case of the present AGCM neglect of some physical processes important in this region (*e.g.* moist convective processes) can explain the result. Present experiments show that the zonal surface temperature variability peaks around 70° and in the mixed layer case, there is a secondary peak around 45°. Polar variability levels are comparable to the mid-latitudes.

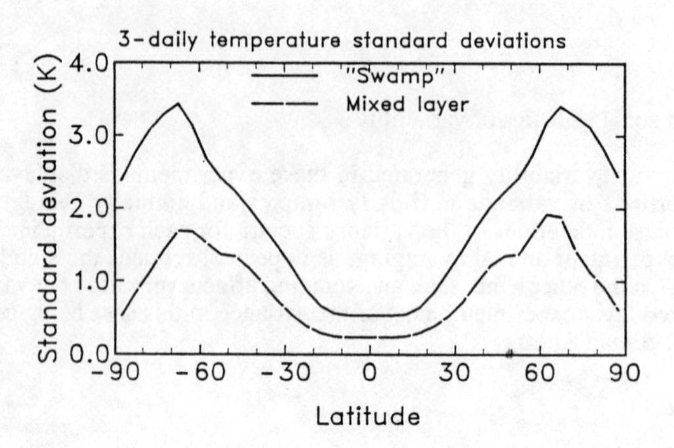

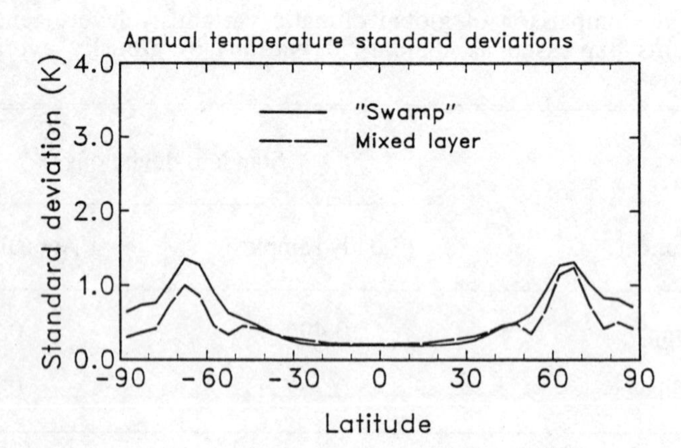

FIG. 1. Meridional profile of zonal surface temperature variability expressed as standard deviations of 3-daily and annually averaged time series.

3.2. Climate spectra: character of the continuum

Hasselman's formalism for stochastic climate models predicts a Lorentzian shape for the simplest case of a linear system with *additive* noise (*i.e.* a Langevin equation). Such a spectrum has an f^{-2} asymptote (where f is the frequency). Later studies (Lemke, 1977; DST) have shown that this result cannot be generalized. Additions of nonlinearities to climate models or changes from the original additive noise scheme to *parametric* noise schemes affect spectral shapes substantially. Also it has been observed that the spectral asymptotes depend on latitude.

Examples of spectra derived from time series generated by the "mixed layer" experiment can be seen on Fig. 2. A clear dependence on latitude is evident. For frequencies less than 1 y^{-1} (*i.e.* "climate" time scales), all spectra have an almost f^{-2} asymptote. For time scales less than one year, spectral shapes differ substantially between different latitudes. In the tropics, it has approximately a f^{-4} dependence. In the polar regions the asymptote has exponent of -3/4. In the mid-latitudes, a broad local maximum

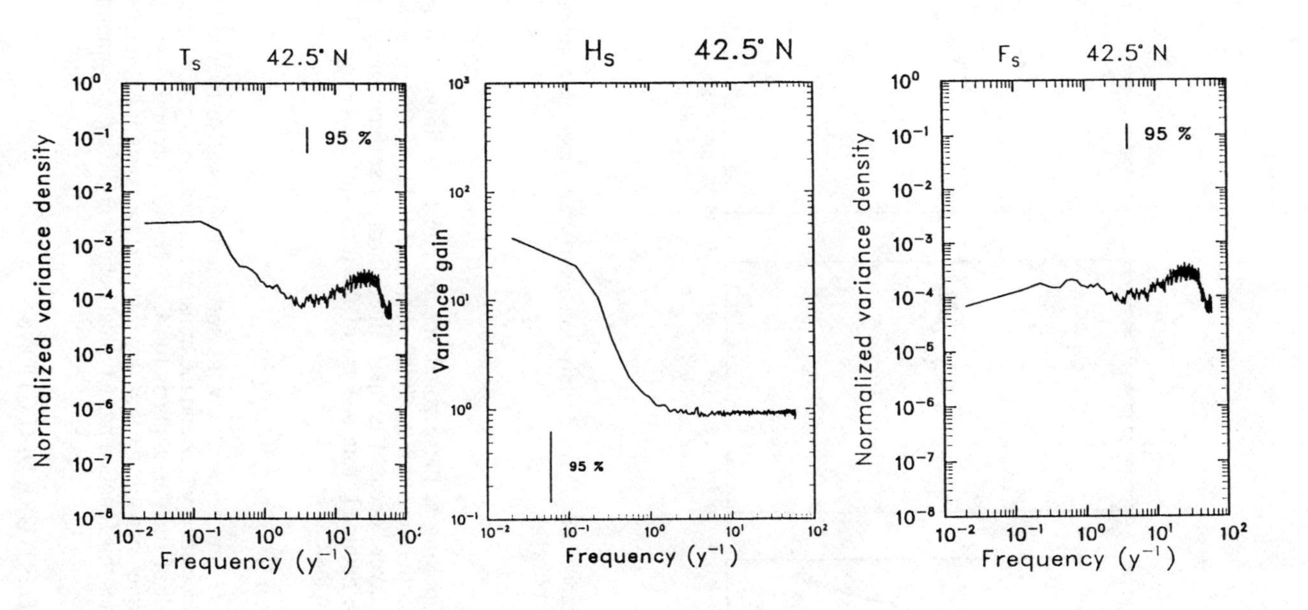

FIG. 3. An input-output analysis of susbsystem interaction at a given latitude band: spectrum of the zonal surface temperature S_T can be analyzed as a product of the input spectrum of the total surface energy flux S_F and of a variance gain function H_s representing the integrating effect of the "slow" subsystem. For frequencies greater than 1 y^{-1}, there is no integrating effect; at lower frequencies the function has the typical behavior with a -2 log-log slope.

centered around 2 y^{-1} is observed.

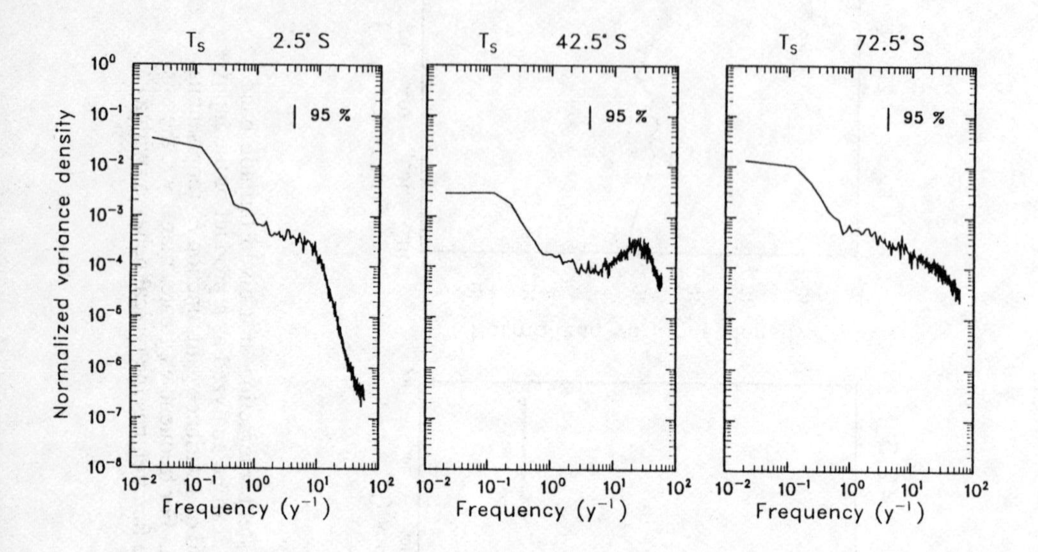

FIG. 2. Normalized variance spectra for 3-daily zonal surface temperature time series for the mixed layer experiment.

Another way of looking at these zonal spectra can be an input-output type of analysis. Since the connection between the "fast" system (*i.e.* the AGCM) and the "slow" system (*i.e.* the mixed layer "ocean") is maintained through surface energy fluxes, a transfer function can be deduced from spectra of the total surface energy flux and of surface temperatures such that:

$$S_T(f) \;=\; H_s(f) \;\; S_F(f)$$

where S_T, S_F and H_s are respectively the spectrum of zonal surface temperature, the spectrum of total surface energy flux and the transfer function for a given latitude band. Fig. 3. displays the result of such an analysis for 42.5° N. As a preliminary conclusion, one can say that, for frequencies greater than 1 y^{-1}, the "slow" system has no integrating effect; at lower frequencies this function has a typical behavior with a f^{-2} asymptote.

4. CONCLUSIONS AND FUTURE DIRECTIONS

The model and experiments described here provide a good example of the usefulness of a hierarchical approach to modeling of climatic change and variability. The use of a model with more complexity and realism than the stochastic models provides a test bed for the assumptions that went into these simpler models.

The climate model described has the advantage of explicitly computing atmospheric circulation and its variations at all time scales of interest; it provides more "realistic" forcings on the "slow" system. In turn, changes in the state of this "slow" system can influence, for example through equator-to-pole temperature gradient, the behavior of the atmosphere, hence the statistics of the surface forcings. As a result, investigations of the effects of two-way feedbacks on the climate variability are possible.

The computational efficiency of this simple AGCM-mixed layer climate model, allows long-term *synchronous* integrations at moderate computational costs on the present generation vector computers. On the other hand, as we have mentioned in several occasions in this paper, certain simplifications of this model limit the significance of its results in some regions (*e.g.* tropics) or time scales.

The present set of experiments seems to compare favorably with stochastic modelling results, though all possibilities of intercomparison have not yet been exhausted. Experiments with more sophisticated "mixed layer" or other type simple ocean models will be useful in this respect. Also, the "all ocean" planet assumption should be relaxed and hemispheric asymmetries should be introduced.

The capability of long integrations at moderate costs will allow applications to those problems where an extensive sampling in the phase space is a requirement. Some of these applications will be presented in forthcoming papers.

5. REFERENCES

Hasselmann, K., 1976: Stochastic climate models, Part 1. Theory. *Tellus*, **28**, 473-485.

Dalfes, H. N., S. H. Schneider and S. L. Thompson, 1983: Numerical experiments with a stochastic zonal climate model. *J. Atmos. Sci.*, **40**, 1648-1658.

Kutzbach, J. E., and R. A. Bryson, 1974: Variance spectrum of holocene climatic fluctuations in the North Atlantic sector. *J. Atmos. Sci.*, **31**, 1958-1963.

Lemke, P., 1977: Stochastic climate models, Part 3. Application to zonally averaged energy models. *Tellus*, **29**, 385-392.

Mitchell, J. M., 1966: Stochastic models of air-sea interaction and climatic fluctuation. *Symp. Arctic Heat Budget and Atmospheric Circulation*, Lake Arrowhead, CA, Memo. RM-5233-NSF, The Rand Corp., 45-74.

Robock, A., 1978: Internally and externally caused climate change. *J. Atmos. Sci.*, **35**, 1112-1122.

Thompson, S. L. and H. N. Dalfes, 1986: *in preparation.*

Author's present address:

Laboratoire de Météorologie Dynamique
École Polytechnique
91128 Palaiseau CEDEX
France

NUCLEAR WINTER: RECENT CLIMATE MODEL RESULTS, UNCERTAINTIES, AND RESEARCH NEEDS

L.D. Danny Harvey
Department of Geography
University of Toronto
100 St. George Street
Toronto, Ontario, M5S 1A1 Canada
and
National Center for Atmospheric Research*
P.O. Box 3000
Boulder, Colorado 80307 USA

*The National Center for Atmospheric Research is operated by the University Corporation for Atmospheric Research and is sponsored by the National Science Foundation.

ABSTRACT. The nuclear winter hypothesis and the associated research effort represent a difficult challenge to the climate modelling community because of the inherently unpredictable nature of a hypothetical nuclear war, uncertainties involving the amount and optical properties of aerosols surviving initial rainout processes, the highly nonlinear nature of the various forcings and interactions involved in determining the climate response to a given aerosol loading and, associated with this nonlinearity, a strong dependence of the climatic response on initial conditions. At the same time, the nuclear winter hypothesis serves as a stimulus for climate model improvements, which should narrow some of the uncertainty in nuclear winter research, but should also aid in the study of other climatic and environmental problems, such as the CO_2 increase, Arctic haze, and natural ice age atmospheric aerosol content increases.

1. INTRODUCTION

In this discussion of nuclear winter we shall be concerned primarily with the value of the nuclear winter hypothesis, and the associated research effort, in elucidating certain features of the climate system and of climate models, and in serving as a stimulus for climate model improvement. The substantive results emerging from the current research effort shall not be discussed *per se*, except to the extent that they serve the above-mentioned objectives. The nuclear winter hypothesis is currently the subject of intensive research, and the reader is referred elsewhere for detailed results (Alexandrov and Stenchikov, 1983; Turco et al., 1983; Covey et al., 1984, 1985; Robock, 1984; Thompson et al., 1984; Warren and Wiscombe, 1985; Ramaswamy and

C. Nicolis and G. Nicolis (eds.), Irreversible Phenomena and Dynamical Systems Analysis in Geosciences, 475–489.

Kiehl, 1985; Thompson, 1985; National Academy of Sciences, 1985; Royal Society of Canada, 1985; Cotton, 1985; Cess et al., 1986; Malone et al., 1986). Similarly, the political and strategic implications of nuclear winter, and the personal responsibility of scientists as citizens, will not be discussed here, as these important issues are also discussed elsewhere (i.e., Sagan, 1984; Malone, 1984; Postol, 1985). Rather, the emphasis here shall be on the methodological issues of (1) how results obtained from deliberately highly simplified models and modelling assumptions have nevertheless served to build intuition concerning the nuclear winter problem; (2) how simple models aid in the interpretation of more complex models; and (3) how the results obtained depend on initial conditions.

The nuclear winter phenomenon represents a massive perturbation to the climate system, which for some variables and processes is much larger than existing models were developed to study. Thus, in some ways the study of nuclear winter stretches our climate models beyond the usual range of conditions for which they are calibrated. In other ways the nuclear winter problem is simpler than more conventional problems such as the CO_2 problem, because for heavy, widely distributed smoke clouds, water droplet cloud feedback processes are almost irrelevant, whereas potential cloud feedback processes represent one of the major sources of uncertainty in estimating the climatic effect of a CO_2 increase (Schlesinger and Mitchell, 1986). The nuclear winter problem has served to focus attention on certain features of existing climate models, and at the same time has forced the climate modelling community to try to improve several deficiencies in existing models. These developments should not only lead to more credible simulations of the climatic impact of a given smoke and dust injection scenario, but also lead to improvements in our ability to model other important (albeit less catastrophic but hopefully more probable) climatic and environmental perturbations, such as acid rain and the climatic impact of the atmospheric CO_2 increase and Arctic haze.

The nuclear winter hypothesis is plagued by major uncertainties at almost every step in the calculation of climatic and physical effects. These uncertainties can be classified as <u>resolvable</u> and <u>unresolvable</u> uncertainties. Many of the unresolvable uncertainties arise from the inherently unpredictable nature of a hypothetical nuclear war, but other uncertainties arise from the highly nonlinear nature of the problem and, associated with this nonlinearity, the strong dependence of the outcome on initial conditions.

The climatic effects of nuclear war arise from the generation of large quantities of smoke and dust aerosols following a nuclear exchange. Smoke aerosols, in which the most important optically-active constituent is elemental carbon, are produced by the burning of forests, grasslands and cities for several days or weeks after a nuclear exchange, and are initially injected largely or entirely into the troposphere. Dust aerosols, on the other hand, are produced by ground bursts at military targets (primarily missile silos according to most scenarios) and might be injected into the stratosphere within a few minutes after the nuclear explosion. Smoke aerosols, particularly those produced by burning cities, are highly absorbing, whereas dust aerosols are largely scattering. The climatic effects predicted by the nuclear winter hypothesis arise largely from smoke aerosols which, for absorption optical depths on the order of 3 or more and a global mean optical path length of 2.0, are sufficient to reduce the amount of solar energy incident at the surface to less than 1% of the insolation incident of the top of the atmosphere. The short and intermediate term climatic effects of a given smoke aerosol loading, however, are highly dependent on the dynamical response of the atmosphere to the initial and subsequent radiative perturbation, as well as on the aerosol lifetime. The radiative perturbation, dynamical response, and aerosol lifetime are highly interdependent, and all three are modulated by the presence or absence of a stratospheric dust layer.

In the following discussion we shall deal primarily in terms of various smoke and dust optical depth scenarios, rather than in terms of actual war scenarios and amounts of smoke and dust produced. The uncertainties in the number and types of targets, area burnt, mass of combustibles per unit area, amount of smoke produced per unit mass of combustibles, fraction of smoke produced which survives initial rainout processes, and the evolution of the optical properties of the smoke (and, to a lesser extent, dust) through time are so large that a given aerosol optical depth could be produced by a large number of war scenarios (Crutzen et al., 1985; NAS, 1985). Since our interest here is primarily in the climate modelling facets of the nuclear winter problem, we begin with specified optical depths. The range of optical depths considered here, however, is well within the range of plausibility. The specific combination of smoke or dust amount and absorption and scattering coefficients which lead to a given optical depth do, however, assume greater importance for interactive aerosol-climate models, in which nonlinearities involving climate response and residence times of the aerosols do depend on the absolute physical amount of aerosol present (Malone et al., 1986).

2. RADIATIVE FORCING OF SMOKE AND DUST AEROSOLS

Some of the most critical factors in determining the climatic impact of a given nuclear war and aerosol generation scenario involve the microphysical and small meteorological scale processes of smoke aerosol coagulation, dispersal, and removal. Coagulation processes alter both the optical and physical properties of smoke aerosols. Most aerosols would quickly become largely restricted to the "accumulation mode" size range, between 0.1 μm and 1.0 μm in radius. According to calculations appropriate for polydispersed spherical particles, for a given index of refraction and fixed total mass the visible optical depth decreases with increasing particle size larger than about 0.1 μm, whereas the infrared optical depth increases with increasing particle size larger than about 0.1 μm (Ramaswamy and Kiehl, 1985). In the size interval 0.1 - 1.0 μm the visible optical depth is greater than the infrared optical depth. A major, put in principle, resolvable uncertainty is the effect of nonspherical particle shape on optical properties. Coagulation processes would thus alter the optical properties of smoke by changing particle size and probably also by changing particle shape.

Coagulation process would also alter the physical properties of aerosols by, for example, transforming a hydrophobic smoke aerosol into a hydroscopic one by coagulating it with sulfate (for example). Thus, the early processes of aerosol coagulation could have a significant impact on the radiative forcing and hence on the magnitude of nuclear winter climatic effects. The extent of early aerosol coagulation would be at least partially dependent on cloud droplet condensation processes, and hence on antecedent meteorological conditions, because cloud droplet formation around aerosol particles followed by droplet evaporation would result in particle coagulation. In the event that the cloud droplet does not evaporate but falls to the ground as rain, on the other hand, the enclosed aerosol particles will be removed from the atmosphere. This introduces another major uncertaintay associated with the initial small-scale processes, namely, the fraction of aerosol originally injected into the atmosphere which is washed out within a few hours after the onset of fires, and before three-dimensional mixing processes can spread the smoke to the typical scale of a General Circulation Model (GCM) grid ($\sim 10^3 \times 10^3$ km). As there is almost no observational basis on which to estimate this quantity, it has generally been assumed that 50% of the smoke aerosols generated by nuclear war fires would be promptly removed before interacting radiatively, as in NAS (1985).

For those smoke aerosols which survive early washout processes, the net radiative forcing depends not only on the intrinsic optical properties of the aerosols, the aerosol size distribution, and the effects of coagulation to other particles and of variable relative humidity, but also depends significantly on the vertical distribution of the aerosols. The main way in which the vertical distribution of aerosols influences the net radiative forcing is by controlling the relative heights of solar absorption and infrared emission. For the simple case of two layers with equal solar absorption and infrared emission, they will tend to be at the same temperature. If solar absorption occurs predominately above infrared emission, the lower layer will tend to be colder (neglecting surface heat sources such as oceans), whereas if solar absorption occurs predominantly below infrared emission, the lower layer will tend to be warmer. Because the solar optical depth tends to be much greater than the infrared optical depth for mode radii between 0.1 and 1.0 μm, the level of solar absorption tends to be much higher than the effective emitting level for smoke aerosols having a constant density with height, leading to surface cooling. For aerosols with an exponential decrease of density with height, the height of solar absorption is closer to the infrared emitting level, thereby reducing the magnitude of surface cooling (Ramaswamy and Kiehl, 1985).

3. CLIMATE SENSITIVITY

A climate sensitivity parameter λ can be defined by

$$\Delta T_s = \frac{G}{\lambda} \qquad (1)$$

where G is the radiative forcing of the surface-troposphere system, ΔT_s is the surface temperature response, and λ is a response parameter which incorporates various feedback processes. For small perturbations of the type usually studied with climate models, such as a CO_2 increase or a solar constant increase, the surface temperature response depends not on the radiative forcing at the surface itself or at the top of the atmosphere, but rather depends on the net surface-troposphere forcing. This dependence arises because, for these perturbations, the surface and troposphere are convectively coupled and so behave as a single system. Thus, both solar constant changes and CO_2 changes result in a similar surface sensitivity parameter, in spite of the fact that the partitioning of the direct radiative forcing between the surface and troposphere is significantly different for these two perturbations. This sensitivity parameter is plotted in Fig. 1 (taken

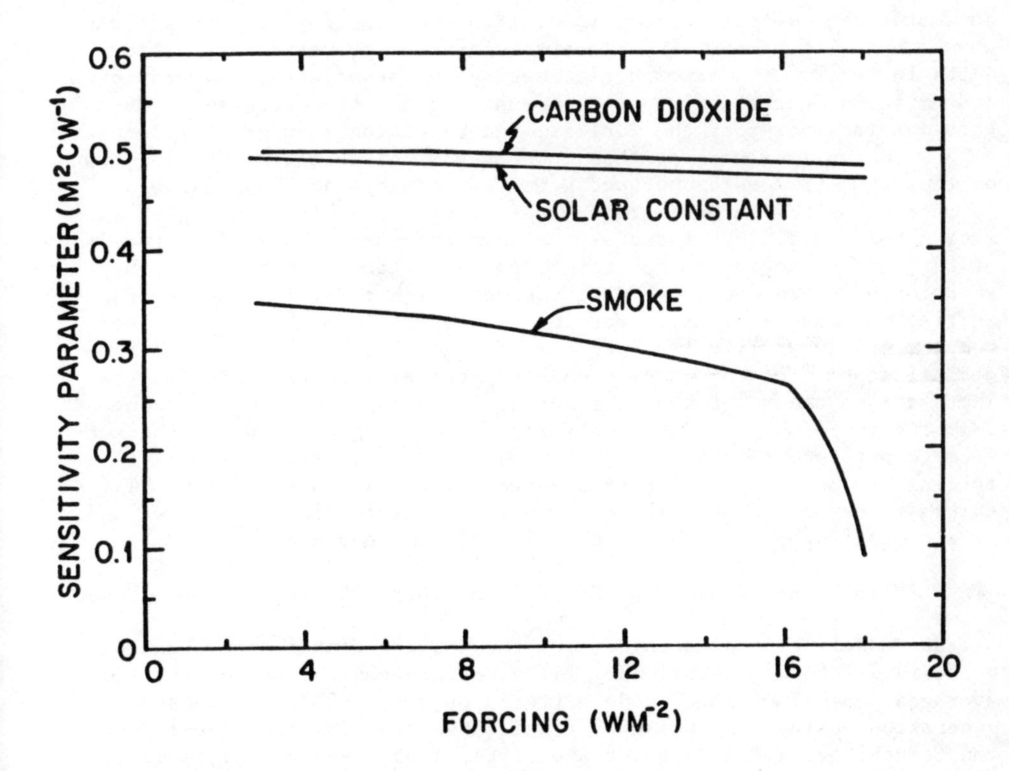

Figure 1. Climate sensitivity parameter for a given radiative-convective climate model as a function of the direct surface-troposphere radiative forcing, for forcings associated with a carbon dioxide increase, solar constant increase, and injection of atmospheric smoke, taken from Cess et al. (1986).

from Cess et al., 1986) for a given radiative-convective climate model as a function of the perturbation in net troposphere-surface forcing for CO_2 and solar constant increases. As seen from Fig. 1, the sensitivity parameter is essentially the same for both types of external forcing, and is largely independent of the magnitude of the forcing, within the range considered here.

In the case of highly absorbing smoke aerosols, in which the direct radiative forcing is one of atmospheric heating and surface cooling, the situation is dramatically different, as seen from Fig. 1. The sensitivity parameter is significantly smaller than for either CO_2 or solar constant forcing and, furthermore, depends on the magnitude of the forcing, with an abrupt downturn at a forcing of 16 Wm^{-2} (corresponding to an absorption optical depth of about 1.0). Reasons for this behavior are explained in detail in Cess et al. (1986). Briefly, for small optical depths less than about 1.0, the surface and troposphere remain convectively coupled but less so than under normal conditions. Hence, net surface-troposphere forcing (one of warming) still dominates over direct surface forcing (one of cooling) so the surface <u>warms</u>, but with a reduced surface sensitivity parameter. As optical depth increases, the atmospheric heating and associated downward infrared emission increase, largely compensating for the decrease in absorbed solar energy at the surface, and resulting in a gradual decrease in λ. The surface warming thus increases with increasing forcing, but by less than if λ were constant. Near an optical depth of unity (depending on the aerosol single scattering albedo, zenith angle, surface albedo, and other factors) the downward infrared emission incident at the surface begins to decrease because the atmospheric solar absorption by smoke occurs at progressively higher levels. Hence, the initial increase in surface warming with optical depth begins to decrease, and eventually leads to a surface cooling. At about the same optical depth (for the model used here) the atmospheric stability is increased to the point that the surface becomes convectively decoupled from the troposphere. Thus, we switch from a regime in which the surface temperature response is governed by net surface-troposphere forcing, to one in which it is governed by net surface forcing. This switch in regime occurs rather abruptly as the forcing increases, and is accompanied by a change in the sign of the response.

4. CLIMATE MODEL STUDIES OF THE NUCLEAR WINTER PROBLEM

Climate modeling studies to date of the nuclear winter phenomenon can be divided into 4 generations: (1) First generation, using globally-averaged, one-dimensional models (Turco et al., 1983); (2) second generation, using 3-D GCMs with fixed smoke distribution (Alexandrov and Stenchikov, 1983; Covey et al., 1984, 1985; Cess et al., 1986); (3) third generation, using a 3-D GCM with interactive smoke transport and radiation, but no smoke removal (Thompson, 1985); and (4) fourth generation, using a 3-D GCM with smoke transport and absorpton as in (3), and smoke removal as well (Malone et al., 1986). In many ways this sequence satisfies the first commandment of climate modelling, "Thou shalt not change more than one thing at a time."

The first generation studies demonstrated the potential importance of nuclear war smoke and dust for climate, and suggested that there would be important differences between land and sea.

The second generation studies used 3-D GCMs, thereby incorporating the effects of land-sea contrasts, regional geography, and propagating, time-dependent storm systems and weather patterns. The imposed smoke cloud interacted with solar radiation, but unlike the first-generation, one-dimensional study of Turco et al. (1983), the smoke amount was constant in time (and fixed in space). That is, although the smoke perturbed the radiative heating and generated anomalous winds, the smoke itself was not transported by these winds. In reality, the smoke would be transported, thereby creating new radiative perturbations beyond the region in which the smoke was initially placed, which would further modify the winds. The second generation studies, although using a fixed smoke distribution (zonally uniform in the case of NCAR), nevertheless demonstrated the rapidity with which local freezing can occur (this could have been demonstrated with 1-D models), and suggested that in models with more realistic smoke patterns that quick, short-term but frequent freezing or near freezing temperatures might occur on a regional basis (individual gridpoints in the NCAR CCM in mid-continents were sometimes colder than that obtained by Turco et al. (1983), even though the mean mid-continent cooling was less). The NCAR study (Covey et al., 1984) suggested that heating of the smoke might perturb flow patterns enough to lift the smoke well above its initial injection heights, suggested that significant cloud amount decreases would occur in the heated part of the atmosphere, thereby reducing precipitation and increasing the lifetime of the smoke that survived initial rainout processes, strongly suggested that a dynamical threshold exists for the response of the Hadley Cell to the smoke and dust induced heating perturbations, and demonstrated the importance of seasonality.

The Cess et al. (1986) study in particular indicated the importance of initial meteorological conditions in determining the temperature response in specific regions. Figure 2, taken from the Cess et al. (1986), shows the zonally-averaged temperature response ten days after a given smoke injection scenario, as a function of the starting day of a control run in which the smoke is injected. The dependence on initial conditions indicated in Fig. 2 is in addition to that which would arise from the dependence of the evolving aerosol optical properties and aerosol amounts on antecedent meteorological conditions discussed earlier, because for each of the experiments shown in Fig. 2 the aerosol loading and optical properties are identical.

The NCAR study (Covey et al., 1984, 1985) is particularly interesting from a methodological point of view because, although deliberately simplified and unrealistic assumptions were made (namely, a zonally-uniform smoke cloud suddenly imposed at time $t = 0$), the NCAR study nevertheless provided many useful insights into the nature of the nuclear winter problem (enumerated above), and provided clear indications of what might happen in more realistic simulations.

In the third generation study of Thompson (1985), transport of smoke by winds is added, thereby permitting interactive feedback between the radiation field and model dynamics, such that once the

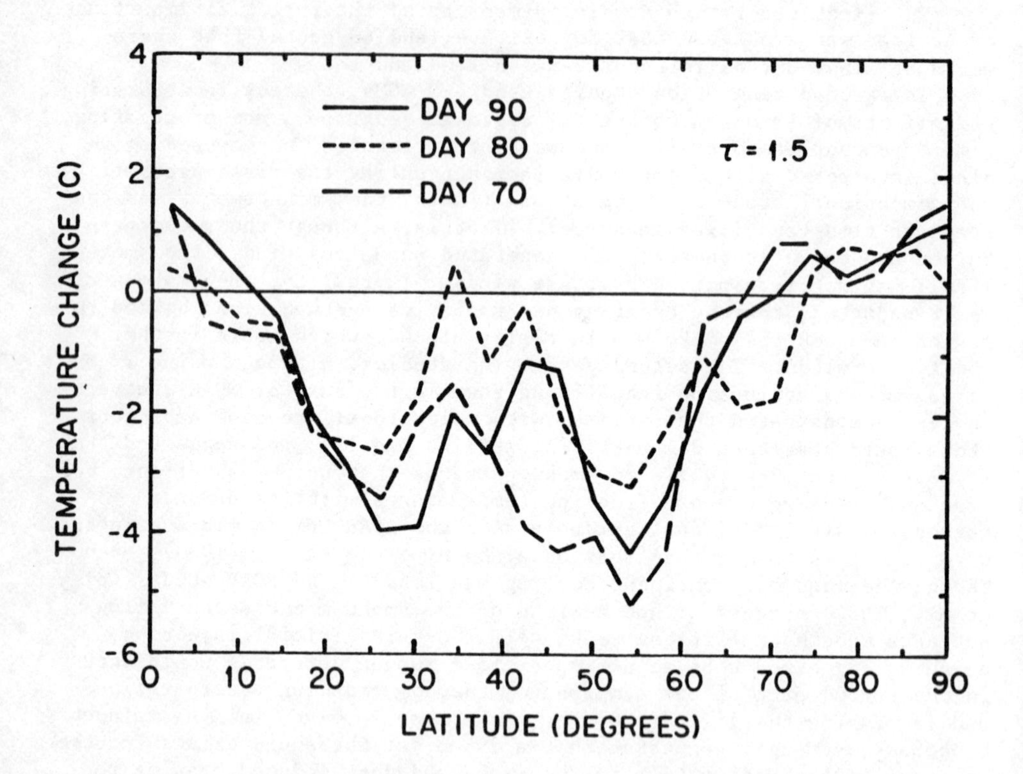

Figure 2. Zonally-averaged changes in July surface air temperature for a smoke optical depth of 1.5 for three different times of a control run in which the smoke is injected, using the GCM of Cess et al. (1986). Temperature changes are for 10 days after the smoke injection, and are based on comparison with the corresponding day of the control run.

radiation field modifies the winds, these winds modify the smoke distribution, which further modifies the radiation field. Removal processes, however, were not included, so that the total smoke content remains constant. The third generation studies confirmed the possibility of quick, short term transient freezing episodes, confirmed that heating of the smoke would lead to upward movement of the smoke beyond its initial injection heights, but did not produce streamers of smoke moving into the Southern Hemisphere (as might have been expected based on the second generation results). Instead, rather patchy blobs that look more like irregular diffusion moved into the subtropics in places, causing 10-15°C quick chills.

The fourth generation studies (Malone et al., 1986, and currently in progress at NCAR) include smoke solar radiative effects and smoke transport, as in the third generation studies, but in addition attempt to account for the long term (days to weeks) subgrid scale processes of aerosol removal from the atmosphere. These studies do not, however, attempt to model the subgrid scale processs of cloud generation, aerosol coagulation, and rainout which would occur in the immediate aftermath of a nuclear war, while fires are still burning; one must still make highly uncertain assumptions as to how much of the smoke initially injected into the atmosphere would survive these processes, spread to GCM grid scales, and participate in the radiative, transport, and removal processes parameterized in the GCM. The fourth generation studies nevertheless allow investigation of the effect of the perturbations in the three-dimensional temperature field induced by the smoke aerosols on smoke removal rates, and hence on the duration of the radiative perturbation and subsequent climatic effects. Results obtained by Malone et al. (1986) indicate that large-scale precipitation is effective in removing smoke in what remains of the troposphere within a few days after the smoke injection, but that above the perturbed atmosphere tropopause smoke scavenging is greatly reduced, thereby increasing the aerosol half-life considerably.

Figure 3a, taken from Malone et al. (1986), shows the model predicted total aerosol mass following on injection of 170 Tg ($1\ Tg = 10^{12}\ g$) of smoke over a seven-day period for a passive tracer case, in which the radiative effects of the smoke aerosol are suppressed so that smoke removal rates are not perturbed. Two injection heights, low (2-5 km) and middle (5-9 km) are considered. The simulated residence times for these two cases agree well with observations. Figure 3b shows the time-dependent smoke mass for the same total mass injection, but with solar radiative effects included, injected according to the NAS (1985) profile of constant smoke density between the surface and 9 km, and injected between 2-5 km (low), as in one of the passive tracer cases of Fig. 3a. Once radiative effects of the smoke are included, the radiative perturbation sets up an inversion in what was originally the troposphere, thereby suppressing processes of aerosol removal associated with precipitation (which is also suppressed) and dramatically increasing the aerosol lifetime. This effect represents an important nonlinearity, inasmuch as a larger initial smoke injection leads to a larger smoke amount at some later time not only for a given smoke removal rate, but also causes a reduction in the removal rate, at least above some threshold injection. The sharp decrease in climate sensitivity near a smoke forcing of about 16 Wm^{-2} seen in Fig. 1, followed by a change in the sign of the sensitivity parameter and surface response, is associated with the same phenomenon which leads to the increase in aerosol lifetime seen in Fig. 3b: namely, establishment of a mid-tropospheric temperature inversion and suppression of vertical convection. It is therefore quite likely that a threshold aerosol optical depth exists for significant aerosol lifetime effects, such that below a given aerosol optical depth an inversion is not established, and hence convection and aerosol removal processes can continue. If such a threshold exists, however, it clearly depends on season, as indicated by the difference in aerosol half-life for the January and July cases shown in Fig. 3b. Height of aerosol injection is also another factor in January (winter), when

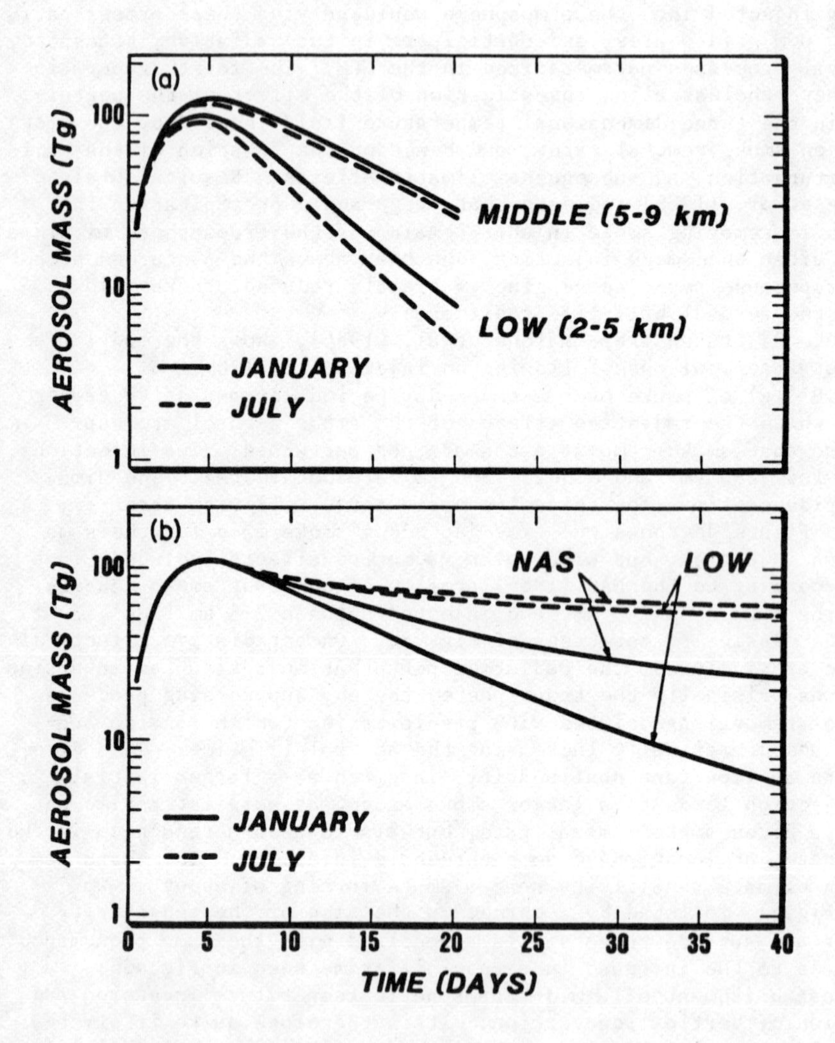

Figure 3. Total mass of aerosol remaining in the atmosphere as a function of time obtained by Malone et al. (1986) for (a) a passive tracer, and (b) a radiatively-active smoke tracer. NAS refers to the NAS (1985) smoke injection profile, whereas LOW and MIDDLE refer to alternate smoke injection profiles (see text).

solar heating of the smoke cloud and resultant self-lofting are much weaker.

An equally important factor likely to significantly influence the aforementioned nonlinearities involving convective coupling, surface temperature response, and aerosol removal rates is the presence of a stratospheric dust layer above the smoke cloud (not included in Malone et al., 1986). Because dust is highly scattering, its effect is to reduce the energy available to the surface-troposphere system, tending to increase surface cooling (as found by Cess et al., 1986). The presence of dust also reduces the mid-tropospheric heating induced by smoke, thereby weakening the resultant inversion. Since the surface cooling is largely dependent upon surface-troposphere decoupling, it is conceivable that under certain highly unpredictable circumstances, the presence of stratospheric dust could lead to a smaller surface cooling. For the same reasons, the presence of dust above smoke is likely to alter the effect of the smoke on its own scavenging rates, at least under some circumstances. Similarly, smoke scattering effects (also not included in Malone et al., 1986) could alter the magnitude of the mid-tropospheric heating and thereby alter the effect of smoke on its own scavenging rates.

In addition to the sequence of modelling efforts outlined above, "off-line" studies with relatively detailed radiative transfer and radiative convective models (Cess, 1985; Ramaswamy and Kiehl, 1985) have served to (1) quantify the combined effects of smoke and dust; (2) indicate the relative importance of smoke scattering and absorption; (3) indicate the importance of smoke single scattering albedo and of assumptions concerning the nature of the vertical smoke distribution; and (4) indicate what simplifying assumptions are acceptable in GCM simulations for a given smoke loading.

Calculations by Cess (1985) indicate that for moderate smoke optical depths ($\tau \sim 1-2$) inclusion of a diurnal cycle in GCM calculations is desirable, since the diurnally-averaged solar radiation reaching the surface can be up to 30 Wm^{-2} larger using a diurnally-varying zenith angle than using the mean diurnal zenith angle. This difference arises as a simple consequence of the nonlinear dependence of light transmission on optical pathlength and the fact that insolation is strongest when the zenith angle is smallest, and indicates that intermittent convection may occur for a few hours each day in a model with a diurnal cycle under circumstances in which it would not occur using a diurnally-averaged model. Intermittent convection in turn could significantly affect the surface temperature response and aerosol removal rates. All of the aforementioned GCM studies except Cess et al. (1986) did not include a diurnal cycle.

Other "off-line" studies involving small-scale models originally designed to study natural cloud dynamics have been used to investigate the dependence of the height to which smoke clouds are injected on antecedent meteorological conditions, and to study aerosol rainout (Cotton, 1985; Manins, 1985). Even the most detailed cloud models or aerosol models currently under development, however, rely upon highly simplified and idealized parameterizations of the microphysical processes involved in droplet formation and particle scavenging. In this respect GCMs and the most detailed cloud models suffer from the same limitation, so that one is not inherently more reliable than the other. We shall have to rely on laboratory studies and experimental fires, but the question will remain (hopefully forever) as to whether aerosol scavenging follows similarity laws in going from the laboratory and experimental scales to nuclear war scales.

5. NONLINEARITIES INVOLVED IN THE NUCLEAR WINTER PHENOMENON

The nuclear winter phenomenon is of particular interest from a strictly scientific point of view, and at the same time poses a number of difficult problems and presents major uncertainties, because of the highly nonlinear nature of the various forcings and interactions involved in determining the climatic response to an initial smoke and dust injection. Some of the important nonlinearities, some of which have already been discussed, include: (1) the transition from direct surface-troposphere forcing to direct surface forcing of the surface temperature response, associated with a reduction in surface-troposphere convective coupling; (2) the change in precipitation and hence in aerosol removal associated with the reduction in surface-troposphere convective coupling; (3) the dynamical response of the Hadley cell, and hence the degree to which smoke would be transported upward, equatorward, and into the Southern Hemisphere; (4) the dependence of radiative transmission on absorption optical depth; (5) the difference in heat capacity and thermal response of land and sea, and resultant land-sea breezes.

6. CONCLUDING COMMENTS ON MODEL STRENGTHS AND WEAKNESSES

Several elements of climate theory needed to study the nuclear winter problem are very well known. These include the transfer of solar and infrared radiation for a given vertical distribution of optically active constituents with given optical properties, the large-scale surface temperature response to large-scale radiative forcing, which can be observationally verified through the diurnal and seasonal cycles, and the large-scale dynamical response to large-scale heating.

Most of the difficult problems concerning nuclear winter pertain to processes which are subgrid scale not only for GCMs, but in many cases also for the most detailed models available. These include the turbulent mixing of smoke and dust, and processes associated with removal of smoke and dust, both in the immediate aftermath of a war and on a longer time scale.

There are nevertheless a number of ways in which existing climate models could be improved in order to more confidently assess the plausible range of climatic effects resulting from nuclear war aerosols. These improvements include (1) inclusion of a diurnal cycle; (2) improved boundary layer subroutines; (3) improved treatment of cloud formation and precipitation, which is related to the processes of aerosol removal; and (4) more accurate treatment of the radiative effects of aerosols--in particular, scattering effects of smoke should be included if not already present, and stratospheric dust should be included in addition to smoke. These improvements should not only serve to narrow the uncertainty in nuclear winter studies, but should also aid in a large number of other areas of atmospheric research.

The first three possible improvements are closely interconnected, and progress in one area will depend in part on progress in the other areas. Inclusion of a diurnal cycle should aid in the development of improved boundary layer and cloud prediction subroutines, since many boundary layer and cloud formation processes have a strong diurnal variation in nature and are strongly nonlinear. Equally importantly, inclusion of a diurnal cycle is required for validation of improved boundary layer and cloud prediction parameterizations. An improved treatment of the boundary layer should in itself lead to improved prediction of cumulus cloud formation, since some cumulus parameterization schemes used in GCMs depend explicitly on boundary layer processes (i.e., Arakawa and Schubert, 1974; Suarez et al., 1983). Some cloud processes in turn feed back on boundary layer processes, both radiatively and through the effects of cumulus processes on capping inversions (Suarez et al., 1983). Furthermore, any improvement in our ability to model cloud processes should help to reduce the uncertainty in predicted CO_2 climate effects, much of which arises from feedback processes involving clouds (Schlesinger and Mitchell, 1986).

The effort to develop fully interactive aerosol radiation, -transport, and -removal models in response to the nuclear winter problem represents another spinoff of nuclear winter research which should aid in the understanding of other atmospheric problems. The same computer codes being developed specifically for nuclear winter applications can also be used, with little or no modification, to study more general problems of atmospheric tracers. Furthermore, there is evidence for large natural increases in the dust and marine aerosol loading of the atmosphere during the last glacial maximum (Thompson and Mosley-Thompson, 1981; Petit et al., 1981; Royer et al., 1983), and much of the research currently in progress involving nuclear war aerosols could be applied to the modelling work which has already started on this large-scale geophysical problem (Joussame et al., 1984).

ACKNOWLEDGMENTS. I wish to thank R. D. Cess and R. C. Malone for permission to use their figures included here. S. H. Schneider and S. L. Thompson critically read an earlier version of this paper and provided useful comments.

REFERENCES

Alexandrov, V. V., and G. L. Stenchikov, 1983: 'On the modelling of the climatic consequences of the nuclear war,' USSR Academy of Sciences, Moscow.

Arakawa, A., and W. H. Schubert, 1974: 'Interaction of a cumulus cloud ensemble with the large-scale environment. Part I,' J. Atmos. Sci., 31, 674-701.

Cess, R. D., 1985: 'Nuclear war: Illustrative effects of atmospheric smoke and dust upon solar radiation,' Clim. Change, 7, 237-251.

_____, G. L. Potter, S. J. Ghan, and W. L. Gates, 1986: 'The climatic effects of large injections of atmospheric smoke and dust: a study of climate feedback mechanisms with one- and three-dimensional climate models,' J. Geophys. Res., in press.

Cotton, W. R., 1985: 'Atmospheric convection and nuclear winter,' Amer. Sci., 73, 275-280.

Covey, C., S. H. Schneider, and S. L. Thompson, 1984: 'Global atmospheric effects of massive smoke injections from a nuclear war: results from general circulation model simulations,' Nature, 308, 21-25.

_____, S. H. Schneider, and S. L. Thompson, 1985: 'Nuclear winter: a diagnosis of atmospheric general circulation model simulations.' J. Geophys. Res., in press.

Crutzen, P. J., I. E. Galbally, and C. Bruhl, 1985: 'Atmospheric effects from post-nuclear fires,' Clim. Change, 6, 323-364.

Joussaume, S., I. Rasool, and R. Sadourny, 1984: 'Simulation of desert dust cycles in an atmospheric general circulation model,' Annals Glac., 5, 208-210.

Malone, R. C., L. H. Auer, G. A. Glatzmaier, and M. C. Wood, 1986: winter: Three dimensional simulations including interactive transport, scavenging and solar heating of smoke,' J. Geophys. Res., in press.

Malone, T. F., 1984: 'What can the scientist do?' in J. London and G. F. White (eds.), The environmental effects of nuclear war, AAAS Selected Symposium 98, Westview Press, Boulder, 151-172.

Manins, P. C., 1985: 'Cloud heights and stratospheric injections resulting from a thermonuclear war,' Atmos. Environment, in press.

National Academy of Sciences, 1985: The effects on the atmosphere of a major nuclear exchange, National Academy Press, Washington, 193 pp.

Postol, T. A., 1985: 'Strategic confusion - with or without nuclear winter,' Bull. Atomic Scientists, 41, 14-17.

Petit, J.-R., M. Briat, and A. Royer, 1981: 'Ice age aerosol content from East Antarctic ice core samples and past wind strength,' Nature, 293, 391-394.

Ramaswamy, V., and J. T. Kiehl, 1985: 'Sensitivities of the radiative forcing due to large loadings of smoke and dust aerosols,' J. Geophys. Res., 90, 5597-5613.

Robock, A., 1984: 'Snow and ice feedbacks prolong effects of nuclear winter,' Nature, 310, 667-670.

Royal Society of Canada, 1985: Nuclear winter and associated effects: A Canadian appraisal of the environmental impact of nuclear war, Royal Society of Canada, Ottawa, 382 pp.

Royer, A., M. De Angelis, and J. R. Petit, 1983: 'A 30,000 year record of physical and optical properties of microparticles from an East Antarctic ice core and implications for paleoclimate reconstruction models,' Clim.Change, 5, 381-412.

Sagan, C., 1983: 'Nuclear war and climatic catastrophe: some policy implications,' Foreign Affairs, 62, 257-292.

Schlesinger, M. E., and J. F. B. Mitchell, 1986: 'Model projections of equilibrium climatic response to increased CO_2 concentration,' in preparation for U.S. Department of Energy State-of-Art Volume on CO_2 Climate Effects.

Suarez, M. J., A. Arakawa, and D. A. Randall, 1983: 'The parameterization of the planetary boundary layer in the UCLA General Circulation Model: Formulation and results,' Mon. Wea. Review, 111, 2224-2243.

Thompson, L. G., and E. Mosley-Thompson, 1981: 'Microparticle concentration variations linked with climatic change: evidence from polar ice cores,' Science, 212, 812-815.

Thompson, S. L., V. V. Aleksandrov, G. L. Stenchikov, S. H. Schneider, C. Covey, and R. M. Chervin, 1984: 'Global climatic consequences of a nuclear war: simulations with three-dimensional models.' Ambio, 13, 236-243.

_____, 1985: 'Global interactive transport simulations of nuclear war smoke,' Nature, in press.

Turco, R. P., O. B. Toon, T. P. Ackerman, J. B. Pollack, and C. Sagan, 1983: 'Nuclear winter: global consequences of multiple nuclear exposions,' Science, 222, 1283-1292.

Warren, S. G., and W. J. Wiscombe, 1985: 'Dirty snow after nuclear war,' Nature, 313, 467-470.

PART V

GEOLOGY

MODELING GEOCHEMICAL SELF ORGANIZATION

Peter J. Ortoleva
Departments of Chemistry and Geology
Indiana University
Bloomington, Indiana
47405

and

Geo-Chem Research Assoc.
400 East Third Street
Bloomington, Indiana
47401

ABSTRACT. Rocks of a wide variety of origins can manifest symmetry breaking instabilities and the development of repetitive and other patterns of mineralization. Models are set forth that couple local processes (aqueous reactions, precipitation/dissolution and nucleation of mineral grains) to a variety of transport processes. Linear stability, bifurcation, matched asymptotic and numerical analysis are used.

1. INTRODUCTION

The same forces that drive atmospheric structures and tectonic changes induce chemical disequilibrium. Since displacement from equilibrium is a necessary condition for self organization through the interaction of reaction and transport,[1] one is led to search for such phenomena in geological systems. In fact there are a great many examples of such phenomena as we shall demonstrate herein.

The patterns of interest in these notes are those which form autonomously - thus seasonal variations in deposition rate are not examples of geochemical self organization because they are driven by an external periodicity. Self organization can take place even in a system whose boundaries are held at constant conditions. Interest in the physico-chemical basis of pattern formation dates back to the nineteenth century.[2,3] In recent years a variety of self organization phenomena in rocks has been identified.[4,5]

Patterns in rocks consist of spatial variations of "textural" variables such as the size, shape, number density and crystallographic orientation of the mineral grains constituting the pattern-bearing rock. Since the redistribution of the texture into patterns generally involves the dissolution of grains, transport of dissolution product molecules and reprecipitation of the original minerals or nucleation of new ones the study of geochemical self organization is the investigation of first order phase transitions coupled to transport in porous media.

In Refs. 4,5 a variety of examples of patterns in rocks of a range of origins from igneous to sedimentary are described. It is quite clear that there is no single mechanism for pattern formation. Mathematical

C. Nicolis and G. Nicolis (eds.), Irreversible Phenomena and Dynamical Systems Analysis in Geosciences, 493–510.

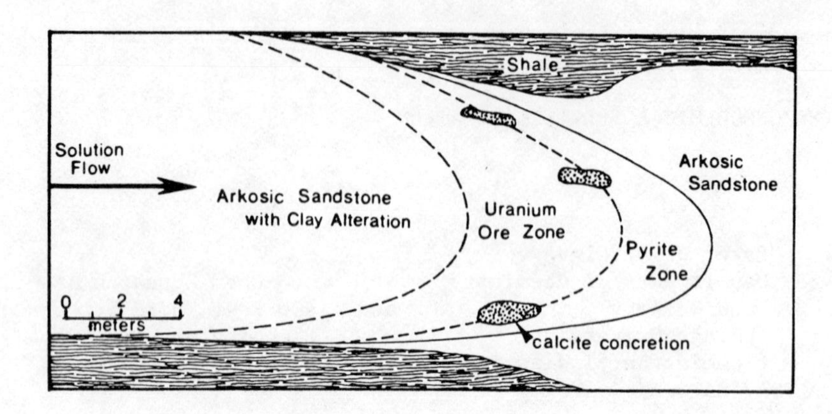

Fig. 1 Sandstone aquifer showing zone of alteration (upflow) and unaltered arkosic sandstone (downflow) and a re-- action zone when uranium ore, calcite and pyrite are deposited, the unaltered rock containing pyrite also. The inlet flow is rich in oxygen and leads to iron oxide mineralization in the altered zone.

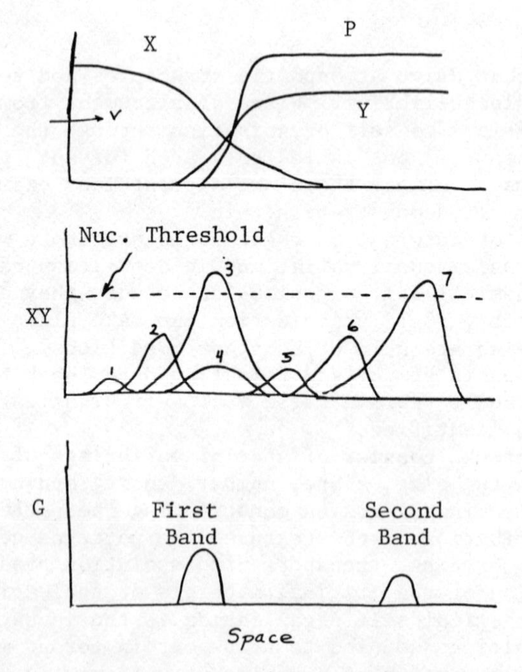

Fig. 2 X-rich fluids are seen to dissolve out mineral P and create mobile species Y. When XY exceeds a nucleation threshold G is P produced.

models of these phenomena should reflect the variety of physico-chemical processes yielding these patterns. The descriptive equations are found to have interesting differences from reaction-diffusion equations traditionally used to study pattern formation as set forth by Turing.[6] The variety of types of equations of geochemical self organization is suggested by the following list of processes operative in various geological contexts. Transport may be by percolation, advection, viscous or plastic rock flow, diffusion, dispersion, and thermal conduction. Local processes include aqueous phase reactions, adsorption, grain growth/dissolution, solidification and nucleation. From the recent experience in working on geochemical systems it seems that one can find self organization phenomena involving essentially all combinations of local and transport processes. It is the purpose of these notes to demonstrate that this conjecture is indeed the case and therefore that geochemical self organization is indeed a rich field for future research.

2. THE OSTWALD-LIESEGANG CYCLE

Since the nineteenth century experiments on the interdiffusion of co-precipitates have shown that banded precipitation can occur. These bands were conjectured by Liesegang (after whom they were named) to have the same mechanism as that leading to many banded mineral occurrences.[2,3] Ostwald suggested that the bands could be formed via a sequence of supersaturation, nucleation and depleation.[7] The Ostwald theory was first given a mathematical formulation by Prager.[8]

A more recent formulation of the Ostwald cycle surmounts some of the difficulties of the Prager theory and, most interesting for our present discussion, has geological interest.[9] Consider the situation shown in Fig. 1 where oxygenated waters flow down an aquifer containing pyrite (FeS_2)(denoted P) leading to the dissolution of P and the precipitation of an iron oxide mineral, denoted simply G here, from the dissolution products of the P reaction. The G deposition is often banded. In Fig. 2 the Ostwald cycle for this system is indicated. This situation may be summarized via the following schematic reaction mechanism: letting X be aqueous O_2 we have

$$X + P \rightarrow Y \tag{1}$$

$$X + Y \rightleftarrows G \tag{2}$$

The overall reaction $2X + P \rightleftarrows G$ is driven forward by the influx of X into the P-bearing domain as seen in Fig. 2.

This has been modeled mathematically via the conservation equations.[9,10] For X we have

$$\frac{\partial X}{\partial t} = \vec{\nabla} \cdot (D_X \vec{\nabla} X - \vec{v} X) + \frac{\partial P}{\partial t} - \frac{\partial G}{\partial t} \tag{3}$$

where X,Y,P,G are molar density per rock volume, D_X is the X diffusion coefficient and $\vec{v}$ is the velocity of the flow. The key to the model

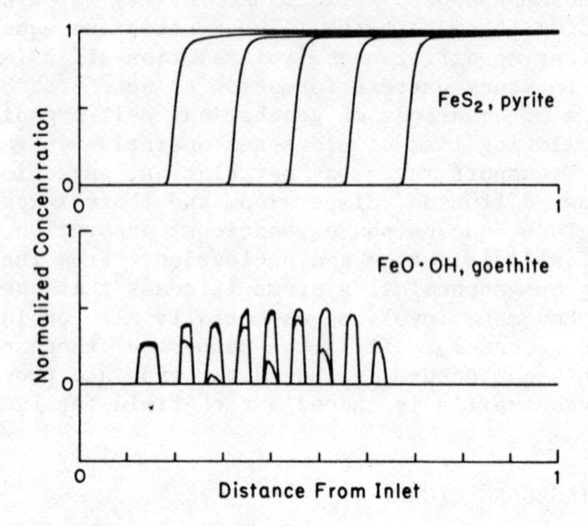

Fig. 3 Numerical simulation of Eqns. (3), (4) and an equation similar to (4) for P showing banded precipitation.

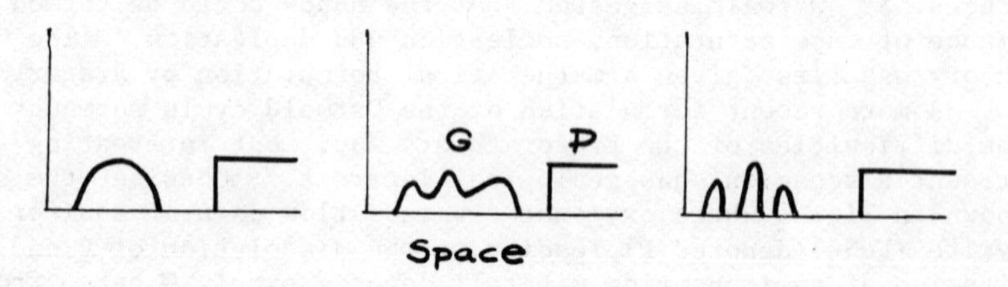

Fig. 4 Solutions of the precipitate model (5)-(7) showing variation of behavior as the growth rate parameter Γ of (7) increases.

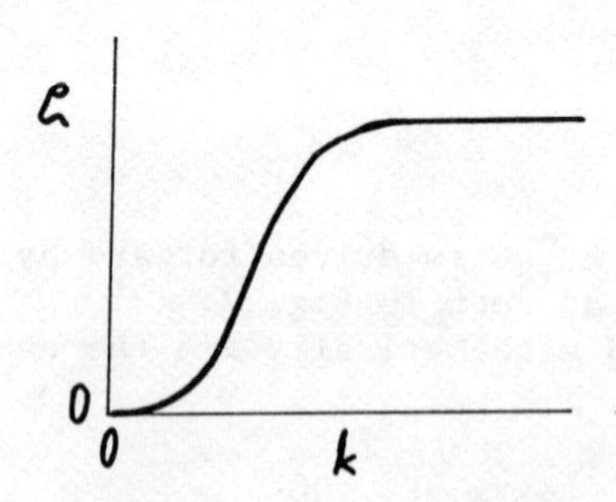

Fig. 5 Dependence of the inverse growth time ζ for patterns of precipitate of wavelength $2\pi/k$ as given in Eqn. (14).

is the nucleation of G. This process has been included by adopting the model

$$\frac{\partial G}{\partial t} = q(G + g)^{2/3} [XY - Q] \tag{4}$$

where q is a rate coefficient and Q is the equilibrium constant for process (2). The G + g factor can be derived by assuming that G forms as a coating on the surface of inert "host" grains of other minerals constituting the rock. In that case the 2/3 exponent reflects that growth occurs at the surface and g represents the molar amount of the host grains per unit rock volume as if they had the same solid molar density as G. To account for nucleation, g is a constant g_o unless the concentration product XY does not exceed a threshold value $Q_n(>Q)$ and G = 0; in the latter situation g vanishes. A mineral growth law similar to that in (4) was adopted for P. The result of a numerical simulation of these equations in Fig. 3 shows the formation of a banded G deposit behind the advancing P front as predicted by the cycle of Fig. 2.

The sparing solubility of most minerals can be used to map the above model onto a free boundary problem (see Chapter XV of Ref. 11). Using a typical value of X divided by a solid molar density as a smallness parameter denoted ε, one arrives at the following free boundary approximation to the above model. The model holds great promise because it allows for the formulation of the problem of band formation as a Hopf bifurcation.

In this "infinite solid density asymptotic" approximation the above model transforms as follows. P experiences a discontinuity at a surface denoted $S(\vec{r},t) = 0$. For S > 0, P is its initial value P_o; P vanishes for S < 0; X vanishes for S > 0. Letting t denote the original time (as in (3),(4)) multiplied by ε, we find as $\varepsilon \to 0$ that

$$\vec{\nabla} \cdot (D_x \vec{\nabla} X - \vec{v} X) - \omega \frac{\partial \tau}{\partial t} = 0, \quad S<0 \tag{5}$$

$$\vec{\nabla} \cdot (D_y \vec{\nabla} Y - \vec{v} Y) - \omega \frac{\partial \tau}{\partial t} = 0 \tag{6}$$

$$\frac{\partial \tau}{\partial t} = \Gamma \Theta(XY,\tau)[XY - Q] \; . \tag{7}$$

We assume process (1) is irreversible, process (2) is slow (with rate coefficient proportional to ε) and define τ to be the thickness of the G coating (taken to be thin) on the host grains. The constant ω measures the surface area of the host grains whereas Γ is proportional to the G rate coefficient. The Θ-factor is one unless $XY < Q_n$ and $\tau = 0$ in which case it vanishes.

To complete the theory we require an equation of motion for the S function. Letting u be the velocity of the interface along the normal $\vec{n}$ to S = 0 pointing in the S > 0 direction, kinematic considerations yield

$$\frac{\partial S}{\partial t} + u|\vec{\nabla} S| = 0 \ . \tag{8}$$

Matched asymptotic analysis reveals that[10,11]

$$\vec{n}\cdot\vec{\nabla} X|_{0^-} = -\lambda u P_o \tag{9}$$

$$-D\vec{n}\cdot[\vec{\nabla} Y|_{0^+} - \vec{\nabla} Y|_{0^-}] = -\lambda u P_o \tag{10}$$

where λ is a constant, $D = D_y/D_x$ and subscripts 0^+, 0^- denote evaluation at $S = \pm 0^+$, 0^+ being a positive infinitesimal. Continuity of X and Y at S = 0 must also be invoked.

The above free boundary problem can be shown to admit constant velocity solutions when Q_n is near Q and the rate parameter Γ is below a certain critical value. It was assumed that the incoming waters are undersaturated with respect to G (i.e. $XY < Q$ in the incoming water). The nature of the constant velocity solution is shown in the first frame of Fig. 4. Also shown are the types of results when the G-rate parameter Γ is increased. The G pulse takes on an undulatory character. A new peak forms to the right as an old peak is dissolved away. In one period the smooth, undulatory G profile attains the same form it had at the beginning of the cycle when it was located further upstream. Thus the steady G-pulse seems to be vulnerable to Hopf bifurcation. Further increase of Γ leads to the formation of gap banding as is also shown in Fig. 4. This sequence of bifurcations and possible onset of chaotic deposition makes the present model ideal for the study of nonlinear phenomena in Liesegang type systems.

3. PRECIPITATE SELF ORGANIZATION VIA COMPETITIVE PARTICLE GROWTH

The equilibrium constant of small precipitate particles can depend strongly on their radius. Thus if a larger particle is near a smaller one, the former will grow at the expense of the latter. Such a surface tension mediated competition is responsible for "Ostwald ripening" whereby a sol always evolves towards a state of fewer but larger particles. In the present section we review results that show how a sol can also evolve into macroscopic patterns of precipitate content. Since precipitation is a common feature of most geochemical processes, it is likely that this self organization via CPG is operative in a variety of geological contexts.

The simplest model of CPG is in terms of the radius $R(\vec{r},t)$ of particles in a macrovolume element about spatial point $\vec{r}$ at time t containing many particles; assuming the particles grow via a surface kinetic limited process we take the following model[12,13]

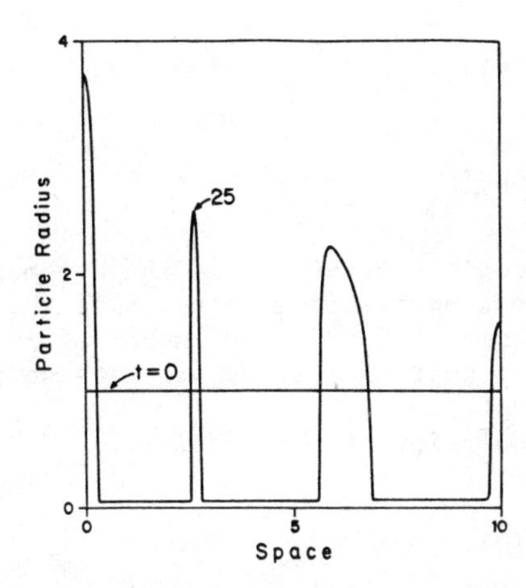

Fig. 6 Numerical solution of (11), (12) showing evolution of a small bump at the left into a large amplitude precipitate pattern.

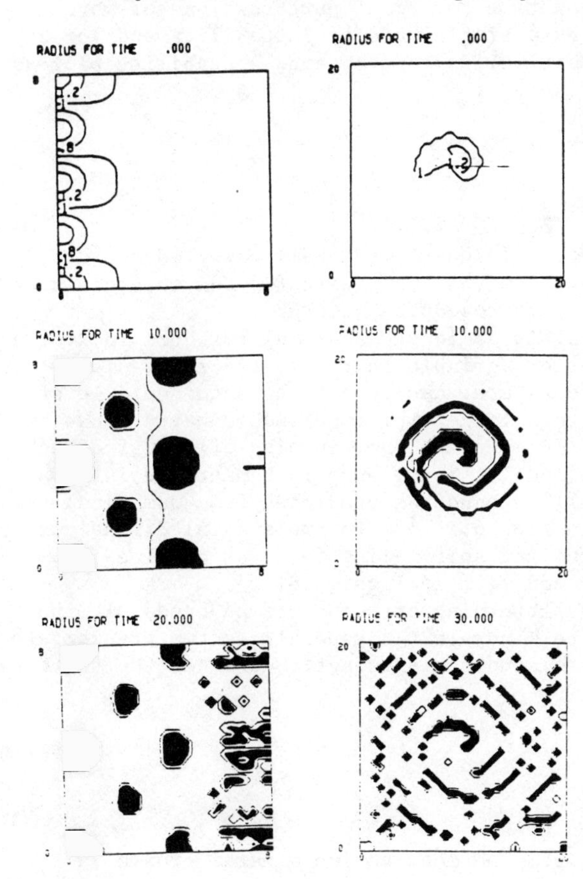

Fig. 7,8 Evolution of spotted and spiral patterns induced by small amplitude, spatially localized perturbations of particle radius from its initially uniform values as per numerical solutions of Eqns. (11), (12).

$$\frac{\partial R}{\partial t} = q[c - c^{eq}(R)] \tag{11}$$

$$\frac{\partial c}{\partial t} = D\nabla^2 c - 4\pi n\rho R^2 \frac{\partial R}{\partial t} \tag{12}$$

where q is the growth kinetic constant, c is the monomer concentration with $c^{eq}(R)$ its R-dependent equilibrium value, ρ is the solid molar density of the particle material and n is the number of particles per unit volume. This model is for post nucleation kinetics so that it is assumed n is independent of time.

The uniform steady solution of the model (11), (12) is, for arbitrary R,

$$R = \bar{R}, \; c = c^{eq}(\bar{R}). \tag{13}$$

Since $dc^{eq}/dR < 0$ for particles larger than a critical nucleus size, one finds that small deviations from the average tend to grow. If ζ(k) is the inverse exponentiation time for small perturbations of wave length 2π/k, then one obtains a single mode of instability and for $c^{eq}(R)/\rho$ small (as it is for most solids) one obtains a stability plot as in Fig. 5: ζ takes the form

$$\zeta = \frac{-(dc^{eq}/dR)qk^2D}{k^2D + 4\pi\rho n\bar{R}^2 q} \tag{14}$$

From this we see that the model is unstable to all wave lengths and that those with smallest wave lengths grow fastest. How then can such a system self organize into a macroscopic pattern?

It is found that a local large length scale disturbance in R or an overall gradient in R (or other variable that can effect the dissolution equilibrium like temperature) can lead to a macroscopic length scale pattern. Once grown to a large amplitude such a pattern can become locked in because $dc^{eq}/dR \to 0$ as $R \to \infty$, turning off the feedback. We see that the growth of a local initial bump in R near the left wall leads to a sequence of satellite bands as predicted from numerical simulation of (11), (12) seen in Fig. 6.[11-13] In two spatial dimensions the patterns can take on speckled and spiral-like forms as obtained experimentally on PbI_2 sols,[12-14] and seen in Figs. 7,8.

In geochemical systems patterning often occurs as bands alternating in the content of two or more minerals. A generalized two precipitate CPG model was investigated to study this effect. Consider the two mineral (A,B) model[15]

$$mX + nY \rightleftarrows A + \lambda Z \tag{15}$$

$$pX + qZ \rightleftarrows B + \mu Y \tag{16}$$

for stoicheometry m,n,λ,p,q,μ. The mass action mineral growth laws

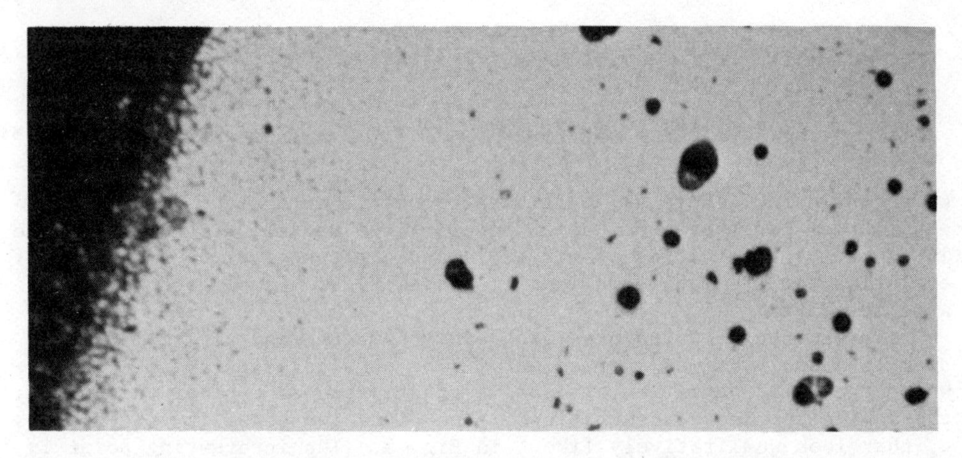

Fig. 9 Micrograph showing edge of precipitate laden region (left) and "greedy gaints" (right) in PbI_2.

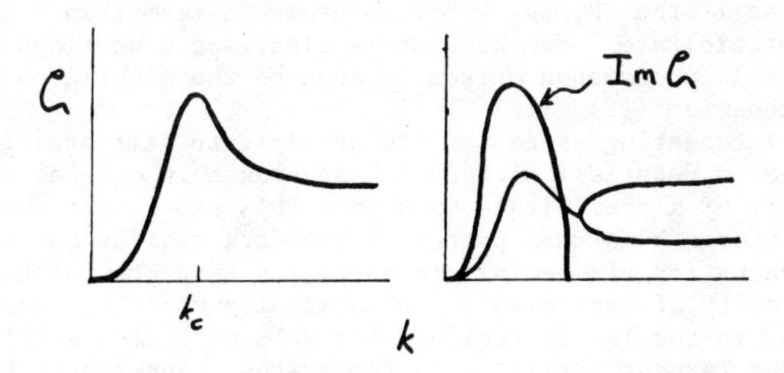

Figs. 10,11 Inverse growth times associated with metamorphic layering. Maxima in Reζ shows the existence of a preferred wavelength.

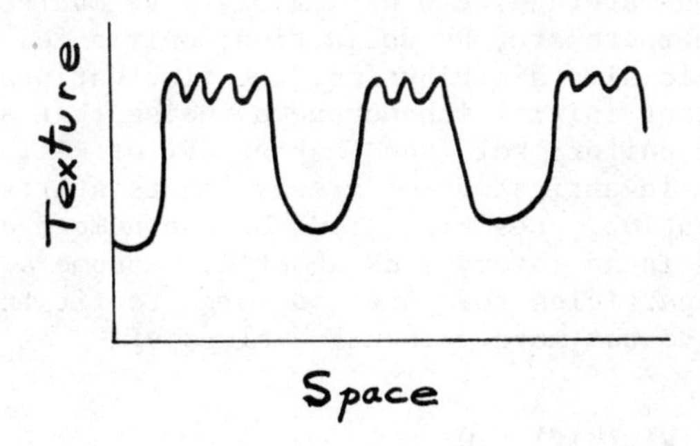

Fig. 12 Multiply periodic metamorphic layering as per Sect. 4.

$$\frac{\partial R_A}{\partial t} = q_A[X^m Y^n - Q_A(R_A)Z^\lambda] \tag{17}$$

$$\frac{\partial R_B}{\partial t} = q_B[X^p Z^q - Q_B(R_B)Y^\mu] \tag{18}$$

were considered. Here q_A, q_B are rate coefficients and Q_A and Q_B are equilibrium constants that depend on particle radius; R_A and R_B are the local radii of particles of mineral A and B respectively. Equations of evolution for the mobile molecular species X, Y, Z analogous to (12) were set forth.

Interesting new features arose in the multi-mineral model. The stability analysis was carried out in the physically relevant limit where the solid densities of A and B greatly exceed typical values of the mobile species concentrations. One finds two stability eigenvalues $\zeta_\pm$ that look qualitatively like ζ in Fig. 5. The interesting point is that for one branch the ratio of perturbations $\delta R_A/\delta R_B$ is positive while $\delta R_A/\delta R_B$ is negative for the other branch. Thus if the branch with $\delta R_A/\delta R_B < 0$ grows fastest then alternating (A,B,A,B,...) banding is predicted whereas if the $\delta R_A/\delta R_B > 0$ mode grows fastest then A and B both grow in the precipitate laden band and a clear space develops between the bands. This phenomenon arises because of the kinetic coupling embedded in the mechanism (15), (16).

In Fig. 9 an interesting aspect of CPG precipitate self organization is illustrated. Seen is a microphotograph taken in our laboratory showing the vicinity of a precipitate band in a PbI_2 experiment wherein a uniform PbI_2 sol forms a mottled pattern. The dark regions constitute a dense concentration of precipitate particles that grew at the expense of the majority of particles in the nominally clear region. Note, however, that in the latter region a few PbI_2 particles survive that are in fact the largest particles in the system. Apparently these "greedy giants" were in the tail of the local particle size distribution in the regions of the roughly initially uniform sol that formed the clear regions of the pattern. Thus the greedy giant phenomenon serves as an example where a statistically improbable event (finding particles much larger than the average) can ultimately have macroscopic consequences. Indeed since there are, by definition, only a few particles in the tail of the particle size distribution, the fluctuations in the tail can be the most important initial inhomogeneous noise that starts pattern development in the uniform sol (see Chapter XVI of Ref. 11).

A theoretical investigation of greedy giants starts with the particle size distribution. Let $F(R,\vec{r},t)dR$ be the number density of particles with radius in an interval dR about R. Assume a single precipitate system with particles that are too large to fit through a gel matrix so that they do not move. Then F evolves via

$$\frac{\partial F}{\partial t} + \frac{\partial}{\partial R}(V(c,R)F) = 0 \tag{19}$$

where V is the particle growth rate ($\partial R/\partial t = V(c,R)$ in the simple monodisperse theory, and c is the concentration of monomer from which the particles are taken to grow; c is taken to satisfy

$$\frac{\partial c}{\partial t} = D\nabla^2 c - \frac{4}{3}\pi\rho\frac{\partial}{\partial t}\int_0^\infty dR\ R^3 F(R,\vec{r},t)\ . \quad (20)$$

To capture the dynamics of the interaction of the tail with the "majority" particles we write F as F = f + g where f represents the majority part of F (a narrowly peaked function of R) and g is the tail (a small amplitude contribution that is nonzero only for large particles). This two population analysis is furthered by a scaling ansatz. Let ε be a smallness parameter. Then let the radius of tail particles be denoted S and scale as $S = \varepsilon^{-1} S'$. Since the tail is small put $g(S,\vec{r},t) = \varepsilon\, g'(S',\vec{r},t')$. The appropriate time scale for evolution turns out to be $t' = \varepsilon t$. Since ρ is large we write $\rho = \varepsilon^{-1}\rho$. With this an asymptotic analysis with $\varepsilon \to 0$ can be carried out to capture the majority-tail interaction. A stability analysis of the above dynamics shows that indeed the tail particles grow where the majority particles dissolve to form the precipitate free domains (see Chapter XVI of Ref. 11 for details).

4. MECHANO-CHEMICAL COUPLING: METAMORPHIC LAYERING AND STYLOLITES

As the stress on a mineral grain increases so does its free energy and hence its solubility. Forces on subsurface rocks leave an individual grain in a state of stress that depends on the hardness of the grains in its vicinity and on the local porosity (i.e. percentage of the rock that is pore space). Since the hardness of a grain varies according to the mineral constituting it, we see that the equilibrium constant of a mineral grain in a rock under stress is a functional of the local texture (i.e. the size, shape, orientation and number densities of the mineral grains in the rock). This texture-solubility coupling can lead to a variety of self organization phenomena in stressed rocks.

Stylolites are dissolution seams that occur in stressed sedimentary rocks. A model for their formation has been set forth based on a porosity-solubility feedback.[16] Consider a monomineralic rock. Let ϕ be the porosity, n be the number of grains per rock volume and L^3 be the volume of each grain; then

$$nL^3 + \phi = 1 \quad (21)$$

since nL^3 is the fraction of the rock occupied by the grains and, by definition, ϕ is the volume fraction that is occupied by pores. Assuming for simplicity that grain growth/dissolution is via a monomer of concentration c then surface kinetic limited growth (with growth constant q) yields

$$\frac{\partial L^3}{\partial t} = qL^2[c - c^{eq}(\phi)] \ . \tag{22}$$

Since the equilibrium concentration c^{eq} increases with stress and hence with ϕ for a given whole rock stress, we take $c^{eq}(\phi)$ to be a monotonically increasing function of ϕ. The theory is completed by a reaction-diffusion equation for c, namely

$$\frac{\partial \phi}{\partial t} = \vec{\nabla} \cdot [\phi D(\phi) \vec{\nabla} c] - \rho n \frac{\partial L^3}{\partial t} \tag{23}$$

where ρ is the solid density. Using (21) to eliminate L from (22), (23) it is clear that the coupled c-ϕ dynamics is essentially identical to that for the simple CPG model. A difference is that ϕc replaces c on the LHS since c is the concentration per pore fluid volume and hence ϕc is the concentration per rock volume. Also the diffusion coefficient $D(\phi)$ is ϕ dependent. The factor of ϕ in front of D comes from the fact that we need the flux per rock cross-section area and the $\vec{\nabla} c$ arises because $\vec{\nabla} c$ and not $\vec{\nabla}(\phi c)$ arises from the chemical potential gradient that drives diffusion.

Using the model (21)-(23) we have shown via numerical simulations that the induction of satellite maxima and minima in ϕ similar to those in R as in Fig. 6 are obtained when an initial small ϕ maximum is introduced into an otherwise uniform system.

The texture-solubility feedback has some surprising new features when more than one mineral is considered. Take the case of "metamorphic" conditions - i.e. pressures and temperatures higher than at the earth's surface but such that the rocks are still solids. In this case the porosity is usually negligible; if n_i and L_i^3 are the number density and average volume of mineral i grains then (21) is replaced by

$$\sum_{i=1}^{N} n_i L_i^3 = 1 \tag{24}$$

for an N mineral system. We now show how rocks under metamorphic conditions can become banded compositionally.

A theory of matamorphic grain growth-transport kinetics has been set forth based on the dependence of the equilibrium constant Q_i of mineral i on texture (i.e. $n_i, L_i^3, i=1,2,\ldots N$).[17-19] Conservation of mass is expressed via equations of motion for C_α (molar density of solute α in moles per rock volume), n_i, L_i^3 (i=1,2,...N) and the rock flow velocity $\vec{v}$ that accounts for the fact that if all grains in a region are growing then a flow is induced as the growing grains push the surrounding medium away; we have

$$\frac{\partial C_\alpha}{\partial t} = - \vec{\nabla} \cdot \vec{J}_\alpha + R_\alpha \tag{25}$$

$$\frac{\partial n_i}{\partial t} = - \vec{\nabla} \cdot (n_i \vec{v}) \tag{26}$$

$$\frac{\partial L_i^3}{\partial t} = - \vec{v} \cdot \vec{\nabla} L_i^3 + G_i \tag{27}$$

where $\vec{J}_\alpha$ is the flux of mobile species α due to diffusion along the grain boundaries, $\vec{J}_\alpha = - \phi D_\alpha \vec{\nabla}(C_\alpha/\phi) + \vec{v}C_\alpha$; R_α represents the net rate of all reaction and grain growth processes and G_i is the rate of growth of volume of mineral i grains. Finally ϕ represents the small porosity associated with the intergranular space along grain-grain contacts through which the mobile species migrate.

The texture-solubility coupling arises from the fact that the equilibrium constant for grain growth depends on local stress and hence on local texture. Let "texture vector" or simply texture be denoted

$$\underset{\sim}{T} = \{n_i, L_i^3, i=1,2,\ldots N\} \tag{28}$$

for the N mineral system. Assume mass action, surface attachement limited kinetics. Mineral i, denoted M_i, grows from the aqueous species $\alpha(=1,2,\ldots N_m)$ via the scheme

$$\sum_{\alpha=1}^{N_m} \nu_{i\alpha} X_\alpha = M_i \tag{29}$$

where X_α denotes an α-molecule and the stoicheometric coefficients $\nu_{i\alpha}$ are $\gtrless 0$ for reactants/products. Then we adopt the law

$$G_i = q_i L_i^2 [\prod_{\substack{\alpha \\ \nu_{i\alpha}>0}} c_\alpha^{\nu_{i\alpha}} - \prod_{\substack{\alpha \\ \nu_{i\alpha}<0}} c_\alpha^{-\nu_{i\alpha}} Q_i(\underset{\sim}{T})] \tag{30}$$

where $c_\alpha = C_\alpha/\phi$ is the concentration per intergranular fluid volume. The feedback is mediated by the $\underset{\sim}{T}$ dependence of the equilibrium constants Q_i.

The linear stability analysis of two and three mineral systems shows that metamorphic systems can manifest a great variety of nonlinear reaction-transport phenomena. In Figs. 10 and 11 we show stability eigenvalues as a function of k (=2π/wavelength) for two interesting cases. In Fig. 10 we see that patterns of wavelength $2\pi/k_c$ grow the fastest and hence a wavelength selectivity is indicated in the linear analysis. This is in sharp constrast to the ζ-plot of Fig. 5 attainable for CPG and stylolite systems or for metamorphic systems under some ranges of parameters. Perhaps the most surprising result is

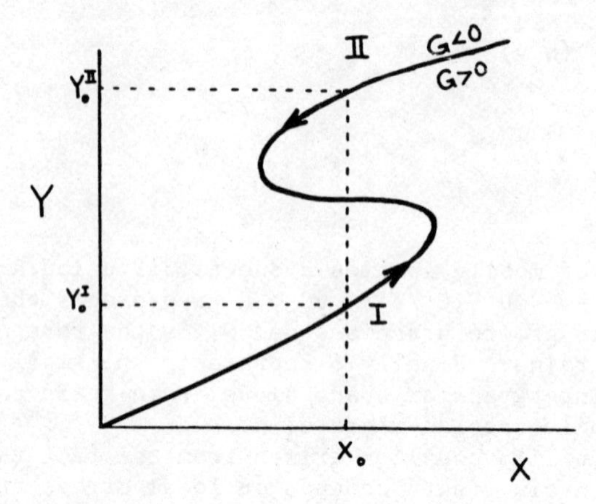

Fig. 13 Folded slow manifold G = 0 for the system (31), (32).

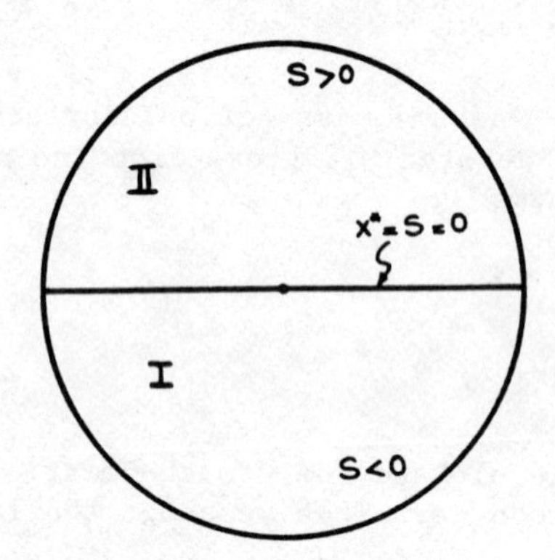

Fig. 14 Static dipolar pattern for model (31), (32) for the special case as per (37) and attending text.

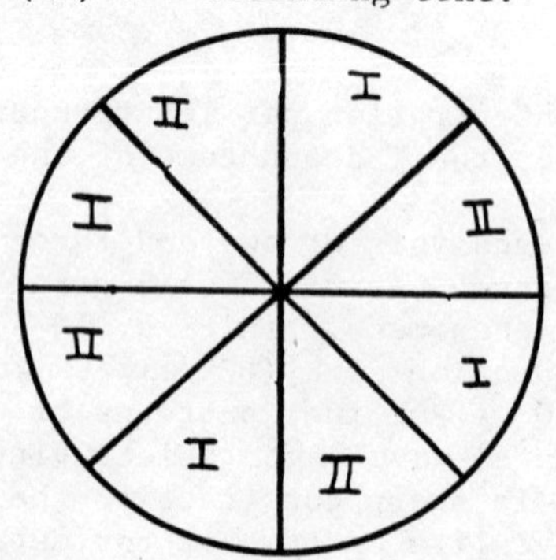

Fig. 15 Static "spoke" pattern exactly calculable for model (35), (36) in special case $F_0^I = -F_0^{II}$.

that shown in Fig. 11 where a pair of complex eigenvalues was found in a kinetic model of a quartz, fedlspar mica system. This implies that propagating wave-like behavior may be obtained in the course of the onset of patterning in some metamorphic rocks. During pattern growth in such a system, $\underset{\sim}{T}$ oscillates both in space and time.

Band formation in metamorphic rocks is in fact a common occurrence. Typically differentiation occurs with elastically stiff minerals like quartz and feldspars segregating from more compliant grains like micas. Banding may also be observed to occur on multiple length scales as suggested in Fig. 12. Such multiple period structures have been observed in our numerical simulation of the metamorphic differentiation equations (24)-(30) for a three mineral system. (See Refs. 17-20 for details.)

5. REACTION TRANSPORT PATTERNS IN MULTIPLE TIME SCALE SYSTEMS

Since periodic wave-like behavior is likely in metamorphic systems as discussed above, it is relevant to comment briefly on two routes to the onset of periodic waves in reaction-transport systems. The route most often cited for rotating or other periodic waves is via a bifurcation of a small amplitude solution as a system parameter passes through a critical value.[1] However we now describe the onset of periodic solutions at finite amplitude as the frequency emerges from zero as a system parameter passes through a critical value.

A simple example in this regard is one involving two species evolving via

$$\frac{\partial X}{\partial t} = D_x \nabla^2 X + \varepsilon^{1/4} F(X,Y) \tag{31}$$

$$\frac{\partial Y}{\partial t} = D_y \nabla^2 Y + \frac{1}{\varepsilon} G(X,Y) \ , \ D_y = \varepsilon^{1/2} \hat{D}_y \ . \tag{32}$$

Here $\varepsilon^{1/4}F$ and G/ε are the net rates of the X and Y reactions and ε is a parameter. We seek the behavior of the system as $\varepsilon \to 0$ for the particular case of the "slow manifold" $G = 0$ as in Fig. 13. In these systems it is well known that discontinuities may propagate corresponding to transition fronts connecting the upper and lower branches of the slow manifold at constant X.[21] However there exists a value X_o of X such that the velocity vanishes and if the transition takes place at X near X_o we have a transition layer speed v given by

$$v \sim - \frac{Q(X-X_o)}{\varepsilon^{1/4}} \quad \text{as } \varepsilon \to 0, \ X \text{ near } X_o \ , \tag{33}$$

where Q is a positive constant and the sign is chosen so that $v > 0$ when the jump is from the upper (II) branch down to the lower (I) branch (see Fig. 13).

The asymptotic ($\varepsilon \to 0$) analysis of the system (31), (32) proceeds much like that in Ref. 21 but because of the $\varepsilon^{1/4}$ factor in front of F in (31) yields an interesting possibility: there may exist solutions such that X never departs far from X_o. As described in Refs. 22,23 the scaling

$$X = X_o + \varepsilon^{1/4} X^* \tag{34}$$

yields the following free boundary problem. Let the jump between branches I and II be indicated by the surface S = 0. When S < 0 $Y = Y_o^I$ and $Y = Y_o^{II}$ for S > 0 (see Fig. 13 for definitions). Furthermore X^* satisfies the piecewise linear problem

$$\frac{\partial X^*}{\partial t} = D_x \nabla^2 X^* + \begin{cases} F_o^I, & S<0 \\ F_o^{II}, & S>0 \end{cases} \tag{35}$$

where $F_o^I = F(X_o, Y_o^I)$ and similarly for F_o^{II}. As seen by the direction of the flows on branches I and II we have $F_o^I > 0$ and $F_o^{II} < 0$.

The description is completed by giving an equation of motion for the interface function S and imposing continuity of X^* at S = 0. The interface equation follows from kinematic considerations (i.e. $\partial S/\partial t + v|\vec{\nabla} S| = 0$) and the expression (33) for v: we obtain

$$\frac{\partial S}{\partial t} = QX^* |\vec{\nabla} S| \ . \tag{36}$$

This free boundary problem yields a host of interesting solutions.

Consider a system in a disc of radius r_o with no-flux boundary at r_o ($\partial X/\partial r = 0$ at r_o, r being the radial distance from the center of the disc). Then for the symmetric case $F_o^I = - F_o^{II} \equiv \beta > 0$ we can show that S = 0 is a straight line as shown in Fig. 14 and that

$$X^*(r,\phi) = - \frac{8\beta r_o^2}{D_x} \sum_{n=1,3,5...} \frac{[2\rho^n - n\rho^2]\sin(n\phi)}{n^2[4-n^2]} \tag{37}$$

where $\rho = r/r_o$ and ϕ is the angular variable (see Fig. 14). As β passes through a critical value β_c this dipolar static structure commences to rotate with a frequency proportional to $(\beta - \beta_c)^{1/2}$.[22]

The above free boundary problem yields a host of exact or approximate bifurcation analytical results in one or more spatial dimensions and holds great promise of generating interesting and complex structures. For example one can calculate static structures as in Fig. 15 exactly. Complex 2 and 3 dimensional textural patterns have been reviewed by Merino [5] for geological systems and in Ref. 24 for this model.

REFERENCES

1. G. Nicolis and I. Prigogine (1977), Self-Organization in Nonequilibrium Systems, Wiley, New York.

2. R.E. Liesegang (1913), Geologische Diffusionen, Stein Kopff, Dresden and Liepzig.
3. E. Hedges and J.E. Myers (1926), The Problem of Physico-Chemical Periodicity, Longmans, Green, New York.
4. P. Ortoleva (1978), Bordeaux Conference on Far From Equilibrium Phenomena, Springer-Verlag, New York.
5. E. Merino (1984) in Chemical Instabilities, G. Nicolis and F. Baras, eds., D. Reidel Pub. Co., Boston, p. 305.
6. A.M. Turing (1952), 'The Chemical Basis of Morphogenesis,' Phil. Trans. Roy. Soc. B, 237, 37.
7. W. Ostwald (1925), 'The Theory of Liesegang Rings,' Kolloid. Z., 36, 330.
8. S. Prager (1956), 'Periodic Precipitations,' J. Chem. Phys. 25, 279.
9. P. Ortoleva (1984), 'From Nonlinear Waves to Spiral and Speckled Patterns: Non-Equilibrium Phenomena in Geological and Biological Systems,' in Proceedings of the Third Annual International Conference of the Center for Nonlinear Studies on Fronts, Interfaces and Patterns, Los Alamos, New Mexico, May 2-6, 1983, edited by A. Bishop, L.J. Campbell, and P.J. Channell, Amsterdam, North-Holland Physics Publishing.
10. P. Ortoleva, G. Auchmuty, J. Chadam, E. Merino, C. Moore, 'Redox Fronts and Their Banding Modalities,' Physica D (to appear).
11. P. Ortoleva (1985), The Variety and Structure of Chemical Waves, a volume in the Synergetics Series, H. Haken, ed., Springer-Verlag, N.Y.
12. D. Feinn, P. Ortoleva, W. Scalf, S. Schmidt and M. Wolff (1978), 'Spontaneous Pattern Formation in Precipitating Systems,' J. Chem. Phys. 69, 27.
13. J. Chadam, R. Feeney, P. Ortoleva, S.L. Schmidt and P. Strickholm (1983), 'Periodic Precipitation and Coarsening Waves: Applications of the Competitive Particle Growth Model,' J. Chem. Phys. 78, 1293.
14. M. Flicker and J. Ross, 'Mechanism of Chemical Instability for Periodic Precipitation Phenomena,' J. Chem. Phys. 60 (1974) 3458.
15. R. Sultan and P. Ortoleva (1985), 'Cooperative Banding Patterns in Two Precipitate Systems,' in preparation.
16. E. Merino, P. Ortoleva and P. Strickholm (1981), 'A Kinetic Theory of Stylolite Generation and Spacing,' Transactions, v. 62, no. 45, p. 1056.
17. P. Ortoleva, E. Merino and P. Strickholm (1982), 'Kinetics of Metamorphic Layering in Anisotropically Stressed Rocks,' Amer. J. of Sci. 282, 617.
18. E. Merino, P. Ortoleva and P. Strickholm (1984), 'The Self Organization of Metamorphic Layering,' in preparation.
19. P. Ortoleva (1983), 'Modeling Nonlinear Wave Propagation and Pattern Formation at Geochemical First Order Phase Transitions," an invited review for the Proceedings of the NATO Advanced Research Workshop on Chemical Instabilities: Applications in Chemistry, Engineering, Geology, and Materials Science, Austin, Texas, March 14-18, 1983.
20. T. Dewers, E. Merino and P. Ortoleva, 'Development of Metamorphic Layering in a Three Mineral Pressure Solution Model,' in preparation.
21. P. Ortoleva and J. Ross (1975), 'Theory of the Propagation of Dis-

continuities in Kinetic Systems with Multiple Time Scales: Fronts, Front Multiplicity, and Pulses," J. Chem. Phys. 63, 3398.

22. R. Sultan and P. Ortoleva (1985), 'Rotating Waves in Reaction-Diffusion Systems with Folded Slow Manifolds,' J. Chem. Phys., submitted for publication.
23. P.C. Fife (1984), 'Propagator-Controller Systems and Chemical Patterns,' in Nonequilibrium Dynamics in Chemistry, C. Vidal and A. Pacault, eds. (Springer, New York), pg. 76.
24. R. Sultan and P. Ortoleva, 'Rotating, Propagating and Static Structures in Reaction-Transport Systems With Folded Slow Manifolds,' in preparation.

NONLINEAR CONVECTION PROBLEMS IN GEOLOGY

BERNARD GUY
Département Géologie
Ecole des Mines
158, cours Fauriel
42023 Saint-Etienne Cedex France

ABSTRACT. Many problems in geology involve the relative displacement of a fluid or liquid with the solid portion of a rock. As far as chemical exchange is concerned, there may be no need to consider two separate classes of functions (i.e. concentrations of components in the fluid and in the solid) provided there is a tendancy toward local equilibrium: both concentrations are then connected by a relation : this one is generally non linear. The simplest transport equation that may be written on such a basis is of the form $c_t + f(c)_x = 0$ where c and f are concentrations (scalars or vectors) ; this equation is very rich : it may particularly provide a framework within which such salient features as the appearance of discontinuities may be understood.

The numerical modeling allows one to simulate several aspects such as front propagation, washing out of heterogeneities, local increase of concentrations and so on. Other features such as oscillatory behaviour need to consider additive terms like diffusion terms $Df(c)_{xx}$ or chemical kinetics terms. The starting problem is set in x and t variables but one is led to find the "stationary" evolution in the concentration space only thanks to a minimum entropy production principle.

1. THE EXCHANGE OF CHEMICAL COMPONENTS BETWEEN SOLIDS AND INTERSTITIAL FLUIDS OR LIQUIDS IN ROCKS : A NON-LINEARITY

In convection - diffusion - reaction problems in general, different kinds of non-linearities may be envisaged : they may lie in thermal, chemical kinetic, inertial terms and so on.

In an homogeneous medium the transport as expressed by an equation of the type

$$c_t + vc_x = 0 \tag{1}$$

where c is the concentration of a component at time t and position x, v the velocity of this component, does not contain by itself - if v is constant - a non-linearity.

C. Nicolis and G. Nicolis (eds.), Irreversible Phenomena and Dynamical Systems Analysis in Geosciences, 511–521.

In the geological problems we will study here, the system is in fact divided into two parts : i) the solid fraction of the rock and ii) an interstitial fluid or liquid that may migrate across it.

The movement of such a fluid is generally very slow and at a geological spatial scale we may consider there is a chemical (or tendancy toward) local equilibrium between the solid and the fluid. As for the partitioning of a component of concentration $c_s = c$ in the solid and c_f in the fluid, this equilibrium may be expressed by a law of the type

$$c_f = f(c_s) \tag{2}$$

which is non linear if the solution is non ideal in either medium (and particularly the solid) ; f is called the isotherm. It comes to considering both media i.e. solid and fluid as a single one. In the usual case where the porosity p of the rock is small, the simple mass balance equation (1) may be replaced by the also rather simple one :

$$c_t + pvf(c)_x = 0 \tag{3}$$

where the flux of component c is a non-linear fonction of c.

2. RELATIVE MOVEMENTS SOLIDS / FLUIDS IN GEOLOGY

The situation depicted in the first section is very common in geology. It may be encountered in several fields, for instance :

i) meteoric alteration : this is especially important in hot and rainy tropical climate : the rain falls on the rocks, goes across them and transforms them.

ii) movements of water in sedimentary basins : the rain infilters in some parts of the basins and may migrate across them for long distances.

iii) magma movements : we may mention the partial fusion (or "anatexy") of rocks and subsequent expell of magma at the basis of the crust or upper mantle (about 50 km deep), the ascent of magmas through the crust and the differential movement crystals / magma, often invoked by geologists (also called fractional crystallisation).

iv) diagenesis of sediments : at the bottom of oceans, the sediments are full of water : this water is progressively expelled during the compaction and goes upward from the deeper levels up to the sea floor where it is substracted from the rock.

v) hydrothermal alteration : in deeper parts of the crust, at higher temperatures and pressures (up to 700°C and a few kilobars) different types of water migration may be inferred, for instance around crystallising granitic intrusions where important chemical exchanges occur.

"Skarns" for instance are "metasomatic" rocks which mainly result from the transformation of carbonatic starting materials by the operation of aqueous fluids rich in silica, iron, manganese, aluminium and so on. The aqueous solutions are thought to have travelled, away from the granite, along lithologic discontinuities that have guided them : they have transformed the different rocks that were on both sides. The size is metric to decametric of irregular lens shape. The duration of formation may be about 10 000 to 100 000 years per meter of skarn.

One conspicuous feature of these phenomena is the separation in space of the chemical transformations due to the movement of the fluid. Starting from a homogeneous situation, discontinuities appear and are maintained, in a case where a dispersion could be expected. A spatial organisation of the concentrations of the components in different domains separated by stong gradients is thus produced. This quantification in space and time is common to the several examples I have given, although because of the scale it may be difficult to see in some cases.

One can see on fig. 1 the different zones that may be encountered in the transformation of a dolostone and the corresponding geochemical profiles for some elements (after [1] and [2]).

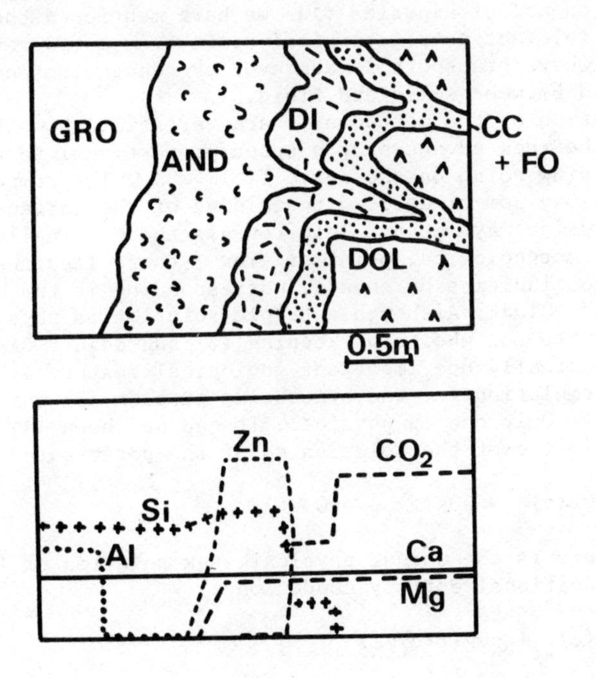

Figure 1. An example of mineralogical and geochemical zoning in skarns.

After i) the dolostone $CaMg(CO_3)_2$, one can meet the following zones : ii) calcite $CaCO_3$ + forsterite Mg_2SiO_4, iii) diopside $CaMgSi_2O_6$ iv) andradite $Ca_3Fe_2Si_3O_{12}$ and v) grossular $Ca_3(Al,Fe)_2Si_3O_{12}$. Several metals may be concentrated within such rocks, such as W, Sn, Cu, Fe, Pb, Sn and so on. The role in concentrating elements is another characteristic feature of these non linear convection problems and may be of economic importance.

It is impossible to give here references for all the fields evoked in this section. One will find for instance in [3] , [4] , [5] , [6] , [7] , [8] , hints for further investigations.

3. THE BASIC NON LINEAR CONVECTION PROBLEM

3.1. Our aim here is thus to simulate the evolution in time and space of the chemical components in the rock following

$$c_t + pvf(c)_x = 0 \tag{3}$$

at first for one chemical component. In eq. (3) we have neglected the roles of diffusion and of kinetics : as we have mentioned the problem as it is set is relevant to a geological scale i.e. a few centimeters to a few meters where transport is achieved by convection and equilibrium is realized between solid and fluid.

The application of the method of characteristics shows that each composition of the rock advances at a speed proportional to the slope of the corresponding point on the curve f(c) [9] . The competition of these velocities may sometimes lead to folding of the surface c(x,t) and a triple solution may appear. This reveals the originality of the situation from a morphological point of view ; there lies the problem of onset of discontinuities or shocks, corresponding in the model to a cusp catastrophe [10] . Although a triple solution is physically meaningless, we make the choice of keeping (3) and adjust discontinuities (which are actually one important geological feature).

The weak formulation for the hyperbolic problem (3) may admit several solutions. Only one is physical. It can be shown [11] , [12] , [13], that the limit c of the solution c_D of the perturbated problem

$$c_t + pvf(c)_x - pDf(c)_{xx} = 0 \tag{4}$$

when D goes to zero is the unique physical weak solution of (3) fulfilling the additional entropy condition

$$\text{Min } (h(c)_t + pvg(f(c))_x) \tag{5}$$

where h and g are concave entropy functions of $c = c_s$ and $c_f = f(c_s)$ and such that

$$dh/dc = dg/df \tag{6}$$

Eq. (6) is equivalent to a condition of thermodynamic equilibrium betwenn the solid and the fluid for an athermic isochore exchange ; eq.(5) is the expression of the second principle of thermodynamics (see the detailed analyses in [14] and [15]). Eq. (3) and (5) are now the basis for a good understanding of the problem of the appearance and propagation of chemical discontinuities or fronts. In particular, the location and velocity of shocks may be derived from them (see further). Thus in the model diffusion appears connected to convection : this latter is responsible of the strong gradients that are characteristic of the phenomenon whereas diffusion has the hidden role allowing to approach the physical solution. This sheds some light in the meaning of a discontinuity : practically (on the field) speaking of a discontinuity is equivalent to saying that a given range of the possible concentrations of the components has a negligible probability to appear within the transformed rocks.

3.2. Condition (5) has an interesting consequence : let us choose for h and g the concave functions $h(c_s) = -|c - f^{-1}(k)|$ and $g(c_f) = -|c_f - k|$ the derivatives of which are $-\,sg(c_s - f^{-1}(k))$ and $-\,sg(c_f - k)$, and put them in the weak formulation of (3) ; we obtain, in the case of a discontinuity between c^- and c^+ the condition

$$\frac{f(c^+) - f(c^-)}{c^+ - c^-} = \mathrm{Max}\ \frac{f(c^+) - f(c)}{c^+ - c} \tag{7}$$

for c in the interval (c^+, c^-), called also the Oleinik condition (use has been made of the condition giving the velocity of a shock, or Rankine-Hugoniot condition [13]). Condition (7) allows one to determine the sequence of compositions found in such a phenomenon and the possible location of shocks : for a Riemann problem corresponding to a fluid of constant composition c_2 entering a rock of constant compositions c_1 the evolution is given by the concave enveloppe f^* of the isotherm (fig. 2).

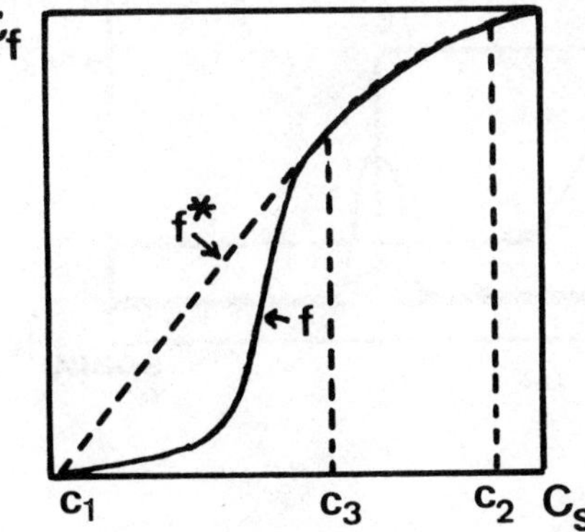

Fig. 2 Evolution path given by the concave enveloppe f* of the isotherm f, making a shock appear between c_3 and c_1.

Eq. (7) rules the overall organisation of the sequence of compositions and is similar to the implicit condition given in [16] and expressing that the fastest velocities go before the slowest. It may look paradoxical to be able to find an evolution by a direct reasoning on the isotherm i.e. with equilibrium data. This has been permitted by the minimum entropy principle (5) : x and t have been dropped and we have found a kind of stationary behaviour in (c_f, c_s) space for a system that increases similarly to itself. Quantification is here related to the second principle and to the definition of a concave function f* (cf. also [17] and [18]).

3.3. The approach of hyperbolic systems is in a way similar to the scalar case (e.g. [12], [19]); the possibility of concentrating components is a new feature due to the coupling between the components ; for two components for instance, eq. (3) may develop in

$$\frac{\partial c_1}{\partial t} + pv\,\frac{\partial f_1}{\partial c_1}\,\frac{\partial c_1}{\partial x} + pv\,\frac{\partial f_1}{\partial c_2}\,\frac{\partial c_2}{\partial x} = 0 \tag{8}$$

for the mass balance relevant to the first component ; $\partial c_1/\partial t$ and $\partial c_1/\partial x$ are no longer of opposite sign as the additive term $\partial f_1/\partial c_2$. $\partial c_2/\partial x$ may have different sign from $\partial f_1/\partial c_1 . \partial c_1/\partial x$ due to the sign of $\partial c_2/\partial x$ or of $\partial f_1/\partial c_2$: this last quantity may be < 0 whereas $\partial f_1/\partial c_1$ is always > 0 for reasons of stability. On figure 3 is represented a numerical modeling for a two component system, after [20].

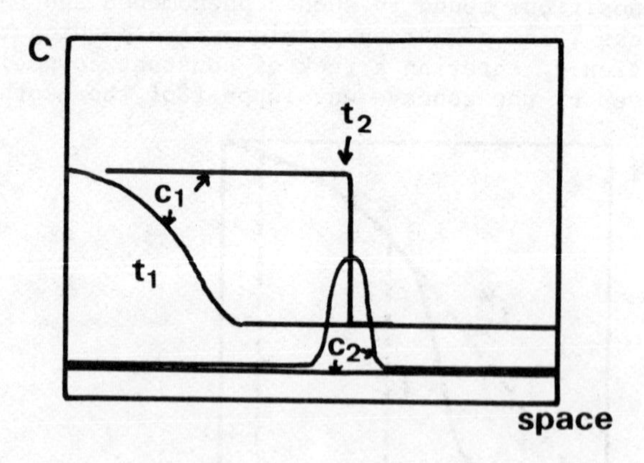

Figure 3. Increase of the concentration of minor component 2 due to front propagation relevant to major component 1.

One can see that minor component 2 is concentrated in the region of strong gradient of component 1, although it was at the beginning at a low grade in the rock. This increase of concentration may either correspond to the sweeping of the component that was downstream, or to the accumulation of the component coming with the inflowing fluid ; the front is then a "chemical dam".

Numerous ore deposits may be as a whole understood in such a way : U roll-type deposits (cf. section 2.ii), Mn, Ni, Pb, Al deposits (cf. 2.i) and so on, beyond the case of the numerous metals concentrated in hydrothermal deposits. The understanding of sea floor polymetallic nodules (cf. 2.iv) may possibly be set in the same framework. For all these examples, the perenniality at geological scale of the strong gradients is a condition for the efficiency of the concentration mechanism. Hyperbolic systems may also be written if we couple chemical exchange with heat and energy exchange.

4. NUMERICAL MODELING

For the scalar case two methods have been used. In the first one, the partial differential equation of the problem is replaced by a finite discretisation scheme (Godunov's method [13]). At each step, one has to solve a Riemann problem in order to determine the value of the unknown function on the sides of the discretisation block ; the choice of the unknown values is done thanks to the entropy condition established in the last paragraph. We are led to

$$c_i^{n+1} = c_i^n - q\,(f(c_i^n) - f(c_{i-1}^n)) \tag{9}$$

if f is increasing and to

$$c_i^{n+1} = c_i^n - q\,(f(c_{i+1}^n) - f(c_i^n)) \tag{10}$$

if f is decreasing where i is the space index and n the time index. Several numerical experiments have been performed : the concentration profiles have been obtained for several times, with different types of isotherms and different initial conditions. We can see the appearance and propagation of fronts, of plateaus, the washing out of heterogeneities and so on, in agreement with what is expected, particularly from a quantitative point of view (e.g. velocity of fronts) [21] , [22] . In order to avoid interfering between adjacent Riemann problems, the algorithm is subjected to a stability condition :

$$\sup f'(c).dt/dx < 1 \tag{11}$$

where dt and dx are the time and space intervals ; if not fulfilled numerical oscillations may appear.

The second method, called the transport-écroulement method ("transport -and -collapse") [23] , is not subjected to such stability condition ; the initial condition is transported on a chosen time interval by the method of characteristics : a multivalued function may thus be obtained ; it is converted into a single valued function by collapsing the overhanging parts. If the initial condition g(x) is transported during the time interval t/n, we obtain after collapse

$$c(x,t/n) = T(t/n).g \tag{12}$$

where T is the "transport-écroulement" operator. It is shown [23] that

$$c(x,t) = \lim_{n \to \infty} T(t/n)^n.g \tag{13}$$

is the unique entropy condition of problem (3) + (5). Numerical experiments have also been performed by this method [24], [25].

5. OSCILLATORY BEHAVIOURS

Oscillatory behaviours are observed at local (cm) scale within the rocks we are dealing here with [26] : spatial alternations of different minerals may form at the expense of homogeneous starting materials. When the concentration of the fluid is around the value where two phases are in equilibrium, the preceding approach as in eq. (2) and (3) ceases to be valid as perturbative effects may have a destabilizing effect ; this may be responsible of the bifurcation toward a periodic regime (cf the Hopf bifurcation). A very simple model can indeed be set forth, with purely chemical kinetics terms [27], [28] and based on the autocatalytic role of surfaces on the growing velocity of crystals [29], [30].

Let us consider the case where an external fluid carrying components A and B has dissolved a rock that contained component C, and where different minerals of composition $A_xB_{1-x}C$ may precipitate ; x is the molar fraction of A in the solid and we will call y the corresponding fraction in the fluid phase. At the surface of the growing solid phase, we may write four reactions with respective speeds v_1 (precipitation of AC on AC), v_2 (AC on BC), v_3 (BC on BC) and v_4 (BC on AC), where the selective autocatalytic effect of each phase is expressed with exponential terms.

For the steady state, $v_2 = v_4$ and we obtain a lay y = g(x) which may have two branches corresponding to two minerals : v_1 predominates on the first branch, v_3 on the second ; starting on the first, AC precipitates and the fluid is enriched in B till the moment when the second is reached ; BC then precipitates and so on, provided the transport is slower than chemical kinetics and imposes the concentration of components around the equilibrium value of the two phases.

We can show semi-quantitatively that the spatial wave-length of the structures is related to the amplitude of the chemical change betwenn the two minerals involved : the greater the chemical change, the greater the wave length, a result in good agreement with natural evidences. On the other hand, if transport is efficient with respect to kinetics the wave length decreases and the oscillation may disappear, in accordance with the situation where the oscillatory structures are found within chemically modified rock systems, e.g. in places where transport is operated by diffusion rather than convection.

A more general equation than (3) could be written like :

$$c_t + pvf(c)_x - pDf(c)_{xx} + g(c) = 0$$
$$\text{(I)} \quad \text{(II)} \quad \text{(III)} \quad \text{(IV)}$$

where c may be a vector $(c_1, c_2 \ldots c_n)$, so as $f = (f_i(c_1 \ldots c_n))$ and g. We have added here the diffusion term n° III which is not an ordinary diffusion term in that it involves the function f(c) ; in g(c,x,t) (term n°IV) we may consider chemical kinetics terms : in g may also appear terms involving derivatives of p and v with respect to x and t if the porosity and the fluid velocity are not constant but are supposed to be some known functions of x and t for instance. The first sections of this text have been concerned with terms (I) and (II) whereas the foregoing lines of the present section have been concerned with terms (I) and (IV). Some authors [33] have shown that the combination of (I) and (III) may also lead to oscillatory patterns. On the whole we may suspect that if we combine terms (I) (II) and (III) or (I) (II) and (IV) or if we take all of them in eq. (14) we will have bifurcations between several types of behaviours (between shocks and oscillatory behaviours for instance), see also [31] and [32] . Such approach will perhaps produce chaotic behaviours which are already suspected in crystal growth within such rock systems (cf observations made by J. Verkaeren, pers. comm.).

6. GENERAL COMMENTS

In this paper, we have taken natural discontinuities or fronts in their mathematical sense. One major interest for geologists of the modeling exposed in section 3 is to give an "intellectual tool" that allows one to consider as a whole a phenomenon that is responsible for the appearance of distinct events in time and space. Many fields in geology show such situation and the temptation is often to invoke separate causes. Of course the logic as given here is not always easy to ascertain on natural examples, particularly if the dimensions are large and if there are disequilibria (to that respect skarns are examplary). But looking for such a logic may be a useful method to understand natural cases.

The association of transport and chemical exchange is fundamental : in that respect current metallogenical models are different for they make a distinction between transport and (or) deposition. I stress here the association of both, needed to maintain strong gradients and concentrative effect ; such features are observed in the several fields I have listed in section 2 : to my sense these fields could be revisited on that ground. The model exposed here desires more work in order to be used for quantitative purposes.

Acknowledgements. This text reviews some works already achieved or in progress in our group in Saint-Etienne. It results from the collaborative efforts of and discussions with several individuals : Mlle Benhadid MM Conrad, Cournil, Fer, Fargue, Fonteilles, Gruffat, Kalaydjian, Valour. It received financial support from the Commission of the European Communities and Centre National de la Recherche Scientifique.

REFERENCES

[1] Guy B. (1979), *Pétrologie et Géochimie isotopique (S, C, O) des skarns à scheelite de Costabonne (Pyrénées Orientales, France)*thèse Ing. Doct., Ecole des Mines, Paris, 270 p.
[2] Le Guyader R. (1983) *Eléments-traces dans les skarns à scheelite et les roches associées à Costabonne (Pyrénées Orientales, France).*
[3] Barnes J.L. ed. (1979) *Geochemistry of Hydrothermal Ore Deposits*, 2d edition, J. Wiley, 798 p.
[4] Berner R.A. (1980) *Early diagenesis : a theoretical approach*, Princeton University Press, Princeton N.J., 241 p.
[5] Einaudi M.T., Meinert L.D. and Newberry R.J. (1981) 'Skarn deposits' *Econ. Geol.*, 75th anniversary volume, 317-391.
[6] Fonteilles M. (1978) 'Les mécanismes de la métasomatose', *Bull. Minéral.*, **101**, 166-194.
[7] Hargraves R.B. (1980) *Physics of magmatic processes*, Princeton University Press, Princeton N.J., 585 p.
[8] Marsilly G. de (1976), *Cours d'Hydrogéologie*, Ecole des Mines, Paris, 273 p.
[9] Guy B. (1984) 'Contribution to the theory of infiltration metasomatic zoning ; the formation of sharp fronts : a geometrical model' *Bull. Minér.*, **107** , 93-105.
[10] Thom R. (1977) *Stabilité structurelle et morphogénèse*, Interéditions, Paris, 351 p.
[11] Krushkov S.N. (1970) 'First order quasilinear equations in several independant variables', *Math USSR Sb.*, **10** , 217-243.
[12] Lax P.D. (1971) 'Shock waves and entropy', in *Contribution to non-linear functional analysis*, E.M. Zarantonello ed., Acad. Press, 603-634.
[13] Le Roux (1979), *Approximation de quelques problèmes hyperboliques non linéaires*, thèse, Rennes, 286 p.
[14] Conrad F., Cournil M. et Guy B. (1983) 'Bilan et "condition" d'entropie dans la métasomatose de percolation', *C.R. Acad. Sc.*, **296** , II, 1655-1658.
[15] Guy B., Cournil M., Conrad F. and Kalaydjian F. (1984) 'Chemical instabilities and "shocks" in a nonlinear convection problem issued from geology' in *Chemical Instabilities*, G. Nicolis and F. Baras ed., D. Reidel, 341-348.
[16] Korzhinskii D.S. (1970), *Theory of Metasomatic Zoning*, Clarendon Press, Oxford, 162 p.
[17] Fer F. (1977) *L'irréversibilité, fondement de la stabilité du monde physique*, Gauthier Villars, 135 p.
[18] Prigogine I. (1980) *Physique, temps et devenir*, Masson, 275 p.

[19] Lax P.D. (1973) 'Hyperbolic systems of conservation laws and the mathematical theory of shock waves', CRMS-NSF, Regional conference series in applied mathematics, SIAM publications.
[20] Benhadid S. (1984) *Simulation numérique des échanges fluides - roches (cas de deux constituants)*, Université de Saint-Etienne, Département Mathématique, Rapport de DEA, 79 p.
[21] Kalaydjian F. (1983) *Etude d'une transformation de granite en endoskarn sur le site tungstifère de Costabonne (Pyrénées Orientales) : pétrographie, géochimie (éléments majeurs, isotopes de l'hydrogène), modèlisation*, Travail Personnel d'Option, Ecole des Mines de Saint-Etienne, 125 p.
[22] Savard M. (1983) *Simulation numérique d'échanges fluides roches*, Ecole des Mines de Saint-Etienne, Département Géologie, 86 p.
[23] Brenier Y. (1981) *Calcul de lois de conservation scalaires par la méthode de transport - écroulement*, Rapp. Inria n° 53.
[24] Valour B. (1983)*Modèlisation et étude numérique d'un problème géologique : la formation des skarns*, thèse Doct. 3e cycle, Université de Saint-Etienne, 250 p.
[25] Fuxova D. (1984) *Numerical modelling of the formation of metasomatic rocks -running Bernard Valour's programs (1983)*- Rapport de stage, Département Géologie, Ecole des Mines de Saint-Etienne, 82 p.
[26] Guy B. (1981) 'Certaines alternances récurrentes rencontrées dans les skarns et les structures dissipatives au sens de Prigogine : un rapprochement ', *C.R. Acad. Sc.*, Paris, **292**, II, 413-416.
[27] Gruffat J.J. et Guy B. (1984) 'Un modèle pour les précipitations alternantes de minéraux dans les roches métasomatiques : l'effet autocatalytique des surfaces', *C.R. Acad. Sc. Paris*, **299**, 961-964.
[28] Guy B. and Gruffat J.J. (1984) 'A model for the oscillatory precipitations of minerals in chemically modified rocks : the autocatalytic role of surfaces', in *Non Equilibrium Dynamics in Chemical Systems*, C. Vidal and A. Pacault editors, Springer Verlag, p. 227.
[29] Chaix J.M. (1983) *Etude d'instabilités et de phénomènes nonlinéaires en réactivité des solides*, thèse Doct. Univ. Dijon, 190 p. annexes 33 p.
[30] Slin'ko M. and Slin'ko M. (1978) 'Self-oscillations of heterogeneous catalytic reaction rates', *Catal. Rev. Sci. Eng.*, **17** (1), 119-153.
[31] Ortoleva P., this volume.
[32] Merino E. (1984) 'Survey of geochemical self-patterning phenomena', *Chemical Instabilities*, G. Nicolis and F. Baras editors, D. Reidel Publishing Company, 305-328.
[33] Hazewinkel M., Kaazhoek J.K. and Leynse B. (1985) *Pattern formation for a one's dimensional evolution equation based on Thom's river basin model*, Report 851 9/B, Econometric Institute, Erasmus University Rotterdam, 24 p.

REACTION-PERCOLATION INSTABILITY

J. Chadam*
Department of Mathematical Sciences
McMaster University
Hamilton, Ontario, Canada, L8S 4K1

ABSTRACT. Reactive waters percolating through a porous medium cause changes in the porosity which, through Darcy's law, alter the flow. This feedback mechanism can cause instabilities in the shape of the porosity level surfaces. In this note we describe the geological problem in a simplified situation. We model the phenomenon mathematically as a coupled system of nonlinear ordinary-partial differential equations and in a certain limiting case as a moving free boundary problem. We formulate the shape stability problem in this context and summarize our analytical and numerical results to date.

1. A SIMPLE REACTION-PERCOLATION MODEL

Consider an aquifer consisting of an insoluble porous matrix (e.g. quartz sandstone) with some soluble mineral (e.g. calcite) partially filling the pores. If water is forced through this porous medium, the soluble component will be dissolved out upstream and the water will become saturated sufficiently far downstream. Between these extremes

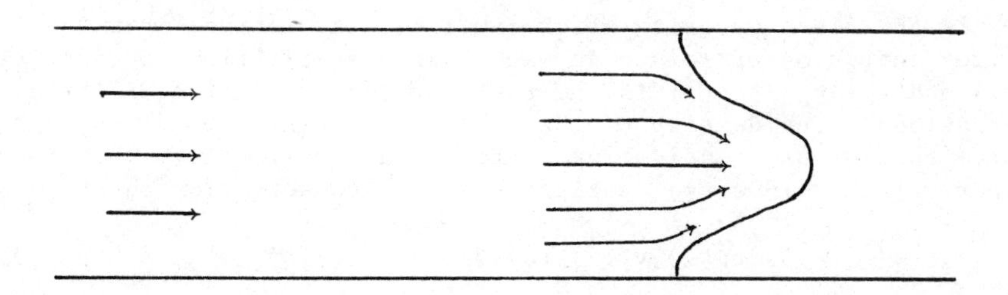

Fig. 1. Focusing of flow to tip of porosity level curve.

*Lecturer and preparer of this report. Work done in collaboration with P. Ortoleva, A. Sen and J. Hettmer.

C. Nicolis and G. Nicolis (eds.), Irreversible Phenomena and Dynamical Systems Analysis in Geosciences, 523–532.

there is a dissolution zone across which the soluble mineral content - and hence the porosity - changes from its original downstream value to the final, altered value upstream. The question of interest is whether the shape of this dissolution zone is stable. Notice that if a bump (in the porosity level curves) in the reaction zone exists at some time, the flow of the undersaturated waters tends to be focused to the tip of the bump via Darcy's law since inside the bump (on the upstream side) the permeability is greater than in the neighbouring regions (see Fig. 1). Thus dissolution is enhanced at the tip causing it to advance more rapidly. This is the porosity change/flow destabilization mechanism. On the other hand, diffusion from the sides of the tip raises the concentration of the solute in the water which is focusing at the tip and hence will decelerate this advancement. The competition between these two processes can lead to decay of the bump, restabilization to a morphologically more complicated dissolution zone or possibly to complete destabilization. A thorough understanding of this phenomenon and the resulting shapes is central to many important applications (e.g. the locating of roll-front ore deposits for which the restabilized shape outlines the deposits; the possible fingering of the flow of reactive nuclear wastes which have broken through their underground confinement).

2. TWO MATHEMATICAL MODELS

In this section we write down without details (c.f. [1;2;3, chap. XIV] for derivations) the models we shall subsequently treat analytically and numerically.

2.1. Coupled ODE/PDE Model

The rate of increase of the porosity ϕ (equivalently the rate of dissolution of the soluble mineral) is proportional to the reaction rate:

$$\varepsilon\frac{\partial\phi}{\partial t} = -(\phi_f - \phi)^{2/3}(\gamma - 1)(= -R(\phi,\gamma)) \tag{2.1}$$

Here ϕ_f is the final porosity after complete dissolution, γ is the scaled concentration of solute in water (with equilibrium concentration being 1) and $\varepsilon = c_{eq}/\rho \ll 1$ is the ratio of the original equilibrium concentration to the density of the soluble mineral. The 2/3-power indicates that we are considering surface reactions. The solute concentration per rock volume, $\phi\gamma$, satisfies a mass conservation equation:

$$\varepsilon\frac{\partial(\phi\gamma)}{\partial t} = \nabla\cdot[\phi D(\phi)\nabla\gamma + \phi\lambda(\phi)\gamma\nabla p] + \frac{1}{\varepsilon}R(\phi,\gamma) \tag{2.2}$$

where $D(\phi)$, $\lambda(\phi)$ are the porosity dependent, scaled diffusion coefficient and permeability respectively, and p is the pressure. Darcy's law has been used in the convective term of (2.2). It is also used in combination with the continuity equation to give:

$$\nabla\cdot[\phi\lambda(\phi)\nabla p] = \frac{\partial\phi}{\partial t} \left(= \frac{1}{\varepsilon}R(\phi,\gamma)\right). \tag{2.3}$$

In addition we impose the asymptotic conditions:

$$\gamma \to 0, \quad \phi \to \phi_f \text{ and } \frac{\partial p}{\partial x} \to \frac{\kappa_f p_f'}{D_f} = -\frac{v_f}{D_f} \quad \text{as } x \to -\infty \tag{2.4}$$

and

$$\gamma \to 1, \quad \phi \to \phi_0, \frac{\partial p}{\partial x} \to ? \quad \text{as } x \to +\infty. \tag{2.5}$$

These indicate that far upstream the water is fresh ($\gamma = 0$) and the mineral has been completely dissolved out ($\phi = \phi_f$). Also the pressure gradient (equivalently the velocity through Darcy's law $v_f = -\kappa_f p_f'$) is specified as in (2.4) with the effects of the scaling appearing explicitly. Far downstream the water is saturated ($\gamma = 1$), the porosity is still at its original, unaltered value ($\phi = \phi_0$) and the pressure gradient (equivalently the velocity) is to be determined. Equations (2.1-5), along with given initial data and periodic boundary conditions on the transverse boundaries, form a complete problem for the unknowns γ, ϕ, p. Unfortunately, nothing can be calculated analytically from these equations except the velocity of a travelling planar dissolution zone. On the other hand, they form the basis of our numerical simulations which will be discussed later.

2.2. Moving Free Boundary Model

In order to obtain an analytically tractable problem, we take the large solid density limit $\varepsilon = c_{eq}/\rho \to 0$. The dissolution zone, typically of width $\varepsilon^{1/2}$, collapses to a dissolution interface located at $x = R(y,t)$, with R unknown. Then, off this interface there is no reaction and the only consistent way to satisfy equations (2.1-3) to all orders of ε is as follows. Upstream of the dissolution interface where from scaling $\lambda(\phi_f)$ and $D(\phi_f) = 1$, one has

$$\left.\begin{aligned} &\nabla\gamma + \nabla\gamma\cdot\nabla p = 0 && (2.6)\\ &\phi = \phi_f && (2.7)\\ &\Delta p = 0 && (2.8)\end{aligned}\right\} \quad \text{in } x < R(y,t),\ 0 \leqslant y \leqslant L.$$

while downstream one obtains

$$\left.\begin{aligned} &\gamma \equiv 1 && (2.9)\\ &\phi = \phi_0 && (2.10)\\ &\Delta p = 0 && (2.11)\end{aligned}\right\} \quad \text{in } x > R(y,t),\ 0 \leqslant y \leqslant L$$

where we have taken $\phi_0(x,y,0) = \phi_0$, constant, to show that the morphological instabilities will even occur in this spatially homogeneous situation. Besides the asymptotic conditions (2.4,5) one also obtains, via matched asymptotics, boundary conditions on the unknown moving dissolution interface. Specifically, one has

$$\gamma = 1 \tag{2.12}$$

$$p- = p+ \qquad \text{on} \tag{2.13}$$

$$\frac{\partial p-}{\partial y} - \frac{\partial p-}{\partial y}\frac{\partial R}{\partial y} = \Gamma\left(\frac{\partial p+}{\partial x} - \frac{\partial p+}{\partial y} \cdot \frac{\partial R}{\partial y}\right) \qquad x = R(y,t) \tag{2.14}$$

$$0 \leqslant y \leqslant L$$

$$\frac{\partial \gamma}{\partial x} - \frac{\partial \gamma}{\partial y} \cdot \frac{\partial R}{\partial y} = (1 - \phi_0/\phi_f)R_t \tag{2.15}$$

where $0 \leqslant \Gamma = \phi_0\kappa_0/\phi_f\kappa_f \leqslant 1$ is a measure of the porosity change. Equation (2.15) relates the rate of advancement of the moving dissolution interface to the flux of the concentration and is called a Stefan condition. A final scaling $x' = \frac{\pi}{L}x$, $y' = \frac{\pi}{L}y$, $t' = (\frac{\pi}{L})^2(1 - \phi_0/\phi_f)^{-1}t$ with $R' = \frac{\pi}{L}R$ (and dropping the primes) makes the transverse dimension $0 \leqslant y \leqslant \pi$, and results in the two changes

$$\gamma \to 0,\ \phi \to \phi_f,\ \frac{\partial p}{\partial x} \to -\frac{v_f L}{D_f \pi} \quad \text{as } x \to -\infty \tag{2.4'}$$

and

$$\frac{\partial \gamma}{\partial x} - \frac{\partial \gamma}{\partial y} \cdot \frac{\partial R}{\partial y} = R_t \quad \text{on } x = R(y,t),\ 0 \leqslant y \leqslant \pi \tag{2.15'}$$

Problem (2.4',5,...14,15') with initial conditions and periodic transverse conditions on $y = 0,\pi$ is the version we shall examine analytically in the next section. Notice that only two essential parameters remain in the problem, the dynamical parameter $\nu_f = v_f L/D_f$ and the measure of the porosity change $\Gamma = \phi_0\kappa_0/\phi_f\kappa_f$.

3. SHAPE INSTABILITIES

In this section we describe our analytical results in the context of the large solid density problem (2.4',5,...14,15'). Here the planar, constant velocity solution can be obtained explicitly and completely, including the concentration and pressure profiles which were not available for the more general coupled ODE/PDE model. The linearized stability of this solution is then described, giving a precise value of the parameter ν_f(in terms of Γ) for which the planar solution looses stability to another, more structured, solution. In the language of G. Nicolis' lectures, we determine the critical parameter value for which the spectrum of the linearized problem changes sign from negative to positive, thus determining the location of a possible bifurcation point. Finally we sketch the bifurcation analysis to show that the linear instabilities are restabilized by the nonlinearities to a morphologically more complicated solution. In the context of G. Nicolis' lectures, we obtain a Landau equation for the amplitude of the linearly unstable mode thus indicating a pitchfork bifurcation diagram.

3.1. Planar Solution

Denoting the planar state quantities with a super bar, one can easily check [2,3] that the following constant velocity solution satisfies problem (2.4',5,...14,15'):

$$\bar{R}(t) = \bar{V}t \tag{3.1}$$

$$\bar{\gamma}(x,t) = \begin{cases} e^{-\bar{\nu}_f(x-\bar{V}t)} & x < \bar{V}t \quad (3.2a) \\ 1 & x > \bar{V}t \quad (3.2b) \end{cases}$$

$$\bar{p}(x,t) = \begin{cases} -\bar{\nu}_f(x-\bar{V}t) & x < \bar{V}t \quad (3.3a) \\ -\bar{\nu}_0(x-\bar{V}t) & x > \bar{V}t. \quad (3.3b) \end{cases}$$

where $\bar{\nu}_f = \nu_f/\pi = v_f L/D_f\pi$, $\bar{\nu}_0 = \phi_f\bar{\nu}_f/\phi_0$ (from (2.5) and (2.14)) and the velocity of the planar interface $\bar{V} = \bar{\nu}_f$ from (2.15').

3.2. Linear Shape Instability

In order to examine the stability of the above planar solutions (3.1,3) with respect to bumps we consider perturbations of the type (i.e. a generic term in the Fourier decomposition)

$$R(y,t) = \bar{V}t + \delta\ r_{1m}(t) \cos my \tag{3.4a}$$

$$\gamma(x,y,t) = \bar{\gamma}(x,t) + \delta\ \gamma_{1m}(x)\ r_{1m}(t) \cos my \tag{3.4b}$$

$$p(x,y,t) = \bar{p}(x,t) + \delta\ p_{1m}(x)\ r_{1m}(t) \cos my. \tag{3.4c}$$

Considering δ to be small, the linearized version of equations (2.4',5, ...14,15') can be derived [2] (i.e. retain terms in first power of δ). These can be solved explicitly for γ_{1m} and p_{1m} and the Stefan condition gives [2,3] the following condition on the amplitude $r_{1m}(t)$ of the cos my bump:

$$r'_{1m}(t) = \frac{\bar{\nu}_f}{1+\Gamma}\,[\bar{\nu}_f - (\bar{\nu}_f^2 + 4m^2)^{1/2} + + (1-\Gamma)|m|]\,r_{1m}(t) \tag{3.5a}$$

This differential equation indicates that the amplitude of the bump grows or decays depending on the sign of the coefficient. The connection with the equivalent, more conventional viewpoint presented in G. Nicolis' lectures follows by expressing $r_{1m}(t) = re^{\sigma(m)t}$ in terms of the spectrum $\sigma(m)$ of the linearized problem and obtaining from (3.5a) the dispersion relation

$$\sigma(m) = \frac{\bar{\nu}_f}{1 + \Gamma} [\bar{\nu}_f - (\bar{\nu}_f^2 + 4m^2)^{1/2} + (1 - \Gamma)|m|]. \qquad (3.5b)$$

The m-dependence of σ is shown in Fig. 2 revealing clearly that the planar solution (3.1,3) is linearly unstable to long wavelength perturbations (because $\Gamma < 1$) and stable to short wavelength perturbations.

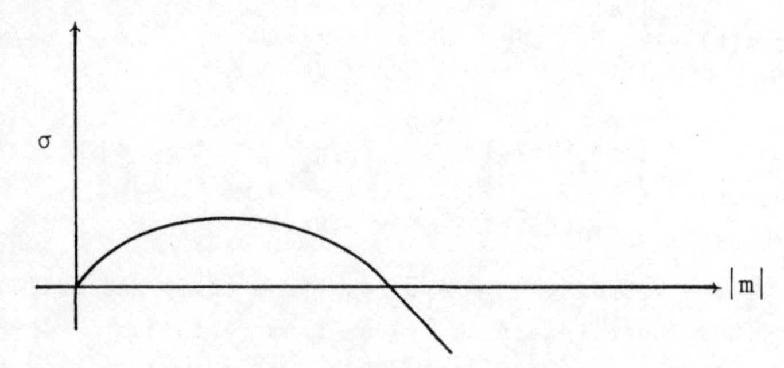

Fig. 2. Graph of dispersion relation (3.5b).

The critical wave number ($|m_0|$ at which $\sigma(m_0) = 0$) is given by

$$|m_0| = \frac{2(1 - \Gamma)}{(3 - \Gamma)(1 + \Gamma)} \bar{\nu}_f \qquad (3.6)$$

Since our channel width has been normalized to π, the first mode which can be carried is $|m_0| = 1$ giving, from (3.6), the critical parameter value (of $\nu_f = v_f L/D_f$)

$$\nu_c = \nu_c(\Gamma) = \frac{(3 - \Gamma)(1 + \Gamma)\pi}{2(1 - \Gamma)}. \qquad (3.7)$$

From this we see that the instability does indeed arise analytically and that, as is physically realistic, larger flow speeds, larger transverse dimensions, larger porosity/permeability changes promote the instability while larger diffusion coefficients inhibit the instability (i.e. diffusion is stabilizing, as mentioned earlier). The limit of $\Gamma \to 1$ (i.e. no porosity change) suggests it is very difficult to produce instabilities. This has been verified by a separate analysis [1]. Thus this instability can occur only if "significant" amounts of the soluble mineral are dissolved.

3.3. Non-linear Restabilization

We begin by scaling the independent variables. Because the instability occurs at finite wavelength none is required for the spatial variables while, as is common for pitchfork bifurcations,

$$t_2 = \varepsilon^2 t. \tag{3.8}$$

(i.e. single terms suffice for the more general series in G. Nicolis' lectures). Additionally, we write

$$\nu_f = \nu_c + \varepsilon\, \nu_1 + \varepsilon^2 \nu_2 + ---. \tag{3.9}$$

We find [3] at $0(\varepsilon^2)$ that $\nu_1 = 0$ (as usual for pitchfork bifurcations) so that the physical significance of the small parameter ε is

$$\varepsilon \simeq (\nu_f - \nu_c)^{1/2} \tag{3.10}$$

where we have taken, without loss of generality, $\nu_2 = 1$. Thus

$$\nu_f = \nu_c + \varepsilon^2 + \varepsilon^3 \nu_3 + ---. \tag{3.11}$$

The stability calculation then proceeds by expanding all of the dependent variables in terms of ε (suppressing the sub-2 in the new t_2 variable):

$$\begin{aligned} R(y,t) = {} & \bar{V}(\varepsilon)t \\ & + \varepsilon(r_{10}(t) + r_{11}(t)\cos y + r_{12}(t)\cos 2y + ---) \\ & + \varepsilon^2(r_{20}(t) + r_{21}\cos y + r_{22}(t)\cos 2y + ---) \\ & + \varepsilon^3(r_{30}(t) + r_{31}\cos y + r_{23}(t)\cos 2y + ---) \\ & + 0(\varepsilon^4), \end{aligned} \tag{3.12a}$$

$$\begin{aligned} \gamma(x,y,t) = {} & \bar{\gamma}(x,t;\varepsilon) \\ & + \varepsilon(\gamma_{10}(x,t) + \gamma_{11}(x,t)\cos y + \gamma_{12}(x,t)\cos 2y + ---) \\ & + \varepsilon^2(\gamma_{20}(x,t) + \gamma_{21}(x,t)\cos y + \gamma_{22}(x,t)\cos 2y + ---) \\ & + \varepsilon^3(\gamma_{30}(x,t) + \gamma_{31}(x,t)\cos y + \gamma_{32}(x,t)\cos 2y + ---) \\ & + 0(\varepsilon^4), \end{aligned} \tag{3.12b}$$

and similarly for $p(x,y,t)$. Following the prescription outlined in G. Nicolis' lectures one obtains a Landau differential equation for the amplitude of the unstable mode, $r_{11}(t)$:

$$r_{11}'(t) = w\, r_{11}(t) - \Lambda\, r_{11}(t)^3 \tag{3.13}$$

where

$$w = w(\Gamma) = \frac{\nu_c}{1+\Gamma} \frac{[\nu_c^2 + 4)^{1/2} - \nu_c]}{(\nu_c^2 + 4)^{1/2}} > 0 \tag{3.14}$$

and the Landau constant, $\Lambda = \Lambda(\Gamma)$, which is algebraically very complicated, is given in Fig. 3. The positivity of Λ indicates that in the

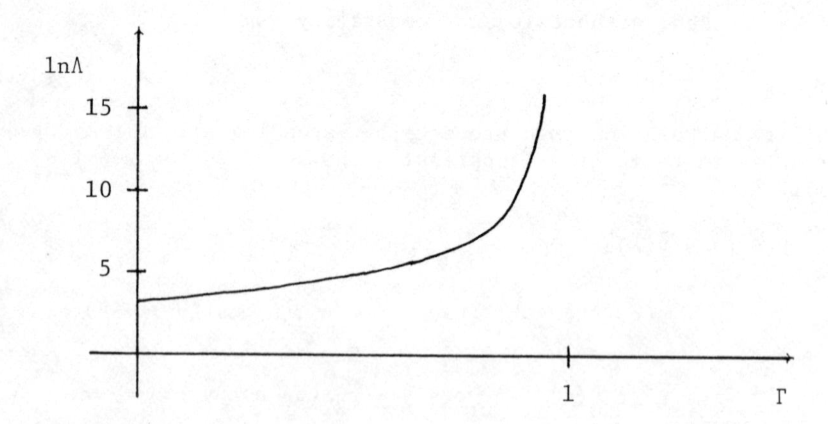

Fig. 3. Graph of the logarithm of the Landau constant versus $\Gamma = \phi_0\kappa_0/\phi_f\kappa_f$, a measure of the porosity change.

vicinity of the critical point the linearized instabilities (from $w > 0$) are restabilized by the nonlinearities at the next highest order, and (from (3.13)) that the bifurcation diagram is the symmetric pitchfork. The asymptotic amplitude of the bump can be obtained from (3.13) to be $(w/\Lambda)^{1/2}$.

4. NUMERICAL SIMULATIONS

The actual shape of the stabilized bump, especially far from the critical point (to which the analysis of the last section does not apply) must be obtained from numerical simulations. Because interface tracking is a difficult problem we return to the coupled ODE/PDE model with $\varepsilon = c_{eq}/\rho$ small (= 0.05) but not zero. Using parameter values suggested by the analytical results of the previous section and standard numerical methods for solving equations (2.1-5) [1], we investigated the three cases $\nu_f \ll \nu_c$, $\nu_f \simeq \nu_c$ and $\nu_f \gg \nu_c$. Figs. 4 a), b), c) depict these

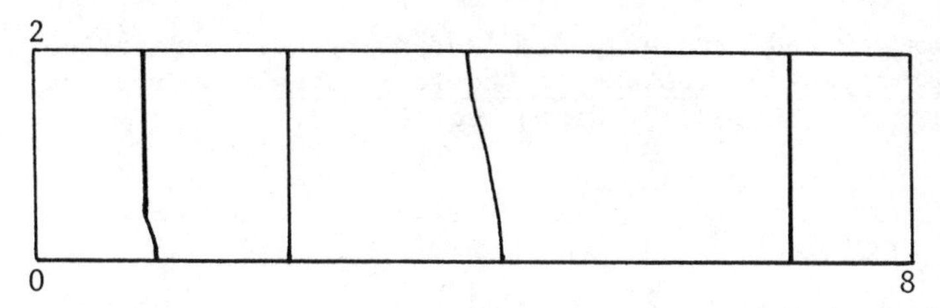

Fig. 4 a). $\nu_f \ll \nu_c$. Evolution of porosity level curve for times 0.0, 1.8, 3.6, 5.4.

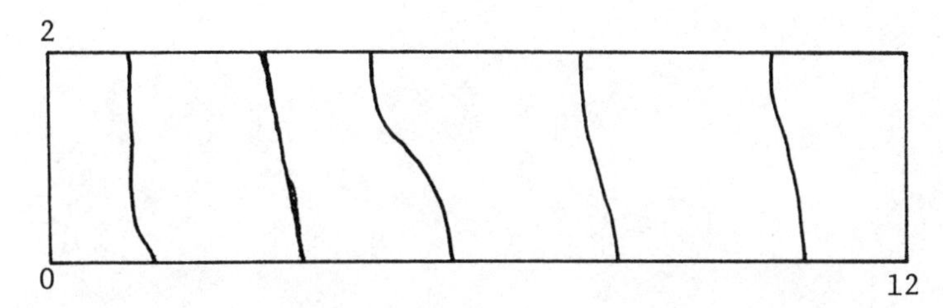

Fig. 4 b). $\nu_f \simeq \nu_c$. Evolution of porosity level curve for times 0.0, 1.5, 3.0, 3.75.

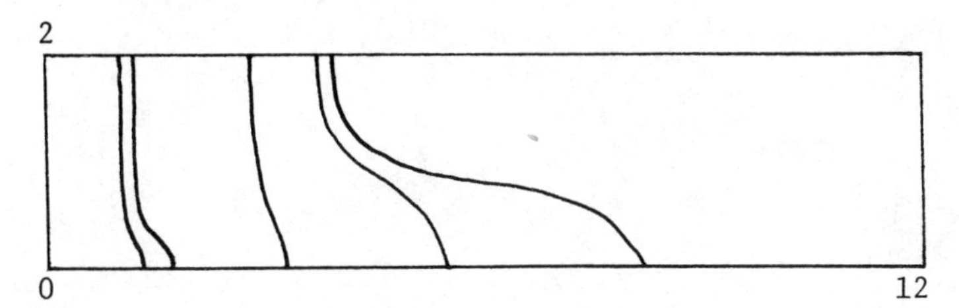

Fig. 4 c). $\nu_f \gg \nu_c$. Evolution of porosity level curves for times 0.0, .24, .48, .72, .96.

cases respectively, indicating stability of the planar front, restabilization to a new shape and a highly unstable dissolution zone respectively.

REFERENCES

1. Auchmuty, G., Chadam, J., Merino, E., Ortoleva, P. and Ripley, E., 'The Structure and Stability of Moving Redox Fronts', accepted for publication in SIAM J. Appl. Math. (1985).

2. Chadam, J., Ortoleva, P. and Sen, A., 'Reactive Percolation Instabilities', 34 mimeo. pages, submitted for publication to JIMA (1985).

3. Chadam, J. and Ortoleva, P., 'Morphological Transitions', chap. XIV of Variety and Structure of Chemical Waves, P. Ortoleva et al., Synergetics Series, H. Haken, ed. (1985).

ON GEOMORPHIC PATTERNS WITH A FOCUS ON STONE CIRCLES VIEWED AS A FREE-CONVECTION PHENOMENON

Bernard Hallet
Quaternary Research Center AK-60
University of Washington
Seattle, Wa. 98195
U.S.A.

ABSTRACT I describe a series of prominent patterns seen in landscapes which constitutes a research area ripe for the application of dynamical systems analysis. Stone circles and polygons in Arctic soils are particularly striking examples of self-organization and, hence, are presented in greater detail. A conceptual model of these features is developed, in which the soil is idealized as a very viscous linear fluid and intermittent free-convection of the soil is inferred. It is likely to result from gradients in the bulk density of the soil arising from an increase in water content with depth due to ice melt. This model is in accord with the observed size, regularity, pattern, texture, micro-relief, surface soil motion and vegetation of stone circles in Spitsbergen.

1. INTRODUCTION

It is hoped that this cursory presentation of geomorphic patterns will inform non-geomorphologists that a variety of orderly patterns, some with striking regularity, occur in landscapes. In addition to being interesting in their own right, considerations of these orderly patterns may pave the way to exploration of the intriguing possibility of chaotic patterning in landscapes (Culling, 1985).

For clarity of presentation geomorphic patterns are arbitrarily subdivided into two groups: closed forms, which are those that enclose an area, and open forms that do not.

2. OPEN FORMS

2.1 Drainage Networks.

The branching network of streams with a progressive convergence of tributaries of different sizes to form larger streams is obvious from any map. The geometry of the drainage network can vary substantially

C. Nicolis and G. Nicolis (eds.), Irreversible Phenomena and Dynamical Systems Analysis in Geosciences, 533–553.

in response to the substrate. In areas of relatively uniform bedrock, dendritic networks are common; they are characterized by sinuous streams with a range of orientations that generally converge in one direction. In contrast, in areas underlain by tilted layers of rocks with differing resistances to erosion drainage tends to form trellis patterns in which angular junctions and abrupt bends in stream courses reflect the orientations of relatively weak zones in the bedrock. In view of the obvious contrasts in drainage form and the irregular geometry of each channel, the topology of these networks is remarkably constant, indicating that the branching patterns are orderly on a statistical basis.

The various segments of a drainage network can be ranked from first-order channels, also termed exterior links, that have no tributaries, to higher-order channels that have successively more tributaries. Interior links are the reaches of channels between adjacent confluences. Horton (1945) discovered that the number of channels of a particular order decreases geometrically as the order increases. This "law of stream numbers" was empirically known to apply generally, regardless of the size of the basin, which can range from 1 to 10^7 km^2, and regardless of the lengths, shapes and orientation of channels. Shreve (1966, p. 36) suggested that "the law of stream numbers is largely a consequence of random development of the topology of channel network according to the laws of chance." His seminal studies are "founded on two basic postulates: (1) Natural channel networks in the absence of strong geologic control are very nearly topologically random and (2) interior and exterior link lengths and associated areas in basins with homogenous climate and bedrock have separate statistical distributions approximately independent of location in the basin" (Shreve, 1975, p. 527). In addition to providing a basis for the law of stream numbers, Shreve's probabilistic topological approach unifies the large number of known quantitative empirical relationships among the orientation-free planimetric properties of channel networks and drainage basins (Shreve, 1975).

The remarkable lack of variation in drainage network topology with basin size reflects the scale-independence of the branching patterns. Indeed, drainage nets seen on maps on vastly different scale often have very similar appearances. Given the fundamental nature of Shreve's postulates, these probabilistic considerations are not likely to be limited to drainage networks; indeed they have obvious applications in studies of other arborescent networks as in trees and other plants, and in circulatory and other physiological systems (e.g. Vogel, 1983). The self-similarity and universality of branching networks is well recognized (Mandelbrot, 1982).

2.2 River Meanders

The sinuous paths of many rivers are often comprised of smoothly curving segments with a rather prominent wavelength that tends to scale with the water discharge and the width of the channel. Spectral analyses of meandering rivers reveal considerable power over a wide range of wavelengths, however, commonly with several spectral peaks (Speight, 1965). A simple prominent geometric property of meandering channels, recognized by Langbein and Leopold (1966), is that changes in direction of channel segments closely approximate a sine function of channel distance. The resulting form is known as a sine-generated curve. They noted further that "the geometry of a meander is that of a random walk whose most frequent form minimizes the sum of squares of the changes in direction in each unit length" of the channel. Thus, in cases where a straight channel is unstable, as discussed below, a river of given sinuosity tends toward a form that minimizes the overall curvature of the channel.

The close geometric resemblance of a buckled elastic beam to the idealized curve of least angular variance offers a suggestive analogy helpful in elucidating overall meander shape (Leopold and Langbein, 1966). The shape of a buckled elastic beam is known to be that which minimizes the overall strain energy compatible with the imposed shortening and other constraints. In this conservative system, since strain energy scales with the square of the curvature (Timoshenko and Goodier, 1951), the overall strain energy of a uniform beam is least when the curvature is as uniform as possible. By analogy, therefore, it might be inferred that the sine-generated curve is closely approximated by a channel because it tends toward a form for which erosion is as uniform along the channel as possible. This form appears likely because areas of relatively high curvature would not persist; bank erosion would naturally be concentrated there, which would tend to decrease the curvature in these areas and to render the overall channel curvature uniform.

Similar considerations have motivated an array of searches for simple extremum principles that would govern the geometry of meanders. Statements of minimum variance, minimum flow resistance, minimum total energy dissipation and minimum rate of energy dissipation (e.g. Ritter, 1978, p.240) have all been developed in this context. At present, they all suffer from the facts that the "system" is difficult to define precisely, and the governing equations for sediment transport and, especially, bank erosion are poorly known. Moreover, the classic "thermodynamic" considerations of landscape evolution (e.g. Leopold and Langbein, 1962) preceded recent advances in dynamical systems analysis and, in particular, preceded the widespread recognition of potentially complex behavior in systems far from equilibrium.

The fundamental cause of meanders has been the subject of a series of enlightening fluid dynamical studies (see Callander, 1978 and references therein). The general thesis is that the mobile bed of a channel with straight banks is unstable, so that a perturbation in the form of a set of alternating bars and pools will grow in amplitude. It is often assumed the perturbations with a wavelength for which the initial growth is fastest will dominate the meander pattern. However, stability analyses say little about meander growth beyond the incipient stage. A recently developed three-dimensional model of water flow and erosion in curved channels with large-amplitude bed relief (Nelson and Smith, 1985) now provides a powerful tool for studying meander evolution and geometry.

2.3 Beach Cusps and Other Coastal Crescentic Landforms

Remarkable patterns of evenly spaced crescentic embayments separated by seaward projections are prominent on many shorelines. The most common features of this type are beach cusps which can form in beach sediments of any size, ranging from fine sand to boulders. The spacing of beach cusps varies from beach to beach from about 3 to 30 m, and appears to increase with wave height.

A variety of origins have been proposed to account for the generation of evenly spaced beach cusps (e.g. Ritter 1978), but no consensus has been reached. According to several prominent hypotheses the cusps reflect periodic variations in wave height along the length of the waves. A popular version involves standing edge waves (e.g. Guza and Inman, 1975), as was mentioned by Dr. Putterman during his discussion of edge wave instabilities at this Institute.

Beach cusps become more perplexing when one realizes that they belong to a complete hierarchy of crescentic shoreline forms ranging from cusplets with a wavelength of less than a meter to major cuspate indentations of the coastline with spacings on the order of a hundred kilometers. Moreover, except for large-scale features with spacings in excess of kilometers, the crescentic forms all tend to display pronounced periodicity (Dolan, et al., 1974). Presently, there appears to be no widely-accepted explanation for this large range of periodic shoreline forms.

2.4 Ripples and Dunes

In many desert and coastal regions of the Earth, winds mold their erodible substrates of cohesionless grains into strikingly regular bedforms of three distinct wavelengths: ripples, dunes and draas (also termed megadunes). The Algodunes dune chain in southern California, U.S.A. provides a spectacular example of bedform superposition. Aerial photographs (e.g. Greeley et al. 1978) reveal the large scale order: a regular spacing, about 1.3 km, of 100 m high megadunes separated by flat areas largely free of sand. The megadunes are themselves comprised of coalescing dunes with a horizontal length scale on the order of 100 m. Moreover, decimeter-sized ripples are characteristic of the dune surfaces. This hierarchy of eolian bedforms, with three prominent wavelengths on the order of 10^{-1}, 10^{2} and 10^{3} m, has been thoroughly documented by Wilson (1972). He has also shown that these bedforms can be grouped into three dimensional categories without overlap by taking into account the grain size of the material. The recognition that, given a particular grain size, bedforms belonging to each wavelength category can form simultaneously and that no transitional forms exist, suggests that distinct processes lead to each geomorphic form.

Eolian ripples form some of the most regular and aesthetically pleasing patterns in nature. Yet, no theory presently explains quantitatively how the complex fluid-bed interactions, including the inherently stochastic motions of coarse grains bouncing energetically along the bed, give rise to this regularity. A recently developed model of sediment transport by wind (Anderson & Hallet, in press) provides the necessary foundation permitting the development of a model for ripples and larger bedforms. This development is likely to parallel closely the successful theoretical analysis of the analogous problem of fluvial bedforms (e.g. Smith and McLean, 1976), in which the turbulent structure associated with finite-amplitude bedforms controls their continued growth, or their decay.

3. CLOSED FORMS

3.1 Polygonal Features

Distinct polygonal patterns, with a predominance of hexagonal forms, are characteristic of crack networks that develop in homogeneously cooling lava and in cooling frozen soils. These patterns belong to a broad class of phenomena called contraction-crack polygons, which form in response to tensions resulting from a decrease in volume such as could be caused by thermal contraction, desiccation, chemical reaction or phase change (Lachenbruch, 1962 p.55). They span a broad range of scale ranging from centimeters in drying mud to tens of meters in permafrost. The size and configuration of polygons depend upon the rheological behavior of the medium and the nature of the induced volume change. The differential contraction at the surface induces horizontal tensile stresses that peak at the surface and cause fractures to form and propagate. As cracks form, they cause a local relief of tension in the surficial material. Progressively, cracks intersect to form polygons that "attain such a size that zones of stress relief of neighboring cracks are superposed at the polygon center so as to keep the stress there below the tensile strength" (Lachenbruch,1962 p.58). The polygonal patterns with common triple junctions of cracks intersecting at nearly 120° angles are typical of conditions of relatively great thermal and mechanical homogeneity. Less homogeneous conditions tend to lead to a sequence of cracks, the first ones forming at loci of low strength or high stress concentration. As they propagate, subsequent cracks are influenced by stress fields around existing cracks; this influence causes cracks to intersect pre-existing ones at right angles to form orthogonal systems of polygons.

Interestingly, compression ridges that appear to result from an increase in volume can give rise to similar patterns of polygons. Such patterns are perhaps most obvious on lake beds where water with high salt content is evaporating. The resulting salt crystallization tends to give rise to compressive horizontal stresses in the surficial material. With continued crystallization, eventually the salt-encrusted surface layer tends to buckle upward thereby forming ridges. As ridges often stand out above the receding water line, they are relatively dry and, hence, fluids tend to migrate osmotically to these areas. This results in continued salt transport and further growth of the ridges.

When conditions are relatively uniform, as in the center of a playa, the ridge pattern is strikingly regular. The ridges almost invariably join in triple junction with intersection angles close to 120° and they subtend equidimensional domains. Interestingly, the resulting patterns are distinct from a hexagonal network because the ridge segments are irregular in plan view. In fact, they closely resemble the fractal Koch curve whose initiator is a regular hexagon (Mandelbrot, 1982, p. 46). An exceptional aspect of this plane-filling form is that each cell can be subdivided into equal sub-cells of identical form. This similarity is absent in hexagons, as several hexagons cannot form a larger hexagon. It is noteworthy that the triple junctions that occur in many natural plane-filling mosaics occur on scales ranging from 10^{-3} m in mudcracks to 10^6 m for oceanic ridges.

3.2 Circular Features

Circular forms occur in a variety of geomorphic settings. Arrays of circular forms are well known to occur in areas where subfreezing temperatures are common. The next section is devoted to these features.

4. STONE CIRCLES

Stone circles are fine-grained soil domains delineated by raised borders of gravel (Figure 1). They form largely in Arctic and alpine areas of permafrost where, below the upper one meter or so, the soil is permanently frozen. Their orderly geometry and the distinct segregation of material according to grain size both represent a

Figure 1. Stone circles averaging 3-5 m in diameter with 1 m wide raised gravel borders in western Spitsbergen.

degree of self-organization that is striking when noting that these structures generally originate from a featureless mixture of mineral material. These stone circles constitute only one of many forms of periglacial patterned ground that have been widely described and illustrated (e.g. Washburn, 1980 and references cited therein). Although numerous hypotheses have have been proposed regarding the formation of patterned ground (Washburn, 1956), there is no consensus and no rigorous assessment of the relative importance of the many invoked processes.

Diverse geologic evidence, including the similarity between plan forms of permafrost patterned ground and of Benard cells, give considerable motivation for viewing these soil patterns as a free convection phenomenon (Nordenskjold, 1907; Low, 1925; Elton, 1927; Gripp, 1926, 1927; Mortensen, 1932). Wasiutynski (1946, p.202) assembled existing observations of stone circles, and presented a strong case for soil convection in the active layer, which is the upper meter or so of soil in permafrost areas that thaws during the summer. Despite this early work, and other studies suggestive of circulatory soil motion (Sorenson, 1935; Pissart, 1966; Nicholson, 1976), the notion of free convection of soil has lost popularity among geomorphologists, partly because of the lack of a mechanism producing a large systematic decrease in soil bulk density with depth.

In view of the current lack of concensus regarding the formation of stone circles (e.g. Washburn, 1956, 1980), it is instructive to review their important characteristics, and to consider the similarities and differences between them and Benard cells. I will first briefly describe stone circles and present measurements of surface soil displacements, drawing largely on recent field studies in Spitsbergen (Hallet and Prestrud, in press). I will then discuss driving mechanisms for the inferred soil convection.

4.1 Description of Patterns

4.1.1. Plan View. The most distinctive soil patterns in our study area, at 79°N 12°E in western Spitsbergen, generally comprise equidimensional domains of essentially unvegetated, fine-grained soil outlined by broad curvilinear ridges of gravel (Figure 1). They resemble closely the exceptionally well-developed sorted circles studied by Czeppe (1961) and Jahn (1963) in the Hornsund area of Spitsbergen. Although they only rarely approach a true circular form, we follow existing usage and refer to these equidimensional figures as "stone circles", also known as "sorted circles" because of the distinct sorting of material according to size. Whereas certain neighboring areas are completely covered with polygonal arrays of coalescing stone circles that share common borders, adjacent circles in our study sites are generally separated by low relief areas lacking the lateral sorting characteristic of stone circles. In extreme cases individual stone circles may be separated from all others.

Considerable variation is apparent in the size of stone circles from one area to another. However, within a single network they tend to be rather uniform in size. Preliminary transects measured from aerial photographs of our study area indicate that outside diameters of stone circles, including the 0.5 to 1 m wide border, range from 3 to 6 m.

4.1.2. Microtopography and Subsurface. Typically the surfaces of fine domains are smooth and convex upward with the highest point near the circle center being 50 to 100 mm higher than the edges of the fine domain. The transition between the fine domain and coarse border is abrupt, both in terms of texture and slope (Figure 2). The surfaces of the fine domains and of the coarse border both dip down toward the sharply defined fine/coarse contact, the latter being generally steeper. The border heights vary considerably from area to area, ranging from a few mm to 0.5 m.

The surface marking the contact between fine and coarse material commonly dips about 20° radially outward near the surface and steepens sharply to near-vertical below about 0.2 m. The depth to frozen ground increases throughout the thaw season and tends to be greatest under the fine centers. The active layer thickness ranges from 1 to 1.5 m in the study areas.

Figure 2. Sharp textural break at the distinct trough marking the contact between domed fine-grained domain and raised borders. Shadow cast by a horizontal bar highlights the micro-relief.

4.1.3. Vertical Displacements. Considerable upward and downward motion due to freezing and thawing have been well documented near the surface of sorted circles at Hornsund (Jahn, 1963 and Czeppe, 1961). Our observations yield similar results with maximum surface heaves on the order of 0.1 m, roughly 10% of the active layer thickness. Thawing of the soil starts shortly before the surface is exposed by snow melt. The surface first settles quickly, particularly in the fine-grained domains. After a few weeks, settling slows to very low rates.

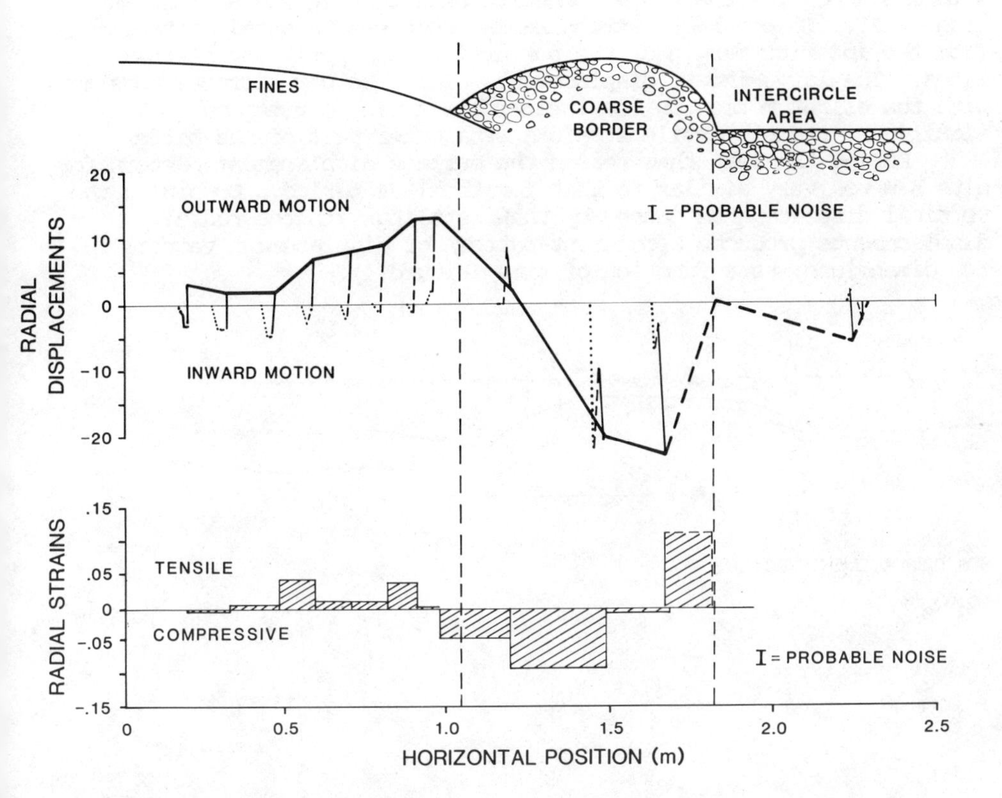

Figure 3. Radial displacements (in mm's) at stone circle site KIW throughout the 1984 thaw season. Inward motions are toward circle center, outward motions are toward circle exterior. Successive measurements are represented by subvertical line segments, slightly offset for illustrative clarity. Heavy solid line represents net radial displacements for entire thaw season. Radial strains are calculated from the net radial displacements; a tensile strain corresponds to an increase in radial displacement with distance from the circle center.

4.1.4 Horizontal Displacements at the Surface. Twelve radial displacement profiles were measured throughout the 1984 and 1985 thaw seasons. In all sorted circles monitored, radial displacements of individual markers at the surface ranged up to 20 mm and averaged a few millimeters. Interestingly, a coherent pattern of motion emerged in each sorted circle. As seen in a representative example in Figure 3, the rate and direction of the radial motions of individual markers during successive time intervals are complicated, but by the end of the thaw season, the markers in the fines moved outward toward the border, and those in the coarse border moved inward. Little or no net motion was recorded in the intercircle areas.

In the fines, the increase in net outward radial displacements with distance from the center constitutes a tensile radial strain (Figure 3). In contrast, strong radial compressive strain extends from the interior margin of the border into the periphery of the fines. The large inward displacements of the border surface, together with the slight motion in the intercircle areas, correspond to significant radial tensile strain in the outer part of the border.

During the 1985 thaw season the surface displacement pattern for site K1W was very similar to that for the 1984 period. Combining the vertical displacements caused by thaw settling with the radial displacements produces a coherent pattern of displacement vectors in two dimensions, as a function of time (Figure 4).

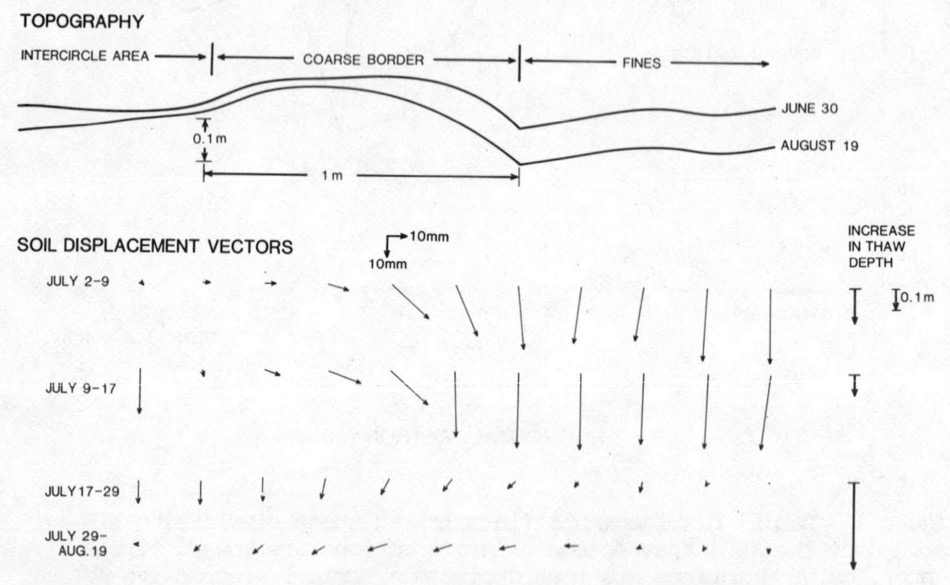

Figure 4. Topography, surface soil displacements and thaw depth (shown on right) as a function of time at stone circle K1W during the 1985 thaw season.

4.2 The Case for Free Convection of Soil

4.2.1. Dynamic Maintenance of the Micro-relief. The bulbous ridges of coarse material, and the sharp troughs that separate them from the domed fine domains, are prominent stone circle characteristics that appear to vary only slowly over periods on the order of 10^3 to 10^4 years (Hallet and Prestrud, in press). This micro-relief is striking in view of the numerous processes that tend to reduce the relief. These include gelifluction, pluvial erosion, and biogenic surface disturbances, all of which result in downslope transfer of material. Because relatively steep slopes converge to form the troughs at the fine/coarse contact, the troughs would tend to fill particularly quickly; yet they persist (Figure 3). A simple diffusional model of the overall erosional response, using the range of plausible effective topographic diffusivities 10^{-2} to 10^{-3} m^2/yr (from tabulations by Kenyon and Turcotte, 1985, p. 1464), suggest that the troughs at the fine/coarse contacts would essentially disappear in decades or, at most, centuries. Thus for this conspicuous micro-relief to persist for periods on the order of 10^3 to 10^4 years, degradational processes must be compensated by subsurface motions; in other words, the micro-relief must be dynamically maintained.

Given the expected lateral divergence of material from topographic highs, these domains can only remain in their raised positions through the upward motion of the underlying soils relative to adjacent areas. Thus both the convex upward fine center and the high coarse border would tend to reflect longterm upwelling of soil in these areas. Conversely the trough at fine/coarse contacts would reflect relative downward soil motion there. The simplest overall pattern of motion compatible with these constraints is illustrated in Figure 5. Note the inferred subsurface motions required to satisfy

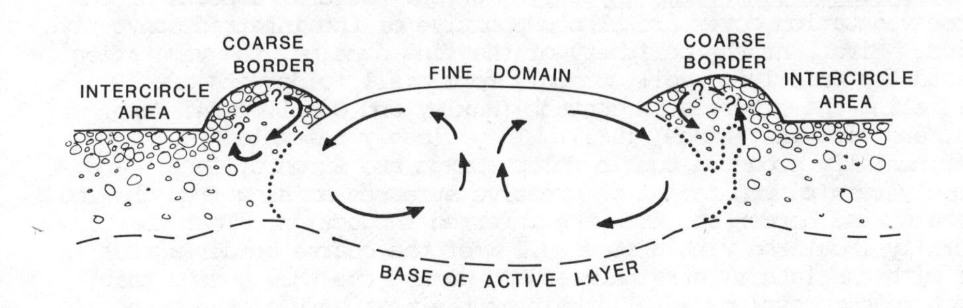

Figure 5. Simplest pattern of soil motion compatible with micro-relief, measured surface displacements, and indications of occasional diapirism under the borders of isolated sorted circles. The motion includes convection in the fine center, "subduction" at the fine/coarse contact, and probably rolling of the border.

the conservation of material. We stress that actual soil motions are likely to be much more complicated as they will reflect large fluctuating displacements associated primarily with freezing and thawing; nevertheless the pattern of net motions over a number of years can be deduced to be as illustrated in Figure 5.

4.2.2. Interpretation of Lateral Motions. The observed net radial strains during both the 1984 and 1985 thaw seasons (Figure 3) displayed identical complicated patterns with weak but pervasive radial extension in the fines, and a juxtaposition of marked compressive and tensile strains, respectively, in the interior and exterior portions of the border. Aside from the unlikely possibility that this strain pattern is systematically reversed after the thaw period, or that the micro-relief of circles is changing rapidly through time, such strains would in the long term reflect a general pattern of upward soil motion in the fine centers, and downward motion increasing from the outward portion of the fines to the interior portion of the coarse borders. Again, continuity requires a transfer of material at depth from under the borders to the center of the circles. The unexpectedly large motion of the border surface toward the center implies circulatory motion in the coarse material as well. The simplest overall pattern of motion compatible with the surface displacements is identical to that inferred from the microtopography (Figure 5).

Radial motions and tilts of dowels initially inserted vertically in the fine domain of sorted circles have also been measured by other workers, notably by Jahn (1963) in the Hornsund area of Spitsbergen, and, more recently, by A.L. Washburn (1985, personal communication) at Resolute, N.W.T., Canada. Markers are generally found to tilt outward in accord with the inferred convective motion of the fine-grained soil.

4.2.3. Inferred Long-term Convective Motion. Several aspects of the sparse vegetation cover are also supportive of the inferred convective motion. First, at the periphery of the fine domains, the vegetation mat and the underlying soil commonly form small folds with axes paralleling the fine/coarse contact (Huxley and Odell, 1924; Jahn, 1963, and Van Vliet-Lanoë, 1983). They closely resemble experimentally produced buckle folds (Dixon and Summers, 1985), and probably result from radial compressive stresses arising from the load of the coarse border and from the inferred subduction. That they are generally localized within about 0.1 m of the coarse border agrees well with the lateral strains measured during the thaw period that switch to compressive radial strain at the periphery of the fine domains (Figure 3). The folded nature of the vegetation mat implies that at least a portion of the convergence recorded during the thaw period near the fine/coarse contact persists in the long term.

Second, subduction under the coarse border, which is inferred from the micro-relief and from the extension of fine-grained material under the border, is expected to transport organic material into the active layer. In accord with other studies of patterned ground (e.g. Ugolini, 1966; Washburn, 1969), we find that organic carbon content does not decrease monotonically with depth, as expected in a stable, nonconvecting soil. Instead, throughout the active layer, the organic carbon content of the soil at the periphery of the fine domain is considerably higher than in the center. As material that has not been at the surface recently will presumably be depleted in biogenic carbon, the very low carbon content of the central portion of the fine domains throughout the entire sampled depth probably reflects upward motion there. That the vegetation cover is often discontinuous or absent in these areas, whereas it is thickest at the periphery of the fines, also reflects the expected radial increase in time available for biotic colonization. These observations all support the inferred pattern of soil motion shown in Figure 5.

Additional geologic data further support elements of long-term convective soil motion. Upward flow of material in the circle centers is compatible with the upward migration of stones and fossils to the surface (e.g. Gripp, 1926, 1927) in places forming islands of coarse material near the centers of the fine domains. Furthermore, the radial flow pattern characteristic of hexagonal convection cells is in accord with the preferred alignment in a radial direction of elongated rock fragments in sorted patterns (Schmertmann and Taylor, 1965).

4.2.4. Geometric Similarity Between Benard Cells and Sorted Circles. As it is primarily the regular geometry of sorted patterns that has long invited comparison with Benard cells (Nordenskjold, 1907; Wasiutynski, 1946), several geometric aspects of sorted circles merit further discussion. In plan view, the similarity between Benard cells and sorted patterns can be quite striking. Figure 6a shows a map of sorted patterns from Cornwallis Island, N.W.T. The fine-grained pattern centers, shown with the stippling, grade from equidimensional domains in a relatively flat area next to a small lake, to elongate forms on a slope that is slightly steeper and higher. Figure 6b shows a very similar cell form found in numerical simulations of free-convection (Bestehorn and Haken, 1983). The geometry of peripheral cells in the simulations reflects the imposed boundary conditions. In sorted patterns, elongated forms are likely to appear where shear forces, which result from the downslope component of gravity, approach the order of the buoyancy forces that drive convection. The transition from circles to stripes in our map area can indeed be shown to correspond to a zone where the estimated ratio of buoyancy force to shear force decreases markedly. The elongation of stone circles due to shear forces was recognized early (e.g. Wasiutynski, 1946, p. 204).

In vertical sections, several further similarities exist between sorted patterns and Benard convection cells. Diameters of stone circles range from 3 to 6 m in areas with active layer depths between 1 and 1.5 m. Assuming that the entire active layer convects, the diameter-to-depth ratio of stone circles is in good agreement with the average aspect ratio found experimentally to be about 3.3 for hexagonal convection cells (Benard, 1901). The central domain of stone circles is invariably convex upward, in accord with theoretical analysis (Jeffreys, 1951) and experimental studies (Ceriser et al., 1984) of the deformation of the free surface by buoyancy-driven convection. That this free surface deflection is opposite to that observed by Benard (1901) is now recognized to result from the fact that Benard cells were primarily caused by a gradient in surface tension, and not by buoyancy (Scriven and Sternling, 1964). Similarly, although Benard found that the highest points on the free surface of the fluid are at the tips of the hexagon, buoyancy-driven convection is likely to result in the lowest points at hexagon tips.

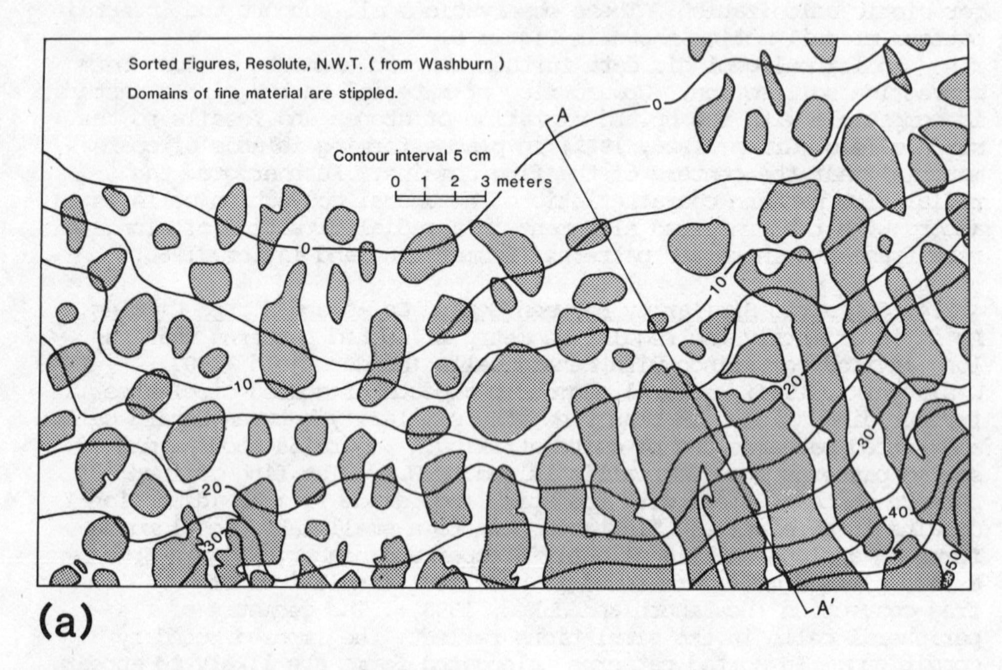

Figure 6. Patterned-ground (a) resembles closely pattern (b) obtained in a numerical simulation of free convection (taken from Bestehorn and Haken, 1983). In accord with the inferred motion, the fine domains correspond to domains of calculated upward velocities; both are stippled. Elongated forms appear in the model due to boundary conditions, and in nature due to increasing slope. The mapped pattern was derived from A.L. Washburn's map, produced by R.S. Anderson, C.G. Gregory, and the author. Topography is shown with 5 cm contours.

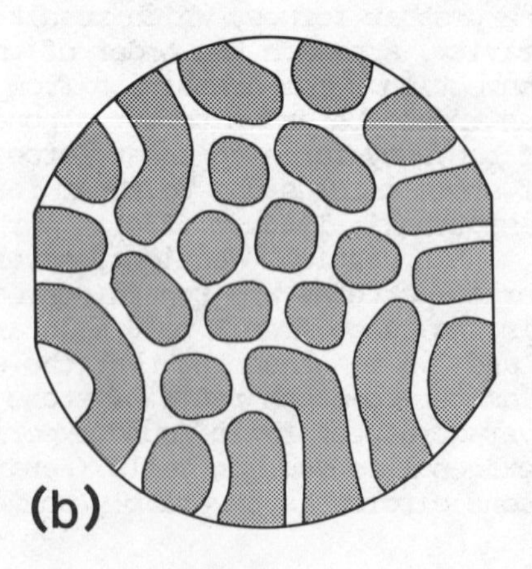

Indeed, the lowest areas in sorted patterns tend to occur between three circles, usually at triple junctions of coarse-grained ridges.

The raised position of the coarse-grained borders appears to result primarily from the accumulation of stones at the border. Freezing and thawing of the soil causes stones to move relative to the soil generally in the direction of heat flow. As heat flow is primarily upward, stones are brought to the surface. This process is termed up-freezing. Stones at the surface will tend to proceed to the borders because surface processes tend to move them downslope toward the fine/coarse contact, and because the soil itself is moving in that direction. Stones accumulate at the circle borders because the rate of stone up-freezing is likely to exceed the rate of downward soil motion there. Thus the up-freezing process "may be considered as a substitute for buoyancy, and the stones behave as if floating on the mud" (Wasiutynski, 1946).

4.2.5. <u>Driving Mechanism for Soil Convection.</u> In view of the fact that the early work on free convection dealt with fluid motion driven by temperature-induced density differences, it is natural that early considerations of soil convection focussed on temperature gradients for a mechanism to drive such motion (e.g. Romanovsky, 1941). It was recognized early that the density of water could be expected to decrease during the thaw season because near the surface it would tend to be a few degrees above 0°C while at the base of the active layer it would remain at 0°C, the ground being frozen below. Simple calculations show, however, that the maximum variation in bulk density of the soil that could result from varying the water temperature from 0°C to 4°C is insufficient to induce convection (Wasiutynski, 1946). Such small bulk density differences, on the order of one part in 10^4, are likely to be completely overshadowed by differences in bulk density arising from realistic variations in soil moisture or texture (Mortensen, 1932, Dzulynski, 1966). Moreover, it was recognized that differences in soil moisture may actually provide the driving mechanism for convective motion of the soil (Mortensen, 1932; Gripp and Simon, 1934; and Sorensen, 1935). A compelling case for soil convection due to vertical gradients in water content was presented by Wasiutynski (1946, p. 202) who clearly recognized that "...if we substitute water content for temperature, the equations of motion in the mud layer will be the same as those of the problem of thermally maintained instability". Convection could be maintained by a supply of water at the base due to ice melting and removal at the soil surface.

A limitation of this work is that no precise mechanism emerged to account for the requisite moisture-induced decrease in soil bulk density with depth. A possible solution becomes apparent through more explicit consideration of active layer processes. Heave records indicate that the soil surface heaves upward on the order of 0.1 m each seasonal freeze period; this implies that the soil in a 1-m thick active layer expands on the average by about 10% seasonally. As this volumetric expansion is due primarily to water intake and subsequent ice lensing, the average bulk density at the end of the freezing season, which is about $2x10^3$ kg/m^3, must be about 5% lower than at the onset of freezing. Note that this expansion will be relatively large in fine-grained soils with abundant moisture as they tend to be particularly favorable for ice lensing. It follows that the average soil bulk density increases throughout the thaw season by about 5% as excess pore water is expulsed; this is over 500 times the maximum soil bulk density difference that would arise from cooling pore water from 4°C to 0°C. Soil densification starts near the surface where soil thaws first, and it progresses downward at a rate determined by the rate of upward water percolation above the thaw front. As illustrated in Figure 7, near-surface consolidation may naturally give rise to denser material near the surface than at depth where water migration has been limited due to the relatively short period during which the material has been in a thawed state. The resulting unstable density gradient would be further accentuated by an increase in ice content with depth, which is often reported (e.g. Smith, 1985). Preliminary measurements of soil bulk density in sorted circles show that the density does decrease with depth under parts of a sorted circle, but more data are needed to substantiate this (Hallet and Prestrud, in press).

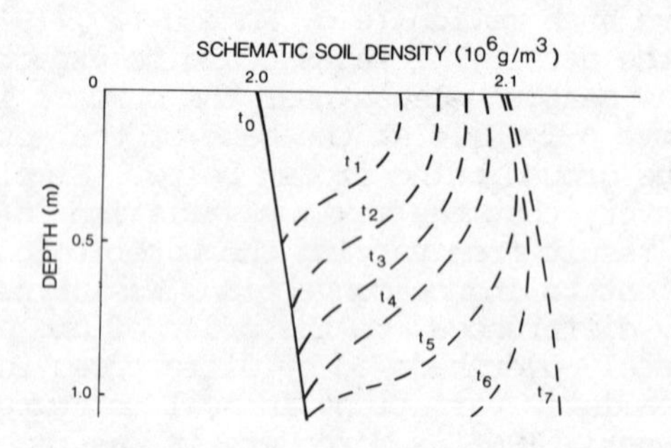

Figure 7. Inferred pattern of soil consolidation during the thaw period. At time t_0, the soil is completely frozen and the soil bulk density is low because of high ice content. At time t_1 the upper 0.5 m of soil has thawed, and the near surface bulk density has increased substantially because the soil consolidates, as water drains from the soil pore space. Near the thaw front, relatively little drainage has occurred because little time has elapsed since the soil was frozen. Hence, consolidation decreases as one approaches the thaw front. The inferred bulk density increases through the thaw season, but at any time from t_1 to t_6, it is inferred to decrease with depth.

4.2.6. Effective Rayleigh Number. Useful insights into the postulated convective motion of periglacial soils can be gained by idealizing soil as a very viscous Newtonian fluid and estimating the effective Rayleigh number:

$$Ra = \frac{\Delta\rho \; g \; h^3}{\eta \; \kappa}$$

where g is the gravitational acceleration, $\Delta\rho$ is the density difference across the fluid layer, h is the depth of the unstable layer, η is the effective viscosity of the material, and κ is a diffusivity that parameterizes the dissipation of density differences with time. As convection appears to be dictated by moisture-related differences in bulk density, κ would naturally be the hydraulic diffusivity, as recognized implicitly by Wasiutynski (1946).

Preliminary estimates of the parameters are 10^2 kg/m^3 for $\Delta\rho$, about 5% of the mean bulk density; 1 m for h, approximating the active layer depth; 10^{-7} m^2/s for κ, a representative value for frost-susceptible soils (Morgenstern and Smith, 1973); and 10^8 Pa-s for η based on measurements of radial velocity at the surface of a sloping layer of thawed fine-grained soil. Considering the order of magnitude uncertainties in the last two parameters, and the difficulty in idealizing cohesive fine-grained soil as a simple Newtonian viscous fluid, it is of interest that the calculated Rayleigh number is of order 10^2, within only one order of magnitude of the critical value. This suggests that soil convection is plausible in soils with slightly lower effective viscosities or hydraulic diffusivities than assumed in this preliminary calculation. Efforts are currently underway to obtain more precise values of $\Delta\rho$, η, and κ, and to examine the implications of our rheological assumptions.

An explanation now emerges for the contrast between the abundance of sorted circles in our study areas in wet swales rich in fine material and their absence from adjacent beach ridges. Convection would be strongly favored in fine grained materials because they would tend to have relatively low effective viscosities, and low hydraulic diffusivities due to their low permeability. Moreover, the moist frost-susceptible fine grained soils are likely to contain considerable ice, leading to a high $\Delta\rho$ during the thaw period.

In contrast, on the beach ridges, the very coarse-grained soils, the relatively low soil moisture and the thinner snow cover are all unfavorable for ice lensing. Pattern initiation on the gravelly ridges is likely to be limited severely by weathering inasmuch as considerable fine-grained material is required to reduce permeability and effective viscosity, and to produce conditions favorable for large frost-induced expansion.

4.3 Free Convection of Water in the Active-Layer?

Although the expected decrease in water density with depth in the active layer is incapable of causing soil convection, Ray et al. (1983) recognized that this density gradient could induce percolative convection of water through the soil. Moreover, they proposed that such water convection could affect the pattern of melting at the bottom of the active layer. The resulting scalloped geometry of the base of the active layer is then inferred to dictate the development of certain patterned ground. Much as is the case for soil convection, pore water convection is an attractive mechanism because it is in accord with the regularity, shape and size of stone circles. However, this mechanism only addresses the geometry of the incipient patterns; the pattern of sorting, the development and maintenance of the features with distinct micro-relief, and the measured surface displacements of the soil cannot be addressed directly. Moreover, whether percolative convection of water can be sufficiently vigorous to affect heat transport significantly in fine-grained soil typical of sorted circles is questionable.

5. CONCLUSIONS

Self-organization appears in many facets of landscapes. The many geomorphic examples of stochastic and deterministic order reviewed here constitute an area of research ripe for the application of dynamical systems analysis because of the inherent complexity of the systems. As we learn more about individual systems, accurate governing equations should emerge. Because they are likely to be difficult to solve directly, qualitative analysis of the equations promises to shed considerable light on system behavior, particularly on transitions from simple to complex behavior. For example, modern analysis of convection instability helps explain how, with an increasing Rayleigh number, orderly cellular convection patterns evolve, and eventually break down to chaotic behavior. Perhaps this evolution is evident in stone circles that grade from circular forms to complicated curvilinear forms with suggestions of segmentation and secondary flows. This would perhaps constitute the first reported example in geomorphology of a transition from orderly to chaotic patterning (Culling, 1985).

6. ACKNOWLEDGEMENTS

I benefitted from many stimulating discussions with R.S. Anderson, who shares my interest in geomorphic patterns, and with A.L. Washburn, who is very knowledgeable about patterned ground. I thank A.L. Washburn for the use of his unpublished map to derive figure 6A; R.S. Anderson and A. J. Heyneman, P. Leaming and A.L. Washburn for their careful reading of the manuscript; and L. Carothers and S. Rasmussen for their typing. The recent work on stone circles emanates from a collaborative effort involving S. Prestrud, C. Gregory and C. Stubbs. It was rendered possible by funding from the National Science Foundation (NSF DPP-8303630) and by logistical support from the Norwegian Polar Research Institute.

7. REFERENCES

Anderson, R.S. and Hallet, B. (1986) 'Sediment transport by wind: Toward a general model'. Geol. Soc. Am. Bull. 97.

Benard, H. (1901) 'Les tourbillons cellulaires dans une nappe liquide'. Rev. Gen. des Sci. 11, 1261-1271, 1309-1328.

Bestehorn, M. and Haken, H. (1983) 'A calculation of transient solutions describing roll and hexagon formation in the convection instability'. Physics Letters. **99A**, 265-267.
Callander, R.A. (1978). 'River Meandering'. Ann. Rev. Fluid Mech. **10**, 129-158.
Cerisier, P., Jamond, C., Pantaloni, J., and Charmet, J.C. (1984) 'Deformation de la surface en convection de Benard - Marangoni'. Jour. Physique, **45**, 405-411.
Culling, W.E.H. (1985) 'Equifinality: Chaos, Dimension and Pattern' New Series Discussion Paper 19, Department of Geography, London School of Econmics, Houghton Street, London WC2A 2AE, 83p.
Czeppe, Z. (1961) 'Annual course of frost ground movements at Hornsund (Spitsbergen) 1957-1958'. Zeszyty Naukowe Uniwersytetu Jagiellońskiego, Prace Geograficzne **3**, 50-63.
Dixon, J.M. and Summers, J.M. (1985) 'Recent developments in centrifuge modelling of tectonic processes: equipment, model construction techniques and rheology of model materials'. Jour. Struc. Geol. **7**, 83-102.
Dolan, R., Vincent, L. and Hayden, B. (1974) 'Crescentic coastal landforms'. Zeit f. Geomorph. **18**, 1-12.
Dzulynski, S. (1966) 'Sedimentary structures resulting from convection-like pattern of motion'. Rocz. Pol. Tow. Geol. **36**, 3-21.
Elton, C. (1927) 'The nature and origin of soil-polygons in Spitsbergen'. Quatr. Jour. Geol. Soc. London **83**, 163-194.
Greeley, R., Womer, M.B., Papson, R P. and Spudis, P.D. eds. (1978). 'Aeolian Features of Southern California: a comparative planetary geology guidebook'. NASA. U.S. Government Printing Office, Wash. D.C.
Gripp, K. (1926) 'Über Frost und Strukturboden auf Spitzbergen'. Ztschr. d. Ges. f. Erdkunde zu Berlin, 351-354.
____ (1927) 'Beitrage zur Geologie von Spitzbergen'. Naturwiss. Ver Hamburg, Abh **21**, 1-38.
Gripp, K. and Simon, W. (1934) 'Die experimentelle Darstellung des Brodelbodens'. Naturwiss. **22**, 8-10.
Guza, R.T. and Inman, D.C. (1975) 'Edge waves and beach cusps'. Jour. Geophys. Res. **80**, 2997-3012.
Hallet, B. and Prestrud, S. 'Dynamics of periglacial sorted circles in Western Spitsbergen'. Submitted to Quaternary Research.
Horton, R.E. (1945) 'Erosional development of streams and their drainage basins: hydrophysical approach to quantitative morphology'. Geol. Soc. America Bull. **56**, 275-370.
Huxley, J.S. and Odell, N.E. (1924) 'Notes on surface markings in Spitsbergen'. Geog. Jour. **63**, 207-229.
Jahn, A. (1963) 'Origin and development of patterned ground in Spitsbergen'. In "Proceedings of the First International Conference on Permafrost", pp. 140-145. Lafayette, Indiana.
Jeffreys, H. (1951) 'The surface elevation in cellular convection'. Quart. Jour. Mech. Appl. Math **4**, 283-288.

Kenyon, P.M. and Turcotte, D.C. (1985) 'Morphology of a delta prograding by bulk sediment transport'. Geol. Soc. Am. Bull. 96, 1457-1467.

Lachenbruch, A.H. (1962) 'Mechanics of thermal contraction cracks and ice-wedge polygons in permafrost'. Geol. Soc. Am. Special Paper 70.

Langbein, W.B. and Leopold, L.B. (1966) 'River meanders - theory of minimum variance'. U.S. Geol. Surv. Prof. Paper 422H, 15 p.

Leopold, L.B. and Langbein, W.B. (1962) 'The concept of entropy in landscape evolution'. U.S. Geol. Surv. Prof. Paper 500A, 20 p.

_____ (1966) 'River meanders'. Sci. Am. 214, June issue, 61-70.

Low, A.R. (1925) 'Instability of viscous fluid motion'. Nature 115, 229-300.

Mandelbrot, B.B. (1982) The Fractal Geometry of Nature, W.H. Freeman, San Francisco, 461 p.

Morgenstern, N.R., and Smith, L.B. (1973) 'Thaw-consolidation tests on remoulded clays'. Can. Geotechn. Jour. 10, 25-40.

Mortensen, H. (1932) 'Uber die physikalische Moglichkeit der "Brodel" Hypothese'. Centralblatt f. Min., Geol. u. Palaont. Abt. B, 417-422.

Nelson, J.M. and Smith, J.D. (1985) 'Numerical prediction of meander evolution', EOS Transactions, Am. Geophys. Union 66, 910.

Nicholson, F.H. (1976) 'Patterned ground formation and description as suggested by Low Arctic and Subarctic examples'. Arctic and Alpine Res. 8, 329-342.

Nordenskjold, O. (1907) 'Uber die Natur der Polarlander'. Geogr. Ztschr., Jhg. 13, 563-566.

Pissart, A. (1966) 'Experiences et observations a propos de la genese des sols polygonaux tries'. Revue Belge de Geographie 90, 55-73.

Ray, R.J., Krantz, W.B., Caine, T. N., and Gunn, R. D. (1983) 'A model for sorted patterned-ground regularity'. Jour. Glac. 29, 317-337.

Ritter, D.F. (1978) Process Geomorphology. Wm. C. Brown, Dubuque, Iowa. 603 p.

Romanovsky, V. (1941) 'Application du criterion de Lord Rayleigh a la formation des tourbillons convectifs dans les sols polygonaux du Spitsberg'. Comptes Rendus Acad. Sciences Paris 211, 877-878.

Schmertmann, J.H. and Taylor, R.S. (1965) 'Quantitative data from a patterned ground site over permafrost'. U.S. Army Cold Regions Research and Laboratory Research Report 96, 76p.

Scriven, L.E. and Sternling, C.V. (1964). 'On cellular convection driven by surface-tension gradients: effects of mean surface tension and surface viscosity'. Jour. Fluid Mech. 19, 321-340.

Shreve, R.L. (1966). 'Statistical law of stream numbers'. Jour. of Geology, 74, 17-37.

_____ (1975) 'The probabilistic-topologic approach to drainage-basin geomorphology'. Geology 3, 527-529.

Smith, J.D. and Mclean, S.R. (1976) 'Spatially averaged flow over a wavy boundary'. Jour. Geophys. Res. 82, 1735-1746.

Smith, M.W. (1985) 'Observations of soil freezing and frost heave at Inuvik, Northwest Territories, Canada'. Can. Jour. Earth Sci. 22, 283-290.

Sorensen, T. (1935) 'Bodenformen und Pflanzendecke in Nordostgronland'. Meddelelser om Gronland, 93.
Speight, J.G. (1965) 'Meander spectra of the Anabunga River'. Jour. Hydrol. 3, 1-15.
Timoshenko, S. and Goodier, J.N. (1951) Theory of Elasticity. McGraw-Hill Book Co., New York, 506 p.
Ugolini, F.C. (1966) 'Soils of the Mesters Vig district, Northeast Greenland: the Arctic Brown and related soils'. Meddelelser om Gronland 176, 1-22.
Van Vliet-Lanoe, B. (1983) 'Etudes cryopedologiques au sud du Kongsfjord - Svalbard'. Centre de Geomorpholgie du C.N.R.S., Caen, France.
Vogel, S. (1983) Life in moving fluids. Princeton Univ. Press, Princeton, New Jersey.
Washburn, A.L. (1956) 'Classification of patterned ground and review of suggested origins'. Geol. Soc. Am. Bull., 67, 823-865.
____ (1969) 'Patterned ground in the Mesters Vig district, Northeast Greenland'. Biuletyn Peryglacjalny 18 , 259-330.
____ (1980) 'Geocryology'. Wiley, New York.
Wasiutynski, J. (1946) 'Studies in hydrodynamics and structure of stars and planets'. Astrophysica Norvegica 4, p. 1-497.
Wilson, R.G. (1972) 'Aeolian bedforms - their development and origins'. Sedimentology 19, 173-210.

PART VI

CONCLUDING REMARKS

LOW-ORDER MODELS AND THEIR USES

Edward N. Lorenz
Center for Meteorology and Physical Oceanography
Massachusetts Institute of Technology
Cambridge, Massachusetts, U.S.A.

1. INTRODUCTION

I have been asked by the organizer of this multidisciplinary gathering to present a few closing remarks. I could attempt to summarize everything that has transpired here, but any such effort would necessarily be superficial. Instead I shall discuss a specific topic which appears to be of fairly general relevance.

In a good number of the communications presented, the results have involved the use of low-order models. The discussions that have followed the presentations have suggested that the rationale for these models has not always been appreciated by the audience. I have therefore decided to speak about the construction and potential use of low-order models. My discussion will deal with the procedures commonly used in meteorology, and the illustrative examples will be drawn from meteorology; however, the methods described should be equally appropriate in any science where the basic laws are well enough known to allow them to be formulated as a system of equations. In particular, they should be applicable throughout the geophysical sciences.

2. CONSTRUCTION OF LOW-ORDER ATMOSPHERIC MODELS

Atmospheric models usually consist of systems of equations that are supposed to approximate to some degree the physical laws governing the atmosphere. This is in contrast to some other disciplines, where the models may consist of empirically determined equations, or simply postulated relationships. In a sense all systems of equations used in atmospheric dynamics are approximations, and are therefore models; I am unaware of any recent studies where, for example, the earth's surface, aside from orographic features, has been treated as an ellipsoid rather than a sphere. However, when the purpose of a study is to obtain qualitative results, or even quantitative results where departures from reality up to a factor of about two are acceptable, much more drastic simplifications are allowable.

C. Nicolis and G. Nicolis (eds.), Irreversible Phenomena and Dynamical Systems Analysis in Geosciences, 557–567.

One of the commonest simplifications is to replace the earth's spherical surface by a plane, thus permitting the use of rectangular coordinates. Topographic features are often omitted altogether. A Coriolis force is introduced to take into account the effect of the earth's rotation. The rationale is that systems that would develop above such a plane are likely not to differ too greatly from those that actually form above the earth. Another common simplification is to treat the atmosphere as an ideal gas, neglecting the presence of water in its various phases. It is assumed that in a dry atmosphere the global-scale currents, once formed, would behave much as they actually do, although such systems as tropical cyclones, which depend upon water for their formation and maintenance, would have no counterparts.

Two other common simplifications are the hydrostatic approximation, which specifies a permanent balance between gravity and the vertical pressure force, and the geostrophic approximation, which balances the Coriolis force with the horizontal pressure force. Each of these approximations replaces a prognostic equation, which expresses the time derivative of one dependent variable in terms of the set of variables, by a diagnostic equation, which expresses the contemporary value of one variable in terms of the others. Each approximation effectively reduces the number of dependent variables.

Whether or not the above simplifications are introduced, the model consists at this point of a system of partial differential equations (PDE's). Numerical methods of dealing with PDE's are becoming increasingly common. Numerical solution requires each dependent variable to be replaced by a number of new variables which are functions of time alone; these are often the values of the original variable at a prechosen grid of points. Each PDE is then replaced by a set of ordinary differential equations (ODE's) governing the new variables. The total number of ODE's serves as a convenient measure of the approximate size of the model. Ultimately the ODE's will be replaced by difference equations.

A low-order model is one where the number of ODE's is very small. Physical simplifications may in some cases reduce the number of equations by almost an order of magnitude, but the greatest savings come from drastic reduction of the horizontal and vertical resolution.

When the resolution is barely sufficient to capture the features of interest, finite differences do not afford good approximations to the partial derivatives that they are supposed to represent. The usual procedure is therefore to transform the PDE's into spectral form; this is done by expressing the field of each dependent variable as a series of orthogonal functions, such as multiple Fourier series or spherical harmonics, and letting the coefficients in these series be the variables in an infinite system of ODE's. This system is then truncated by discarding all but a finite number of variables and equations; for a low-order model this number is very small. Usually the retained variables are the coefficients of the orthogonal functions of largest spatial scale, although selective truncation is sometimes used. Partial-derivative fields are obtained by differentiating the orthogonal functions, and no spatial differencing is needed.

in magnitude to the wind speed. We treat the atmosphere as an ideal gas. We introduce the hydrostatic and geostrophic approximations. We omit all thermal and mechanical forcing and damping; by so doing we forgo the possibility of explaining the presence of the westerlies, and simply take their existence for granted.

We next confine our attention to flow patterns in which there are no vertical variations of the wind. In this way we effectively eliminate the vertical coordinate as an independent variable. We then find that the two-dimensional flow is free of divergence, so that it may be expressed in terms of a stream function ψ, while individual values of the vorticity ζ are conserved, i.e., ζ remains fixed at any point moving with the flow. Introducing x- and y-axes pointing eastward and northward, so that

$$\zeta = \frac{\partial^2 \psi}{\partial x^2} + \frac{\partial^2 \psi}{\partial y^2} = \nabla^2 \psi, \tag{1}$$

and letting t denote time, we find that the system of governing equations reduces to a single PDE--the familiar vorticity equation

$$\frac{\partial \nabla^2 \psi}{\partial t} = - \frac{\partial \psi}{\partial x} \frac{\partial \nabla^2 \psi}{\partial y} + \frac{\partial \psi}{\partial y} \frac{\partial \nabla^2 \psi}{\partial x} \tag{2}$$

containing a single dependent variable ψ. In a slightly modified form, the vorticity equation was actually used at one time for operational weather forecasting, despite its obvious shortcomings.

We next transform the equation into spectral form by letting

$$\psi(x,y,t) = \sum_{k,\ell=-\infty}^{\infty} \psi_{k,\ell}(t) \exp i(kx + \ell y) ; \tag{3}$$

ψ will be real if $\psi_{k,\ell}$ and $\psi_{-k,-\ell}$ are complex conjugates. We obtain the infinite system of ODE's

$$\frac{d\psi_{k,\ell}}{dt} = (k^2 + \ell^2)^{-1} \sum_{m,n=-\infty}^{\infty} (m\ell - nk)(m^2 + n^2)\psi_{k-m,\ell-n}\psi_{m,n}. \tag{4}$$

We now note that parallel belts of westerly and easterly winds may be described by the terms in eq. (3) containing $\psi_{0,L}$ and $\psi_{0,-L}$, where L is any single value of ℓ. Likewise, variations with longitude are captured by the terms containing $\psi_{K,0}$ and $\psi_{-K,0}$, where K is a single value of k. If we choose initial conditions for eq. (4) in which the only nonvanishing variables are $\psi_{K,0}$ and $\psi_{0,L}$ and their complex conjugates, we find tht the only variables whose time derivatives differ from zero are $\psi_{K,L}$ and $\psi_{K,-L}$ and their complex conjugates.

We can therefore convert eq. (4) into a non-trivial low-order model by retaining only the variables $\psi_{K,0}$, $\psi_{0,L}$, $\psi_{K,L}$, and $\psi_{K,-L}$ and

Low-order models cannot be expected to produce good weather forecasts in real situations, and their main use is in theoretical work. Their most obvious advantage is the large saving in computation time which they afford--this is especially important when the more powerful computers are not available--but, because they also minimize the numerical output, they can make the subsequent interpretation much easier. Ideally a low-order model should be tailored to fit the particular phenomenon, such as the intensification of middle-latitude cyclonic storms, to which it is to be applied. Physical processes that are patently irrelevant are best omitted. Afterward, only enough variables need be retained for an adequate representation of the phenomenon.

If the purpose of the model is simply to describe an already understood phenomenon, perhaps for instructional purposes, the number of variables may be the minimum needed for its description. If instead the model is to be used in an attempt to explain a phenomenon, or, better, to test the hypothesis that a particular process is responsible for the phenomenon of interest, less extreme simplification is generally called for. Care must be taken not only that the process being tested is unambiguously described, but also that alternative processes, which might be the ones actually responsible for the phenomenon, are included, since otherwise the model would be unable to choose among the various processes, and might be forced to accept the hypothesis.

We shall present two examples of low-order atmospheric models. The first model is used to examine the influence of superposed large-scale vortices on a globe-encircling westerly wind current. Its purpose is descriptive, so it need not include other processes that might affect the current. The second model is designed to investigate the maintenance of approximate geostrophic balance in middle and high latitudes. Clearly a model using the geostrophic approximation, which does not permit geostrophic unbalance, is not suitable for the purpose, and a so-called primitive-equation model, where the wind and pressure fields are not diagnostically related, is used instead.

3. A DESCRIPTIVE MODEL

A prominent feature of the atmospheric circulation is the presence of a belt of westerly winds in the middle latitudes of either hemisphere. These winds undergo continual fluctuations in intensity. Observations indicate that a major process in producing these fluctuations is the horizontal transport of eastward momentum into or out of these belts by the large-scale superposed vortices. We shall describe the construction of a low-order model which displays the working of this process. Since we are not attempting to explain why other processes, such as large-scale overturning, are not equally important in producing the fluctuations, we do not require a model that includes these other processes.

We begin by introducing some of the commonly used physical simplifications. We replace the earth's spherical surface by an infinite plane, and introduce a horizontal Coriolis force proportional

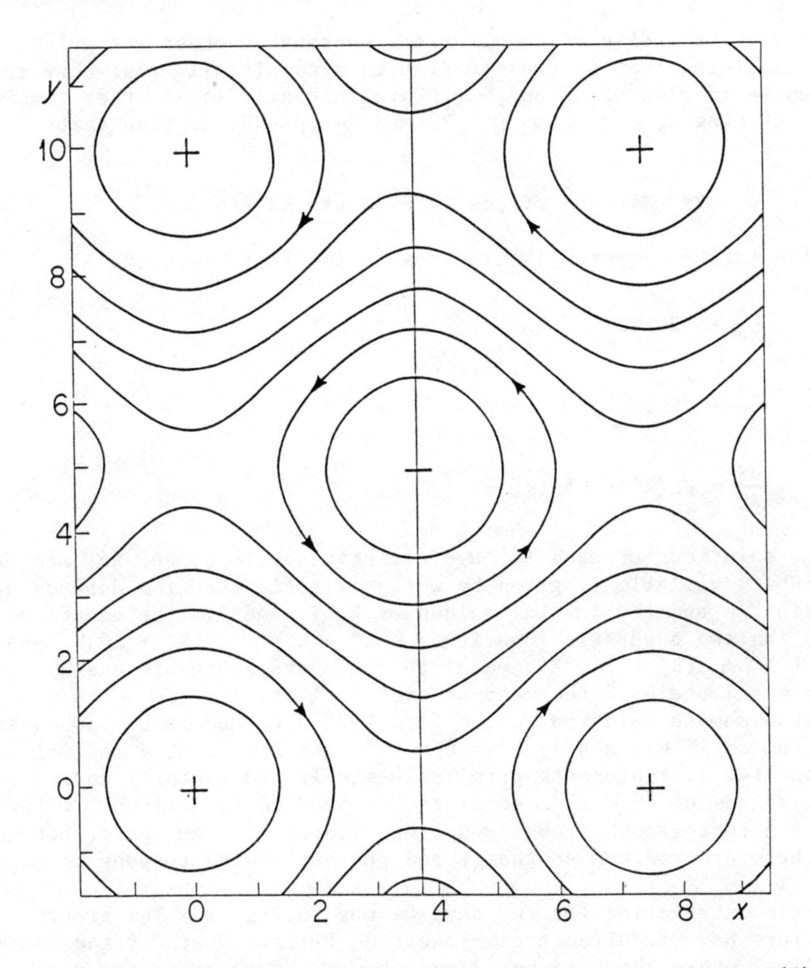

Fig. 1. Initial streamlines for the solution of eqs. (6) described in the text. The arrowheads indicate the direction of flow. The stream-function interval is $4 \times 10^6 \text{ m}^2\text{s}^{-1}$. The thin line extending from the bottom to the top is a trough line. The x- and y-scales are in thousands of km.

their conjugates, discarding all terms containing other variables. We note in addition that if these variables are initially real they remain real, while if also $\psi_{K,-L}$ and $-\psi_{K,L}$ are initially equal, they remain equal. Letting $\psi_{K,0} = X$, $\psi_{0,L} = Y$, and $\psi_{K,-L} = Z$, we find that eq. (3) becomes

$$\psi = 2X \cos Kx + 2Y \cos Ly + 4Z \sin Kx \sin Ly \ , \qquad (5)$$

while the infinite system (4) reduces to the finite system

$$\frac{dX}{dt} = -2KLYZ, \qquad (6a)$$

$$\frac{dY}{dt} = 2KLXZ, \qquad (6b)$$

$$\frac{dZ}{dt} = KL(K^2 - L^2)(K^2 + L^2)^{-1}XY. \qquad (6c)$$

The solutions of eqs. (6) are elliptic functions sn, cn, and dn of time. Which variable is given by which elliptic function depends upon the ratio K/L and the initial values of X, Y, and Z. The equations possess the two quadratic invariants $K^2X^2 + L^2Y^2 + 2(K^2 + L^2)Z^2$ and $K^4X^2 + L^4Y^2 + 2(K^2 + L^2)^2Z^2$, equal to the average kinetic energy per unit mass and one half the mean-square vorticity.

For a sample solution we let $2\pi/K = 7500$ km and $2\pi/L = 10000$ km, and we choose $KX = 6$ m s^{-1}, $LY = 8$ m s^{-1}, and $Z = 0$. The initial state, shown in Fig. 1, represents parallel westerly and easterly currents with central speeds of 16 m s^{-1}, separated by 5000 km in latitude, with superposed north-south trough and ridge lines, 3750 km apart, between which there are maximum southerly and northerly wind components of 12 m s^{-1}.

After integrating for two days we obtain Fig. 2. The trough and ridge lines have acquired cross-longitude tilts. South of the maximum westerlies, where the arrowheads are shown, the contours are more closely spaced in the northward flowing air than the southward flowing air, while north of the maximum westerlies they are more closely spaced in the southward flowing air. There is thus a net transport of eastward momentum into the belt of westerlies. As a consequence, the westerlies have increased in strength. Farther north, the easterlies have also become stronger. This is revealed by the numerical values $KX = 3.75$, $LY = 8.74$, and $(K^2 + L^2)^{1/2}Z = 2.19$ m s^{-1}; the westerly and easterly currents, represented by Y, now account for a larger fraction of the kinetic energy.

Continued integration reveals that the westerlies reach their maximum speed of 18.36 m s^{-1} after 3.52 days and return to their original strength after 7.04 days, thereafter repeating the cycle. Integrations with other initial values of X, Y, and Z or other values of K and L reveal how these values affect the period and amplitude of the fluctuations. A detailed description of these aspects of the model is given elsewhere [1].

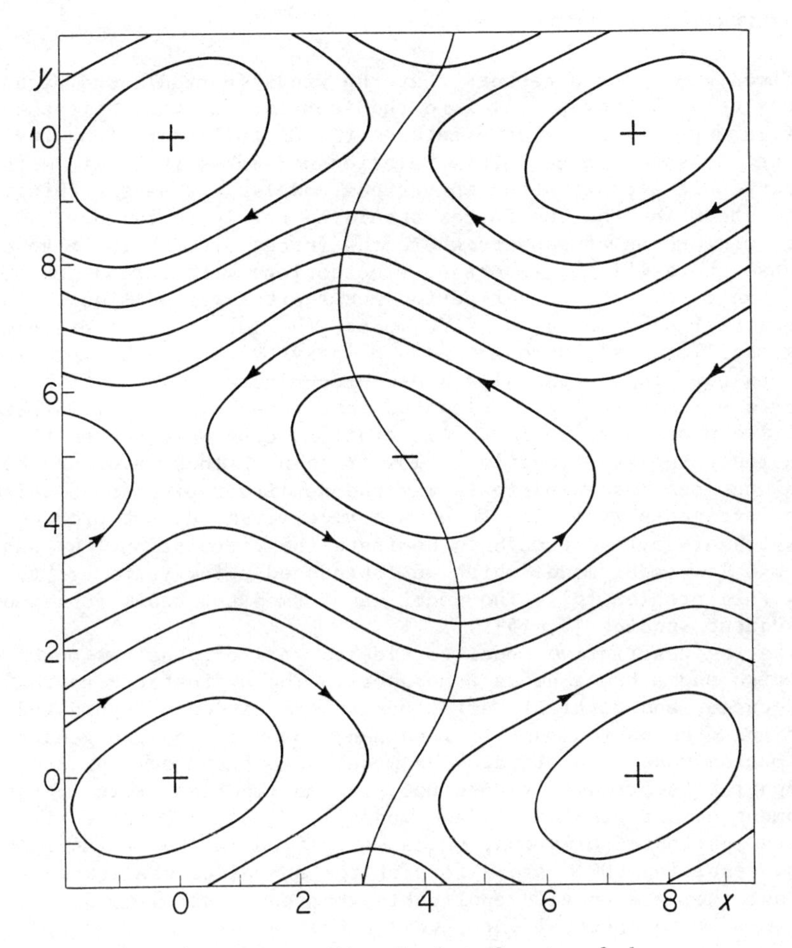

Fig. 2. The same as Fig. 1, but for t = 2 days.

4. AN INVESTIGATIVE MODEL

It has been known for a century that the winds in middle and high latitudes tend to blow parallel to the isobars, and that this phenomenon implies an approximate balance between the Coriolis force and the horizontal pressure force. This relationship makes it feasible to use the geostrophic approximation in various models. It says nothing, however, about why the two forces should be nearly in balance. One can, in fact, picture an atmosphere where the forces are not in balance, in which case there will be important fluctuations with periods of hours rather than days. The longer-period and shorter-period fluctuations are sometimes call slow modes and fast modes; the latter are inertial-gravity oscillations, commonly known as gravity waves. The problem at hand is to explain why the slow modes predominate.

It is evident that both slow and fast modes will be diminished by dissipative processes, while the external forcing which keeps the circulation going is primarily of low frequency, thus favoring the slow modes. What one must explain is why the nonlinear processes, which can produce fast modes from slow modes and vice versa, do not produce fast modes which are strong enough to dominate the circulation. We shall describe a low-order model which we introduced a few years ago to address this problem [2]. The model has formed the basis for a number of subsequent studies [3,4,5].

Like the descriptive model considered earlier, the new model uses a plane earth and a homogeneous atmosphere. The hydrostatic approximation is introduced, and vertical variations of the wind are suppressed. Forcing and dissipation must be retained, however, and the geostrophic approximation must be avoided. The model also possesses surface topography; this presumably does not play an important role in the development or suppression of fast modes, but it can be effective in producing aperiodic solutions, which we desired in the original study.

The resulting PDE's are a form of the so-called shallow-water equations; they are equally applicable to a gas and a liquid. To transform them to spectral form, we let

$$\chi = \chi_0 \sum_{j=1}^{\infty} x_j \exp i(k_j x + \ell_j y), \tag{7c}$$

$$\psi = \psi_0 \sum_{j=1}^{\infty} y_j \exp i(k_j x + \ell_j y), \tag{7b}$$

$$p = p_0 \sum_{j=1}^{\infty} z_j \exp i(k_j x + \ell_j y), \tag{7c}$$

where χ, ψ, and p are the velocity potential, stream function, and pressure, χ_0, ψ_0, and p_0 are dimensional constants, and the values of k_j and ℓ_j are chosen so that $k_1 + k_2 + k_3 = 0$, $\ell_1 + \ell_2 + \ell_3 = 0$, and $k_1\ell_2 - \ell_1 k_2 = 0$ but are otherwise unrestricted. We then truncate the

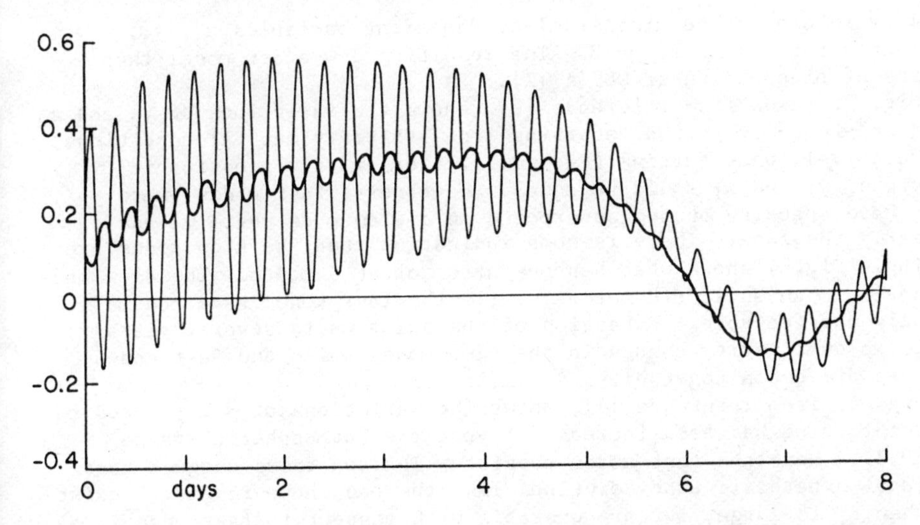

Fig. 3. Variations of y_1 (heavy curve) and z_1 (thin curve) during the first 8 days of a numerical solution of the 9-variable primitive-equation model, with $F_1 = 0.1$. The values are dimensionless.

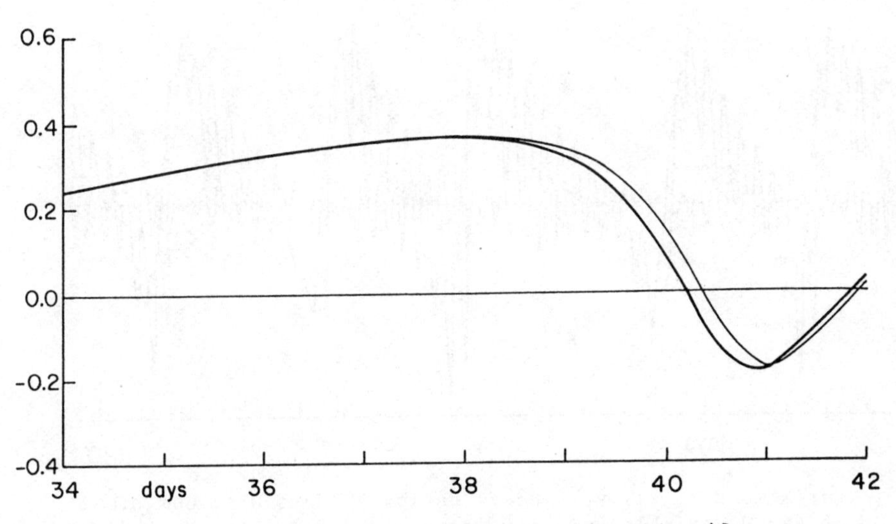

Fig. 4 The same as Fig. 3, but from day 34 to day 42.

system by retaining the dimensionless dependent variables x_j, y_j, and z_j only for j = 1, 2, or 3. The resulting low-order model then consists of nine nonlinear ODE's [2].

Fig. 3, taken from reference [2], shows the variations of y_1 and z_1 during an eight-day period, starting from arbitrary initial conditions, with relatively weak forcing ($F_1 = 0.1$). For exactly geostrophic conditions, y_1 and z_1 would be equal. We observe fast modes where y_1 and z_1 have opposite phases superposed on a slow mode where y_1 and z_1 are nearly the same. There is some indication that the slow modes are decaying. Fig. 4 shows what happens after about a month. The slow mode is behaving much as it did before, while the fast modes have become virtually undetectable. Extension of the solution to several years reveals no qualitative change in the slow mode, while the fast modes appear to die out altogether.

Fig. 5, from reference [4], shows the variations of x_1, y_1, and z_1 when the forcing has been increased beyond the "atmospheric" range ($F_1 = 0.3$). Here the fast modes persist. The two cases suggest the additional hypothesis that solutions like the one shown in Fig. 4 exist even when F_1 is large, but are unstable with respect to fast-mode perturbations when F_1 exceeds a critical value. However, further investigation [4] has failed to reveal any abrupt transition from one type of behavior to the other, and has suggested the alternative hypothesis that fast-mode activity varies more like a fairly high power of F_1. If this is the case, some fast-mode activity should always be present, except when all fluctuations die out. Clearly we are dealing with a problem where not all questions have been answered, and where something more may yet be learned from the model.

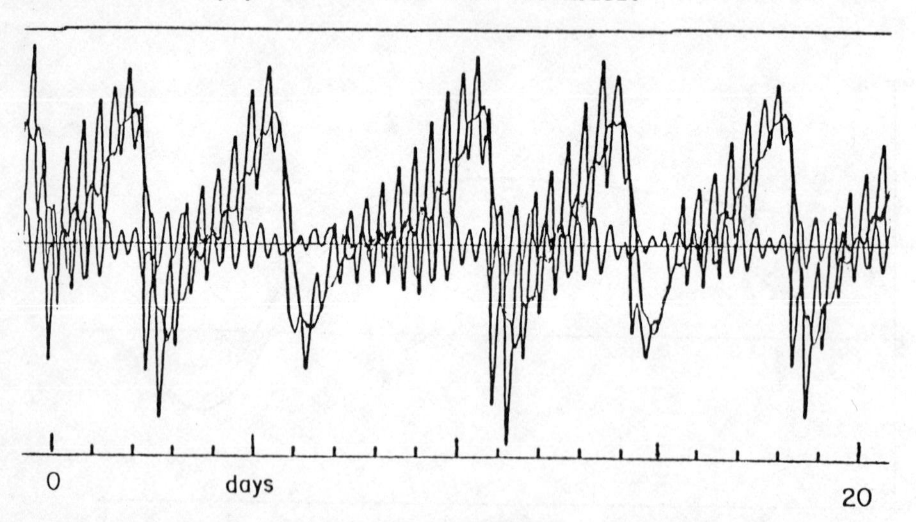

Fig. 5. Variations of x_1, y_1, and z_1 during 20 days of a numerical solution of the 9-variable primitive-equation model, with $F_1 = 0.3$, after transient effects have died out. The central horizontal line is the zero line. The variable remaining closest to the zero line is x_1. The variable with the most intense short-period fluctuations is z_1.

5. CONCLUDING COMMENTS

The cases which I have discussed illustrate only a few of the potentialities of low-order models. New uses, in fact, are continually being found. In a recent review article [6] I have described the general aspects of low-order models in considerable detail. Here one will also find a much larger selection of examples.

ACKNOWLEDGMENT. This work has been supported by the GARP Program of the Atmospheric Sciences Section, National Science Foundation, under Grant 82-14582 ATM.

REFERENCES

1. Lorenz, E. N., 1960: 'Maximum simplification of the dynamic equations' Tellus, 12, 243-254.
2. Lorenz, E. N., 1980: 'Attractor sets and quasi-geostrophic equilibrium' J. Atmos. Sci., 37, 1685-1699.
3. Gent, P. P., and McWilliams, J. C., 1982: 'Intermediate model solutions to the Lorenz equations: strange attractors and other phenomena' J. Atmos. Sci., 39,. 3-13.
4. Krishnamurthy, V., 1985: The Slow Manifold and the Persisting Gravity Waves. Ph.D. Thesis, Dept. of Earth, Atmospheric, and Planetary Sciences, Mass. Inst. Technology.
5. Vautard, R., and Legras, B., 1986: 'Invariant manifolds, quasi-geostrophy, and initialization' J. Atmos. Sci., 43, (in press).
6. Lorenz, E. N., 1982: 'Low-order models of atmospheric circulations' J. Meteor. Soc. Japan, 60, 255-267.

INDEX